Applied Mathematical Sciences
Volume 97

Springer Science+Business Media, LLC

Applied Mathematical Sciences

(continued following index)

Andrzej Lasota   Michael C. Mackey

Chaos, Fractals, and Noise
Stochastic Aspects of Dynamics

Second Edition

With 48 Illustrations

Springer

Andrzej Lasota
Institute of Mathematics
Silesian University
ul. Bankowa 14
Katowice 40-058, Poland

Michael C. Mackey
Center of Nonlinear Dynamics
McGill University
Montreal, Quebec H3G 1Y6
Canada

Editors

J.E. Marsden
Control and Dynamical Systems, 107-81
California Institute of Technology
Pasadena, CA 91125
USA

L. Sirovich
Division of Applied Mathematics
Brown University
Providence, RI 02912
USA

Mathematics Subject Classifications (1991): 60Gxx, 60Bxx, 58F30

Library of Congress Cataloging-in-Publication Data
Lasota, Andrzej, 1932–
    Chaos, fractals, and noise : stochastic aspects of dynamics /
    Andrzej Lasota, Michael C. Mackey.
        p.   cm. — (Applied mathematical sciences ; v. 97)
    Rev. ed. of: Probabilistic properties of deterministic systems.
1985.
    Includes bibliographical references and index.
    ISBN 978-1-4612-8723-0   ISBN 978-1-4612-4286-4 (eBook)
    DOI 10.1007/978-1-4612-4286-4
    1. System analysis. 2. Probabilities. 3. Chaotic behavior in
systems. I. Mackey, Michael C., 1942–   II. Lasota, Andrzej,
1932   . Probabilistic properties of deterministic systems.
III. Title. IV. Series:Applied mathematical sciences
(Springer-Verlag New York Inc.) ; v.97.
QA1.A647   vol. 97
[QA402]
510s—dc20
[003'.75]   93-10432

Printed on acid-free paper.

© 1994 Springer Science+Business Media New York
Originally published by Springer-Verlag New York Inc. in 1994
Softcover reprint of the hardcover 2nd edition 1994

Production managed by Hal Henglein; manufacturing supervised by Vincent R. Scelta.
Photocomposed copy prepared from a TeX file.

9 8 7 6 5 4 3

To the memory of

Maria Ważewska-Czyżewska

Preface to the Second Edition

The first edition of this book was originally published in 1985 under the title "Probabilistic Properties of Deterministic Systems." In the intervening years, interest in so-called "chaotic" systems has continued unabated but with a more thoughtful and sober eye toward applications, as befits a maturing field. This interest in the serious usage of the concepts and techniques of nonlinear dynamics by applied scientists has probably been spurred more by the availability of inexpensive computers than by any other factor. Thus, computer experiments have been prominent, suggesting the wealth of phenomena that may be resident in nonlinear systems. In particular, they allow one to observe the interdependence between the deterministic and probabilistic properties of these systems such as the existence of invariant measures and densities, statistical stability and periodicity, the influence of stochastic perturbations, the formation of attractors, and many others. The aim of the book, and especially of this second edition, is to present recent theoretical methods which allow one to study these effects.

We have taken the opportunity in this second edition to not only correct the errors of the first edition, but also to add substantially new material in five sections and a new chapter. Thus, we have included the additional dynamic property of sweeping (Chapter 5) and included results useful in the study of semigroups generated by partial differential equations (Chapters 7 and 11) as well as adding a completely new Chapter 12 on the evolution of distributions. The material of this last chapter is closely related to the subject of iterated function systems and their attractors-fractals. In addition, we have added a set of exercises to increase the utility of the work for graduate courses and self-study.

In addition to those who helped with the first edition, we would like to thank K. Alligood (George Mason), P. Kamthan, J. Losson, I. Nechayeva, N. Provatas (McGill), and A. Longtin (Ottawa) for their comments.

A.L.
M.C.M.

Preface to the First Edition

This book is about densities. In the history of science, the concept of densities emerged only recently as attempts were made to provide unifying descriptions of phenomena that appeared to be statistical in nature. Thus, for example, the introduction of the Maxwellian velocity distribution rapidly led to a unification of dilute gas theory; quantum mechanics developed from attempts to justify Planck's ad hoc derivation of the equation for the density of blackbody radiation; and the field of human demography grew rapidly after the introduction of the Gompertzian age distribution.

From these and many other examples, as well as the formal development of probability and statistics, we have come to associate the appearance of densities with the description of large systems containing inherent elements of uncertainty. Viewed from this perspective one might find it surprising to pose the questions: "What is the smallest number of elements that a system must have, and how much uncertainty must exist, before a description in terms of densities becomes useful and/or necessary?" The answer is surprising, and runs counter to the intuition of many. A one-dimensional system containing only one object whose dynamics are completely deterministic (no uncertainty) can generate a density of states! This fact has only become apparent in the past half-century due to the pioneering work of E. Borel [1909], A. Rényi [1957], and S. Ulam and J. von Neumann. These results, however, are not generally known outside that small group of mathematicians working in ergodic theory.

The past few years have witnessed an explosive growth in interest in physical, biological, and economic systems that could be profitably studied using densities. Due to the general inaccessibility of the mathematical literature to the nonmathematician, there has been little diffusion of the concepts and techniques from ergodic theory into the study of these "chaotic" systems. This book attempts to bridge that gap.

Here we give a unified treatment of a variety of mathematical systems generating densities, ranging from one-dimensional discrete time transformations through continuous time systems described by integro-partial-differential equations. We have drawn examples from a variety of the sciences to illustrate the utility of the techniques we present. Although the range of these examples is not encyclopedic, we feel that the ideas presented here may prove useful in a number of the applied sciences.

This book was organized and written to be accessible to scientists with a knowledge of advanced calculus and differential equations. In various places, basic concepts from measure theory, ergodic theory, the geometry of manifolds, partial differential equations, probability theory and Markov processes, and stochastic integrals and differential equations are introduced. This material is presented only as needed, rather than as a discrete unit at the beginning of the book where we felt it would form an almost insurmountable hurdle to all but the most persistent. However, in spite of our presentation of all the necessary concepts, we have not attempted to offer a compendium of the existing mathematical literature.

The one mathematical technique that touches every area dealt with is the use of the lower-bound function (first introduced in Chapter 5) for proving the existence and uniqueness of densities evolving under the action of a variety of systems. This, we feel, offers some partial unification of results from different parts of applied ergodic theory.

The first time an important concept is presented, its name is given in bold type. The end of the proof of a theorem, corollary, or proposition is marked with a ■; the end of a remark or example is denoted by a □.

A number of organizations and individuals have materially contributed to the completion of this book.

In particular the National Academy of Sciences (U.S.A.), the Polish Academy of Sciences, the Natural Sciences and Engineering Research Council (Canada), and our home institutions, the Silesian University and McGill University, respectively, were especially helpful.

For their comments, suggestions, and friendly criticism at various stages of our writing, we thank J. Bélair (Montreal), U. an der Heiden (Bremen), and R. Rudnicki (Katowice). We are especially indebted to P. Bugiel (Krakow) who read the entire final manuscript, offering extensive mathematical and stylistic suggestions and improvements. S. James (McGill) has cheerfully, accurately, and tirelessly reduced several rough drafts to a final typescript.

Contents

1
Introduction

We begin by showing how densities may arise from the operation of a one-dimensional discrete time system and how the study of such systems can be facilitated by the use of densities.

If a given system operates on a density as an initial condition, rather than on a single point, then successive densities are given by a linear integral operator, known as the Frobenius–Perron operator. Our main objective in this chapter is to offer an intuitive interpretation of the Frobenius–Perron operator. We make no attempt to be mathematically precise in either our language or our arguments.

The precise definition of the Frobenius–Perron operator is left to Chapter 3, while the measure-theoretic background necessary for this definition is presented in Chapter 2.

1.1 A Simple System Generating a Density of States

One of the most studied systems capable of generating a density of states is that defined by the quadratic map

$$S(x) = \alpha x(1 - x) \qquad \text{for } 0 \le x \le 1. \tag{1.1.1}$$

We assume that $\alpha = 4$ so S maps the closed unit interval $[0,1]$ onto itself. This is also expressed by the saying that the **state** (or **phase**) **space** of the system is $[0,1]$. The graph of this transformation is shown in Fig. 1.1.1a.

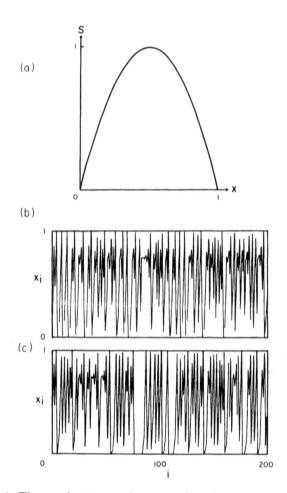

FIGURE 1.1.1. The quadratic transformation (1.1.1) with $\alpha = 4$ is shown in (a). In (b) we show the trajectory (1.1.2) determined by (1.1.1) with $x^0 = \pi/10$. Panel (c) illustrates the sensitive dependence of trajectories on initial conditions by using $x^0 = (\pi/10) + 0.001$. In (b) and (c), successive points on the trajectories have been connected by lines for clarity of presentation.

Having defined S we may pick an initial point $x^0 \in [0, 1]$ so that the successive states of our system at times $1, 2, \ldots$ are given by the trajectory

$$x^0, S(x^0), S^2(x^0) = S(S(x^0)), \ldots . \tag{1.1.2}$$

A typical trajectory corresponding to a given initial state is shown in Figure 1.1.1b. It is visibly erratic or chaotic, as is the case for almost all x^0. What is even worse is that the trajectory is significantly altered by a slight change

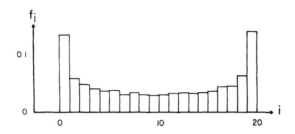

FIGURE 1.1.2. The histogram constructed according to equation (1.1.3) with $n = 20$, $N = 5000$, and $x^0 = \pi/10$.

in the initial state, as shown in Figure 1.1.1c for an initial state differing by 10^{-3} from that used to generate Figure 1.1.1b. Thus we are seemingly faced with a real problem in characterizing systems with behaviors like that of (1.1.1).

By taking a clue from other areas, we might construct a histogram to display the frequency with which states along a trajectory fall into given regions of the state space. This is done in the following way. Imagine that we divide the state space $[0, 1]$ into n discrete nonintersecting intervals so the ith interval is (we neglect the end point 1)

$$[(i - 1)/n, i/n) \qquad i = 1, \ldots, n.$$

Next we pick an initial system state x^0 and calculate a long trajectory

$$x^0, S(x^0), S^2(x^0), \ldots, S^N(x^0)$$

of length N where $N >> n$. Then it is straightforward to determine the fraction, call it f_i, of the N system states that is in the ith interval form

$$f_i = \frac{n}{N}\{\text{number of } S^j(x^0) \in [(i - 1)/n, i/n), \ j = 1, \ldots, N\}. \qquad (1.1.3)$$

We have carried out this procedure for the initial state used to generate the trajectory of Figure 1.1.1b by taking $n = 20$ and using a trajectory of length $N = 5000$. The result is shown in Figure 1.1.2. There is a surprising symmetry in the result, for the states are clearly most concentrated near 0 and 1 with a minimum at $\frac{1}{2}$. Repeating this process for other initial states leads, in general, to the same result. Thus, in spite of the sensitivity of trajectories to initial states, this is not *usually* reflected in the distribution of states within long trajectories.

However, for certain select initial states, different behaviors may occur. For some initial conditions the trajectory might arrive at one of the fixed points of equation (1.1.1), that is, a point x_* satisfying

$$x_* = S(x_*).$$

(a)

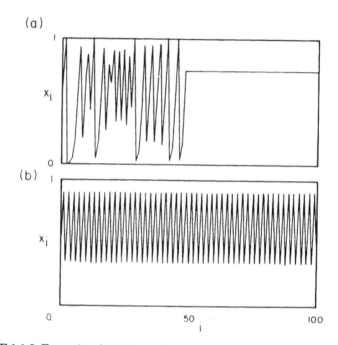

(b)

FIGURE 1.1.3. Exceptional initial conditions may confound the study of transformations via trajectories. In (a) we show how an initial condition on the quadratic transformation (1.1.1) with $\alpha = 4$ can lead to a fixed point x_* of S. In (b) we see that another initial condition leads to a period 2 trajectory, although all other characteristics of S are the same.

(For the quadratic map with $\alpha = 4$ there are two fixed points, $x_* = 0$ and $x_* = \frac{3}{4}$.) If this happens the trajectory will then have the constant value x_* forever after, as illustrated in Figure 1.1.3a. Alternately, for some other initial states the trajectory might become periodic (see Figure 1.1.3b) and also fail to exhibit the irregular behavior of Figures 1.1.1b and c. The worst part about these exceptional behaviors is that we have no a priori way of predicting which initial states will lead to them.

In the next section we illustrate an alternative approach to avoid these problems.

Remark 1.1.1. Map (1.1.1) has attracted the attention of many mathematicians, Ulam and von Neumann [1947] examined the case when $\alpha = 4$, whereas Ruelle [1977], Jakobson [1978], Pianigiani [1979], Collet and Eckmann [1980] and Misiurewicz [1981] have studied its properties for values of $\alpha < 4$. May [1974], Smale and Williams [1976], and Lasota and Mackey [1980], among others, have examined the applicability of (1.1.1) and similar maps to biological population growth problems. Interesting properties related to the existence of periodic orbits in the transformation (1.1.1) follow from the classical results of Šarkovskiĭ [1964].   □

1.2 The Evolution of Densities: An Intuitive Point of View

The problems that we pointed out in the previous section can be partially circumvented by abandoning the study of individual trajectories in favor of an examination of the flow of densities. In this section we give a heuristic introduction to this concept.

Again we assume that we have a transformation $S : [0,1] \rightarrow [0,1]$ (a shorthand way of saying S maps $[0,1]$ onto itself) and pick a large number N of initial states

$$x_1^0, x_2^0, \ldots, x_N^0.$$

To each of these states we apply the map S, thereby obtaining N new states denoted by

$$x_1^1 = S(x_1^0), x_2^1 = S(x_2^0), \ldots, x_N^1 = S(x_N^0).$$

To define what we mean by the densities of the initial and final states, it is helpful to introduce the concept of the **characteristic** (or **indicator**) **function** for a set Δ. This is simply defined by

$$1_\Delta(x) = \begin{cases} 1 & \text{if } x \in \Delta \\ 0 & \text{if } x \notin \Delta . \end{cases}$$

Loosely speaking, we say that a function $f_0(x)$ is the **density function** for the initial states $x_1^0, \ldots, x_N^0$ if, for every (not too small) interval $\Delta_0 \subset [0,1]$, we have

$$\int_{\Delta_0} f_0(u)du \simeq \frac{1}{N} \sum_{j=1}^N 1_{\Delta_0}(x_j^0). \tag{1.2.1}$$

Likewise, the density function $f_1(x)$ for the states $x_1^1, \ldots, x_N^1$ satisfies, for $\Delta \subset [0,1]$,

$$\int_\Delta f_1(u)du \simeq \frac{1}{N} \sum_{j=1}^N 1_\Delta(x_j^1). \tag{1.2.2}$$

We want to find a relationship between f_1 and f_0.

To do this it is necessary to introduce the notion of the **counterimage** of an interval $\Delta \subset [0,1]$ under the operation of the map S. This is the set of all points that will be in Δ after one application of S, or

$$S^{-1}(\Delta) = \{x : S(x) \in \Delta\}.$$

As illustrated in Figure 1.2.1, for the quadratic map considered in Section 1.1, the counterimage of an interval will be the union of two intervals.

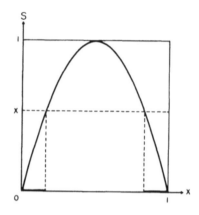

FIGURE 1.2.1. The counterimage of the set $[0, x]$ under the quadratic transformation consists of the union of the two sets denoted by the heavy lines on the x-axis.

Now note that for any $\Delta \subset [0, 1]$

$$x_j^1 \in \Delta \qquad \text{if and only if } x_j^0 \in S^{-1}(\Delta).$$

We thus have the very useful relation

$$1_\Delta(S(x)) = 1_{S^{-1}(\Delta)}(x). \tag{1.2.3}$$

With (1.2.3) we may rewrite equation (1.2.2) as

$$\int_\Delta f_1(u)\, du \simeq \frac{1}{N} \sum_{j=1}^{N} 1_{S^{-1}(\Delta)}(x_j^0). \tag{1.2.4}$$

Because Δ_0 and Δ have been arbitrary up to this point, we simply pick $\Delta_0 = S^{-1}(\Delta)$. With this choice the right-hand sides of (1.2.1) and (1.2.4) are equal and therefore

$$\int_\Delta f_1(u)\, du = \int_{S^{-1}(\Delta)} f_0(u)\, du. \tag{1.2.5}$$

This is the relationship that we sought between f_0 and f_1, and it tells us how a density of initial states f_0 will be transformed by a given map S into a new density f_1.

If Δ is an interval, say $\Delta = [a, x]$, then we can obtain an explicit representation for f_1. In this case, equation (1.2.5) becomes

$$\int_a^x f_1(u)\, du = \int_{S^{-1}([a,x])} f_0(u)\, du,$$

and differentiating with respect to x gives

$$f_1(x) = \frac{d}{dx} \int_{S^{-1}([a,x])} f_0(u)\, du. \tag{1.2.6}$$

It is clear that f_1 will depend on f_0. This is usually indicated by writing $f_1 = Pf_0$, so that (1.2.6) becomes

$$Pf(x) = \frac{d}{dx} \int_{S^{-1}([a,x])} f(u)\, du \tag{1.2.7}$$

(we have dropped the subscript on f_0 as it is arbitrary). Equation (1.2.7) explicitly defines the **Frobenius–Perron operator** P corresponding to the transformation S; it is very useful for studying the evolution of densities.

To illustrate the utility of (1.2.7) and, incidentally, the Frobenius–Perron operator concept, we return to the quadratic map $S(x) = 4x(1-x)$ of the preceding section. To apply (1.2.7) it is obvious that we need an analytic formula for the counterimage of the interval $[0, x]$. Reference to Figure 1.2.1 shows that the end points of the two intervals constituting $S^{-1}([0, x])$ are very simply calculated by solving a quadratic equation. Thus

$$S^{-1}([0,x]) = \left[0, \tfrac{1}{2} - \tfrac{1}{2}\sqrt{1-x}\right] \cup \left[\tfrac{1}{2} + \tfrac{1}{2}\sqrt{1-x}, 1\right].$$

With this, equation (1.2.7) becomes

$$Pf(x) = \frac{d}{dx} \int_0^{1/2 - 1/2\sqrt{1-x}} f(u)\, du + \frac{d}{dx} \int_{1/2 + 1/2\sqrt{1-x}}^1 f(u)\, du,$$

or, after carrying out the indicated differentiation,

$$Pf(x) = \frac{1}{4\sqrt{1-x}} \left\{ f\left(\tfrac{1}{2} - \tfrac{1}{2}\sqrt{1-x}\right) + f\left(\tfrac{1}{2} + \tfrac{1}{2}\sqrt{1-x}\right) \right\}. \tag{1.2.8}$$

This equation is an explicit formula for the Frobenius–Perron operator corresponding to the quadratic transformation and will tell us how S transforms a given density f into a new density Pf. Clearly the relationship can be used in an iterative fashion.

To see how this equation works, pick an initial density $f(x) \equiv 1$ for $x \in [0, 1]$. Then, since both terms inside the braces in (1.2.8) are constant, a simple calculation gives

$$Pf(x) = \frac{1}{2\sqrt{1-x}}. \tag{1.2.9}$$

Now substitute this expression for Pf in place of f on the right-hand side of (1.2.8) to give

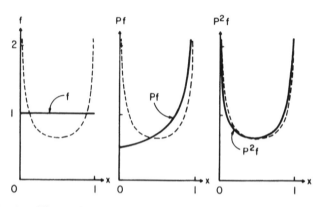

FIGURE 1.2.2. The evolution of the constant density $f(x) = 1$, $x \in [0, 1]$, by the Frobenius–Perron operator corresponding to the quadratic transformation. Compare the rapid and regular approach of $P^n f$ to the density given in equation (1.2.11) (shown as a dashed line) with the sustained irregularity shown by the trajectories in Figure 1.1.1.

$$P(Pf(x)) = P^2 f(x)$$

$$= \frac{1}{4\sqrt{1-x}} \left\{ \frac{1}{2\sqrt{1 - \frac{1}{2} + \frac{1}{2}\sqrt{1-x}}} + \frac{1}{2\sqrt{1 - \frac{1}{2} - \frac{1}{2}\sqrt{1-x}}} \right\}$$

$$= \frac{\sqrt{2}}{8\sqrt{1-x}} \left\{ \frac{1}{\sqrt{1 + \sqrt{1-x}}} + \frac{1}{\sqrt{1 - \sqrt{1-x}}} \right\}. \qquad (1.2.10)$$

In Figure 1.2.2 we have plotted $f(x) \equiv 1$, $Pf(x)$ given by (1.2.9), and $P^2 f(x)$ given by (1.2.10) to show how rapidly they seem to approach a limiting density. Actually, this limiting density is given by

$$f_*(x) = \frac{1}{\pi\sqrt{x(1-x)}}. \qquad (1.2.11)$$

If f_* is really the ultimate limit of $P^n f$ as $n \to \infty$, then we should find that $Pf_* \equiv f_*$ when we substitute into equation (1.2.8) for the Frobenius–Perron operator. A few elementary calculations confirm this. Note also the close similarity between the graph of f_* in Figure 1.2.2 and the histogram of Figure 1.1.2. Later we will show that for the quadratic map the density of states along a trajectory approaches the same unique limiting density f_* as the iterates of densities approach.

Example 1.2.1. Consider the transformation $S : [0, 1] \to [0, 1]$ given by

$$S(x) = rx \qquad (\text{mod } 1), \qquad (1.2.12)$$

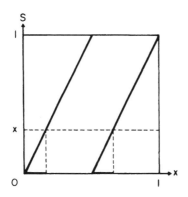

FIGURE 1.2.3. The dyadic transformation is a special case of the r-adic transformation. The heavy lines along the x-axis mark the two components of the counterimage of the interval $[0, x]$.

where r is an integer. The notation rx (mod 1) means $rx - n$, where n is the largest integer such that $rx - n \geq 0$. This expression is customarily called the r-**adic transformation** and is illustrated in Figure 1.2.3 for $r = 2$ (the **dyadic transformation**)..

Pick an interval $[0, x] \subset [0, 1]$ so that the counterimage of $[0, x]$ under S is given by

$$S^{-1}([0, x]) = \bigcup_{i=0}^{r-1} \left[\frac{i}{r}, \frac{i}{r} + \frac{x}{r} \right]$$

and the Frobenius–Perron operator is thus

$$Pf(x) = \frac{d}{dx} \sum_{i=0}^{r-1} \int_{i/r}^{i/r + x/r} f(u) \, du = \frac{1}{r} \sum_{i=0}^{r-1} f\left(\frac{i}{r} + \frac{x}{r} \right). \qquad (1.2.13)$$

This formula for the Frobenius–Perron operator corresponding to the r-adic transformation (1.2.12) shows again that densities f will be rapidly smoothed by P, as can be seen in Figure 1.2.4a for an initial density $f(x) = 2x$, $x \in [0, 1]$. It is clear that the density $P^n f(x)$ rapidly approaches the constant distribution $f_*(x) \equiv 1$, $x \in [0, 1]$. Indeed, it is trivial to show that $P1 \equiv 1$. This behavior should be contrasted with that of a typical trajectory (Figure 1.2.4b).   □

1.3   Trajectories Versus Densities

In closing this chapter we offer a qualitative examination of the behavior of two transformations from both the flow of trajectories and densities viewpoints.

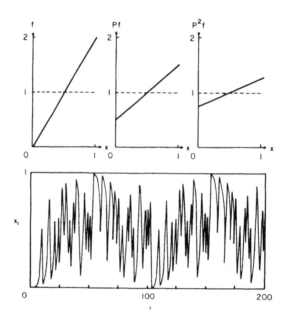

FIGURE 1.2.4. Dynamics of the dyadic transformation. (a) With an initial density $f(x) = 2x$, $x \in [0,1]$, successive applications of the Frobenius–Perron operator corresponding to the dyadic transformation result in densities approaching $f_* = 1$, $x \in [0,1]$. (b) A trajectory calculated from the dyadic transformation with $x^0 \cong 0.0005$. Compare the irregularity of this trajectory with the smooth approach of the densities in (a) to a limit.

Let R denote the entire real line, that is, $R = \{x: -\infty < x < \infty\}$, and consider the transformation $S: R \to R$ defined by

$$S(x) = \alpha x, \qquad \alpha > 0. \tag{1.3.1}$$

Our study of transformations confined to the unit interval of Section 1.2 does not affect expression (1.2.7) for the Frobenius–Perron operator. Thus (1.3.1) has the associated Frobenius–Perron operator

$$Pf(x) = (1/\alpha)f(x/\alpha).$$

We first examine the behavior of S for $\alpha > 1$. Since $S^n(x) = \alpha^n x$, we see that, for $\alpha > 1$,

$$\lim_{n \to \infty} |S^n(x)| = \infty, \qquad x \neq 0,$$

and thus the iterates $S^n(x)$ escape from any bounded interval.

This behavior is in total agreement with the behavior deduced from the flow of densities. To see this note that

$$P^n f(x) = (1/\alpha^n)f(x/\alpha^n).$$

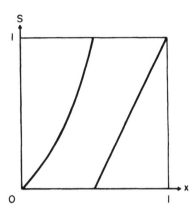

FIGURE 1.3.1. The transformation $S(x)$, defined by equation (1.3.2), has a single weak repelling point at $x = 0$.

By the qualitative definition of the Frobenius–Perron operator of the previous section, we have, for any bounded interval $[-A, A] \subset R$,

$$\int_{-A}^{A} P^n f(x) \, dx = \int_{-A/\alpha^n}^{A/\alpha^n} f(x) \, dx.$$

Since $\alpha > 1$,

$$\lim_{n \to \infty} \int_{-A}^{A} P^n f(x) \, dx = 0$$

and so, under the operation of S, densities are reduced to zero on every finite interval when $\alpha > 1$.

Conversely, for $\alpha < 1$,

$$\lim_{n \to \infty} |S^n(x)| = 0$$

for every $x \in R$, and therefore all trajectories converge to zero. Furthermore, for every neighborhood $(-\varepsilon, \varepsilon)$ of zero, we have

$$\lim_{n \to \infty} \int_{-\varepsilon}^{\varepsilon} P^n f(x) \, dx = \lim_{n \to \infty} \int_{-\varepsilon/\alpha^n}^{\varepsilon/\alpha^n} f(x) \, dx = \int_{-\infty}^{\infty} f(x) \, dx = 1,$$

so in this case all densities are concentrated in an arbitrarily small neighborhood of zero. Thus, again, the behaviors of trajectories and densities seem to be in accord.

However, it is not always the case that the behavior of trajectories and densities seem to be in agreement. This may be simply illustrated by what we call the **paradox of the weak repellor**. In Remark 6.2.1 we consider the transformation $S: [0, 1] \to [0, 1]$ defined by

$$S(x) = \begin{cases} x/(1 - x) & \text{for } x \in \left[0, \tfrac{1}{2}\right] \\ 2x - 1 & \text{for } x \in \left(\tfrac{1}{2}, 1\right], \end{cases} \tag{1.3.2}$$

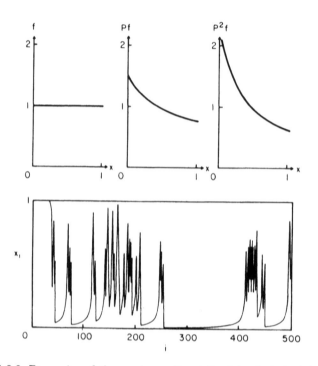

FIGURE 1.3.2. Dynamics of the weak repellor defined by (1.3.2). (a) The evolution $P^n f$ of an initial distribution $f(x) = 1$, $x \in [0, 1]$. (b) The trajectory originating from an initial point $x^0 \approx 0.25$.

(see Figure 1.3.1). There we prove that, for every $\varepsilon > 0$,

$$\lim_{n \to \infty} \int_\varepsilon^1 P^n f(x)\, dx = 0.$$

Thus, since $P^n f$ is a density,

$$\lim_{n \to \infty} \int_0^\varepsilon P^n f(x)\, dx = 1,$$

and all densities are concentrated in an arbitrarily small neighborhood of zero. This behavior is graphically illustrated in Figure 1.3.2a.

If one picks an initial point $x_0 > 0$ very close to zero (see Figure 1.3.2b), then, as long as $S^n(x_0) \in (0, \frac{1}{2}]$, we have

$$S^n(x_0) = x_0/(1 - nx_0) \geq \alpha^n x_0$$

where $\alpha = 1/(1 - x_0) > 1$. Thus initially, for small x_0, this transformation behaves much like transformation (1.3.1), and the behavior of the trajectory near zero apparently contradicts that expected from the behavior of the densities.

This paradox is more apparent than real and may be easily understood. First, note that even though all trajectories are repelled from zero (zero is a repellor), once a trajectory is ejected from $\left(0, \frac{1}{2}\right]$ it is quickly reinjected into $\left(0, \frac{1}{2}\right]$ from $\left(\frac{1}{2}, 1\right]$. Thus zero is a "weak repellor." The second essential point to note is that the speed with which any trajectory leaves a small neighborhood of zero is small; it is given by

$$S^n(x_0) - S^{n-1}(x_0) = \frac{x_0^2}{(1 - nx_0)[1 - (n - 1)x_0]}.$$

Thus, starting with many initial points, as n increases we will see the progressive accumulation of more and more points near zero. This is precisely the behavior predicted by examining the flow of densities.

Although our comments in this chapter lack mathematical rigor, they offer some insight into the power of looking at the evolution of densities under the operation of deterministic transformations. The next two chapters are devoted to introducing the mathematical concepts required for a precise treatment of this problem.

Exercises

Simple numerical experiments can greatly clarify the material of this and subsequent chapters. Consequently, the first five exercises of this chapter involve the writing of simple utility programs to study the quadratic map (1.1.1) from several perspectives. Exercises in subsequent chapters will make use of these programs to study other maps. If you have access to a personal computer (preferably with a math coprocessor), a workstation, or a microcomputer with graphics capabilities, we strongly urge you to do these exercises.

1.1. Write a program to numerically generate a sequence of iterates $\{x_n\}$ from $x_{n+1} = S(x_n)$, where S is the quadratic map (1.1.1). Write your program in such a way that the map S is called from a subroutine (so it may be changed easily) and include graphics to display x_n versus n. When displaying the sequence $\{x_n\}$ graphically, you will find it helpful to connect successive values by a straight line so you can keep track of them. Save this program under the name TRAJ so you can use it for further problems.

1.2. Using TRAJ study the behavior of (1.1.1) for various values of α satisfying $3 \le \alpha \le 4$, and for various initial conditions x_0. (You can include an option to generate x_0 using the random number generator if you wish, but be careful to use a different seed number for each run.) At a given value of α what can you say about the temporal behavior of the sequence

$\{x_n\}$ for different x_0? What can you say concerning the qualitative and quantitative differences in the trajectories $\{x_n\}$ for different values of α?

1.3. To increase your understanding of the results in Exercise 1.2, write a second program called BIFUR. This program will plot a large number of iterates of the map (1.1.1) as α is varied between 3 and 4, and the result will approximate the bifurcation diagram of (1.1.1). Procedurally, for each value of α, use the random number generator (don't forget about the seed) to select an initial x_0, discard the first 100 or so values of x_n to eliminate transients, and then plot a large number (on the order of 1000 to 5000) of the x_n vertically above the value of α. Then increment α and repeat the process successively until you have reached the maximal value of α. A good incremental value of α is $\Delta\alpha = 0.01$ to 0.05, and obviously the smaller $\Delta\alpha$ the better the resolution of the details of the bifurcation diagram at the expense of increased computation time. Use the resulting bifurcation diagram you have produced, in conjunction with your results of Exercise 1.2, to more fully discuss the dynamics of (1.1.1). You may find it helpful to make your graphics display flexible enough to "window" various parts of the bifurcation diagram so you can examine fine detail.

1.4. Write a program called DENTRAJ (Density from a Trajectory) to display the histogram of the location of the iterates $\{x_n\}$ of (1.1.1) for various values of α satisfying $3 \leq \alpha \leq 4$ as was done in Figure 1.1.2 for $\alpha = 4$. [Constructing histograms from "data" like this is always a bit tricky because there is a tradeoff between the number of points and the number of bins in the histogram. However, a ratio of 200–300 of point number to bin number should provide a satisfactory result, so, depending on the speed of your computer (and thus the number of iterations that can be carried out in a given time), you can obtain varying degrees of resolution.] Compare your results with those from Exercise 1.3. Note that at a given value of α, the bands you observed in the bifurcation diagram correspond to the histogram supports (the places where the histogram is not zero).

1.5. Redo Exercise 1.4 by writing a program called DENITER (Density Iteration) that takes a large number N of initial points $\{x_i^0\}_{i=1}^N$ distributed with some density $f_0(x)$, e.g., $f_0(x)$ could be uniform on $[0,1]$ for (1.1.1), or $f_0(x) = 2x$, etc., and iterates them sequentially to give $\{x_i^1\}_{i=1}^N = \{S(x_i^0)\}_{i=1}^N$, $\{x_i^2\}_{i=1}^N = \{S(x_i^1)\}_{i=1}^N$, etc. Construct your program to display the histogram of the $\{x_i^j\}_{i=1}^N$ for the initial ($j = 0$) and successive iterations. Do the histograms appear to converge to an invariant histogram? How does the choice of the initial histogram affect the result after many iterations? Discuss the rate of convergence of the sequence of histograms.

1.6. Prove that f_* given by (1.2.11) is a solution of the equation $Pf = f$, where P, given by (1.2.8), is the Frobenius–Perron operator corresponding to the quadratic map (1.1.1) with $\alpha = 4$.

1.7. This exercise illustrates that there can sometimes be a danger in drawing conclusions about the behavior of even simple systems based on numerical experiments. Consider the Frobenius–Perron operator (1.2.13) corresponding to the r-adic transformation (1.2.12) when r is an integer. (a) For every integer r show that $f_*(x) = 1_{[0,1]}(x)$ is a solution of $Pf = f$. Can you prove that it is the unique solution? (b) For $r = 2$ and $r = 3$ use TRAJ, DENTRAJ, and DENITER to study (1.2.12). What differences do you see in the behaviors for $r = 2$ and $r = 3$? Why do these differences exist? Discuss your numerical results in light of your computations in (a).

1.8. Consider the example of the weak repellor (1.3.2). (a) Derive the Frobenius–Perron operator corresponding to the weak repellor without looking in Chapter 6. Calculate a few terms of the sequence $\{P^n f\}$ for $f(x) = 1_{[0,1]}(x)$. (b) Use TRAJ, DENTRAJ and DENITER to study the weak repellor (1.3.2). Discuss your results. Based on your observations, what conjectures can you formulate about the behavior of the weak repellor? In what way do these differ from the properties of the quadratic map (1.1.1) that you saw in Exercises 1.1–1.5?

2
The Toolbox

In this and the following chapter, we introduce basic concepts necessary for understanding the flow of densities. These concepts may be studied in detail before continuing on to the core of our subject matter, which starts in Chapter 4, or, they may be skimmed on first reading to fix the location of important concepts for later reference.

We briefly outline here some essential concepts from measure theory, the theory of Lebesgue integration, and from the theory of the convergence of sequences of functions. This material is in no sense exhaustive; those desiring more detailed treatments should refer to Halmös [1974] and Royden [1968].

2.1 Measures and Measure Spaces

We start with the definition of a σ-algebra.

Definition 2.1.1. A collection $\mathcal{A}$ of subsets of a set X is a σ-**algebra** if:

(a) When $A \in \mathcal{A}$ then $X \setminus A \in \mathcal{A}$;

(b) Given a finite or infinite sequence $\{A_k\}$ of subsets of X, $A_k \in \mathcal{A}$, then the union $\bigcup_k A_k \in \mathcal{A}$; and

(c) $X \in \mathcal{A}$.

From this definition it follows immediately, by properties (a) and (c), that the empty set $\emptyset$ belongs to $\mathcal{A}$, since $\emptyset = X \setminus X$. Further, given a sequence $\{A_k\}$, $A_k \in \mathcal{A}$, then the intersection $\bigcap_k A_k \in \mathcal{A}$. To see this, note that

$$\bigcap_k A_k = X \setminus \bigcup_k (X \setminus A_k)$$

and then apply properties (a) and (b). Finally, the difference $A \setminus B$ of two sets A and B that belong to $\mathcal{A}$ also belongs to $\mathcal{A}$ because

$$A \setminus B = A \cap (X \setminus B).$$

Definition 2.1.2. A real-valued function μ defined on a σ-algebra $\mathcal{A}$ is a measure if:

(a) $\mu(\emptyset) = 0$;

(b) $\mu(A) \geq 0$ for all $A \in \mathcal{A}$; and

(c) $\mu(\bigcup_k A_k) = \sum_k \mu(A_k)$ if $\{A_k\}$ is a finite or infinite sequence of pairwise disjoint sets from $\mathcal{A}$, that is, $A_i \cap A_j = \emptyset$ for $i \neq j$.

We do not exclude the possibility that $\mu(A) = \infty$ for some $A \in \mathcal{A}$.

Remark 2.1.1. This definition of a measure and the properties of a σ-algebra $\mathcal{A}$ as detailed in Definition 2.1.1 ensure that (1) if we know the measure of a set X and a subset A of X we can determine the measure of $X \setminus A$; and (2) if we know the measure of each disjoint subset A_k of $\mathcal{A}$ we can calculate the measure of their union. □

Definition 2.1.3. If $\mathcal{A}$ is a σ-algebra of subsets of X and if μ is a measure on $\mathcal{A}$, then the triple $(X, \mathcal{A}, \mu)$ is called a **measure space**. The sets belonging to $\mathcal{A}$ are called **measurable sets** because, for them, the measure is defined.

Remark 2.1.2. A simple example of a measure space is the finite set $X = \{x_1, \ldots, x_N\}$, in which the σ-algebra is all possible subsets of X and the measure is defined by ascribing to each element $x_i \in X$ a nonnegative number, say p_i. From this it follows that the measure of a subset $\{x_{\alpha_1}, \ldots, x_{\alpha_k}\}$ of X is just $p_{\alpha_1} + \cdots + p_{\alpha_k}$. If $p_i = 1$, then the measure is called a **counting measure** because it counts the number of elements in the set. □

Remark 2.1.3. If $X = [0, 1]$ or R, the real line, then the most natural σ-algebra is the σ-algebra $\mathcal{B}$ of Borel sets (the **Borel σ-algebra**), which, by definition, is the smallest σ-algebra containing intervals. (The word smallest means that any other σ-algebra that contains intervals also contains any set contained in $\mathcal{B}$.) It can be proved that on the Borel σ-algebra there exists a unique measure μ, called the **Borel measure**, such that $\mu([a, b]) = b - a$. Whenever considering spaces $X = R$ or $X = R^d$ or subsets of these (intervals, squares, etc.) we *always* assume the Borel measure and will not repeat this assumption again.   □

As presented, Definition 2.1.3 is extremely general. In almost all applications a more specific measure space is adequate, as follows:

Definition 2.1.4. A measure space $(X, \mathcal{A}, \mu)$ is called σ-**finite** if there is a sequence $\{A_k\}$, $A_k \in \mathcal{A}$, satisfying

$$X = \bigcup_{k=1}^{\infty} A_k \quad \text{and} \quad \mu(A_k) < \infty \qquad \text{for all } k.$$

Remark 2.1.4. If $X = R$, the real line, and μ is the Borel measure, then the A_k may be chosen as intervals of the form $[-k, k]$. In the d-dimensional space R^d, the A_k may be chosen as balls of radius k.   □

Definition 2.1.5. A measure space $(X, \mathcal{A}, \mu)$ is called **finite** if $\mu(X) < \infty$. In particular, if $\mu(X) = 1$, then the measure space is said to be **normalized** or **probabilistic**.

Remark 2.1.5. We have defined a hierarchy of measure spaces from the most general (Definition 2.1.3) down to the most specific (Definition 2.1.5). *Throughout this book, unless it is specifically stated to the contrary, a measure space will always be understood to be σ-finite.*   □

Remark 2.1.6. If a certain property involving the points of a measure space is true except for a subset of that space having measure zero, then we say that property is true **almost everywhere** (abbreviated as a.e.). □

2.2 Lebesgue Integration

In the material we deal with it is often necessary to use a type of integration more general than the customary Riemann integration. In this section we introduce the Lebesgue integral, which is defined for abstract measure spaces in which no other structures except a σ-algebra $\mathcal{A}$ and a measure μ must be introduced.

Definition 2.2.1. Let $(X, \mathcal{A}, \mu)$ be a measure space. A real-valued function $f: X \to R$ is **measurable** if $f^{-1}(\Delta) \in \mathcal{A}$ for every interval $\Delta \subset R$.

In developing the concept of the Lebesgue integral, we need the notation

$$f^+(x) = \max(0, f(x)) \quad \text{and} \quad f^-(x) = \max(0, -f(x))$$

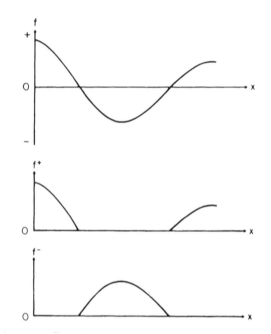

FIGURE 2.2.1. Illustration of the notation $f^+(x)$ and $f^-(x)$.

(see Figure 2.2.1). Observe that

$$f(x) = f^+(x) - f^-(x) \quad \text{and} \quad |f(x)| = f^+(x) + f^-(x).$$

Before presenting the formal definitions for the Lebesgue integral of a function, consider the following. Let $f: X \to R$ be a bounded, nonnegative measurable function, $0 \le f(x) < M < \infty$. Take the partition of the interval $[0, M]$, $0 = a_0 < a_1 < \cdots < a_n = M$, $a_i = Mi/n$, $i = 0, \ldots, n$, and define the sets A_i by

$$A_i = \{x: f(x) \in [a_i, a_{i+1})\}, \qquad i = 0, \ldots, n-1.$$

Then it is clear that the 1_{A_i} are measurable and

$$\left| f(x) - \sum_{i=0}^{n-1} a_i 1_{A_i}(x) \right| \le \frac{M}{n}.$$

Therefore, we must conclude that every bounded nonnegative measurable function can be approximated by a finite linear combination of characteristic functions. This observation is crucial to our development of the Lebesgue integral embodied in the following four definitions.

Definition 2.2.2. Let $(X, \mathcal{A}, \mu)$ be a measure space, and the sets $A_i \in \mathcal{A}$,

$i = 1, \ldots, n$ be such that $A_i \cap A_j = \emptyset$ for all $i \neq j$. Then the Lebesgue integral of the function

$$g(x) = \sum_i \lambda_i 1_{A_i}(x), \qquad \lambda_i \in R, \tag{2.2.1}$$

is defined as

$$\int_X g(x)\,\mu(dx) = \sum_i \lambda_i \mu(A_i).$$

A function g of the form (2.2.1) is called a **simple function**.

Definition 2.2.3. Let $(X, \mathcal{A}, \mu)$ be a measure space, $f: X \to R$ an arbitrary nonnegative bounded measurable function, and $\{g_n\}$ a sequence of simple functions converging uniformly to f. Then the Lebesgue integral of f is defined as

$$\int_X f(x)\,\mu(dx) = \lim_{n \to \infty} \int_X g_n(x)\,\mu(dx).$$

Remark 2.2.1. It can be shown that the limit in Definition 2.2.3 exists and is independent of the choice of the sequence of simple functions $\{g_n\}$ as long as they converge uniformly to f.   □

Definition 2.2.4. Let $(X, \mathcal{A}, \mu)$ be a measure space, $f: X \to R$ a nonnegative unbounded measurable function, and define

$$f_M(x) = \begin{cases} f(x) & \text{if } 0 \le f(x) \le M \\ M & \text{if } M < f(x). \end{cases}$$

Then the Lebesgue integral of f is defined by

$$\int_X f(x)\,\mu(dx) = \lim_{M \to \infty} \int_X f_M(x)\,\mu(dx).$$

Remark 2.2.2. Note that $\int_X f_M(x)\,\mu(dx)$ is an increasing function of M so that the limit in Definition 2.2.4 always exists even though it might be infinite.   □

Definition 2.2.5. Let $(X, \mathcal{A}, \mu)$ be a measure space and $f: X \to R$ a measurable function. Then the Lebesgue integral of f is defined by

$$\int_X f(x)\,\mu(dx) = \int_X f^+(x)\,\mu(dx) - \int_X f^-(x)\,\mu(dx)$$

if at least one of the terms

$$\int_X f^+(x)\,\mu(dx), \quad \int_X f^-(x)\,\mu(dx)$$

is finite. If both of these terms are finite then the function f is called **integrable**.

Remark 2.2.3. The four Definitions 2.2.2–2.2.5 are for the Lebesgue integral of f over the entire space X. For $A \in \mathcal{A}$ we have, by definition,

$$\int_A f(x)\, \mu(dx) = \int_X f(x) 1_A(x)\, \mu(dx). \qquad \square$$

The Lebesgue integral has some important properties that we will often use. We state them without proof. Throughout a measure space $(X, \mathcal{A}, \mu)$ is assumed.

(L1) If $f, g: X \to R$ are measurable, g is integrable, and $|f(x)| \leq g(x)$, then f is integrable and

$$\left| \int_X f(x)\, \mu(dx) \right| \leq \int_X g(x)\, \mu(dx).$$

(L2) $\int_X |f(x)|\, \mu(dx) = 0$ if and only if $f(x) = 0$ a.e.

(L3) If $f_1, f_2: X \to R$ are integrable functions, then for $\lambda_1, \lambda_2 \in R$ the linear combination $\lambda_1 f_1 + \lambda_2 f_2$ is integrable and

$$\int_X [\lambda_1 f_1(x) + \lambda_2 f_2(x)]\, \mu(dx)$$

$$= \lambda_1 \int_X f_1(x)\, \mu(dx) + \lambda_2 \int_X f_2(x)\, \mu(dx).$$

(L4) Let $f, g: X \to R$ be measurable functions and $f_n: X \to R$ be measurable functions such that $|f_n(x)| \leq g(x)$ and $\{f_n(x)\}$ converges to $f(x)$ almost everywhere. If g is integrable, then f and f_n are also integrable and

$$\lim_{n \to \infty} \int_X f_n(x)\, \mu(dx) = \int_X f(x)\, \mu(dx).$$

The last formula is also true if the assumption $|f_n(x)| \leq g(x)$ with an integrable g is replaced by $0 \leq f_1(x) \leq f_2(x) \ldots$. In this case, however, the integrals could be infinite.

Remark 2.2.4. The properties described in (L4) are often referred to as the **Lebesgue dominated convergence theorem** ($|f_n(x)| \leq g(x)$) and the **Lebesgue monotone convergence theorem** ($0 \leq f_1(x) \leq \cdots$). $\square$

(L5) Let $f: X \to R$ be an integrable function and the sets $A_i \in \mathcal{A}$, $i = 1, 2, \ldots$, be disjoint. If $A = \bigcup_i A_i$, then

$$\sum_i \int_{A_i} f(x)\, \mu(dx) = \int_A f(x)\, \mu(dx).$$

Remark 2.2.5. Observe that f is integrable if and only if $|f|$ is integrable. This is easy to see since $|f| = f^+ + f^-$. If f is integrable, f^+ and f^- are also and thus

$$\int_X |f(x)| \, \mu(dx) = \int_X f^+(x) \, \mu(dx) + \int_X f^-(x) \, \mu(dx)$$

is finite. Hence $|f|$ is integrable. The converse is equally easy to prove. □

Remark 2.2.6. Our definition of the Lebesgue integral was stated in four distinct steps. It should be evident from this construction that for every integrable function f there is a sequence of simple functions

$$f_n(x) = \sum_i \lambda_{i,n} 1_{A_{i,n}}(x)$$

such that

$$\lim_{n \to \infty} f_n(x) = f(x) \text{ a.e.} \quad \text{and} \quad |f_n(x)| \le |f(x)|.$$

Thus, by the Lebesgue dominated convergence theorem (L4), we have

$$\lim_{n \to \infty} \int_X f_n(x) \, \mu(dx) = \int_X f(x) \, \mu(dx).$$

This observation will be used many times in simplifying proofs since it enables us to reduce our arguments to two steps: First, we must only verify some formula for simple functions and then, second, pass to the limit. □

Remark 2.2.7. The notion of the Lebesgue integral is quite important since it is defined for very abstract measure spaces $(X, \mathcal{A}, \mu)$ in which no other structures are introduced except for the existence of a σ-algebra $\mathcal{A}$ and a measure μ. In calculus the definition of the Riemann integral is intimately related to the algebraic properties of the real line, and it is easy to establish a connection between the Lebesgue and Riemann integrals. For example, if we define μ as in Remark 2.1.3, then

$$\int_{[a,b]} f(x) \, \mu(dx) = \int_a^b f(x) \, dx$$

where the left-hand side is the Lebesgue integral and the right-hand side is the Riemann integral. This equality is true for any Riemann integrable function f since any Riemann integrable function is automatically Lebesgue integrable. An analogous connection exists in higher dimensions. □

From the properties of the Lebesgue integral it is easy to demonstrate that if $f \colon X \to R$ is a nonnegative integrable function then $\mu_f(A)$, defined by

$$\mu_f(A) = \int_A f(x) \, \mu(dx),$$

is a finite measure. In fact, by the definition of the Lebesgue integral it is clear that $\mu_f(A)$ is nonnegative and finite, and from property (L5) it is also additive. Further, from (L2) if $\mu(A) = 0$ then

$$\mu_f(A) = \int_A 1_A(x) f(x) \mu(dx) = 0$$

since $1_A(x) f(x) = 0$ a.e. Thus $\mu_f(A)$ satisfies all the properties of a measure as detailed in Definition 2.1.2, and $\mu_f(A) = 0$ whenever $\mu(A) = 0$. This observation that every integrable nonnegative function defines a finite measure can be reversed by the following theorem, which is of fundamental importance for the development of the Frobenius–Perron operator.

Theorem 2.2.1. (Radon–Nikodym theorem). *Let $(X, \mathcal{A}, \mu)$ be a measure space and let ν be a second finite measure with the property that $\nu(A) = 0$ for all $A \in \mathcal{A}$ such that $\mu(A) = 0$. Then there exists a nonnegative integrable function $f: X \to R$ such that*

$$\nu(A) = \int_A f(x) \mu(dx) \qquad \text{for all } A \in \mathcal{A}.$$

Remark 2.2.8. It should be observed that we have not explicitly stated that $(X, \mathcal{A}, \mu)$ is a σ-finite measure space, which is an important assumption in the Radon–Nikodym theorem. Once again we wish to stress our earlier assumption that *all measure spaces are taken to be σ-finite* unless a contrary assumption is made.   □

Although we omit the proof of the Radon–Nikodym theorem, it is easy to show that the function f is in some sense unique. To see this, assume that there are two functions $f_1, f_2: X \to R$ such that

$$\nu(A) = \int_A f_1(x) \mu(dx) \quad \text{and} \quad \nu(A) = \int_A f_2(x) \mu(dx).$$

Then for all $A \in \mathcal{A}$ we have

$$\int_A [f_1(x) - f_2(x)] \mu(dx) = 0.$$

Define two sets A_1 and A_2 by

$$A_1 = \{x: f_1(x) > f_2(x)\} \quad \text{and} \quad A_2 = \{x: f_1(x) \le f_2(x)\}.$$

Then

$$0 = \int_{A_1} [f_1(x) - f_2(x)] \mu(dx) - \int_{A_2} [f_1(x) - f_2(x)] \mu(dx)$$

$$= \int_{A_1 \cup A_2} |f_1(x) - f_2(x)| \mu(dx)$$

$$= \int_X |f_1(x) - f_2(x)| \mu(dx).$$

Hence, from property (L2) of Lebesgue integrals, we have $|f_1(x) - f_2(x)| = 0$ a.e., so that $f_1(x)$ and $f_2(x)$ differ only on a set of measure zero.

Observe that our argument is quite general, and we have in fact proved the following.

Proposition 2.1.1. *If f_1 and f_2 are integrable functions such that*

$$\int_A f_1(x)\,\mu(dx) = \int_A f_2(x)\,\mu(dx) \qquad \text{for } A \in \mathcal{A}$$

then $f_1 = f_2$ a.e.

Also from property (L2) of the Lebesgue integral it is clear that two measurable functions, f_1 and f_2, differing from each other only on a set of measure zero, cannot be distinguished by calculating integrals. Thus we say that in the **space of measurable functions**, every two functions f_1, f_2, differing only on a set of measure zero, represent the same **element** of that space. However, to simplify our notation, we will often write "measurable function" instead of "an element of the space of measurable functions." Because of property (L2) this should not lead to any confusion.

With these remarks in mind, we now introduce the concept of an L^p space.

Definition 2.2.6. Let $(X, \mathcal{A}, \mu)$ be a measure space and p a real number, $1 \leq p < \infty$. The family of all possible real-valued measurable functions $f \colon X \to R$ satisfying

$$\int_X |f(x)|^p\,\mu(dx) < \infty \qquad (2.2.2)$$

is the $L^p(X, \mathcal{A}, \mu)$ **space.** Here we use the term "measurable function" to mean "an element of the space of measurable functions."

We shall sometimes write L^p instead of $L^p(X, \mathcal{A}, \mu)$ if the measure space is understood, or $L^p(X)$ if $\mathcal{A}$ and μ are understood. Note that if $p = 1$ then the L^1 space consists of all possible integrable functions.

The integral appearing in (2.2.2) is very important for an element $f \in L^p$. Thus it is assigned the special notation

$$\|f\|_{L^p} = \left[\int_X |f(x)|^p\,\mu(dx) \right]^{1/p} \qquad (2.2.3)$$

and is called the L^p **norm** of f. When property (L2) of the Lebesgue integral is applied to $|f|^p$, it immediately follows that the condition $\|f\|_{L^p} = 0$ is equivalent to $f(x) = 0$ a.e. Or, more precisely, $\|f\|_{L^p} = 0$ if and only if f is a zero element in L^p (which is an element represented by all functions equal to zero almost everywhere).

Two other important properties of the norm are

$$\|\alpha f\|_{L^p} = |\alpha| \cdot \|f\|_{L^p} \qquad \text{for } f \in L^p,\ \alpha \in R \qquad (2.2.4)$$

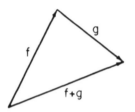

FIGURE 2.2.2. A geometric interpretation of the triangle inequality (2.2.5).

and

$$\|f + g\|_{L^p} \le \|f\|_{L^p} + \|g\|_{L^p} \qquad \text{for } f, g \in L^p. \qquad (2.2.5)$$

The first condition, (2.2.4), simply says that the norm is homogeneous. The second is called the **triangle inequality**. As shown in Figure 2.2.2, if we think of f, g, and $f + g$ as vectors, we can consider a triangle with sides f, g, and $f + g$. Then, by equation, (2.2.5), the length of the side $(f + g)$ is shorter than the sum of the lengths of the other two sides.

From (2.2.4) it follows that for every $f \in L^p$ and real α, the product αf belongs to L^p. Further, from (2.2.5) it follows that for every $f, g \in L^p$ the sum $f + g$ is also an element of L^p. This is denoted by saying that L^p is a **vector space**.

Because the value of $\|f\|_{L^p}$ is interpreted as the length of f, we say that

$$\|f - g\|_{L^p} = \left[\int_X |f(x) - g(x)|^p \, \mu(dx) \right]^{1/p}$$

is the L^p **distance** between f and g.

It is important to note that the product fg of two functions $f, g \in L^p$ is not necessarily in L^p, for example, $f(x) = x^{-1/2}$ is integrable on $[0, 1]$ but $[f(x)]^2 = x^{-1}$ is not.

This leads us to define the space adjoint to L^p.

Definition 2.2.7. Let $(X, \mathcal{A}, \mu)$ be a measure space. The **space adjoint** to $L^p(X, \mathcal{A}, \mu)$ is $L^{p'}(X, \mathcal{A}, \mu)$, where

$$(1/p) + (1/p') = 1.$$

Remark 2.2.9. If $p = 1$, Definition 2.2.7 of adjoint space fails. The adjoint space, in the case $p = 1$, by definition, consists of all bounded almost everywhere measurable functions and is denoted by L^∞. Functions that differ only on a set of measure zero are considered to represent the same element. □

It is well known that if $f \in L^p$ and $g \in L^{p'}$, then fg is integrable, and hence we define the **scalar product** of two functions by

$$\langle f, g \rangle = \int_X f(x)g(x)\,\mu(dx).$$

An important relation we will often use is the **Cauchy–Hölder inequality**. Thus, if $f \in L^p$ and $g \in L^{p'}$, then

$$|\langle f, g \rangle| \le \|f\|_{L^p} \cdot \|g\|_{L^{p'}}.$$

For this inequality to make sense when $f \in L^1$, $g \in L^\infty$, we take the L^∞ norm of g to be the smallest constant c such that

$$|g(x)| \le c$$

for almost all $x \in X$. This constant is denoted by ess sup $|g(x)|$, called the **essential supremum** of g.

Remark 2.2.10. As we almost always work in L^1, we will not indicate the space in which the norm is taken unless it is not L^1. Thus we will write $\|f\|$ instead of $\|f\|_{L^1}$. Observe that in L^1 the norm has the exceptional property that the triangle inequality is sometimes an equality. To see this, note from property (L3) that

$$\|f + g\| = \|f\| + \|g\| \qquad \text{for } f \ge 0,\ g \ge 0;\ f, g \in L^1.$$

Thus geometrical intuition in some abstract spaces may be misleading. □

The concept of the L^1 space simplifies the Radon–Nikodym theorem as shown by the following corollary.

Corollary 2.2.1. *If $(X, \mathcal{A}, \mu)$ is a measure space and ν is a finite measure on $\mathcal{A}$ such that $\nu(A) = 0$ whenever $\mu(A) = 0$, then there exists a unique element $f \in L^1$ such that*

$$\nu(A) = \int_A f(x)\,\mu(dx) \qquad \text{for } A \in \mathcal{A}.$$

One of the most important notions in analysis, measure theory, and topology, as well as other areas of mathematics, is that of the Cartesian product. To introduce this concept we start with a definition.

Definition 2.2.8. Given two arbitrary sets A_1 and A_2, the **Cartesian product** of A_1 and A_2 (note that the order is important) is the set of all pairs (x_1, x_2) such that $x_1 \in A_1$ and $x_2 \in A_2$. This is customarily written as

$$A_1 \times A_2 = \{(x_1, x_2) : x_1 \in A_1,\ x_2 \in A_2\}.$$

In a natural way this concept may be extended to more than two sets. Thus the Cartesian product of the sets $A_1, \ldots, A_d$ is the set of all sequences $(x_1, \ldots, x_d)$ such that $x_i \in A_i$, $i = 1, \ldots, d$, or

$$A_1 \times \cdots \times A_d = \{(x_1, \ldots, x_d): x_i \in A_i \text{ for } i = 1, \ldots, d\}.$$

An important consequence following from the concept of the Cartesian product is that if a structure is defined on each of the factors A_i, for example, a measure, then it is possible to extend that property to the Cartesian product. Thus, given d measure spaces (X_i, A_i, μ_i), $i = 1, \ldots, d$, we define

$$X = X_1 \times \cdots \times X_d, \tag{2.2.6}$$

A to be the smallest σ-algebra of subsets of X containing all sets of the form

$$A_1 \times \cdots \times A_d \quad \text{with } A_i \in A_i, \ i = 1, \ldots, d, \tag{2.2.7}$$

and

$$\mu(A_1 \times \cdots \times A_d) = \mu_1(A_1) \cdots \mu_d(A_d) \quad \text{for } A_i \in A_i. \tag{2.2.8}$$

Unfortunately, by themselves they do not define a measure space (X, A, μ). There is no problem with either X or A, but μ is defined only on special sets, namely $A = A_1 \times \cdots \times A_d$, that do not form a σ-algebra. To show that μ, as defined by (2.2.8), can be extended to the entire σ-algebra A requires the following theorem.

Theorem 2.2.2. *If measure spaces* (X_i, A_i, μ_i), $i = 1, \ldots, d$ *are given and* X, A, *and* μ *are defined by equations* (2.2.6), (2.2.7), *and* (2.2.8), *respectively, then there exists a unique extension of* μ *to a measure defined on* A.

The measure space (X, A, μ) whose existence is guaranteed by Theorem 2.2.2, is called the **product of the measure spaces** $(X_1, A_1, \mu_1), \ldots, (X_d, A_d, \mu_d)$, or more briefly a **product space**. The measure μ is called the **product measure**.

Observe that from equation (2.2.8) it follows that

$$\mu(X_1 \times \cdots \times X_d) = \mu(X_1) \cdots \mu(X_d).$$

Thus, if all the measure spaces (X_i, A_i, μ_i) are finite or probabilistic, then (X, A, μ) will also be finite or probabilistic.

Theorem 2.2.2 allows us to define integration on the product space (X, A, μ) since it is also a measure space. A function $f: X \to R$ may be written as a function of d variables because every point $x \in X$ is a sequence $x = (x_1, \ldots, x_d)$, $x_i \in X_i$. Thus it is customary to write integrals on X either as

$$\int_X f(x) \, \mu(dx),$$

where it is implicitly understood that $x = (x_1, \ldots, x_d)$ and $X = X_1 \times \cdots \times X_d$, or in the more explicit form

$$\int_{X_1} \cdots \int_{X_d} f(x_1, \ldots, x_d) \, \mu(dx_1 \cdots dx_d).$$

Integrals on the product of measure spaces are related to integrals on the individual factors by a theorem associated with the name of Fubini. For simplicity, we first formulate it for product spaces containing only two factors.

Theorem 2.2.3 (Fubini's theorem). *Let $(X, \mathcal{A}, \mu)$ be the product space formed by $(X_1, \mathcal{A}_1, \mu_1)$ and $(X_2, \mathcal{A}_2, \mu_2)$, and let a μ integrable function $f: X \to R$ be given. Then, for almost every x_1, the function $f(x_1, x_2)$ is μ_2 integrable with respect to x_2. Furthermore the function*

$$\int_{X_2} f(x_1, x_2) \, \mu_2(dx_2)$$

of the variable x_1 is μ_1 integrable and

$$\int_{X_1} \left\{ \int_{X_2} f(x_1, x_2) \mu_2(dx_2) \right\} \mu_1(dx_1) = \int\int_X f(x_1, x_2) \, \mu(dx_1 dx_2). \quad (2.2.9)$$

Theorem 2.2.3 extends, in a natural way, to product spaces with an arbitrary number of factors. If $(X, \mathcal{A}, \mu)$ is the product of the measure spaces $(X_i, \mathcal{A}_i, \mu_i)$, $i = 1, \ldots, d$, and $f: X \to R$ is μ integrable, then

$$\int_X \cdots \int f(x_1, \ldots, x_d) \, \mu(dx_1 \cdots dx_d) \qquad (2.2.10)$$

$$= \int_{X_1} \left\{ \cdots \int_{X_{d-1}} \left[\int_{X_d} f(x_1, \ldots, x_d) \mu_d(dx_d) \right] \mu_{d-1}(dx_{d-1}) \cdots \right\} \mu_1(dx_1).$$

Remark 2.2.11. As we noted in Remark 2.1.3, the "natural" Borel measure on the real line R is defined on the smallest σ-algebra $\mathcal{B}$ that contains all intervals. For every interval $[a, b]$ this measure satisfies $\mu([a, b]) = b - a$. Having the structure $(R, \mathcal{B}, \mu)$, we define by Theorem 2.2.2 the product space $(R^d, \mathcal{B}^d, \mu^d)$, where

$$R^d = R \times \overset{d}{\cdots} \times R,$$

$\mathcal{B}^d$ is the smallest σ-algebra containing all sets of the form

$$A_1 \times \cdots \times A_d \qquad \text{with } A_i \in \mathcal{B},$$

and

$$\mu^d(A_1 \times \cdots \times A_d) = \mu(A_1) \cdots \mu(A_d). \tag{2.2.11}$$

The measure μ^d is again called the Borel measure. It is easily verified that $\mathcal{B}^d$ may be alternately defined as either the smallest σ-algebra containing all the rectangles

$$[a_1, b_1] \times \cdots \times [a_d, b_d],$$

or as the smallest σ-algebra containing all the open subsets of R^d. From (2.2.11) it follows that

$$\mu^d([a_1, b_1] \times \cdots \times [a_d, b_d]) = (b_1 - a_1) \cdots (b_d - a_d),$$

which is the classical formula for the volume of an n-dimensional box.

The same construction may be repeated by starting, not from the whole real line R, but from the unit interval $[0, 1]$ or from any other finite interval. Thus, from Theorem 2.2.2, we will obtain the Borel measure on the unit square $[0, 1] \times [0, 1]$ or on the d-dimensional cube

$$[0, 1]^d = [0, 1] \times \overset{d}{\cdots} \times [0, 1].$$

In all cases (R^d, $[0, 1]^d$, etc.) we will omit the superscript d on $\mathcal{B}^d$ and μ^d and write $(R^d, \mathcal{B}, \mu)$ instead of $(R^d, \mathcal{B}^d, \mu^d)$. Furthermore, in all cases when the space is R, R^d, or any subset of these ($[0, 1], [0, 1]^d, R^+ = [0, \infty)$, etc.) and the measure and σ-algebra are not specified, we will assume that the measure space is taken with the Borel σ-algebra and Borel measure. Finally, all the integrals on R or R^d taken with respect to the Borel measure will be written with dx instead of $\mu(dx)$. $\square$

Remark 2.2.12. From the additivity property of a measure (Definition 2.1.2c) it follows that every measure is **monotonic**, that is, if A and B are measurable sets and $A \subset B$ then $\mu(A) \leq \mu(B)$. This follows directly from

$$\mu(B) = \mu(A \cup (B \setminus A)) = \mu(A) + \mu(B \setminus A).$$

Thus, if $\mu(B) = 0$ and $A \subset B$, then $\mu(A) = 0$. However, it could happen that $A \subset B$ and B is a measurable set while A is not. In this case, if $\mu(B) = 0$, then it does not follow that $\mu(A) = 0$, because $\mu(A)$ is not defined, which is a peculiar situation.

It is rather natural, therefore, to require that a "good" measure have the property that subsets of measurable sets of measure zero should also be measurable with, of course, measure zero. If a measure has this property it is called **complete**. Indeed, it can be proved that, if $(X, \mathcal{A}, \mu)$ is a measure space, then there exists a smallest σ-algebra $\mathcal{A}_1 \supset \mathcal{A}$ and a measure μ_1 on $\mathcal{A}_1$ identical with μ on $\mathcal{A}$ such that $(X_1, \mathcal{A}_1, \mu_1)$ is complete.

Every Borel measure on R (or R^d, $[0, 1]$, $[0, 1]^d$, etc.) can be completed. This complete measure is called the **Lebesgue measure**. However, when working in R (or R^d, etc.), we will use the Borel measure and *not* the Lebesgue measure, because, with the Lebesgue measure, we encounter problems with the measurability of the composition of measurable functions that are avoided with the Borel measure. $\quad\square$

2.3   Convergence of Sequences of Functions

Having defined L^p spaces and introduced the notions of norms and scalar products, we now consider three different types of convergence for a sequence of functions.

Definition 2.3.1. A sequence of functions $\{f_n\}$, $f_n \in L^p$, $1 \le p < \infty$, is (weakly) **Cesàro convergent** to $f \in L^p$ if

$$\lim_{n \to \infty} \frac{1}{n} \sum_{k=1}^{n} \langle f_k, g \rangle = \langle f, g \rangle \qquad \text{for all } g \in L^{p'}. \tag{2.3.1}$$

Definition 2.3.2. A sequence of functions $\{f_n\}$, $f_n \in L^p$, $1 \le p < \infty$, is **weakly convergent** to $f \in L^p$ if

$$\lim_{n \to \infty} \langle f_n, g \rangle = \langle f, g \rangle \qquad \text{for all } g \in L^{p'}. \tag{2.3.2}$$

Definition 2.3.3. A sequence of functions $\{f_n\}$, $f_n \in L^p$, $1 \le p \le \infty$, is **strongly convergent** to $f \in L^p$ if

$$\lim_{n \to \infty} \|f_n - f\|_{L^p} = 0. \tag{2.3.3}$$

From the Cauchy–Hölder inequality, we have

$$|\langle f_n - f, g \rangle| \le \|f_n - f\|_{L^p} \cdot \|g\|_{L^{p'}},$$

and, thus, if $\|f_n - f\|_{L^p}$ converges to zero, so must $\langle f_n - f, g \rangle$. Hence strong convergence implies weak convergence, and the condition for strong convergence is relatively straightforward to check. However, the condition for weak convergence requires a demonstration that it holds for all $g \in L^{p'}$, which seems difficult to do at first glance. In some special and important spaces, it is sufficient to check weak convergence for a restricted class of functions, defined as follows.

Definition 2.3.4. A subset $K \subset L^p$ is called **linearly dense** if for each $f \in L^p$ and $\varepsilon > 0$ there are $g_1, \ldots, g_n \in K$ and constants $\lambda_1, \ldots, \lambda_n$, such that

$$\|f - g\|_{L^p} < \varepsilon,$$

where

$$g = \sum_{i=1}^{n} \lambda_i g_i.$$

By using the notion of linearly dense sets, it is possible to simplify the proof of weak convergence. If the sequence $\{f_n\}$ is bounded in norm, that is, $\|f_n\|_{L^p} \leq c < \infty$, and if K is linearly dense in $L^{p'}$, then it is sufficient to check weak convergence in Definition 2.3.2 for any $g \in K$.

It is well known that in the space $L^p([0,1])$ $(1 \leq p < \infty)$ the following sets are linearly dense:

$K_1 = \{$the set of characteristic functions $1_\Delta(x)$ of the Borel sets
    $\Delta \subset [0,1]\}$,
$K_2 = \{$the set of continuous functions on $[0,1]\}$,
$K_3 = \{\sin(n\pi x); n = 1, 2, \ldots\}$.

In K_1 it is enough to take a family of sets Δ that are generators of Borel sets on $[0,1]$, for example, $\{\Delta\}$ could be the family of subintervals of $[0,1]$. Observe that the linear density of K_3 follows from the Fourier expansion theorem. In higher dimensions, for instance on a square in the plane, we may take analogous sets K_1 and K_2 but replace K_3 with

$$K_3' = \{\sin(m\pi x)\sin(n\pi y): n, m = 1, 2, \ldots\}.$$

Example 2.3.1. Consider the sequence of functions $f_n(x) = \sin(nx)$ on $L^2([0,1])$. We are going to show that $\{f_n\}$ converges weakly to $f \equiv 0$. First observe that

$$\|f_n\|_{L^2} = \left(\int_0^1 \sin^2 nx\, dx\right)^{1/2} = \left|\frac{1}{2} - \frac{\sin 2n}{4n}\right|^{1/2} \leq 1,$$

and hence the sequence $\{\|f_n\|_{L^2}\}$ is bounded. Now take an arbitrary function $g(x) = \sin(m\pi x)$ from K_3. We have

$$\langle f_n, g \rangle = \int_0^1 \sin(nx)\sin(m\pi x)\, dx$$

$$= \frac{\sin(n - m\pi)}{2(n - m\pi)} - \frac{\sin(n + m\pi)}{2(n + m\pi)}$$

so that

$$\lim_{n\to\infty} \langle f_n, g \rangle = \langle 0, g \rangle = 0, \qquad \text{for } g \in K_3$$

and $\{f_n\}$ thus converges weakly to $f = 0$. $\square$

We have seen that, in a given L^p space, strong convergence implies weak convergence. It also turns out that we may compare convergence in different L^p spaces using the following proposition.

Proposition 2.3.1. *If $(X, \mathcal{A}, \mu)$ is a finite measure space and $1 \le p_1 < p_2 \le \infty$, then*

$$\|f\|_{L^{p_1}} \le c\|f\|_{L^{p_2}} \qquad \text{for every } f \in L^{p_2} \tag{2.3.4}$$

where c depends on $\mu(X)$. Thus every element of L^{p_2} belongs to L^{p_1}, and strong convergence in L^{p_2} implies strong convergence in L^{p_1}.

Proof. Let $f \in L^{p_2}$ and let $p_2 < \infty$. By setting $g = |f|^{p_1}$, we obtain

$$\|f\|_{L^{p_1}}^{p_1} = \int_X |f|^{p_1} \mu(dx) = \langle 1, g \rangle.$$

Setting $p' = p_2/p_1$ and denoting by p the number adjoint to p', that is, $(1/p) + (1/p') = 1$, we have

$$\langle 1, g \rangle \le \|1\|_{L^p} \cdot \|g\|_{L^{p'}} = \left[\int_X \mu(dx)\right]^{1/p} \left[\int_X |f|^{p_1 p'} \mu(dx)\right]^{1/p'}$$
$$= \mu(X)^{1/p} \|f\|_{L^{p_2}}^{p_1}$$

and, consequently,

$$\|f\|_{L^{p_1}}^{p_1} \le \mu(X)^{1/p} \|f\|_{L^{p_2}}^{p_1},$$

which proves equation (2.3.4). Hence, if $\|f\|_{L^{p_2}}$ is finite, then $\|f\|_{L^{p_1}}$ is also finite, proving that L^{p_2} is contained in L^{p_1}. Furthermore, the inequality

$$\|f_n - f\|_{L^{p_1}} \le c\|f_n - f\|_{L^{p_2}}$$

implies that strong convergence in L^{p_2} is stronger than strong convergence in L^{p_1}. If $p_2 = \infty$, the inequality (2.3.4) is obvious, and thus the proof is complete. ∎

Observe that the strong convergence of f_n to f in L^1 (with arbitrary measure) as well as the strong convergence of f_n to f in L^p ($p > 1$) with finite measure both imply

$$\lim_{n \to \infty} \int_X f_n \mu(dx) = \int_X f \mu(dx).$$

To see this simply note that

$$\left| \int_X f_n \mu(dx) - \int_X f \mu(dx) \right| \le \int_X |f_n - f| \, \mu(dx) = \|f_n - f\|_{L^1} \le c\|f_n - f\|_{L^p}.$$

It is often necessary to define a function as a limit of a convergent sequence and/or as a sum of a convergent series. Thus the question arises how to show that a sequence $\{f_n\}$ is convergent if the limit is unknown. The famous **Cauchy condition for convergence** provides such a tool. To understand this condition, first assume that $\{f_n\}$, $f_n \in L^p$, is strongly convergent to f. Take $\varepsilon > 0$. Then there is an integer n_0 such that

$$\|f_n - f\|_{L^p} \leq \tfrac{1}{2}\varepsilon \qquad \text{for } n \geq n_0$$

and, in particular,

$$\|f_{n+k} - f\|_{L^p} \leq \tfrac{1}{2}\varepsilon \qquad \text{for } n \geq n_0 \text{ and } k \geq 0.$$

From this and the triangle inequality, we obtain

$$\|f_{n+k} - f_n\|_{L^p} \leq \|f_{n+k} - f\|_{L^p} + \|f - f_n\|_{L^p} \leq \varepsilon.$$

Thus we have proved that, if $\{f_n\}$ is strongly convergent in L^p to f, then

$$\lim_{n \to \infty} \|f_{n+k} - f_n\|_{L^p} = 0 \qquad \text{uniformly for all } k \geq 0. \tag{2.3.5}$$

This is the Cauchy condition for convergence.

It can be proved that all L^p spaces $(1 \leq p \leq \infty)$ have the property that condition (2.3.5) is also sufficient for convergence. This is stated more precisely in the following theorem.

Theorem 2.3.1. *Let $(X, \mathcal{A}, \mu)$ be a measure space and let $\{f_n\}$, $f_n \in L^p(X, \mathcal{A}, \mu)$ be a sequence such that equation (2.3.5) holds. Then there exists an element $f \in L^p(X, \mathcal{A}, \mu)$ such that $\{f_n\}$ converges strongly to f, that is, condition (2.3.3) holds.*

The fact that Theorem 2.3.1 holds for L^p spaces is referred to by saying that L^p spaces are **complete**.

Theorem 2.3.1 enables us to prove the convergence of series by the use of a **comparison series**. Suppose we have a sequence $\{g_n\} \subset L^p$ and we know the series of norms $\|g_n\|_{L^p}$ is convergent, that is,

$$\sum_{n=0}^{\infty} \|g_n\|_{L^p} < \infty. \tag{2.3.6}$$

Then, using Theorem 2.3.1, it is easy to verify that the series

$$\sum_{n=0}^{\infty} g_n \tag{2.3.7}$$

is also strongly convergent and that its sum is an element of L^p.

To see this note that the convergence of (2.3.7) simply means that the sequence of partial sums

$$s_n = \sum_{m=0}^{n} g_m$$

is convergent. To verify that $\{s_n\}$ is convergent, set

$$\sigma_n = \sum_{m=0}^{n} \|g_m\|_{L^p}.$$

From equation (2.3.6) the sequence of real numbers $\{\sigma_n\}$ is convergent and, therefore, the Cauchy condition holds for this sequence. Thus

$$\lim_{n \to \infty} |\sigma_{n+k} - \sigma_n| = 0 \qquad \text{uniformly for } k \geq 0.$$

Further

$$\|s_{n+k} - s_n\|_{L^p} = \left\| \sum_{m=n+1}^{n+k} g_m \right\|_{L^p} \leq \sum_{m=n+1}^{n+k} \|g_m\|_{L^p} = |\sigma_{n+k} - \sigma_n|$$

so finally

$$\lim_{n \to \infty} \|s_{n+k} - s_n\|_{L^p} = 0 \qquad \text{uniformly for } k \geq 0,$$

which is the Cauchy condition for $\{s_n\}$.

Exercises

2.1. Using Definition 2.1.2 prove the following "continuity properties" of the measure:

(a) If $\{A_n\}$ is a sequence of sets belonging to $\mathcal{A}$ and $A_1 \subset A_2 \subset \ldots$, then

$$\lim_{n \to \infty} \mu(A_n) = \mu \left(\bigcup_{n=1}^{\infty} A_n \right).$$

(b) If $\{A_n\}$ is a sequence of sets belonging to $\mathcal{A}$ and $A_1 \supset A_2 \supset \ldots$, then

$$\lim_{n \to \infty} \mu(A_n) = \mu \left(\bigcap_{n=1}^{\infty} A_n \right).$$

2.2. Let $X = \{1, 2, \ldots\}$ be the set of positive integers. For each $A \subset X$ define

$$k(n, A) = \text{the number of elements of the set } A \cap \{1, \ldots, n\}.$$

Let $\mathcal{A}$ be the family of all $A \subset X$ for which there exists "the average density of A in X" given by

$$\mu(A) = \lim_{n \to \infty} \frac{1}{n} k(n, A).$$

Is μ a measure? [More precisely, is $(X, \mathcal{A}, \mu)$ a measure space?]

2.3. Let $X = [a, b]$ be a compact interval and μ the standard Borel measure. Prove that for a continuous $f : [a, b] \to R$ the values of the Lebesgue and the Riemann integral coincide.

2.4. Let $X = R^+$ and μ be the standard Borel measure. Prove that a continuous function $f : R^+ \to R$ is Lebesgue integral if and only if

$$\lim_{a \to \infty} \int_0^a |f(x)| dx < \infty,$$

and that

$$\int_{R^+} f(x)\mu(dx) = \lim_{n \to \infty} \int_0^a f(x)\, dx.$$

2.5. Consider the space $(X, \mathcal{A}, \mu)$ where $X = \{1, 2, \ldots\}$ is the set of positive integers, $\mathcal{A}$ all subsets of X and μ the counting measure. Prove that a function $f : X \to R$ is integrable if and only if

$$\sum_{k=1}^{\infty} |f(x)| < \infty,$$

and that

$$\int_X f(x)\mu(dx) = \sum_{k=1}^{\infty} f(k).$$

[**Remark.** $L^1(X, \mathcal{A}, \mu)$ is therefore identical with the space of all absolutely convergent sequences. It is denoted by l^1.]

2.6. From Proposition 2.3.1 we have derived the statement: if $1 \le p_1 < p_2 \le \infty$ and $\mu(X) < \infty$, then the strong convergence of $\{f_n\}$ to f in $L^{p_2}(f_n, f \in L^{p_2})$ implies the strong convergence of $\{f_n\}$ to f in L^{p_1}. Construct an example showing that this statement is false when $\mu(X) = \infty$ even if f_n, $f \in L^{p_1} \cap L^{p_2}$.

2.7. Let $(X, \mathcal{A}, \mu)$ be a finite measure space and let $f \in L^\infty(X)$ be fixed. Show that the function

$$\varphi(p) = \|f\|_{L^p}, \qquad 1 \le p < \infty$$

is continuous and that

$$\lim_{n \to \infty} \varphi(p) = \text{ess sup} |f|.$$

2.8. The spaces $L^p(X, \mathcal{A}, \mu)$ are seldom considered for $0 < p < 1$ because in this case an important property of the norm $\| \cdot \|_{L^p}$ given by formulas (2.2.2) is not satisfied. Which property?

3
Markov and Frobenius–Perron Operators

Taking into account the concepts of the preceding chapter, we are now ready to formally introduce the Frobenius–Perron operator, which, as we saw in Chapter 1, is of considerable use in studying the evolution of densities under the operation of deterministic systems.

However, as a preliminary step, we develop the more general concept of the Markov operator and derive some of its properties. Our reasons for this approach are twofold: First, as will become clear, many concepts concerning the asymptotic behavior of densities may be equally well formulated for both deterministic and stochastic systems. Second, many of the results that we develop in later chapters concerning the behavior of densities evolving under the influence of deterministic systems are simply special cases of more general results for stochastic systems.

The theory of Markov operators is extremely rich and varied, and we have chosen an approach particularly suited to an examination of the eventual behavior of densities in dynamical systems. Foguel [1969] contains an exhaustive survey of the asymptotic properties of Markov operators.

3.1 Markov Operators

We define the Markov operator as follows.

Definition 3.1.1. Let $(X, \mathcal{A}, \mu)$ be a measure space. Any linear operator $P: L^1 \to L^1$ satisfying

(a) $Pf \geq 0$ for $f \geq 0, f \in L^1$; and

(b) $\|Pf\| = \|f\|,$ for $f \geq 0, f \in L^1$ (3.1.1)

is called a **Markov operator**.

Remark 3.1.1. In conditions (a) and (b), the symbols f and Pf denote elements of L^1 represented by functions that can differ on a set of measure zero. Thus, for any such function, properties $f \geq 0$ and $Pf \geq 0$ hold almost everywhere. When it is clear that we are dealing with elements of L^1 (or L^p), we will drop the "almost everywhere" notation. $\square$

Markov operators have a number of properties that we will have occasion to use. First, if $f, g \in L^1$, then

$$Pf(x) \geq Pg(x) \quad \text{whenever } f(x) \geq g(x). \tag{3.1.2}$$

Any operator P satisfying (3.1.2) is said to be **monotonic**. To show the monotonicity of P is trivial, since $(f - g) \geq 0$ implies $P(f - g) \geq 0$.

To demonstrate further inequalities that Markov operators satisfy, we offer the following proposition.

Proposition 3.1.1. *If $(X, \mathcal{A}, \mu)$ is a measure space and P is a Markov operator, then, for every $f \in L^1$,*

(M1) $(Pf(x))^+ \leq Pf^+(x)$ (3.1.3)

(M2) $(Pf(x))^- \leq Pf^-(x)$ (3.1.4)

(M3) $|Pf(x)| \leq P|f(x)|$ (3.1.5)

and

(M4) $\|Pf\| \leq \|f\|$.   (3.1.6)

Proof. These inequalities are straightforward to derive. To obtain (3.1.3), note that from the definition of f^+ and f^-, it follows that

$$(Pf)^+ = (Pf^+ - Pf^-)^+ = \max(0, Pf^+ - Pf^-)$$
$$\leq \max(0, Pf^+) = Pf^+;$$

and inequality (3.1.4) is obtained in an analogous fashion. Inequality (3.1.5) follows from (M1) and (M2), namely,

$$|Pf| = (Pf)^+ + (Pf)^- \leq Pf^+ + Pf^-$$
$$= P(f^+ + f^-) = P|f|.$$

Finally, by integrating (3.1.5) over X, we have

$$\|Pf\| = \int_X |Pf(x)|\mu(dx) \le \int_X P|f(x)|\mu(dx)$$
$$= \int_X |f(x)|\mu(dx) = \|f\|,$$

which confirms (3.1.6). ∎

Inequality (3.1.6) is extremely important, and any operator P that satisfies it is called a **contraction**. The actual inequality (3.1.6) is known as the **contractive property** of P. To illustrate its power note that for any $f \in L^1$ we have

$$\|P^n f\| = \|P(P^{n-1}f)\| \le \|P^{n-1}f\|$$

and, thus, for any two $f_1, f_2 \in L^1, f_1 \ne f_2$,

$$\|P^n f_1 - P^n f_2\| = \|P^n(f_1 - f_2)\|$$
$$\le \|P^{n-1}(f_1 - f_2)\| = \|P^{n-1}f_1 - P^{n-1}f_2\|. \quad (3.1.7)$$

Inequality (3.1.7) simply states that during the process of iteration of two individual functions the distance between them can decrease, but it can never increase. This is referred to as the **stability property** of iterates of Markov operators.

By the **support** of a function g we simply mean the set of all x such that $g(x) \ne 0$, that is,

$$\text{supp } g = \{x : g(x) \ne 0\}. \quad (3.1.8)$$

Remark 3.1.2. This is not the customary definition of the support of a function, which is usually defined by

$$\text{supp } g = \text{closure}\{x : g(x) \ne 0\}. \quad (3.1.9)$$

But, because the customary definition (3.1.9) requires the introduction of topological notions not used elsewhere, we have presented the slightly unusual definition (3.1.8). □

Remark 3.1.3. If g is an element of L^p, then the set (3.1.8) is not defined in a completely unique manner, since g may be represented by functions that differ on a set of measure zero. This inaccuracy never leads to any difficulties in calculating measures and integrals. Thus, it is customary to simplify the terminology and to talk about the supports of elements from L^p as if we were speaking of sets. However, if we want to emphasize that a relation between sets does not hold precisely but may be violated on a set of measure zero, we say that it holds **modulo zero**. Thus $A = B$ modulo zero means that the set of x in A that does not belong to B, or vice versa, has measure zero. $\square$

One might wonder under what conditions the contractive property (3.1.6) is a strong inequality. The answer is quite simple.

Proposition 3.1.2. $\|Pf\| = \|f\|$ *if and only if* Pf^+ *and* Pf^- *have disjoint supports.*

Proof. We start from the inequality

$$|Pf^+(x) - Pf^-(x)| \le |Pf^+(x)| + |Pf^-(x)|.$$

Clearly the inequality will be strong if both $Pf^+(x) > 0$ and $Pf^-(x) > 0$, while the equality holds if $Pf^+(x) = 0$ or $Pf^-(x) = 0$. Thus, by integrating over the space X, we have

$$\int_X |Pf^+(x) - Pf^-(x)|\mu(dx) = \int_X |Pf^+(x)|\mu(dx) + \int_X |Pf^-(x)|\mu(dx)$$

if and only if there is no set $A \in \mathcal{A}$, $\mu(A) > 0$, such that $Pf^+(x) > 0$ and $Pf^-(x) > 0$ for $x \in A$, that is, $Pf^+(x)$ and $Pf^-(x)$ have disjoint support. Since $f = f^+ - f^-$, the left-hand integral is simply $\|Pf\|$. Further, the right-hand side is $\|Pf^+\| + \|Pf^-\| = \|f^+\| + \|f^-\| = \|f\|$, so the proposition is proved. $\blacksquare$

Having developed some of the more important elementary properties of Markov operators, we now introduce the concept of a fixed point of P.

Definition 3.1.2. If P is a Markov operator and, for some $f \in L^1$, $Pf = f$ then f is called a **fixed point** of P.

From Proposition 3.1.1 it is easy to show the following.

Proposition 3.1.3. *If* $Pf = f$ *then* $Pf^+ = f^+$ *and* $Pf^- = f^-$.

Proof. Note that from $Pf = f$ we have

$$f^+ = (Pf)^+ \le Pf^+ \quad \text{and} \quad f^- = (Pf)^- \le Pf^-,$$

hence

$$\int_X [Pf^+(x) - f^+(x)]\mu(dx) + \int_X [Pf^-(x) - f^-(x)]\mu(dx)$$
$$= \int_X [Pf^+(x) + Pf^-(x)]\mu(dx) - \int_X [f^+(x) + f^-(x)]\mu(dx)$$
$$= \int_X P|f(x)|\mu(dx) - \int_X |f(x)|\mu(dx)$$
$$= \|P|f|\| - \|\,|f|\,\|.$$

However, by the contractive property of P we know that

$$\| P|f| \| - \| |f| \| \leq 0.$$

Since both the integrands $(Pf^+ - f^+)$ and $(Pf^- - f^-)$ are nonnegative, this last inequality is possible only if $Pf^+ = f^+$ and $Pf^- = f^-$. ∎

Definition 3.1.3. Let $(X, \mathcal{A}, \mu)$ be a measure space and the set $D(X, \mathcal{A}, \mu)$ be defined by $D(X, \mathcal{A}, \mu) = \{f \in L^1(X, \mathcal{A}, \mu) : f \geq 0 \text{ and } \|f\| = 1\}$. Any function $f \in D(X, \mathcal{A}, \mu)$ is called a **density**.

Definition 3.1.4. If $f \in L^1(X, \mathcal{A}, \mu)$ and $f \geq 0$, then the measure

$$\mu_f(A) = \int_A f(x)\mu(dx)$$

is said to be **absolutely continuous** with respect to μ, and f is called the **Radon–Nikodym derivative** of μ_f with respect to μ. In the special case that $f \in D(X, \mathcal{A}, \mu)$, then we also say that f is the **density** of μ_f and that μ_f is a **normalized measure**.

From Corollary 2.2.1 it follows that a normalized measure ν is absolutely continuous with respect to μ if $\nu(A) = 0$ whenever $\mu(A) = 0$. This property is often used as the definition of an absolutely continuous measure.

Using the notion of densities we may extend the concept of a fixed point of a Markov operator with the following definition.

Definition 3.1.5. Let $(X, \mathcal{A}, \mu)$ be a measure space and P be a Markov operator. Any $f \in D$ that satisfies $Pf = f$ is called a **stationary density** of P.

The concept of a stationary density of an operator is extremely important and plays a central role in many of the sections that follow.

3.2   The Frobenius-Perron Operator

Having developed the concept of Markov operators and some of their properties, we are in a position to examine a special class of Markov operators, the Frobenius–Perron operator, which we introduced intuitively in Chapter 1.

We start with the following definitions.

Definition 3.2.1. Let $(X, \mathcal{A}, \mu)$ be a measure space. A transformation $S: X \to X$ is **measurable** if

$$S^{-1}(A) \in \mathcal{A} \qquad \text{for all } A \in \mathcal{A}.$$

Definition 3.2.2. A measurable transformation $S: X \to X$ on a measure space $(X, \mathcal{A}, \mu)$ is **nonsingular** if $\mu(S^{-1}(A)) = 0$ for all $A \in \mathcal{A}$ such that $\mu(A) = 0$.

Before stating a precise definition of the Frobenius–Perron operator, consider the following. Assume that a nonsingular transformation $S: X \to X$ on a measure space is given. We define an operator $P: L^1 \to L^1$ in two steps.

1. Let $f \in L^1$ and $f \geq 0$. Write

$$\int_{S^{-1}(A)} f(x)\mu(dx). \tag{3.2.1}$$

Because

$$S^{-1}\left(\bigcup_i A_i\right) = \bigcup_i S^{-1}(A_i),$$

it follows from property (L5) of the Lebesgue integral that the integral (3.2.1) defines a finite measure. Thus, by Corollary 2.2.1, there is a unique element in L^1, which we denote by Pf, such that

$$\int_A Pf(x)\mu(dx) = \int_{S^{-1}(A)} f(x)\mu(dx) \qquad \text{for } A \in \mathcal{A}.$$

2. Now let $f \in L^1$ be arbitrary, that is, not necessarily nonnegative. Write $f = f^+ - f^-$ and define

$$Pf = Pf^+ - Pf^-.$$

From this definition we have

$$\int_A Pf(x)\mu(dx) = \int_{S^{-1}(A)} f^+(x)\mu(dx) - \int_{S^{-1}(A)} f^-(x)\mu(dx)$$

or, more completely,

$$\int_A Pf(x)\mu(dx) = \int_{S^{-1}(A)} f(x)\mu(dx), \qquad \text{for } A \in \mathcal{A}. \tag{3.2.2}$$

From Proposition 2.2.1 and the nonsingularity of S, it follows that equation (3.2.2) uniquely defines P.

We summarize these comments as follows.

Definition 3.2.3. Let $(X, \mathcal{A}, \mu)$ be a measure space. If $S: X \to X$ is a nonsingular transformation the unique operator $P: L^1 \to L^1$ defined by equation (3.2.2) is called the **Frobenius–Perron operator** corresponding to S.

It is straightforward to show from (3.2.2) that P has the following properties:

(FP1) $P(\lambda_1 f_1 + \lambda_2 f_2) = \lambda_1 P f_1 + \lambda_2 P f_2$ (3.2.3)

     for all $f_1, f_2 \in L^1$, $\lambda_1, \lambda_1 \in R$, so P is a linear operator;

(FP2) $Pf \geq 0$ if $f \geq 0$;   and   (3.2.4)

(FP3) $\displaystyle \int_X Pf(x)\mu(dx) = \int_X f(x)\mu(dx)$ (3.2.5)

(FP4) If $S_n = S \circ \overset{n}{\ldots} \circ S$ and P_n is the Frobenius–Perron operator corresponding to S_n, then $P_n = P^n$, where P is the Frobenius–Perron operator corresponding to S.

Remark 3.2.1. Although the definition of the Frobenius–Perron operator P by (3.2.2) is given by a quite abstract mathematical theorem of Radon-Nikodym, it should be realized that it describes the evolution of f by a transformation S. Properties (3.2.4–3.2.5) of the transformed distribution $Pf(x)$ are exactly what one would expect on intuitive grounds.   □

Remark 3.2.2. From the preceding section, the Frobenius–Perron operator is also a Markov operator.

As we wish to emphasize the close connection between the behavior of stochastic systems and the chaotic behavior of deterministic systems, we will formulate concepts and results for Markov operators wherever possible. The Frobenius–Perron operator is a particular Markov operator, and thus any property of Markov operators is immediately applicable to the Frobenius–Perron operator.

In some special cases equation (3.2.2) allows us to obtain an explicit form for Pf. As we showed in Chapter 1, if $X = [a, b]$ is an interval on the real line R, and $A = [-a, x]$, then (3.2.2) becomes

$$\int_a^x Pf(x)\,ds = \int_{S^{-1}([a,x])} f(s)\,ds,$$

and by differentiating

$$Pf(x) = \frac{d}{dx} \int_{S^{-1}([a,x])} f(s)\,ds. \qquad (3.2.6)$$

It is important to note that in the special case where the transformation S is differentiable and invertible, an explicit form for Pf is available. If S is differentiable and invertible, then S must be monotone. Suppose S is an increasing function and S^{-1} has a continuous derivative. Then

$$S^{-1}([a, x]) = [S^{-1}(a), S^{-1}(x)],$$

and from (3.2.6)

$$Pf(x) = \frac{d}{dx} \int_{S^{-1}(a)}^{S^{-1}(x)} f(s)\,ds = f(S^{-1}(x)) \frac{d}{dx}[S^{-1}(x)].$$

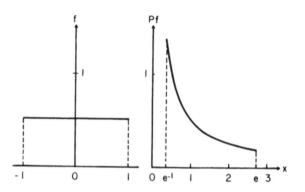

FIGURE 3.2.1. Operation of the Frobenius–Perron operator corresponding to $S(x) = e^x$, $x \in R$. (a) An initial density $f(x) = \frac{1}{2}1_{[-1,1]}(x)$ is transformed into the density $Pf(x) = (2x)^{-1}1_{[e^{-1},e]}(x)$ by S as shown in (b).

If S is decreasing, then the sign of the right-hand side is reversed. Thus, in the general one-dimensional case, for S differentiable and invertible with continuous dS^{-1}/dx,

$$Pf(x) = f(S^{-1}(x))\left|\frac{d}{dx}[S^{-1}(x)]\right|. \tag{3.2.7}$$

Example 3.2.1. To see how the Frobenius–Perron operator works, pick $S(x) = \exp(x)$. Thus from (3.2.7), we have

$$Pf(x) = (1/x)f(\ln x).$$

Consider what happens to an initial f given by

$$f(x) = \frac{1}{2}1_{[-1,1]}(x),$$

and shown in Figure 3.2.1a. Under the action of P, the function f is carried into

$$Pf(x) = (1/2x)1_{[e^{-1},e]}(x)$$

as shown in Figure 3.2.1b. $\quad\square$

Two important points are illustrated by this example. The first is that for an initial f supported on a set $[a,b]$, Pf will be supported on $[S(a), S(b)]$. Second, Pf is small where (dS/dx) is large and vice versa.

We generalize the first observation as follows.

Proposition 3.2.1. Let $S: X \to X$ be a nonsingular transformation and P the associated Frobenius–Perron operator. Assume that an $f \geq 0$, $f \in L^1$, is given. Then

$$\text{supp}f \subset S^{-1}(\text{supp}Pf) \tag{3.2.8}$$

and, more generally, for every set $A \in \mathcal{A}$ the following equivalence holds: $Pf(x) = 0$ for $x \in A$ if and only if $f(x) = 0$ for $x \in S^{-1}(A)$.

Proof. The proof is straightforward. By the definition of the Frobenius–Perron operator, we have

$$\int_A Pf(x)\mu(dx) = \int_{S^{-1}(A)} f(x)\mu(dx)$$

or

$$\int_X 1_A(x)Pf(x)\mu(dx) = \int_X 1_{S^{-1}(A)}(x)f(x)\mu(dx).$$

Thus $Pf(x) = 0$ on A implies, by property (L2) of the Lebesgue integral, that $f(x) = 0$ for $x \in S^{-1}(A)$, and vice versa. Now setting $A = X \setminus \text{supp}(Pf)$, we have $Pf(x) = 0$ for $x \in A$ and, consequently, $f(x) = 0$ for $x \in S^{-1}(A)$, which means that $\text{supp } f \subset X \setminus S^{-1}(A)$. Since $S^{-1}(A) = X \setminus S^{-1}(\text{supp}(Pf))$, this completes the proof. ■

Remark 3.2.3. In the case of arbitrary $f \in L^1$, then, in Proposition 3.2.1 we only have: If $f(x) = 0$ for all $x \in S^{-1}(A)$, then $Pf(x) = 0$ for all $x \in A$. That the converse is not true can be seen by the following example. Take $S(x) = 2x \pmod 1$ and let

$$f(x) = \begin{cases} 1 & 0 \le x < \frac{1}{2} \\ -1 & \frac{1}{2} \le x \le 1. \end{cases}$$

Then from (1.2.13) $Pf(x) = 0$ for all $x \in [0,1)$ but $f(x) \ne 0$ for any $x \in [0,1]$. □

For a second important case consider the rectangle $X = [a,b] \times [c,d]$ in the plane R^2. Set $A = [a,x] \times [c,y]$ so that (3.2.2) now becomes

$$\int_a^x ds \int_b^y Pf(s,t)\, dt = \int\int_{S^{-1}([a,x]\times[c,y])} f(s,t)\, ds\, dt.$$

Differentiating first with respect to x and then with respect to y, we have immediately that

$$Pf(x,y) = \frac{\partial^2}{\partial y\, \partial x} \int\int_{S^{-1}([a,x]\times[c,y])} f(s,t)\, ds\, dt.$$

Analogous formulas can be derived in the case of $X \subset R^d$.

In the general case, where $X = R^d$ and $S: X \to X$ is invertible, we can derive an interesting and useful generalization of equation (3.2.7). To do this we first state and prove a change of variables theorem based on the Radon–Nikodym theorem.

Theorem 3.2.1. Let $(X, \mathcal{A}, \mu)$ be a measure space, $S: X \rightarrow X$ a non-singular transformation, and $f: X \rightarrow X$ a measurable function such that $f \circ S \in L^1(X, \mathcal{A}, \mu)$. Then for every $A \in \mathcal{A}$,

$$\int_{S^{-1}(A)} f(S(x))\mu(dx) = \int_A f(x)\mu S^{-1}(dx) = \int_A f(x)J^{-1}(x)\mu(dx)$$

where μS^{-1} denotes the measure

$$\mu S^{-1}(B) = \mu(S^{-1}(B)), \qquad \text{for } B \in \mathcal{A},$$

and J^{-1} is the density of μS^{-1} with respect to μ, that is,

$$\mu(S^{-1}(B)) = \int_B J^{-1}(x)\mu(dx) \qquad \text{for } B \in \mathcal{A}.$$

Remark 3.2.4. We use the notation $J^{-1}(x)$ to draw the connection with differentiable invertible transformations on R^d, in which case $J(x)$ is the determinant of the Jacobian matrix:

$$J(x) = \left|\frac{dS(x)}{dx}\right| \quad \text{or} \quad J^{-1}(x) = \left|\frac{dS^{-1}(x)}{dx}\right|. \quad \square$$

Proof of Theorem 3.2.1. To prove this change of variables theorem, we recall Remark 2.2.6 and first take $f(x) = 1_B(x)$ so that $f(S(x)) = 1_B(S(x)) = 1_{S^{-1}(B)}(x)$ and, hence,

$$\int_{S^{-1}(A)} f(S(x))\mu(dx) = \int_X 1_{S^{-1}(A)}(x)f(S(x))\mu(dx)$$

$$= \int_X 1_{S^{-1}(A)}(x)1_{S^{-1}(B)}(x)\mu(dx)$$

$$= \mu(S^{-1}(A) \cap S^{-1}(B)) = \mu(S^{-1}(A \cap B)).$$

The second integral of the theorem may be written as

$$\int_A f(x)\mu S^{-1}(dx) = \int_X 1_A(x)1_B(x)\mu S^{-1}(dx) = \mu(S^{-1}(A \cap B))$$

whereas the third and last integral has the form

$$\int_A f(x)J^{-1}(x)\mu(dx) = \int_A 1_B(x)J^{-1}(x)\mu(dx)$$

$$= \int_{A \cap B} J^{-1}(x)\mu(dx) = \mu(S^{-1}(A \cap B)).$$

Thus we have shown that the theorem is true for functions of the form $f(x) = 1_B(x)$. To complete the proof we need only to repeat it for simple functions $f(x)$, which will certainly be true by linearity [property (L3)] of the Lebesgue integral. Finally, we may pass to the limit for arbitrary bounded and integrable function f. [Note that f bounded is required for the integrability of $f(x)J^{-1}(x)$.]   ∎

With this change of variables theorem it is easy to prove the following extension of equation (3.2.7).

Corollary 3.2.1. Let $(X, \mathcal{A}, \mu)$ be a measure space, $S: X \to X$ an invertible nonsingular transformation $(S^{-1}$ nonsingular) and P the associated Frobenius–Perron operator. Then for every $f \in L^1$

$$Pf(x) = f(S^{-1}(x))J^{-1}(x). \tag{3.2.9}$$

Proof. By the definition of P, for $A \in \mathcal{A}$ we have

$$\int_A Pf(x)\mu(dx) = \int_{S^{-1}(A)} f(x)\mu(dx).$$

Change the variables in the right-hand integral with $y = S(x)$, so that

$$\int_{S^{-1}(A)} f(x)\mu(dx) = \int_A f(S^{-1}(y))J^{-1}(y)\mu(dy)$$

by Theorem 3.2.1. Thus we have

$$\int_A Pf(x)\mu(dx) = \int_A f(S^{-1}(x))J^{-1}(x)\mu(dx)$$

with the result that, by Proposition 2.2.1,

$$Pf(x) = f(S^{-1}(x))J^{-1}(x). \quad ∎$$

3.3 The Koopman Operator

To close this chapter, we define a third type of operator closely related to the Frobenius–Perron operator.

Definition 3.3.1. Let $(X, \mathcal{A}, \mu)$ be a measure space, $S: X \to X$ a nonsingular transformation, and $f \in L^\infty$. The operator $U: L^\infty \to L^\infty$ defined by

$$Uf(x) = f(S(x)) \tag{3.3.1}$$

is called the **Koopman operator** with respect to S.

This operator was first introduced by Koopman [1931]. Due to the non-singularity of S, U is well defined since $f_1(x) = f_2(x)$ a.e. implies $f_1(S(x)) = f_2(S(x))$ a.e. Operator U has some important properties:

(K1) $U(\lambda_1 f_1 + \lambda_2 f_2) = \lambda_1 U f_1 + \lambda_2 U f_2$ (3.3.2)

  for all $f_1, f_2 \in L^\infty, \lambda_1, \lambda_2 \in R$;

(K2) For every $f \in L^\infty$,

$$\|Uf\|_{L^\infty} \le \|f\|_{L^\infty},$$ (3.3.3)

  that is, U is a contraction of L^∞;

(K3) For every $f \in L^1$, $g \in L^\infty$,

$$\langle Pf, g \rangle = \langle f, Ug \rangle$$ (3.3.4)

  so that U is adjoint to the Frobenius–Perron operator P.

Property (K1) is trivial to check. Further, property (K2) follows immediately from the definition of the norm since $|f(x)| \le \|f\|_{L^\infty}$ a.e. implies $|f(S(x))| \le \|f\|_{L^\infty}$ a.e. The latter inequality gives equation (3.3.3) since, by (3.3.1), $Uf(x) = f(S(x))$.

Finally, to obtain (K3) we first check it with $g = 1_A$. Then the left-hand side of (3.3.4) becomes

$$\langle Pf, g \rangle = \int_X Pf(x) 1_A(x) \mu(dx) = \int_A Pf(x) \mu(dx),$$

while the right-hand side becomes

$$\langle f, Ug \rangle = \int_X f(x) U 1_A(x) \mu(dx)$$

$$= \int_X f(x) 1_A(S(x)) \mu(dx) = \int_{S^{-1}(A)} f(x) \mu(dx).$$

Thus (K3) is equivalent to

$$\int_A Pf(x) \mu(dx) = \int_{S^{-1}(A)} f(x) \mu(dx)$$

which is the equation defining Pf. Because (K3) is true for $g(x) = 1_A(x)$ it is true for any simple function $g(x)$. Thus, by Remark 2.2.6, property (K3) must be true for all $g \in L^\infty$.

With the Koopman operator it is easy to prove that the Frobenius–Perron operator is weakly continuous. Precisely, this means that for every sequence $\{f_n\} \subset L^1$ the condition

$$f_n \to f \text{ weakly}$$ (3.3.5)

implies

$$Pf_n \to Pf \text{ weakly.} \tag{3.3.6}$$

To show this note that by property (K3) we have

$$\langle Pf_n, g \rangle = \langle f_n, Ug \rangle \qquad \text{for } g \in L^\infty.$$

Furthermore, from (3.3.5) it follows that $\langle f_n, Ug \rangle$ converges to $\langle f, Ug \rangle = \langle Pf, g \rangle$, which means that Pf_n converges weakly to Pf.

The same proof can be carried out for an arbitrary Markov operator P (or even more generally for every bounded linear operator). In this case we must use the fact that for every Markov operator there exists a unique adjoint operator $P^*: L^\infty \to L^\infty$ that satisfies

$$\langle Pf, g \rangle = \langle f, P^* g \rangle \qquad \text{for } f \in L^1, g \in L^\infty.$$

Exercises

3.1. The differential equation

$$u'' - u + f(x) = 0, \qquad 0 \le x \le 1$$

with the boundary value conditions

$$u'(0) = u'(1) = 0$$

for every $f \in L^1([0,1])$ has a unique solution $u(x)$ defined for $0 \le x \le 1$. Show that the mapping that adjoins the solution u to f is a Markov operator on $L^1([0,1])$. This can be done without looking for the explicit formula for u.

3.2. Find the Frobenius–Perron operator P corresponding to the following transformations:

(a) $S: [0,1] \to [0,1]$, $S(x) = 4x^2(1 - x^2)$;

(b) $S: [0,1], \to [0,1]$, $S(x) = \sin \pi x$;

(c) $S: R \to R$, $S(x) = a \tan(bx + c)$.

In (c) observe that the value of $S(x)$ for $bx + c = n\pi$ are irrelevant for the calculation of P.

3.3. Consider the set $X = \{1, \ldots, N\}$ with the counting measure. Prove that any Markov operator $P: L^1(X) \to L^1(X)$ is given by a formula

$$(Pf)_i = \sum_{i=1}^{N} p_{ij} f_i, \qquad i = 1, \ldots, N,$$

where (p_{ij}) is a stochastic matrix, i.e.,

$$p_{ij} \geq 0, \qquad \sum_{i=1}^{N} p_{ij} = 1,$$

and f_i stands for $f(i)$.

3.4. A mapping $S: [0, 1] \to [0, 1]$ is called a **generalized tent transformation** if $S(x) = S(1 - x)$ for $0 \leq x \leq 1$ and if $S(x)$ is strictly increasing for $0 \leq x \leq \frac{1}{2}$. Show that there is a unique generalized tent transformation [given by (6.5.9)] for which the standard Borel measure is invariant.

3.5. Generalize the previous result showing that for every absolutely continuous measure μ on $[0,1]$ with positive density $(d\mu/dx > 0$ a.e.) there is a unique generalized tent transformation S such that μ is invariant with respect to S.

3.6. Let $(X, \mathcal{A}, \mu)$ be a measure space. A Markov operator $P: L^1(X) \to L^1(X)$ is called **deterministic** if its adjoint $U = P^*$ has the following property: For every $A \in \mathcal{A}$ the function $U1_A$ is a characteristic function, i.e., $U1_A = 1_B$ for some $B \in \mathcal{A}$. Show that the Frobenius–Perron operator is a deterministic operator.

3.7. Let $X = \{1, \ldots, N\}$ be a measure space with the counting measure considered in Exercise 3.3. Describe a general form of the matrix (p_{ij}) which corresponds to a deterministic operator.

3.8. Let $P_i: L^1 \to L^1$, $i = 1, 2$, denote deterministic Markov operators. Are the operators $P_1 P_2$ and $\alpha P_1 + (1 - \alpha) P_2$, $0 < \alpha < 1$, also deterministic?

3.9. Let $X = [0, 1]$. Show that $P: L^1([0, 1]) \to L^1([0, 1])$ given by the formula

$$Pf(x) = \frac{1}{2} f(x) + \frac{1}{4} f\left(\frac{x}{2}\right) + \frac{1}{4} f\left(\frac{x}{2} + \frac{1}{2}\right)$$

is not a deterministic Markov operator.

3.10. Let $P: L^1 \to L^1$ be a Markov operator. Prove that for every nonnegative $f, g \in L^1$ the condition supp $f \subset$ supp g implies supp $Pf \subset$ supp Pg.

4
Studying Chaos with Densities

Here we introduce the concept of measure-preserving transformations and then define and illustrate three levels of irregular behavior that such transformations can display. These three levels are known as ergodicity, mixing, and exactness. The central theme of the chapter is to show the utility of the Frobenius–Perron and Koopman operators in the study of these behaviors.

All these basic notions arise in ergodic theory. Roughly speaking, preservation of an initial measure μ by a transformation corresponds to the fact that the constant density $f(x) = 1$ is a stationary density of the Frobenius–Perron operator, $P1 = 1$. Ergodicity corresponds to the fact that $f(x) \equiv 1$ is the unique stationary density of the Frobenius–Perron operator. Finally, mixing and exactness correspond to two different kinds of stability of the stationary density $f(x) \equiv 1$.

In Section 4.5, we briefly introduce Kolmogorov automorphisms, which are closely related to exact transformations. This section is only of a reference nature, and, therefore, all proofs are omitted and the examples are treated superficially.

4.1 Invariant Measures and Measure-Preserving Transformations

We start with a definition.

Definition 4.1.1. Let $(X, \mathcal{A}, \mu)$ be a measure space and $S: X \to X$ a measurable transformation. Then S is said to be **measure preserving** if

$$\mu(S^{-1}(A)) = \mu(A) \qquad \text{for all } A \in \mathcal{A}.$$

Since the property of measure preservation is dependent on S as well as μ, we will alternately say that the measure μ is **invariant** under S if S is measure preserving. Note that every measure-preserving transformation is necessarily nonsingular.

Theorem 4.1.1. *Let $(X, \mathcal{A}, \mu)$ be a measure space, $S \colon X \to X$ a nonsingular transformation, and P the Frobenius–Perron operator associated with S. Consider a nonnegative $f \in L^1$. Then a measure μ_f given by*

$$\mu_f(A) = \int_A f(x)\mu(dx)$$

is invariant if and only if f is a fixed point of P.

Proof. First we show the "only if" portion. Assume μ_f is invariant. Then, by the definition of an invariant measure,

$$\mu_f(A) = \mu_f(S^{-1}(A)) \qquad \text{for all } A \in \mathcal{A},$$

or

$$\int_A f(x)\mu(dx) = \int_{S^{-1}(A)} f(x)\mu(dx) \qquad \text{for } A \in \mathcal{A}. \tag{4.1.1}$$

However, by the very definition of the Frobenius–Perron operator, we have

$$\int_{S^{-1}(A)} f(x)\mu(dx) = \int_A Pf(x)\mu(dx), \qquad \text{for } A \in \mathcal{A}. \tag{4.1.2}$$

Comparing (4.1.1) with (4.1.2) we immediately have $Pf = f$.

Conversely, if $Pf = f$ for some $f \in L^1$, $f \geq 0$, then from the definition of the Frobenius–Perron operator equation (4.1.1) follows and thus μ_f is invariant. ∎

Remark 4.1.1. Note that the original measure μ is invariant if and only if $P1 = 1$. □

Example 4.1.1. Consider the r-adic transformation originally introduced in Example 1.2.1,

$$S(x) = rx \qquad (\text{mod } 1),$$

where $r > 1$ is an integer, on the measure space $([0,1], \mathcal{B}, \mu)$ where $\mathcal{B}$ is the Borel σ-algebra and μ is the Borel measure (cf. Remark 2.1.3). As we have shown in Example 1.2.1, for any interval $[0, x] \subset [0, 1]$

$$S^{-1}([0, x]) = \bigcup_{i=0}^{r-1} \left[\frac{i}{r}, \frac{i}{r} + \frac{x}{r} \right]$$

and the Frobenius–Perron operator P corresponding to S is given by equation (1.2.13):

$$Pf(x) = \frac{1}{r}\sum_{i=0}^{r-1} f\left(\frac{i}{r} + \frac{x}{r}\right).$$

Thus

$$P1 = \frac{1}{r}\sum_{i=0}^{r-1} 1 = 1$$

and by our previous remark the Borel measure is invariant under the r-adic transformation. □

Remark 4.1.2. It should be noted that, as defined, the r-adic transformation is not continuous at $\frac{1}{2}$. However, if instead of defining the r-adic transformation on the interval $[0,1]$ we define it on the unit circle (circle with circumference of 1) obtained by identifying 0 with 1 on the interval $[0,1]$, then it is continuous and differentiable throughout. □

Example 4.1.2. Again consider the measure space $([0,1], \mathcal{B}, \mu)$ where μ is the Borel measure. Let $S\colon [0,1] \to [0,1]$ be the quadratic map $(S(x) = 4x(1-x)$ of Chapter 1). As was shown there, for $[0,x] \subset [0,1]$,

$$S^{-1}([0,x]) = [0, \tfrac{1}{2} - \tfrac{1}{2}\sqrt{1-x}] \cup [\tfrac{1}{2} + \tfrac{1}{2}\sqrt{1-x}, 1]$$

and the Frobenius–Perron operator is given by

$$Pf(x) = \frac{1}{4\sqrt{1-x}}\left\{f\left(\tfrac{1}{2} - \tfrac{1}{2}\sqrt{1-x}\right) + f\left(\tfrac{1}{2} + \tfrac{1}{2}\sqrt{1-x}\right)\right\}.$$

Clearly,

$$P1 = \frac{1}{2\sqrt{1-x}},$$

so that the Borel measure μ is not invariant under S by Remark 4.1.1. To find an invariant measure we must find a solution to the equation $Pf = f$ or

$$f(x) = \frac{1}{4\sqrt{1-x}}\left\{f\left(\tfrac{1}{2} - \tfrac{1}{2}\sqrt{1-x}\right) + f\left(\tfrac{1}{2} + \tfrac{1}{2}\sqrt{1-x}\right)\right\}.$$

This problem was first solved by Ulam and von Neumann [1947] who showed that the solution is given by

$$f_*(x) = \frac{1}{\pi\sqrt{x(1-x)}}, \tag{4.1.3}$$

which justifies our assertion in Section 1.2. It is straightforward to show that f_* as given by (4.1.3) does, indeed, constitute a solution to $Pf = f$. Hence the measure

$$\mu_{f_*}(A) = \int_A \frac{dx}{\pi\sqrt{x(1-x)}}$$

is invariant under the quadratic transformation $S(x) = 4x(1-x)$. □

Remark 4.1.3. The factor of π appearing in equation (4.1.3) ensures that f_* is a density and thus that the measure μ_{f_*} is normalized.   □

Example 4.1.3. (The baker transformation). Now let X be the unit square in a plane, which we denote by $X = [0, 1] \times [0, 1]$ (see Section 2.2). The Borel σ-algebra $\mathcal{B}$ is now generated by all possible rectangles of the form $[0, a] \times [0, b]$ and the Borel measure μ is the unique measure on $\mathcal{B}$ such that

$$\mu([0, a] \times [0, b]) = ab.$$

(Thus the Borel measure is a generalization of the concept of the area.) We define a transformation $S: X \to X$ by

$$S(x, y) = \begin{cases} (2x, \frac{1}{2}y) & 0 \le x < \frac{1}{2}, 0 \le y \le 1 \\ (2x - 1, \frac{1}{2}y + \frac{1}{2}) & \frac{1}{2} \le x \le 1, 0 \le y \le 1. \end{cases} \tag{4.1.4}$$

To understand the operation of this transformation, examine Figure 4.1.1, where X is shown in Figure 4.1.1a. The first operation of S involves a compression of X in the y direction by $\frac{1}{2}$ and a stretching of X in the x direction by a factor of 2 (Figure 4.1.b). The transformation S is completed by vertically dividing the compressed and stretched rectangle, shown in Figure 4.1.1b, into two equal parts and then placing the right-hand part on top of the left-hand part (Figure 4.1.1c). This transformation has become known as the baker transformation because it mimics some aspects of kneading dough. From Figure 4.1.1 it is obvious that the counterimage of any rectangle is again a rectangle or a pair of rectangles with the same total area. Thus the baker transformation is measurable.

Now we calculate the Frobenius–Perron operator for the baker transformation. It will help to refer to Figure 4.1.2 and to note that two cases must be distinguished: $0 \le y < \frac{1}{2}$ and $\frac{1}{2} \le y \le 1$. Thus, for the simpler case of $0 \le y < \frac{1}{2}$ and $0 \le x < 1$ we have

$$S^{-1}([0, x] \times [0, y]) = [0, \tfrac{1}{2}x] \times [0, 2y]$$

so from equation (3.2.9)

$$Pf(x, y) = \frac{\partial^2}{\partial x\, \partial y} \int_0^{x/2} ds \int_0^{2y} f(s, t)dt$$

$$= f\left(\frac{1}{2}x, 2y\right), \qquad 0 \le y < \frac{1}{2}.$$

In the second case, for $\frac{1}{2} \le y \le 1$, we find that

$$S^{-1}([0, x] \times [0, y]) = \left([0, \tfrac{1}{2}x] \times [0, 1]\right) \cup \left([\tfrac{1}{2}, \tfrac{1}{2} + \tfrac{1}{2}x] \times [0, 2y - 1]\right)$$

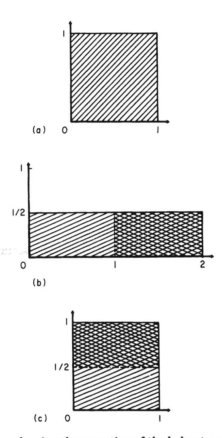

FIGURE 4.1.1. Steps showing the operation of the baker transformation given in equation (4.1.4).

hence

$$Pf(x,y) = \frac{\partial^2}{\partial x\, \partial y} \left\{ \int_0^{x/2} ds \int_0^1 f(s,t)\, dt + \int_{1/2}^{(1/2)+(x/2)} ds \int_0^{2y-1} f(s,t)\,dt \right\}$$
$$= f\left(\tfrac{1}{2} + \tfrac{1}{2}x, 2y-1\right), \qquad \tfrac{1}{2} \le y \le 1.$$

Thus, finally,

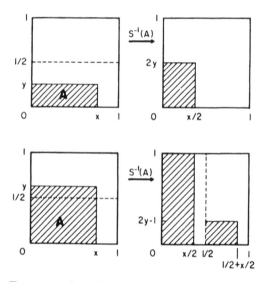

FIGURE 4.1.2. Two cases for calculating the counterimage of a set A by the baker transformation.

$$Pf(x,y) = \begin{cases} f\left(\tfrac{1}{2}x, 2y\right), & 0 \le y < \tfrac{1}{2} \\ f\left(\tfrac{1}{2} + \tfrac{1}{2}x, 2y - 1\right), & \tfrac{1}{2} \le y \le 1 \end{cases} \qquad (4.1.5)$$

so that $P1 = 1$, and the Borel measure is, therefore, invariant under the baker transformation. $\square$

Remark 4.1.4. Note that the transformation of the x-coordinate in the baker transformation is the dyadic transformation. However, the dyadic transformation is not $1 - 1$, whereas the baker transformation is a.e. Given an $X \subset R$ and any (not necessarily invertible) one-dimensional transformation $S: X \to X$, we may construct a two-dimensional invertible transformation $T: X \times X \to X \times X$ with $0 < \beta$ and

$$T(x,y) = (S(x) + y, \beta x).$$

As an example let $S: [0, 1] \to [0, 1]$ be the quadratic map $S(x) = 4x(1 - x)$. Then T is given by

$$T(x,y) = (4x(1 - x) + y, \beta x),$$

which is equivalent to the **Henon map** first studied by Henon [1976]. $\square$

Remark 4.1.5. Our derivation of the Frobenius–Perron operator (4.1.5) corresponding to the baker transformation is longer than it need be. Since

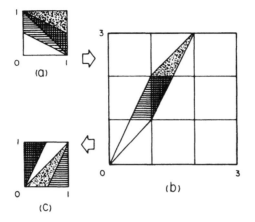

FIGURE 4.1.3. Operation of the Anosov diffeomorphism [equation (4.1.6)].

the baker transformation is invertible (except on the line $y = \frac{1}{2}$), and indeed

$$S^{-1}(x,y) = \begin{cases} (\frac{1}{2}x, 2y) & 0 \le x < 1, 0 \le y < \frac{1}{2} \\ (\frac{1}{2} + \frac{1}{2}x, 2y - 1) & 0 \le x < 1, \frac{1}{2} < y < 1, \end{cases}$$

equation (4.1.5) may be immediately obtained from Corollary 3.2.1. $\square$

Example 4.1.4 (Anosov diffeomorphisms). The baker transformation of the previous example may be considered to be a prototype of a very important class of transformations originally introduced by Anosov [1963]. One of the simplest of the Anosov diffeomorphisms is given by

$$S(x,y) = (x + y, x + 2y) \qquad (\text{mod } 1). \qquad (4.1.6)$$

To see the effect of this transformation consult Figure 4.1.3. In the first part (a) of the figure we depict the unit square in the plane and divide it into four triangular ares. In Figure 4.1.3b we show how the unit square is transformed after one application of $(x, y) \to (x+y, x+2y)$, whereas Figure 4.1.3c shows the result of the full Anosov diffeomorphism. It is clear that the effect of this transformation will be to very quickly scramble, or mix, various regions of the unit square. This property of mixing, also shared by the baker transformation, is most important and is dealt with in more detail in Section 4.3.

The determinant of the Jacobian of transformation (4.1.6) is given by

$$J = \det \begin{vmatrix} 1 & 1 \\ 1 & 2 \end{vmatrix} = 1,$$

so that the transformation is measure preserving; we already noted this result on geometric grounds in Figure 4.1.3. The two eigenvectors associated with S have eigenvalues

$$\lambda_1 = \frac{3}{2} - \frac{\sqrt{5}}{2} \quad \text{and} \quad \lambda_2 = \frac{3}{2} + \frac{\sqrt{5}}{2},$$

hence $0 < \lambda_1 < 1 < \lambda_2$. Thus, as for the baker transformation, the Anosov diffeomorphism involves a stretching in one direction and a corresponding compression in the orthogonal direction.

With some patience it is possible to derive an explicit formula for the Frobenius–Perron operator corresponding to the Anosov diffeomorphism (4.1.6) using a technique analogous to that employed for the baker transformation of the previous example. However, we can obtain this result immediately from Corollary 3.2.1 since the Anosov diffeomorphism is invertible. An easy calculation gives

$$S^{-1}(x, y) = (2x - y, y - x) \qquad (\text{mod } 1)$$

and thus

$$Pf(x, y) = f(2x - y, y - x), \qquad (4.1.7)$$

where, as in (4.1.6), the terms $2x - y$ and $y - x$ should be interpreted modulo 1. From (4.1.7) it is clear that $P1 = 1$, which corresponds to the fact that S preserves the Borel measure.   □

Remark 4.1.6. Observe that, if we replace the unit square $[0, 1] \times [0, 1]$ with the torus, that is, if we identify points $(x, 1)$ with $(x, 0)$ and $(1, y)$ with $(0, y)$, then this example of an Anosov diffeomorphism becomes continuous and differentiable just as the r-adic transformation does when the unit interval is replaced by the unit circle. The word diffeomorphism comes from the fact that the Anosov transformation is invertible, and that both the transformation and its inverse are differentiable.   □

4.2 Ergodic Transformations

Because a transformation S has an invariant measure or because the Frobenius–Perron operator P associated with S has a stationary density does not imply that S has interesting statistical properties. For example, if S is the identity on X, that is, $S(x) = x$ for every $x \in X$, then

$$S^{-1}(A) = A \qquad (4.2.1)$$

for every $A \subset X$, and, consequently, $Pf = f$ for every $f \in L^1$. This is, of course, not an interesting transformation. However, even if (4.2.1) holds for just one subset A of X, then the transformation S may be studied on the sets A and $X \setminus A$ separately. To see this, assume that A is fixed and condition (4.2.1) holds. Consider a trajectory

$$x^0, S(x^0), S^2(x^0), \ldots.$$

Equality (4.2.1) implies that S maps A into itself and no element of $X \setminus A$ is mapped into A. Thus, if $x^0 \in A$ then $S^n(x^0) \in A$ for all n, and if $x^0 \notin A$ then $S^n(x^0) \notin A$ for all n.

Example 4.2.1. A simple example is

$$S(k) = \begin{cases} k+2 & \text{for } k = 1, \ldots, 2(N-1) \\ 1 & \text{for } k = 2N-1 \\ 2 & \text{for } k = 2N \end{cases}$$

operating on the space $X = \{1, \ldots, 2N\}$ with the counting measure. This transformation can be studied separately on the sets $A = \{1, 3, \ldots, 2N-1\}$ and $X \setminus A = \{2, 4, \ldots, 2N\}$ of odd and even interiors. $\square$

Any set A satisfying (4.2.1) is called **invariant**. We require this equality to be satisfied modulo zero (see Remark 3.1.3). Then we can make the following definition.

Definition 4.2.1. Let $(X, \mathcal{A}, \mu)$ be a measure space and let a nonsingular transformation $S: X \to X$ be given. Then S is called **ergodic** if every invariant set $A \in \mathcal{A}$ is such that either $\mu(A) = 0$ or $\mu(X \setminus A) = 0$; that is, S is ergodic if all invariant sets are **trivial** subsets of X.

From this definition it follows that any ergodic transformation S must be studied on the entire space X. Determining ergodicity on the basis of Definition 4.2.1 is, in general, difficult except for simple examples on finite spaces. Thus, for example, the transformation in Example 4.2.1 is not ergodic on the space X of integers, but it is ergodic on the sets of even and odd integers.

In studying more interesting examples the following theorem may be of use.

Theorem 4.2.1. *Let $(X, \mathcal{A}, \mu)$ be a measure space and $S: X \to X$ a nonsingular transformation. S is ergodic if and only if, for every measurable function $f: X \to R$,*

$$f(S(x)) = f(x) \qquad \text{for almost all } x \in X \tag{4.2.2}$$

implies that f is constant almost everywhere.

Proof. We first show that ergodicity implies f is constant. Assume that, as in Figure 4.2.1, we have a function f satisfying (4.2.2), which is not

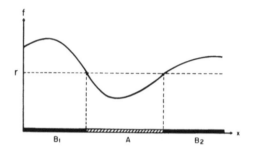

FIGURE 4.2.1. Definition of the sets A and $B = B_1 \cup B_2$.

constant almost everywhere, and that S is ergodic. Then there is some r such that the sets

$$A = \{x\colon f(x) \leq r\} \quad \text{and} \quad B = \{x\colon f(x) > r\}$$

have positive measure. These sets are also invariant because

$$S^{-1}(A) = \{x\colon S(x) \in A\} = \{x\colon f(S(x)) \leq r\}$$
$$= \{x\colon f(x) \leq r\} = A$$

and similarly for B. Because sets A and B are invariant, S is not ergodic, which is a contradiction. Thus, every f satisfying (4.2.2) must be constant.

To prove the converse, assume that S is not ergodic. Then, by Definition 4.2.1, there is a nontrivial set $A \in \mathcal{A}$ that is invariant. Set $f = 1_A$, and since A is nontrivial, f is not a constant function. Moreover, since $A = S^{-1}(A)$ we have

$$f(S(x)) = 1_A(S(x)) = 1_{S^{-1}(A)}(x) = 1_A(x) = f(x) \text{ a.e.}$$

and (4.2.2) is satisfied by a nonconstant function. ∎

Remark 4.2.1. It is clear from the proof that it is sufficient to verify only (4.2.2) for bounded measurable functions since in the last part of the proof we used characteristic functions that are bounded. □

An immediate consequence of Theorem 4.2.1 in combination with the definition of the Koopman operator is the following corollary.

Corollary 4.2.1. *Let $(X, \mathcal{A}, \mu)$ be a measure space, $S\colon X \to X$ a nonsingular transformation, and U the Koopman operator with respect to S. Then S is ergodic if and only if all the fixed points of U are constant functions.*

In addition to Theorem 4.2.1 and the preceding corollary, another result of use in checking the ergodicity of S using the Frobenius–Perron operator is contained in the following theorem.

Theorem 4.2.2. *Let* $(X, \mathcal{A}, \mu)$ *be a measure space,* $S: X \to X$ *a nonsingular transformation, and* P *the Frobenius–Perron operator associated with* S. *If* S *is ergodic, then there is at most one stationary density* f_* *of* P. *Further, if there is a unique stationary density* f_* *of* P *and* $f_*(x) > 0$ *a.e., then* S *is ergodic.*

Proof. To prove the first part of the theorem assume that S is ergodic and that f_1 and f_2 are different stationary densities of P. Set $g = f_1 - f_2$, so that $Pg = g$ by the linearity of P. Thus, by Proposition 3.1.3, g^+ and g^- are both stationary densities of P:

$$Pg^+ = g^+ \quad \text{and} \quad Pg^- = g^-. \tag{4.2.3}$$

Since, by assumption, f_1 and f_2 are not only different but are also densities we have

$$g^+ \not\equiv 0 \quad \text{and} \quad g^- \not\equiv 0.$$

Set

$$A = \operatorname{supp} g^+ = \{x: g^+(x) > 0\}.$$

and

$$B = \operatorname{supp} g^- = \{x: g^-(x) > 0\}.$$

It is evident that A and B are disjoint sets and both have positive (nonzero) measure. By equality (4.2.3) and Proposition 3.2.1, we have

$$A \subset S^{-1}(A) \quad \text{and} \quad B \subset S^{-1}(B).$$

Since A and B are disjoint sets, $S^{-1}(A)$ and $S^{-1}(B)$ are also disjoint. By induction we, therefore, have

$$A \subset S^{-1}(A) \subset S^{-2}(A) \cdots \subset S^{-n}(A)$$

and

$$B \subset S^{-1}(B) \subset S^{-2}(B) \cdots \subset S^{-n}(B),$$

where $S^{-n}(A)$ and $S^{-n}(B)$ are also disjoint for all n. Now define two sets by

$$\bar{A} = \bigcup_{n=0}^{\infty} S^{-n}(A) \quad \text{and} \quad \bar{B} = \bigcup_{n=0}^{\infty} S^{-n}(B).$$

These two sets $\bar{A}$ and $\bar{B}$ are also disjoint and, furthermore they are invariant because

$$S^{-1}(\bar{A}) = \bigcup_{n=1}^{\infty} S^{-n}(A) = \bigcup_{n=0}^{\infty} S^{-n}(A) = \bar{A}$$

and

$$S^{-1}(\bar{B}) = \bigcup_{n=1}^{\infty} S^{-n}(B) = \bigcup_{n=0}^{\infty} S^{-n}(B) = \bar{B}.$$

Neither $\bar{A}$ nor $\bar{B}$ are of measure zero since A and B are not of measure zero. Thus, $\bar{A}$ and $\bar{B}$ are nontrivial invariant sets, which contradicts the ergodicity of S. Thus, the first portion of the theorem is proved.

To prove the second portion of the theorem, assume that $f_* > 0$ is the unique density satisfying $Pf_* = f_*$ but that S is not ergodic. If S is not ergodic, then there exists a nontrivial set A such that

$$S^{-1}(A) = A$$

and with $B = X \setminus A$

$$S^{-1}(B) = B.$$

With these two sets A and B, we may write $f_* = 1_A f_* + 1_B f_*$, so that

$$1_A f_* + 1_B f_* = P(1_A f_*) + P(1_B f_*). \qquad (4.2.4)$$

The function $1_B f_*$ is equal to zero in the set $X \setminus B = A = S^{-1}(A)$. Thus, by Proposition 3.2.1, $P(1_B f_*)$ is equal to zero in $A = X \setminus B$ and, likewise, $P(1_A f_*)$ is equal to zero in $B = X \setminus A$. Thus, equality (4.2.4) implies that

$$1_A f_* = P(1_A f_*) \quad \text{and} \quad 1_B f_* = P(1_B f_*).$$

Since f_* is positive on A and B, we may replace $1_A f_*$ by $f_A = 1_A f_* / \|1_A f_*\|$, and $1_B f_*$ by $f_B = 1_B f_* / \|1_B f_*\|$ in the last pair of equalities to obtain

$$f_A = Pf_A \quad \text{and} \quad f_B = Pf_B.$$

This implies that there exist two stationary densities of P, which is in contradiction to our assumption. Thus, if there is a unique positive stationary density f_* of P, then S is ergodic. ∎

Example 4.2.2. Consider a circle of radius 1, and let S be a rotation through an angle ϕ. This transformation is equivalent to the map $S: [0, 2\pi) \rightarrow [0, 2\pi)$ defined by

$$S(x) = x + \phi \qquad (\text{mod } 2\pi).$$

If ϕ is commensurate with 2π (that is, $\phi/2\pi$ is rational), then S is evidently nonergodic. For example, if $\phi = \pi/3$, then the sets A and B of Figure 4.2.2 are invariant. For any $\phi = 2\pi(k/n)$, where k and n are integers, we will still find two invariant sets A and B, each containing n parts. As n becomes large the intermingling of the two sets A and B becomes more complicated and suggests that the rotational transformation S may be ergodic for $(\phi/2\pi)$ irrational. This does in fact hold, but it will be proved later when we have more techniques at our disposal.

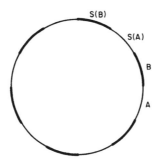

FIGURE 4.2.2. The two disjoint sets A (containing all the arcs denoted by thin lines) and B (containing arcs marked by heavy lines) are invariant under the rotational transformation when $\phi/2\pi$ is rational.

In this example the behavior of the trajectories is moderately regular and insensitive to changes in the initial value. Thus, independent of whether or not $(\phi/2\pi)$ is rational, if the value of ϕ is known precisely but the initial condition is located between α and β, $x_0 \in (\alpha, \beta)$, then

$$S^n(x_0) \in (\alpha + n\phi, \beta + n\phi) \qquad (\text{mod } 2\pi)$$

and all of the following points of the trajectory are known with the same accuracy, $(\beta - \alpha)$. □

Before closing this section we state, without proof, the **Birkhoff individual ergodic theorem** [Birkhoff, 1931a,b].

Theorem 4.2.3. Let $(X, \mathcal{A}, \mu)$ be a measure space, $S: X \to X$ a measurable transformation, and $f: X \to R$ an integrable function. If the measure μ is invariant, then there exists an integrable function f^* such that

$$f^*(x) = \lim_{n \to \infty} \frac{1}{n} \sum_{k=0}^{n-1} f(S^k(x)) \qquad \text{for almost all } x \in X. \qquad (4.2.5)$$

Without additional assumptions the limit $f^*(x)$ is generally difficult to determine. However, it can be shown that $f^*(x)$ satisfies

$$f^*(x) = f^*(S(x)) \qquad \text{for almost all } x \in X, \qquad (4.2.6)$$

and when $\mu(X) < \infty$

$$\int_X f^*(x)\mu(dx) = \int_X f(x)\mu(dx). \qquad (4.2.7)$$

Equation (4.2.6) follows directly from (4.2.5) if x is replaced by $S(x)$. The second property, (4.2.7), follows from the invariance of μ and equation

(4.2.5). Thus, by Theorem 3.2.1,

$$\int_X f(x)\mu(dx) = \int_X f(S(x))\mu(dx) = \cdots$$

so that integrating equation (4.2.5) over X and passing to the limit yields (4.2.7) by the Lebesque-dominated convergence theorem when f is bounded. When f is not bounded the argument is more difficult.

Remark 4.2.2. Theorem 4.2.3 is known as the individual ergodic theorem because it may be used to give information concerning the asymptotic behavior of trajectories starting from a given point $x \in X$. As our emphasis is on densities and not on individual trajectories, we will seldom use this theorem. □

With the notion of ergodicity we may derive an important and often quoted extension of the Birkhoff individual ergodic theorem.

Theorem 4.2.4. *Let $(X, \mathcal{A}, \mu)$ be a finite measure space and $S: X \to X$ be measure preserving and ergodic. Then, for any integrable f, the average of f along the trajectory of S is equal almost everywhere to the average of f over the space X; that is*

$$\lim_{n\to\infty} \frac{1}{n} \sum_{k=0}^{n-1} f(S^k(x)) = \frac{1}{\mu(X)} \int_X f(x)\mu(dx) \quad \text{a.e.} \tag{4.2.8}$$

Proof. From (4.2.6) and Theorem 4.2.1 it follows that f^* is constant almost everywhere. Hence, from (4.2.7), we have

$$\int_X f^*(x)\mu(dx) = f^* \int_X \mu(dx) = f^*\mu(X) = \int_X f(x)\mu(dx),$$

so that

$$f^*(x) = \frac{1}{\mu(X)} \int_X f(x)\mu(dx) \quad \text{a.e.}$$

Thus equation (4.2.5) of the Birkhoff theorem and the preceding formula imply (4.2.8), and the theorem is proved. ∎

One of the most quoted consequences of this theorem is the following.

Corollary 4.2.2. *Let $(X, \mathcal{A}, \mu)$ be a finite measure space and $S: X \to X$ be measure preserving and ergodic. Then for any set $A \in \mathcal{A}, \mu(A) > 0$, and almost all $x \in X$, the fraction of the points $\{S^k(x)\}$ in A as $k \to \infty$ is given by $\mu(A)/\mu(X)$.*

Proof. Using the characteristic function 1_A of A, the fraction of points from $\{S^k(x)\}$ in A is

$$\lim_{n\to\infty} \frac{1}{n} \sum_{k=0}^{n-1} 1_A(S^k(x)).$$

However, from (4.2.8) this is simply $\mu(A)/\mu(X)$. ∎

Remark 4.2.3. Corollary 4.2.2 says that every set of nonzero measure is visited infinitely often by the iterates of almost every $x \in X$. This result is a special case of the **Poincaré recurrence theorem.**   □

4.3   Mixing and Exactness

Mixing Transformations

The examples of the previous section show that ergodic behavior per se need not be very complicated and suggests the necessity of introducing another concept, that of mixing.

Definition 4.3.1. Let $(X, \mathcal{A}, \mu)$ be a normalized measure space, and $S: X \to X$ a measure-preserving transformation. S is called **mixing** if

$$\lim_{n \to \infty} \mu(A \cap S^{-n}(B)) = \mu(A)\mu(B) \qquad \text{for all } A, B \in \mathcal{A}. \qquad (4.3.1)$$

Condition (4.3.1) for mixing has a very simple interpretation. Consider points x belonging to the set $A \cap S^{-n}(B)$. These are the points such that $x \in A$ and $S^n(x) \in B$. Thus, from (4.3.1), as $n \to \infty$ the measure of the set of such points is just $\mu(A)\mu(B)$. This can be interpreted as meaning that the fraction of points starting in A that ended up in B after n iterations (n must be a large number) is just given by the product of the measures of A and B and is independent of the position of A and B in X.

It is easy to see that any mixing transformation must be ergodic. Assume that $B \in \mathcal{A}$ is an invariant set, so that $B = S^{-1}(B)$ and, even further, $B = S^{-n}(B)$ by induction. Take $A = X \setminus B$ so that $\mu(A \cap B) = \mu(A \cap S^{-n}(B)) = 0$. However, from (4.3.1), we must have

$$\lim_{n \to \infty} \mu(A \cap S^{-n}(B)) = \mu(A)\mu(B) = (1 - \mu(B))\mu(B),$$

and thus $\mu(B)$ is either 0 or 1, which proves ergodicity.

Many of the transformations considered in our examples to this point are mixing, for example, the baker, quadratic, Anasov, and r-adic. (The rotation transformation is not mixing according to our foregoing discussion.) To illustrate the mixing property we consider the baker and r-adic transformations in more detail.

Example 4.3.1. (See also Example 4.1.3.) In considering the baker transformation, it is relatively easy to check the mixing condition (4.3.1) for generators of the σ-algebra $\mathcal{B}$, namely, for rectangles. Although the transformation is simple, writing the algebraic expressions for the counterimages is tedious, and the property of mixing is easier to see pictorially. Consider

Figure 4.3.1a, where two sets A and B are represented with $\mu(B) = \frac{1}{2}$. We take repeated counterimages of the set B by the baker transformation and find that after n such steps, $S^{-n}(B)$ consists of 2^{n-1} vertical rectangles of equal area. Eventually the measure of $A \cap S^{-n}(B)$ approaches $\mu(A)/2$, and condition (4.3.1) is evidently satisfied. The behavior of any pair of sets A and B is similar.

It is interesting that the baker transformation behaves in a similar fashion if, instead of examining $S^{-n}(B)$, we look at $S^n(B)$ as shown in Figure 4.3.1b. Now we have 2^n horizontal rectangles after n steps and all of our previous comments apply. So, for the baker transformation the behavior of images and counterimages is very similar and illustrates the property of mixing. This is not true for our next example, the dyadic transformation. □

In general, proving that a given transformation is mixing via Definition 4.3.1 is difficult. In the next section, Theorem 4.4.1 and Proposition 4.4.1, we introduce easier and more powerful techniques for this purpose.

Example 4.3.2. (Cf. Examples 1.2.1 and 4.1.1.) To examine the mixing property (4.3.1) for the dyadic transformation, consider Figure 4.3.2a. Now we take the set $B = [0, b]$ and find that the nth counterimage of B consists of intervals on $[0,1]$ each of the same length. Eventually, as before $\mu(A \cap S^{-n}(B)) \to \mu(A)\mu(B)$.

As for the baker transformation let us consider the behavior of images of a set B under the dyadic transformation (cf. Figure 4.3.2b). In this case, if $B = [0, b]$, then $S(B) = [0, 2b]$ and after a finite number of iterations $S^n(B) = [0, 1)$. The same procedure with any arbitrary set $B \subset [0, 1]$ of positive measure will show that $\mu(S^n(B)) \to 1$ and thus the behavior of images of the dyadic transformation is different from the baker transformation. □

Exact Transformations

The behavior illustrated by images of the dyadic transformation is called exactness, and is made precise by the following definition due to Rochlin [1964].

Definition 4.3.2. Let $(X, \mathcal{A}, \mu)$ be a normalized measure space and $S: X \to X$ a measure-preserving transformation such that $S(A) \in \mathcal{A}$ for each $A \in \mathcal{A}$. If

$$\lim_{n \to \infty} \mu(S^n(A)) = 1 \qquad \text{for every } A \in \mathcal{A}, \mu(A) > 0, \qquad (4.3.2)$$

then S is called **exact**.

It can be proved, although it is not easy to do so from the definition, that exactness of S implies that S is mixing. As we have seen from the

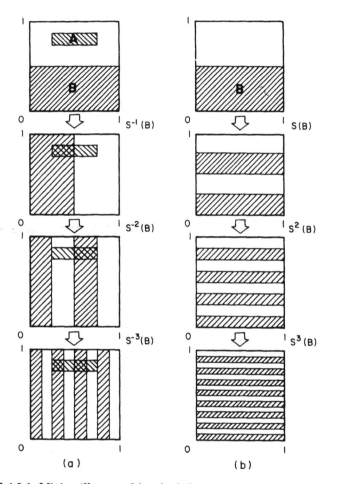

FIGURE 4.3.1. Mixing illustrated by the behavior of counterimages and images of a set B by the baker transformation. (a) The nth counterimage of the set B consists of 2^{n-1} vertical rectangles, each of equal area. (b) Successive iterates of the same set B results in 2^n horizontal rectangles after n iterations.

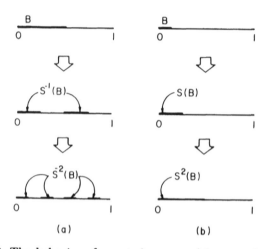

FIGURE 4.3.2. The behavior of counterimages and images of a set B by the dyadic transformation. (a) Successive counterimages of a set B that result after n such counterimages, in 2^n disjoint sets on $[0,1]$. (b) The behavior of images of a set B generated by the dyadic transformation, which is quite different than that for the baker transformation. (See the text for further details.)

baker transformation the converse is not true. We defer the proof until the next section when we have other tools at our disposal.

Condition (4.3.2) has a very simple interpretation. If we start with a set A of initial conditions of nonzero measure, then after a large number of iterations of an exact transformation S the points will have spread and completely filled the space X.

Remark 4.3.1. It cannot be emphasized too strongly that invertible transformations cannot be exact. In fact, for any invertible measure-preserving transformation S, we have $\mu(S(A)) = \mu(S^{-1}(S(A))) = \mu(A)$ and by induction $\mu(S^n(A)) = \mu(A)$, which violates (4.3.2). □

In this and the previous section we have defined and examined a hierarchy of "chaotic" behaviors. However, by themselves the definitions are a bit sterile and may not convey the full distinction between the behaviors of ergodic, mixing, and exact transformations. To remedy this we present the first six successive iterates of a random distribution of 1000 points in the set $X = [0,1] \times [0,1]$ by the ergodic transformation

$$S(x,y) = \left(\sqrt{2} + x, \sqrt{3} + y\right) \qquad (\text{mod } 1) \qquad (4.3.3)$$

in Figure 4.3.3; by the mixing transformation

$$S(x,y) = (x + y, x + 2y) \qquad (\text{mod } 1) \qquad (4.3.4)$$

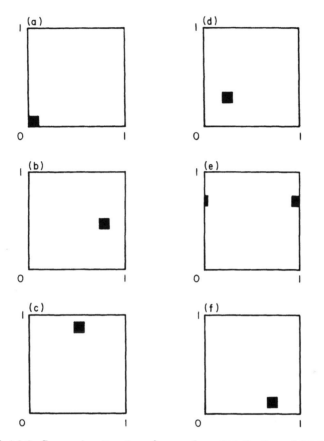

FIGURE 4.3.3. Successive iterates of a random distribution of 1000 points in $[0, 0.1] \times [0, 0.1]$ by the ergodic transformation (4.3.3). Note how the distribution moves about in the space $[0, 1] \times [0, 1]$.

in Figure 4.3.4; and by the exact transformation

$$S(x, y) = (3x + y, x + 3y) \qquad (\text{mod } 1) \tag{4.3.5}$$

in Figure 4.3.5. Techniques to prove these assertions will be developed in the next two chapters.

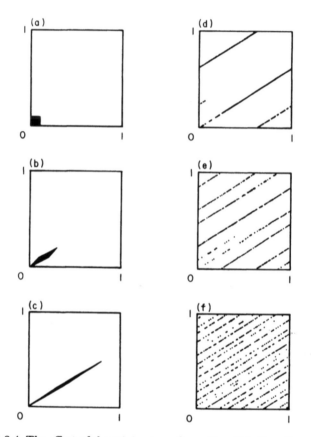

FIGURE 4.3.4. The effect of the mixing transformation (4.3.4) on the same initial distribution of points used in Figure 4.3.3.

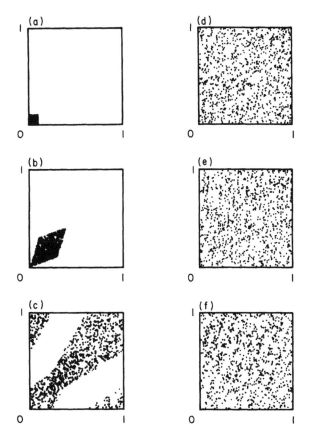

FIGURE 4.3.5. Successive applications of the exact transformation [equation (4.3.5)]. Note the rapid spread of the initial distribution of points throughout the phase space.

4.4   Using the Frobenius–Perron and Koopman Operators for Classifying Transformations

The concepts developed in the previous two sections for classifying various degrees of irregular behaviors (ergodicity, mixing, and exactness) were stated in terms of the behavior of sequences of sets. The proof of ergodicity, mixing, or exactness using these definitions is difficult. Indeed, in all the examples we gave to illustrate these concepts, no rigorous proofs were ever given, although it is possible to do so.

In this section we reformulate the concepts of ergodicity, mixing, and

exactness in terms of the behavior of sequences of iterates of Frobenius–
Perron and Koopman operators and show how they can be used to determine whether a given transformation S with an invariant measure is
ergodic, mixing, or exact. The techniques of this chapter rely heavily on
the notions of Cesàro, weak and strong convergences, which were developed
in Section 2.3.

We will first state and prove the main theorem of this section and then
show its utility.

Theorem 4.4.1. *Let $(X, \mathcal{A}, \mu)$ be a normalized measure space, $S: X \to X$
a measure preserving transformation, and P the Frobenius–Perron operator
corresponding to S. Then*

(a) *S is ergodic if and only if the sequence $\{P^n f\}$ is Cesàro convergent
to 1 for all $f \in D$;*

(b) *S is mixing if and only if $\{P^n f\}$ is weakly convergent to 1 for all
$f \in D$;*

(c) *S is exact if and only if $\{P^n f\}$ is strongly convergent to 1 for all
$f \in D$.*

Before giving the proof of Theorem 4.4.1, we note that, since P is linear,
convergence of $\{P^n f\}$ to 1 for $f \in D$ is equivalent to the convergence of
$\{P^n f\}$ to $\langle f, 1 \rangle$ for every $f \in L^1$. This observation is, of course, valid for
all types of convergence: Cesàro, weak, and strong. Thus we may restate
Theorem 4.4.1 in the equivalent form.

Corollary 4.4.1. *Under the assumptions of Theorem 4.4.1, the following
equivalences hold:*

(a) *S is ergodic if and only if*

$$\lim_{n \to \infty} \frac{1}{n} \sum_{k=0}^{n-1} \langle P^k f, g \rangle = \langle f, 1 \rangle \langle 1, g \rangle \qquad \text{for } f \in L^1, g \in L^\infty;$$

(b) *S is mixing if and only if*

$$\lim_{n \to \infty} \langle P^n f, g \rangle = \langle f, 1 \rangle \langle 1, g \rangle \qquad \text{for } f \in L^1, g \in L^\infty;$$

(c) *S is exact if and only if*

$$\lim_{n \to \infty} \| P^n f - \langle f, 1 \rangle \| = 0 \qquad \text{for } f \in L^1.$$

Proof of Theorem 4.4.1. The proof of part (a) follows easily from Corollary 5.2.3.

Next consider the mixing portion of the theorem. Assume S is mixing, which, by definition, means

$$\lim_{n\to\infty} \mu(A \cap S^{-n}(B)) = \mu(A)\mu(B) \qquad \text{for all } A, B \in \mathcal{A}.$$

The mixing condition can be rewritten in integral form as

$$\lim_{n\to\infty} \int_X 1_A(X)1_B(S^n(x))\mu(dx) = \int_X 1_A(x)\mu(dx) \int_X 1_B(x)\mu(dx).$$

By applying the definitions of the Koopman operator and the scalar product to this equation, we obtain

$$\lim_{n\to\infty} \langle 1_A, U^n 1_B \rangle = \langle 1_A, 1 \rangle \langle 1, 1_B \rangle. \qquad (4.4.1)$$

Since the Koopman operator is adjoint to the Frobenius–Perron operator, equation (4.4.1) may be rewritten as

$$\lim_{n\to\infty} \langle P^n 1_A, 1_B \rangle = \langle 1_A, 1 \rangle \langle 1, 1_B \rangle$$

or

$$\lim_{n\to\infty} \langle P^n f, g \rangle = \langle f, 1 \rangle \langle 1, g \rangle$$

for $f = 1_A$ and $g = 1_B$. Since this relation holds for characteristic functions it must also hold for the simple functions

$$f = \sum_i \lambda_i 1_{A_i} \quad \text{and} \quad g = \sum_i \sigma_i 1_{B_i}.$$

Further, every function $g \in L^\infty$ is the uniform limit of simple functions $g_k \in L^\infty$, and every function $f \in L^1$ is the strong (in L^1 norm) limit of a sequence of simple functions $f_k \in L^1$. Obviously,

$$\begin{aligned} |\langle P^n f, g \rangle - \langle f, 1 \rangle \langle 1, g \rangle| &\le |\langle P^n f, g \rangle - \langle P^n f_k, g_k \rangle| \\ &+ |\langle P^n f_k, g_k \rangle - \langle f_k, 1 \rangle \langle 1, g_k \rangle| \\ &+ |\langle f_k, 1 \rangle \langle 1, g_k \rangle - \langle f, 1 \rangle \langle 1, g \rangle|. \quad (4.4.2) \end{aligned}$$

If $\|f_k - f\| \le \epsilon$ and $\|g_k - g\|_{L^\infty} \le \epsilon$, then the first and last terms on the right-hand side of (4.4.2) satisfy

$$\begin{aligned} &|\langle P^n f, g \rangle - \langle P^n f_k, g_k \rangle| \\ &\le |\langle P^n f, g \rangle - \langle P^n f_k, g \rangle| + |\langle P^n f_k, g \rangle - \langle P^n f_k, g_k \rangle| \\ &\le \epsilon\|g\|_{L^\infty} + \epsilon\|f_k\| \le \epsilon(\|g\|_{L^\infty} + \|f\| + \epsilon) \end{aligned}$$

and analogously

$$|\langle f_k, 1 \rangle \langle 1, g_k \rangle - \langle f, 1 \rangle \langle 1, g \rangle| \le \epsilon(\|g\|_{L^\infty} + \|f\| + \epsilon).$$

Thus these terms are arbitrarily small for small ϵ. Finally, for fixed k the middle term of (4.4.2),

$$|\langle P^n f_k, g_k \rangle - \langle f_k, 1 \rangle \langle 1, g_k \rangle|$$

converges to zero as $n \to \infty$, which shows that the right-hand side of inequality (4.4.2) can be as small as we wish it to be for large n. This completes the proof that mixing implies the convergence of $\langle P^n f, g \rangle$ to $\langle f, 1 \rangle \langle 1, g \rangle$ for all $f \in L^1$ and $g \in L^\infty$. Conversely, this convergence implies the mixing condition (4.4.1) if we set $f = 1_A$ and $g = 1_B$.

Lastly, we show that the strong convergence of $\{P^n f\}$ to $\langle f, 1 \rangle$ implies exactness. Assume $\mu(A) > 0$ and define

$$f_A(x) = (1/\mu(A))1_A(x).$$

Clearly, f_A is a density. If the sequence $\{r_n\}$ is defined by

$$r_n = \|P^n f_A - 1\|,$$

then it is also clear that the sequence is convergent to zero. By the definition of r_n, we have

$$\mu(S^n(A)) = \int_{S^n(A)} \mu(dx)$$

$$= \int_{S^n(A)} P^n f_A(x)\mu(dx) - \int_{S^n(A)} (P^n f_A(x) - 1)\mu(dx)$$

$$\geq \int_{S^n(A)} P^n f_A(x)\mu(dx) - r_n. \qquad (4.4.3)$$

From the definition of the Frobenius–Perron operator, we have

$$\int_{S^n(A)} P^n f_A(x)\mu(dx) = \int_{S^{-n}(S^n(A))} f_A(x)\mu(dx)$$

and, since $S^{-n}(S^n(A))$ contains A, the last integral is equal to 1. Thus inequality (4.4.3) gives

$$\mu(S^n(A)) \geq 1 - r_n,$$

which completes the proof that the strong convergence of $\{P^n f\}$ to $\langle f, 1 \rangle$ implies exactness. ∎

We omit the proof of the converse (that exactness implies the strong convergence of $\{P^n f\}$ to $\langle f, 1 \rangle$) since we will never use this fact and its proof is based on quite different techniques (see Lin [1971]).

Because the Koopman and Frobenius–Perron operators are adjoint, it is possible to rewrite conditions (a) and (b) of Corollary 4.4.1 in terms of the Koopman operator. The advantage of such a reformulation lies in the fact that the Koopman operator is much easier to calculate than the Frobenius–Perron operator. Unfortunately, this reformulation cannot be extended to condition (c) for exactness of Corollary 4.4.1 since it is not expressed in terms of a scalar product.

Thus, from Corollary 4.4.1, the following proposition can easily be stated.

Proposition 4.4.1. *Let* $(X, \mathcal{A}, \mu)$ *be a normalized measure space,* $S: X \rightarrow X$ *a measure-preserving transformation, and* U *the Koopman operator corresponding to* S. *Then*

(a) S *is ergodic if and only if*

$$\lim_{n \to \infty} \frac{1}{n} \sum_{k=0}^{n-1} \langle f, U^k g \rangle = \langle f, 1 \rangle \langle 1, g \rangle \qquad \text{for } f \in L^1, g \in L^\infty;$$

(b) S *is mixing if and only if*

$$\lim_{n \to \infty} \langle f, U^n g \rangle = \langle f, 1 \rangle \langle 1, g \rangle \qquad \text{for } f \in L^1, g \in L^\infty.$$

Proof. The proof of this proposition is trivial since, according to equation (3.3.4), we have

$$\langle f, U^n g \rangle = \langle P^n f, g \rangle \qquad \text{for } f \in L^1, g \in L^\infty, n = 1, 2, \dots,$$

which shows that conditions (a) and (b) of Corollary 4.4.1 and Proposition 4.4.1 are identical. ∎

Remark 4.4.1. We stated Theorem 4.4.1 and Corollary 4.4.1 in terms of L^1 and L^∞ spaces to underline the role of the Frobenius–Perron operator as a transformation of densities. The same results can be proved using adjoint spaces L^p and $L^{p'}$ instead of L^1 and L^∞, respectively. Moreover, when verifying conditions (a) through (c) of Theorem 4.4.1 and Corollary 4.4.1, or conditions (a) and (b) of Proposition 4.4.1, it is not necessary to check for their validity for all $f \in L^p$ and $g \in L^{p'}$. Due to special properties of the operators P and U, which are linear contractions, it is sufficient to check these conditions for f and g belonging to linearly dense subsets of L^p and $L^{p'}$, respectively (see Section 2.3). □

Example 4.4.1. In Example 4.2.2 we showed that the rotational transformation

$$S(x) = x + \phi \quad (\text{mod } 2\pi)$$

is not ergodic when $\phi/2\pi$ is rational. Here we prove that it is ergodic when $\phi/2\pi$ is irrational.

It is straightforward to show that S preserves the Borel measure μ and the normalized measure $\mu/2\pi$. We take as our linearly dense set in $L^p([0, 2\pi])$

that consisting of the functions $\{\sin kx, \cos lx: k, l = 0, 1, \ldots\}$. We will show that, for each function g belonging to this set,

$$\lim_{n \to \infty} \frac{1}{n} \sum_{k=0}^{n-1} U^k g(x) = \langle 1, g \rangle \qquad (4.4.4)$$

uniformly for all x, thus implying that condition (a) of Proposition 4.4.1 is satisfied for all f. To simplify the calculations, note that

$$\sin kx = \frac{e^{ikx} - e^{-ikx}}{2i}, \quad \cos kx = \frac{e^{ikx} + e^{-ikx}}{2}$$

where $i = \sqrt{-1}$. Consequently, it is sufficient to verify (4.4.4) only for $g(x) = \exp(ikx)$ with k an arbitrary (not necessarily positive) integer.

We have, for $k \neq 0$,

$$U^l g(x) = g(S^l(x)) = e^{ik(x+l\phi)},$$

so that

$$u_n(x) = \frac{1}{n} \sum_{l=0}^{n-1} U^l g(x)$$

obeys

$$u_n(x) = \frac{1}{n} \sum_{l=0}^{n-1} e^{ik(x+l\phi)}$$

$$= \frac{1}{n} e^{ikx} \frac{e^{ink\phi} - 1}{e^{ik\phi} - 1}.$$

and

$$\|u_n(x)\|_{L^2} \le \frac{1}{n|e^{ik\phi} - 1|} \left\{ \int_0^{2\pi} |e^{ikx} \left[e^{ink\phi} - 1 \right]|^2 \frac{dx}{2\pi} \right\}^{1/2}$$

$$\le \frac{2}{n|e^{ik\phi} - 1|}.$$

Thus $u_n(x)$ converges in L^2 to zero. Also, however, with our choice of $g(x)$,

$$\langle 1, g \rangle = \int_0^{2\pi} e^{ikx} \frac{dx}{2\pi} = \frac{1}{ik} [e^{2\pi ik} - 1] = 0$$

and condition (a) of Proposition 4.4.1 for ergodicity is satisfied with $k \neq 0$. When $k = 0$ the calculation is even simpler, since $g(x) \equiv 1$ and thus

$$u_n(x) \equiv 1.$$

Noting also that

$$\langle 1, g \rangle = \int_0^{2\pi} \frac{dx}{2\pi} \equiv 1,$$

we have again that $u_n(x)$ converges to $\langle 1, g \rangle$.   □

Example 4.4.2. In this example we demonstrate the exactness of the r-adic transformation

$$S(x) = rx \qquad \text{(mod 1)}.$$

From Corollary 4.4.1 it is sufficient to demonstrate that $\{P^n f\}$ converges strongly to $\langle f, 1 \rangle$ for f in a linearly dense set in $L^p([0,1])$. We take that linearly dense set to be the set of continuous functions.

From equation (1.2.13) we have

$$Pf(x) = \frac{1}{r} \sum_{i=0}^{r-1} f\left(\frac{i}{r} + \frac{x}{r}\right),$$

and thus by induction

$$P^n f(x) = \frac{1}{r^n} \sum_{k=0}^{r^n - 1} f\left(\frac{i}{r^n} + \frac{x}{r^n}\right).$$

However, in the limit as $n \to \infty$, the right-hand side of this equation approaches the Riemann integral of f over [0,1], that is,

$$\lim_{n \to \infty} P^n f(x) = \int_0^1 f(s)ds, \qquad \text{uniformly in } x,$$

which, by definition, is just $\langle f, 1 \rangle$. Thus the condition for exactness is fulfilled.   □

Example 4.4.3. Here we show that the Anosov diffeomorphism

$$S(x, y) = (x + y, x + 2y) \qquad \text{(mod 1)}$$

is mixing. For this, from Proposition 4.4.1, it is sufficient to show that $U^n g(x, y) \equiv g(S^n(x, y))$ converges weakly to $\langle 1, g \rangle$ for g in a linearly dense set in $L^p([0, 1] \times [0, 1])$.

Observe that for $g(x, y)$ periodic in x and y with period 1, $g(S(x, y)) = g(x + y, x + 2y), g(S^2(x, y)) = g(2x + 3y, 3x + 5y)$, and so on. By induction we easily find that

$$U^n g(x, y) = g(a_{2n-2}x + a_{2n-1}y, a_{2n-1}x + a_{2n}y),$$

where the a_n are the Fibonacci numbers given by $a_0 = a_1 = 1$, $a_{n+1} = a_n + a_{n-1}$. Thus, if we take $g(x, y) = \exp[2\pi i(kx + ly)]$ and $f(x, y) =$

$\exp[-2\pi i(px+qy)]$, then we have

$$\langle f, U^n g \rangle = \int_0^1 \int_0^1 \exp\{2\pi i[ka_{2n-2}+la_{2n-1}-p)x$$
$$+(ka_{2n-1}+la_{2n}-q)y]\}dx\,dy,$$

and it is straightforward to show that

$$\langle f, U^n g \rangle = \begin{cases} 1 & \text{if } (ka_{2n-2}+la_{2n-1}-p)=(ka_{2n-1}+la_{2n}-q)=0 \\ 0 & \text{otherwise.} \end{cases}$$

Now we show that for large n either

$$ka_{2n-2}+la_{2n-1}-p \quad \text{or} \quad ka_{2n-1}+la_{2n}-q$$

is different from zero if at least one of k, l, p, q is different from zero. If $k = l = 0$ but $p \neq 0$ or $q \neq 0$ this is obvious. We may suppose that either k or l is not zero. Assume $k \neq 0$ and that $ka_{2n-2}+la_{2n-1}-p = 0$ for infinitely many n. Thus,

$$k\frac{a_{2n-2}}{a_{2n-1}}+l-\frac{p}{a_{2n-1}}=0.$$

It is well known [Hardy and Wright, 1959] that

$$\lim_{n\to\infty}\frac{a_{2n-2}}{a_{2n-1}}=\frac{2}{1+\sqrt{5}} \quad \text{and} \quad \lim_{n\to\infty}a_n=\infty,$$

hence

$$\lim_{n\to\infty}\left[k\left(\frac{a_{2n-2}}{a_{2n-1}}\right)+l-\frac{p}{a_{2n-1}}\right]=\frac{2k}{1+\sqrt{5}}+l.$$

However, this limit can never be zero because k and l are integers. Thus, $ka_{2n-2}+la_{2n-1}-p \neq 0$ for large n. Therefore, for large n,

$$\langle f, U^n g \rangle = \begin{cases} 1 & \text{if } k = 1 = p = q = 0 \\ 0 & \text{otherwise.} \end{cases}$$

But

$$\langle 1, g \rangle = \int_0^1 \int_0^1 \exp[2\pi i(kx+ly)]dx\,dy$$
$$= \begin{cases} 1 & k = l = 0 \\ 0 & k \neq 0 \quad \text{or} \quad l \neq 0, \end{cases}$$

so that

$$\langle f, 1 \rangle \langle 1, g \rangle = \int_0^1 \int_0^1 \langle 1, g \rangle \exp[-2\pi i(px+qy)]dx\,dy$$
$$= \begin{cases} \langle 1, g \rangle & \text{if } p = q = 0 \\ 0 & \text{if } p \neq 0 \quad \text{or} \quad q \neq 0 \end{cases}$$
$$= \begin{cases} 1 & \text{if } k = l = p = q = 0 \\ 0 & \text{otherwise.} \end{cases}$$

Thus

$$\langle f, U^n g \rangle = \langle f, 1 \rangle \langle 1, g \rangle$$

for large n and, as a consequence, $\{U^n g\}$ converges weakly to $\langle 1, g \rangle$. There-fore, mixing of the Anosov diffeomorphism is demonstrated. $\square$

In this chapter we have shown how the study of ergodicity, mixing, and exactness for transformations S can be greatly facilitated by the use of the Frobenius–Perron operator P corresponding to S (cf. Theorem 4.4.1 and Corollary 4.4.1). Since the Frobenius–Perron operator is a special type of Markov operator, there is a certain logic to extending the notions of ergodicity, mixing, and exactness for transformations to Markov operators in general. Thus, we close this section with the following definition.

Definition 4.4.1. Let $(X, \mathcal{A}, \mu)$ be a normalized measure space and P: $L^1(X, \mathcal{A}, \mu) \to L^1(X, \mathcal{A}, \mu)$ be a Markov operator with stationary density 1, that is, $P1 = 1$. Then we say:

(a) The operator P is **ergodic** if $\{P^n f\}$ is Cesàro convergent to 1 for all $f \in D$;

(b) The operator P is **mixing** if $\{P^n f\}$ is weakly convergent to 1 for all $f \in D$; and

(c) The operator P is **exact** if $\{P^n f\}$ is strongly convergent to 1 for all $f \in D$.

4.5 Kolmogorov Automorphisms

Until now we have considered three types of transformations exhibiting gradually stronger chaotic properties: ergodicity, mixing, and exactness. This is not a complete list of possible behaviors. These three types are probably the most important, but it is possible to find some intermediate types and some new unexpected connections between them. For example, between ergodic and mixing transformations, there is a class of **weakly mixing** transformations that, by definition, are measure preserving [on a normalized measure space $(X, \mathcal{A}, \mu)$] and satisfy the condition

$$\lim_{n \to \infty} \frac{1}{n} \sum_{k=0}^{n-1} |\mu(A \cap S^{-k}(B)) - \mu(A)\mu(B)| = 0 \qquad \text{for } A, B \in \mathcal{A}.$$

It is not easy to construct an example of a weakly mixing transformation that is not mixing. Interesting comments on this problem can be found in Brown [1976].

However, Kolmogorov automorphisms, which are invertible and there-fore cannot be exact, are stronger than mixing. As we will see later, to some extent they are parallel to exact transformations. Schematically this situation ca be visualized as follows:

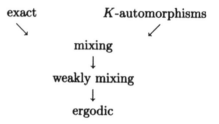

where K-automorphism is the usual abbreviation for a Kolmogorov automorphism and the arrows indicate that the property above implies the one below. Before giving the precise definition of K-automorphisms, we introduce two simple notations.

If $S: X \to X$ is a given transformation and $\mathcal{A}$ is a collection of subsets of X (e.g., a σ-algebra), then $S(\mathcal{A})$ denotes the collection of sets of the form $S(A)$ for $A \in \mathcal{A}$, and $S^{-1}(\mathcal{A})$ the collection of $S^{-1}(A)$ for $A \in \mathcal{A}$. More generally,

$$S^n(\mathcal{A}) = \{S^n(A): A \in \mathcal{A}\}, \qquad n = 0, \pm 1, \pm 2, \ldots.$$

Definition 4.5.1. Let $(X, \mathcal{A}, \mu)$ be a normalized measure space and let $S: X \to X$ be an invertible transformation such that S and S^{-1} are measurable and measure preserving. The transformation S is called a **K-automorphism** if there exists a σ-algebra $\mathcal{A}_0 \subset \mathcal{A}$ such that the following three conditions are satisfied:

(i) $S^{-1}(\mathcal{A}_0) \subset \mathcal{A}_0$;

(ii) the σ-algebra

$$\bigcap_{n=0}^{\infty} S^{-n}(\mathcal{A}_0) \tag{4.5.1}$$

is **trivial**, that is, it contains only sets of measure 1 or 0; and

(iii) the smallest σ-algebra containing

$$\bigcup_{n=0}^{\infty} S^n(\mathcal{A}_0) \tag{4.5.2}$$

is identical to $\mathcal{A}$.

The word automorphism comes from algebra and in this case it means that the transformation S is invertible and measure preserving (analogously the word endomorphism is used for measure preserving but not necessarily invertible transformations).

Examples 4.5.1. The baker transformation is a K-automorphism. For $\mathcal{A}_0$ we can take all the sets of the form

$$\mathcal{A}_0 = \{A \times [0,1]: A \subset [0,1], A \text{ is a Borel set}\}.$$

It is easy to verify condition (i) of Definition 4.5.1. Thus, if $B = A \times [0,1]$, then $B_1 = S^{-1}(B)$ has the form $B_1 = A_1 \times [0,1]$, where

$$A_1 = \tfrac{1}{2}A \cup \left(\tfrac{1}{2} + \tfrac{1}{2}A\right) \tag{4.5.3}$$

and thus condition (i) is satisfied. From this follows a hint of how to prove condition (ii). Namely, from (4.5.3) it follows that the basis A_1 of the set $B_1 = S^{-1}(B)$ is the union of two sets of equal measure that are contained in the intervals $\left[0, \tfrac{1}{2}\right]$ and $\left[\tfrac{1}{2}, 1\right]$, respectively. Furthermore, set $B_2 = S^{-2}(B)$ has the form $A_1 \times [0,1]$ and its basis A_2 is the union of four sets of equal measure contained in the intervals $\left[0, \tfrac{1}{4}\right], \ldots, \left[\tfrac{3}{4}, 1\right]$. Finally, every set B_∞ belonging to the σ-algebra (4.5.1) is of the form $A_\infty \times [0,1]$ and A_∞ has the property that for each integer n the measure of the intersection of A_∞ with $[k/2^n, (k+1)/2^n]$ does not depend on k. From this it is easy to show that the measure of the intersection of A_∞ with $[0, x]$ is a linear function of x or

$$\int_0^x 1_{A_\infty}(y)dy = cx,$$

where c is a constant. Differentiation gives

$$1_{A_\infty}(x) = c \quad \text{a.e.} \qquad \text{for } 0 \leq x \leq 1.$$

Since 1_{A_∞} is a characteristic function, either $c = 1$ or $c = 0$. In the first case, A_∞ as well as B_∞ have measure 1, and if $c = 0$, then A_∞ and B_∞ have measure 0. Thus condition (ii) is verified.

To verify (iii), observe that $\mathcal{A}_0 \cup S(\mathcal{A}_0)$ contains not only sets of the form $A \times [0,1]$ but also the sets of the form $A \times \left[0, \tfrac{1}{2}\right]$ and $A \times \left[\tfrac{1}{2}, 1\right]$. Further, $\mathcal{A}_0 \cup S(\mathcal{A}_0) \cup S^2(\mathcal{A}_0)$ also contains the sets $A \times \left[0, \tfrac{1}{4}\right], \ldots, A \times \left[\tfrac{3}{4}, 1\right]$ and so on. Thus, by using the sets from the family (4.5.2), we can approximate every rectangle contained in $[0,1] \times [0,1]$. Consequently, the smallest σ-algebra containing (4.5.2) is the σ-algebra of Borel sets. $\square$

Example 4.5.2. The baker transformation considered in the previous example has an important geometrical property. At every point it is contracting in one direction and expanding in the orthogonal one. The transformation

$$S(x, y) = (x + y, x + 2y) \pmod 1$$

considered in Example 4.1.4 has the same property. As we have observed the Jacobian of S has two eigenvalues λ_1, λ_2 such that $0 < \lambda_1 < 1 < \lambda_2$. To these eigenvalues correspond the eigenvectors

$$\xi_1 = \left(1, \tfrac{1}{2} - \tfrac{1}{2}\sqrt{5}\right), \quad \xi_2 = \left(1, \tfrac{1}{2} + \tfrac{1}{2}\sqrt{5}\right).$$

Thus, S contracts in the direction ξ_1 and expands in the direction ξ_2. With this fact it can be verified that S is also a K-automorphism. The construction of $\mathcal{A}_0$ is related with vectors ξ_1 and ξ_2; that is, $\mathcal{A}_0$ may be defined as a σ-algebra generated by a class of rectangles with sides parallel to vectors ξ_1 and ξ_2. The precise definition of $\mathcal{A}_0$ requires some technical details, which can be found in an article by Arnold and Avez [1968]. □

As we observed in Remark 4.1.4, the first coordinate in the baker transformation is transformed independently of the second, which is the dyadic transformation. The baker transformation is a K-automorphism and the dyadic is exact. This fact is not a coincidence. It may be shown that every exact transformation is, in some sense, a restriction of a K-automorphism. To make this statement precise we need the following definition.

Definition 4.5.2. Let $(X, \mathcal{A}, \mu)$ and $(Y, \mathcal{B}, \nu)$ be two normalized measure spaces and let $S: X \to X$ and $T: Y \to Y$ be two measure-preserving transformations. If there exists a transformation $F: Y \to X$ that is also measure preserving, namely,

$$\nu(F^{-1}(A)) = \mu(A) \quad \text{for } A \in \mathcal{A}$$

and such that $S \circ F = F \circ T$, then S is called a **factor** of T.

The situation described by Definition 4.5.2 can be visualized by the diagram

$$Y \xrightarrow{T} Y$$

$$\downarrow F \qquad \downarrow F \qquad\qquad\qquad (4.5.4)$$

$$X \xrightarrow{S} X$$

and the condition $S \circ F = F \circ T$ may be expressed by saying that the diagram (4.5.4) **commutes**. We have the following theorem due to Rochlin [1961].

Theorem 4.5.1. *Every exact transformation is a factor of a K-automorphism.*

The relationship between K-automorphisms and mixing transformations is much simpler; it is given by the following theorem.

Theorem 4.5.2. *Every K-automorphism is mixing.*

The proofs and more information concerning K-automorphisms can be found in the books by Walters [1982] and by Parry [1981].

Exercises

4.1. Study the rotation on the circle transformation (Examples 4.2.2 and 4.4.1) numerically. Is the behavior a consequence of the properties of the transformation or of the computer? Why?

4.2. Write a series of programs, analogous to those you wrote in the exercises of Chapter 1, to study the behavior of two-dimensional transformations. In particular, write a program to examine the successive locations of an initial cluster of initial conditions as presented in our study of the baker transformation and of equations (4.3.3)–(4.3.5).

4.3. Let $(X, \mathcal{A}, \mu)$ be a finite measure space and let $S: X \to X$ be a measurable transformation such that

$$\mu(A) \leq \mu(S^{-1}(A)) \qquad \text{for } A \in \mathcal{A}.$$

Show that μ is invariant with respect to S. Is the assumption $\mu(X) < \infty$ essential?

4.4. Consider the space $(X, \mathcal{A}, \mu)$ where

$$X = \{\dots, -2, -1, 0, 1, 2, \dots\}$$

is the set of all integers, $\mathcal{A}$ the family of all subsets of X and μ is the counting measure. Let $S(X) = x + k$ for $x \in X$ where k is an integer. For which k is the transformation S ergodic?

4.5. Prove that the baker transformation of Examples 4.1.3 and 4.3.1 is mixing by using the mixing condition (4.3.1).

4.6. Let $X = [0, 1) \times [0, 1)$ be the unit square with the standard Borel measure. Let $r \geq 2$ be an integer. Consider the following generalization of the baker transformation

$$S(x, y) = \left(rx \,(\text{mod } 1), \frac{y}{r} + \frac{k}{r} \right)$$

for

$$\frac{k}{r} \leq x < \frac{k+1}{r}, k = 0, \dots, r - 1.$$

Prove that S is mixing.

4.7. Let $(X, \mathcal{A}, \mu)$ be a normalized measure space and let $P: L^1(X) \to L^1(X)$ be a Markov operator such that $P1 = 1$. Fix an integer $k \geq 1$. Prove that the following statements are true:

(a) P^k is ergodic $\Rightarrow$ P is ergodic,

(b) P^k is mixing $\Rightarrow$ P is mixing,

(c) P^k is exact $\Rightarrow$ P is exact,

where the arrow, as usual, means "implies that." Where may the arrow be reversed?

5
The Asymptotic Properties of Densities

The preceding chapter was devoted to an examination of the various degrees of "chaotic" behavior (ergodicity, mixing, and exactness) that measure-preserving transformations may display. In particular, we saw the usefulness of the Koopman and Frobenius–Perron operators in answering these questions.

Theorem 4.1.1 reduced the problem of finding an invariant measure to one of finding solutions to the equation $Pf = f$. Perhaps the most obvious, although not the simplest, way to find these solutions is to pick an arbitrary $f \in D$ and examine the sequence $\{P^n f\}$ of successive iterations of f by the Frobenius–Perron operator. If $\{P^n f\}$ converges to f_*, then clearly $\{P^{n+1} f\} = \{P(P^n f)\}$ converges simultaneously to f_* and Pf_* and we are done. However, to prove that $\{P^n f\}$ converges (weakly or strongly) to a function f_* is difficult.

In this chapter we first examine the convergence of the sequence $\{A_n f\}$ of averages defined by

$$A_n f = \frac{1}{n} \sum_{k=0}^{n-1} P^k f$$

and show how this may be used to demonstrate the existence of a stationary density of P. We then show that under certain conditions $\{P^n f\}$ can display a new property, namely, asymptotic periodicity. Finally, we introduce the concept of asymptotic stability for Markov operators, which is a generalization of exactness for Frobenius–Perron operators. We then show how the lower-bound function technique may be used to demonstrate asymptotic stability. This technique is used throughout the remainder of the book.

5.1 Weak and Strong Precompactness

In calculus one of the most important observations, originally due to Weierstrass, is that any bounded sequence of numbers contains a convergent subsequence. This observation can be extended to spaces of any finite dimension. Unfortunately, for more complicated objects, such as densities, this is not the case. One example is

$$f_n(x) = n1_{[0,1/n]}(x), \quad 0 \le x \le 1$$

which is bounded in L^1 norm, that is, $\|f_n\| = 1$, but which does not converge weakly or strongly in $L^1([0,1])$ to any density. In fact as $n \to \infty$, $f_n(x) \to \delta(x)$, the Dirac delta function that is supported on a single point, $x = 0$.

One of the great achievements in mathematical analysis was the discovery of sufficient conditions for the existence of convergent subsequences of functions, which subsequently found applications in the calculus of variations, optimal control theory, and proofs for the existence of solutions to ordinary and partial differential equations and integral equations.

To make these comments more precise we introduce the following definitions. Let $(X, \mathcal{A}, \mu)$ be a measure space and $\mathcal{F}$ a set of functions in L^p.

Definition 5.1.1. The set $\mathcal{F}$ will be called **strongly precompact** if every sequence of functions $\{f_n\}, f_n \in \mathcal{F}$, contains a strongly convergent subsequence $\{f_{a_n}\}$ that converges to an $\bar{f} \in L^p$.

Remark 5.1.1. The prefix "pre-" is used because we take $\bar{f} \in L^p$ rather than $\bar{f} \in \mathcal{F}$. $\square$

Definition 5.1.2. The set $\mathcal{F}$ will be called **weakly precompact** if every sequence of functions $\{f_n\}, f_n \in \mathcal{F}$, contains a weakly convergent subsequence $\{f_{a_n}\}$ that converges to an $\bar{f} \in L^p$.

Remark 5.1.2. These two definitions are often applied to sets consisting simply of sequences of functions. In this case the precompactness of $\mathcal{F} = \{f_n\}$ simply means that every sequence $\{f_n\}$ contains a convergent subsequence $\{f_{a_n}\}$. $\square$

Remark 5.1.3. From the definitions it immediately follows that any subset of a weakly or strongly precompact set is itself weakly or strongly precompact. $\square$

There are several simple and general criteria useful for demonstrating the weak precompactness of sets in L^p [see Dunford and Schwartz, 1957]. The three we will have occasion to use are as follows:

1 Let $g \in L^1$ be a nonnegative function. Then the set of all functions $f \in L^1$ such that

$$|f(x)| \leq g(x) \qquad \text{for } x \in X \text{ a.e.} \qquad (5.1.1)$$

is weakly precompact in L^1.

2 Let $M > 0$ be a positive number and $p > 1$ be given. If $\mu(X) < \infty$, then the set of all functions $f \in L^1$ such that

$$\|f\|_{L^p} \leq M \qquad (5.1.2)$$

is weakly precompact in L^1.

3 A set of functions $\mathcal{F} \subset L^1$, $\mu(X) < \infty$, is weakly precompact if and only if:

(a) There is an $M < \infty$ such that $\|f\| \leq M$ for all $f \in \mathcal{F}$; and

(b) For every $\epsilon > 0$ there is a $\delta > 0$ such that

$$\int_A |f(x)|\mu(dx) < \epsilon \qquad \text{if } \mu(A) < \delta \quad \text{and} \quad f \in \mathcal{F}.$$

Remark 5.1.4. If the measure is not finite these two conditions must be supplemented by

(c) For every $\epsilon > 0$ there is a set $B \in \mathcal{A}$, $\mu(B) < \infty$, such that

$$\int_{X \setminus B} |f(x)|\mu(dx) < \epsilon. \quad \square$$

Strong precompactness is generally more difficult to demonstrate than weak precompactness. One of the simplest criteria, which we present only for one-dimensional spaces, is as follows:

4 Let $\mathcal{F}$ be a set of functions defined on a bounded interval Δ of the real line. $\mathcal{F}$ is strongly precompact in $L^1(\Delta)$ if and only if:

(a) There exists a constant $M > 0$ independent of f such that

$$\|f\| \leq M \qquad \text{for all } f \in \mathcal{F}; \qquad (5.1.3a)$$

(b) For all $\epsilon > 0$ there exists a $\delta > 0$ such that

$$\int_\Delta |f(x+h) - f(x)| \, dx < \epsilon \qquad (5.1.3b)$$

for all $f \in \mathcal{F}$ and all h such that $|h| < \delta$. To ensure that this integral is well defined we assume $f(x+h) - f(x) = 0$ for $x + h \notin \Delta$.

Remark 5.1.5. This necessary and sufficient condition for strong precompactness is valid for unbounded intervals Δ if, in addition, for every $\epsilon > 0$ there is an $r > 0$ such that

$$\int_{|x| \geq r} |f(x)| \, dx < \epsilon \qquad \text{for all } f \in \mathcal{F}. \quad \square \qquad (5.1.4)$$

Remark 5.1.6. In practical situations it is often difficult to verify inequality (5.1.3b). However, if the functions $f \in \mathcal{F}$ have uniformly bounded derivatives, that is, if there is a constant K such that $|f'(x)| \leq K$, then the condition is automatically satisfied. To see this, note that

$$|f(x+h) - f(x)| \leq Kh$$

implies

$$\int_\Delta |f(x+h) - f(x)| \, dx \leq Kh\mu(\Delta)$$

and thus if, for a given ϵ, we pick

$$\delta = \epsilon / K\mu(\Delta)$$

the condition (5.1.3b) is satisfied. Clearly this will not work for unbounded intervals because for $\mu(\Delta) \to \infty$, $\delta \to 0$. $\quad \square$

To close this section we state the following corollary.

Corollary 5.1.1. *For every $f \in L^1$, Δ bounded or not,*

$$\lim_{h \to 0} \int_\Delta |f(x+h) - f(x)| \, dx = 0. \qquad (5.1.5)$$

Proof. To see this note that set $\{f\}$ consisting of only one function f is obviously strongly precompact since the sequence $\{f, f, \dots\}$ is always convergent. Thus equation (5.1.5) follows from the foregoing condition (4b) for strong precompactness. $\quad \blacksquare$

5.2 Properties of the Averages $A_n f$

In this section we assume a measure space $(X, \mathcal{A}, \mu)$ and a Markov operator

$P: L^1 \rightarrow L^1$. We are going to demonstrate some simple properties of the averages defined by

$$A_n f = \frac{1}{n} \sum_{k=0}^{n-1} P^k f, \qquad \text{for } f \in L^1. \qquad (5.2.1)$$

We then state and prove a special case of the Kakutani–Yosida abstract ergodic theorem as well as two corollaries to that theorem.

Proposition 5.2.1. *For all $f \in L^1$,*

$$\lim_{n \to \infty} \|A_n f - A_n P f\| = 0.$$

Proof. By the definition of $A_n f$ (5.2.1) we have

$$A_n f - A_n P f = (1/n)(f - P^n f)$$

and thus

$$\|A_n f - A_n P f\| \leq (1/n)(\|f\| + \|P^n f\|).$$

Since it is an elementary property of Markov operators that $\|P^n f\| \leq \|f\|$, we have

$$\|A_n f - A_n P f\| \leq (2/n)\|f\| \rightarrow 0$$

as $n \rightarrow \infty$, which completes the proof. ■

Proposition 5.2.2. *If, for $f \in L^1$, there is a subsequence $\{A_{a_n} f\}$ of the sequence $\{A_n f\}$ that converges weakly to $f_* \in L^1$, then $P f_* = f_*$.*

Proof. First, since $P A_{a_n} f = A_{a_n} P f$, then $\{A_{a_n} P f\}$ converges weakly to $P f_*$. Since $\{A_{a_n} P f\}$ has the same limit as $\{A_{a_n} f\}$, we have $P f_* = f_*$. ■

The following theorem is a special case of an abstract ergodic theorem originally due to Kakutani and Yosida (see Dunford and Schwartz [1957]). The usefulness of the theorem lies in the establishment of a simple condition for the existence of a fixed point for a given Markov operator P.

Theorem 5.2.1. *Let $(X, \mathcal{A}, \mu)$ be a measure space and $P: L^1 \rightarrow L^1$ a Markov operator. If for a given $f \in L^1$ the sequence $\{A_n f\}$ is weakly precompact, then it converges strongly to some $f_* \in L^1$ that is a fixed point of P, namely, $P f_* = f_*$. Furthermore, if $f \in D$, then $f_* \in D$, so that f_* is a stationary density.*

Proof. Because $\{A_n f\}$ is weakly precompact by assumption, there exists a subsequence $\{A_{a_n} f\}$ that converges weakly to some $f_* \in L^1$. Further, by Proposition 5.2.2, we know $P f_* = f_*$.

Write $f \in L^1$ in the form

$$f = (f - f_*) + f_* \tag{5.2.2}$$

and assume for the time being that for every $\epsilon > 0$ the function $f - f_*$ can be written in the form

$$f - f_* = Pg - g + r, \tag{5.2.3}$$

where $g \in L^1$ and $\|r\| < \epsilon$. Thus, from equation (5.2.2) and (5.2.3), we have

$$A_n f = A_n(Pg - g) + A_n r + A_n f_*.$$

Because $Pf_* = f_*$, $A_n f_* = f_*$, and we obtain

$$\|A_n f - f_*\| = \|A_n(f - f_*)\| \le \|A_n(Pg - g)\| + \|A_n r\|.$$

By Proposition 5.2.1 we know that $\|A_n(Pg - g)\|$ is strongly convergent to zero as $n \to \infty$, and by our assumptions $\|A_n r\| \le \|r\| < \epsilon$. Thus, for sufficiently large n, we must have

$$\|A_n f - f_*\| \le \epsilon.$$

Since ϵ is arbitrary, this proves that $\{A_n f\}$ is strongly convergent to f_*.

To show that if $f \in D$, then $f_* \in D$, recall from Definition 3.1.3 that $f \in D$ means that

$$f \ge 0 \quad \text{and} \quad \|f\| = 1.$$

Therefore $Pf \ge 0$ and $\|Pf\| = 1$ so that $P^n f \ge 0$ and $\|P^n f\| = 1$. As a consequence, $A_n f \ge 0$ and $\|A_n f\| = 1$ and, since $\{A_n f\}$ is strongly convergent to f_*, we must have $f_* \in D$. This completes the proof under the assumption that representation (5.2.3) is possible for every ϵ.

In proving this assumption, we will use a simplified version of the Hahn–Banach theorem (see Remark 5.2.1). Suppose that for some ϵ there does not exist an r such that equation (5.2.3) is true. If this were the case, then $f - f_* \notin \text{closure}(P - I)L^1(X)$ and, thus, by the Hahn–Banach theorem, there must exist a $g_0 \in L^\infty$ such that

$$\langle f - f_*, g_0 \rangle \ne 0 \tag{5.2.4}$$

and

$$\langle h, g_0 \rangle = 0 \quad \text{for all } h \in \text{closure}(P - I)L^1(X).$$

In particular

$$\langle (P - I)P^j f, g_0 \rangle = 0.$$

Thus

$$\langle P^{j+1} f, g_0 \rangle = \langle P^j f, g_0 \rangle \quad \text{for } j = 0, 1, \ldots,$$

and by induction we must, therefore, have

$$\langle P^j f, g_0 \rangle = \langle f, g_0 \rangle. \tag{5.2.5}$$

As a consequence

$$\frac{1}{n} \sum_{j=0}^{n-1} \langle P^j f, g_0 \rangle = \frac{1}{n} \sum_{j=0}^{n-1} \langle f, g_0 \rangle = \langle f, g_0 \rangle$$

or

$$\langle A_n f, g_0 \rangle = \langle f, g_0 \rangle. \tag{5.2.6}$$

Since $\{A_{a_n} f\}$ was assumed to converge weakly to f_*, we have

$$\lim_{n \to \infty} \langle A_{a_n} f, g_0 \rangle = \langle f_*, g_0 \rangle$$

and, by (5.2.6),

$$\langle f, g_0 \rangle = \langle f_*, g_0 \rangle,$$

which gives

$$\langle f - f_*, g_0 \rangle = 0.$$

However, this result contradicts (5.2.4), and therefore we conclude that the representation (5.2.3) is, indeed, always possible. ∎

Remark 5.2.1. The Hahn–Banach theorem is one of the classical results of functional analysis. Although it is customarily stated as a general property of some linear topological spaces (e.g., Banach spaces and locally convex spaces), here we state it for L^p spaces. We need only two concepts. A set $E \subset L^p$ is a **linear subspace** of L^p if $\lambda_1 f_1 + \lambda_2 f_2 \in E$ for all $f_1, f_2 \in E$ and all scalars λ_1, λ_2. A linear subspace is **closed** if $\lim f_n \in E$ for every strongly convergent sequence $\{f_n\} \subset E$. □

Next we state a simple consequence of the Hahn–Banach theorem in the language of L^p spaces [see Dunford and Schwartz, 1957].

Proposition 5.2.3. *Let $1 \le p < \infty$ and let p' be adjoint to p, that is, $(1/p) + (1/p') = 1$ for $p > 1$ and $p' = \infty$ for $p = 1$. Further, let $E \subset L^p$ be a linear closed subspace. If $f_0 \in L^p$ and $f_0 \notin E$, then there is a $g_0 \in L^{p'}$ such that $\langle f_0, g_0 \rangle \ne 0$ and $\langle f, g_0 \rangle = 0$ for $f \in E$.*

Geometrically, this proposition means that, if we have a closed subspace E and a vector $f_0 \notin E$, then we can find another vector g_0 orthogonal to E but not orthogonal to f_0 (see Figure 5.2.1).

Remark 5.2.2. By proving Theorem 5.2.1 we have reduced the problem of demonstrating the existence of a stationary density f_* for the operator P, that is, $P f_* = f_*$, to the simpler problem of demonstrating the weak

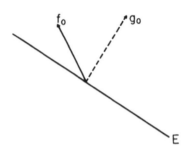

FIGURE 5.2.1. Diagram showing that, for $f_0 \notin E$, we can find a g_0, such that g_0 is not orthogonal to f_0 but it is orthogonal to all $f \in E$. Since g_0 belongs to $L^{p'}$, but not necessarily to L^p, it is drawn as a dashed line.

precompactness of the sequence $\{A_n f\}$. In the special case that P is a Frobenius–Perron operator this also suffices to demonstrate the existence of an invariant measure. $\square$

There are two simple and useful corollaries to Theorem 5.2.1.

Corollary 5.2.1. *Let* $(X, \mathcal{A}, \mu)$ *be a measure space and* $P: L^1 \to L^1$ *a Markov operator. If, for some* $f \in D$ *there is a* $g \in L^1$ *such that*

$$P^n f \leq g \tag{5.2.7}$$

for all n, *then there is an* $f_* \in D$ *such that* $P f_* = f_*$, *that is,* f_* *is a stationary density.*

Proof. By assumption, $P^n f \leq g$ so that

$$0 \leq A_n f = \frac{1}{n} \sum_{k=0}^{n-1} P^k f \leq g$$

and, thus, $|A_n f| \leq g$. By applying our first criterion for weak precompactness (Section 5.1), we know that $\{A_n f\}$ is weakly precompact. Then Theorem 5.2.1 completes the argument. ∎

Corollary 5.2.2. *Again let* $(X, \mathcal{A}, \mu)$ *be a finite measure space and* $P: L^1 \to L^1$ *a Markov operator. If some* $f \in D$ *there exists* $M > 0$ *and* $p > 1$ *such that*

$$\|P^n f\|_{L^p} \leq M \tag{5.2.8}$$

for all n, *then there is an* $f_* \in D$ *such that* $P f_* = f_*$.

Proof. We have

$$\|A_n f\|_{L^p} = \left\| \frac{1}{n} \sum_{k=0}^{n-1} P^k f \right\|_{L^p} \leq \frac{1}{n} \sum_{k=0}^{n-1} \|P^k f\|_{L^p} \leq \frac{1}{n}(nM) = M.$$

Hence, by our second criterion for weak precompactness, $\{A_n f\}$ is weakly precompact, and again Theorem 5.2.1 completes the proof. ∎

Remark 5.2.3. The conditions $P^n f \leq g$ or $\|P^n f\|_{L^p} \leq M$ of these two corollaries guaranteeing the existence of a stationary density f_* rely on the properties of $\{P^n f\}$ for large n. To make this clearer suppose $P^n f \leq g$ only for $n > n_0$. Then, of course, $P^{n+n_0} f \leq g$ for all n, but this can be rewritten in the alternate form $P^n P^{n_0} f = P^n \tilde{f} \leq g$, where $\tilde{f} = P^{n_0} f$. The same argument holds for $\|P^n f\|_{L^p}$, thus demonstrating that it is sufficient for some n_0 to exist such that for all $n > n_0$ either (5.2.7) or (5.2.8) holds. □

We have proved that either convergence or precompactness of $\{A_n f\}$ implies the existence of a stationary density. We may reverse the question to ask whether the existence of a stationary density gives any clues to the asymptotic properties of sequences $\{A_n f\}$. The following theorem gives a partial answer to this question.

Theorem 5.2.2. *Let* $(X, \mathcal{A}, \mu)$ *be a measure space and* $P: L^1 \to L^1$ *a Markov operator with a unique stationary density* f_*. *If* $f_*(x) > 0$ *for all* $x \in X$, *then*

$$\lim_{n \to \infty} A_n f = f_* \qquad \text{for all } f \in D.$$

Proof. First assume f/f_* is bounded. By setting $c = \sup(f/f_*)$, we have

$$P^n f \leq P^n c f_* = c P^n f_* = c f_* \quad \text{and} \quad A_n f \leq c A_n f_* = c f_*.$$

Thus the sequence $\{A_n f\}$ is weakly precompact and, by Theorem 5.2.1, is convergent to a stationary density. Since f_* is the unique stationary density, $\{A_n f\}$ must converge to f_*. Thus the theorem is proved when f/f_* is bounded.

In the general case, write $f_c = \min(f, c f_*)$. We then have

$$f = \frac{1}{\|f_c\|} f_c + r_c, \tag{5.2.9}$$

where

$$r_c = \left(1 - \frac{1}{\|f_c\|}\right) f_c + f - f_c.$$

Since $f_*(x) > 0$ we also have

$$\lim_{c \to \infty} f_c(x) = f(x) \qquad \text{for all } x$$

and, evidently, $f_c(x) \le f(x)$. Thus, by the Lebesgue dominated convergence theorem, $\|f_c - f\| \to 0$ and $\|f_c\| \to \|f\| = 1$ as $c \to \infty$. Thus the remainder r_c is strongly convergent to zero as $c \to \infty$. By choosing $\epsilon > 0$ we can find a value c such that $\|r_c\| < \epsilon/2$. Then

$$\|A_n r_c\| \le \|r_c\| < \frac{\epsilon}{2}. \tag{5.2.10}$$

However, since $f_c/\|f_c\|$ is a density bounded by $c\|f_c\|^{-1} f_*$, according to the first part of the proof,

$$\left\| A_n \left(\frac{1}{\|f_c\|} f_c \right) - f_* \right\| \le \frac{\epsilon}{2} \tag{5.2.11}$$

for sufficiently large n. Combining inequalities (5.2.10) and (5.2.11) with the decomposition (5.2.9), we immediately obtain

$$\|A_n f - f_*\| \le \epsilon$$

for sufficiently large n. ∎

In the case that P is the Frobenius–Perron operator corresponding to a nonsingular transformation S, Theorem 5.2.2 offers a convenient criterion for ergodicity. As we have seen in Theorem 4.2.2, the ergodicity of S is equivalent to the uniqueness of the solution to $Pf = f$. Using this relationship, we can prove the following corollary.

Corollary 5.2.3. *Let (X, A, μ) be a normalized measure space, $S: X \to X$ a measure-preserving transformation, and P the corresponding Frobenius–Perron operator. Then S is ergodic if and only if*

$$\lim_{n \to \infty} \frac{1}{n} \sum_{k=0}^{n-1} P^k f = 1 \qquad \text{for every } f \in D. \tag{5.2.12}$$

Proof. The proof is immediate. Since S is measure preserving, we have $P1 = 1$. If S is ergodic, then by Theorem 4.2.2, $f_*(x) \equiv 1$ is the unique stationary density of P and, by Theorem 5.2.2, the convergence of (5.2.12) follows. Conversely, if the convergence of (5.2.12) holds, applying (5.2.12) to a stationary density f gives $f = 1$. Thus $f_*(x) = 1$ is the unique stationary density of P and again, by Theorem 4.2.2, the transformation S is ergodic. ∎

5.3   Asymptotic Periodicity of $\{P^n f\}$

In the preceding section we reduce the problem of examining the asymptotic properties of the averages $A_n f$ to one of determining the precompactness of $\{A_n f\}$. This, in turn, was reduced by Corollaries 5.2.1 and 5.2.2 to the problem of finding an upper-bound function for $P^n f$ or an upper bound for $\|P^n f\|_{L^p}$. In this section we show that if conditions similar to those in Corollaries 5.2.1 and 5.2.2 are satisfied for Frobenius–Perron operators, then the surprising result is that $\{P^n f\}$ is asymptotically periodic. Even more generally, we will show that almost any kind of upper bound on the iterates $P^n f$ of a Markov operator P suffices to establish that $\{P^n f\}$ will also have very regular (asymptotically periodic) behavior.

Definition 5.3.1. Let $(X, \mathcal{A}, \mu)$ be a finite measure space. A Markov operator P is called **constrictive** if there exists a $\delta > 0$ and $\kappa < 1$ such that for every $f \in D$ there is an integer $n_0(f)$ for which

$$\int_E P^n f(x) \mu(dx) \le \kappa \quad \text{for} \quad n \ge n_0(f) \quad \text{and} \quad \mu(E) \le \delta. \quad (5.3.1)$$

Note that for every density f the integral in inequality (5.3.1) is bounded above by one. Thus condition (5.3.1) for constrictiveness means that eventually $[n \ge n_0(f)]$ this integral cannot be close to one for sufficiently small sets E. This clearly indicates that constrictiveness rules out the possibility that $P^n f$ is eventually concentrated on a set of very small or vanishing measure.

If the space X is not finite, we wish to have a definition of constrictiveness that also prevents $P^n f$ from being dispersed throughout the entire space. To accomplish this we extend Definition 5.3.1.

Definition 5.3.2. Let $(X, \mathcal{A}, \mu)$ be a (σ-finite) measure space. A Markov operator P is called **constrictive** if there exists $\delta > 0$, and $\kappa < 1$, and a measurable set B of finite measure, such that for every density f there is an integer $n_0(f)$ for which

$$\int_{(X \setminus B) \cup E} P^n f(x) \mu(dx) \le \kappa \quad \text{for} \quad n \ge n_0(f) \quad \text{and} \quad \mu(E) \le \delta. \quad (5.3.2)$$

Clearly this definition reduces to that of Definition 5.3.1 when X is finite and we take $X = B$.

Remark 5.3.1. Observe that in inequality (5.3.2) we may always assume that $E \subset B$. To see this, take $F = B \cap E$. Then $(X \setminus B) \cup E = (X \setminus B) \cup F$ and, as a consequence,

$$\int_{(X \setminus B) \cup E} P^n f(x) \mu(dx) = \int_{(X \setminus B) \cup F} P^n f(x) \mu(dx),$$

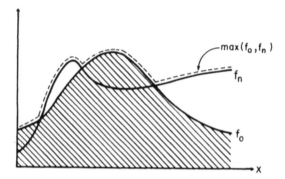

FIGURE 5.3.1. Graph showing convergence of a sequence of functions $\{f_n\}$ to a set $\mathcal{F}$, where the hatched region contains all possible functions drawn from $\mathcal{F}$. (See Example 5.3.1 for details.)

and $\mu(F) \leq \mu(E)$.

From the definition, one might think that verifying constrictiveness is difficult since it is required to find two constants δ and κ as well as a set B with rather specific properties. However, it is often rather easy to verify constrictiveness using one of the two following propositions.

Proposition 5.3.1. *Let $(X, \mathcal{A}, \mu)$ be a finite measure space and $P: L^1(X) \to L^1(X)$ be a Markov operator. Assume there is a $p > 1$ and $K > 0$ such that for every density $f \in D$ we have $P^n f \in L^p$ for sufficiently large n, and*

$$\limsup_{n \to \infty} \|P^n f\|_{L^p} \leq K. \tag{5.3.3}$$

Then P is constrictive.

Proof. From (5.3.3) there is an integer $n_0(f)$ such that

$$\|P^n f\|_{L^p} \leq K + 1 \qquad \text{for } n \geq n_0(f).$$

Thus, by criteria 2 of Remark 5.1.3 the family $\{P^n f\}$, for $n \geq n_0(f)$, $f \in D$, is weakly precompact. Finally, for a fixed $\epsilon \in (0, 1)$, criteria 3 of the same remark implies there is a $\delta > 0$ such that

$$\int_E P^n f(x)\mu(dx) < \epsilon \qquad \text{if } \mu(E) < \delta.$$

Thus weak constrictiveness following from (5.3.3) is demonstrated.   ∎

Our next proposition may be even more useful in demonstrating the constrictiveness of an operator P.

Proposition 5.3.2. *Let $(X, \mathcal{A}, \mu)$ be a σ-finite measure space and $P: L^1(X) \to L^1(X)$ be a Markov operator. If there exists an $h \in L^1$ and $\lambda < 1$ such* *that*

$$\limsup_{n\to\infty} \|(P^n f - h)^+\| \le \lambda \qquad \text{for } f \in D, \tag{5.3.4}$$

then P is constrictive.

Proof. Let $\epsilon = \frac{1}{4}(1-\lambda)$ and take $\mathcal{F} = \{h\}$. Since $\mathcal{F}$, which contains only one element, is evidently weakly precompact (it is also strongly precompact, but this property is not useful to us here), then by criterion 3 of Remark 5.1.3 there exists a $\delta > 0$ such that

$$\int_E h(x)\mu(dx) < \epsilon \qquad \text{for } \mu(E) < \delta. \tag{5.3.5}$$

Furthermore, by Remark 5.1.4 there is a measurable set B of finite measure for which

$$\int_{X\setminus B} h(x)\mu(dx) < \epsilon. \tag{5.3.6}$$

Now fix $f \in D$. From (5.3.4) we may choose an integer $n_0(f)$ such that

$$\|(P^n f - h)^+\| \le \lambda + \epsilon \qquad \text{for } n \ge n_0(f),$$

and, as a consequence,

$$\int_C P^n f(x)\mu(dx) \le \int_C h(x)\mu(dx) + \lambda + \epsilon \tag{5.3.7}$$

for an arbitrary set $C \in \mathcal{A}$. Setting $C = (X \setminus B) \cup E$ in (5.3.7) and using (5.3.5) and (5.3.6) we have

$$\int_{(X\setminus B)\cup E} P^n f(x)\mu(dx) \le \int_{X\setminus B} h(x)\mu(dx) + \int_E h(x)\mu(dx) + \lambda + \epsilon$$

$$< 3\epsilon + \lambda = 1 - \epsilon.$$

this completes the proof. ∎

The interpretation of Proposition 5.3.2 is quite straightforward. Namely, for those regions where $P^n f > h$, if the area of the difference between $P^n f$ and h is bounded above by $\lambda < 1$, then P is constrictive.

In checking conditions (5.3.1)–(5.3.4), it is not necessary to verify them for all $f \in D$. Rather, it is sufficient to verify them for an arbitrary class of densities $f \in D_0 \subset D$ where the set D_0 is dense in D. To be more precise, we give the following definition.

Definition 5.3.3. A set $D_0 \subset D(X)$ is called **dense** in $D(X)$ if, for every $h \in D$ and $\epsilon > 0$, there is a $g \in D_0$ such that $\|h - g\| < \epsilon$.

If X is an interval of the real line R or, more generally, and open set in R^d, then, for example, the following subsets of $D(X)$ are dense:

$D_1 = \{\text{nonnegative continuous functions on } X\} \cap D(X)$

$D_2 = \{\text{nonnegative continuous functions with compact support in } X\} \cap D(X)$

$D_3 = \{\text{nonnegative differentiable functions on } X\} \cap D(X)$

$D_4 = \{\text{positive differentiable functions on } X\} \cap D(X)$

If a set $D_0 \subset D(X)$ is dense in $D(X)$, one need only verify inequality (5.3.1) for $f \in D_0$ when checking for constrictiveness. Then, for any other $f \in D(X)$ this inequality will be automatically satisfied with κ replaced by $\kappa_1 = \frac{1}{2}(1 + \kappa)$. To show this choose an $f \in D$. Then there is another density $f_0 \in D_0$ such that $\|f - f_0\| \leq \kappa_1 - \kappa$. Since, by assumption, (5.3.1) holds for $f_0 \in D_0$, we have

$$\int_E P^n f_0(x)\mu(dx) \leq \kappa \qquad \text{for } n \geq n_0(f_0)$$

and

$$\int_E P^n f(x)\mu(dx) = \int_E P^n f_0(x)\mu(dx) + \int_E [P^n f(x) - P^n f_0(x)]\mu(dx)$$
$$\leq \int_E P^n f_0(x)\mu(dx) + \|f - f_0\|$$
$$\leq \kappa_1.$$

Thus, when (5.3.1) holds for $f_0 \in D_0$ it holds for all densities $f \in D(X)$.

Precisely the same argument shows that it is also sufficient to verify (5.3.2) for densities drawn from dense sets. As a consequence of these observations, in verifying either (5.3.3) or (5.3.4) of Propositions 5.3.1 and 5.3.2 we may confine our attention to $f \in D_0$.

The main result of this section—which is proved in Komornik and Lasota ([1987]; see also Lasota, Li and Yorke [1984]; Schaefer [1980] and Keller [1982])—is as follows.

Theorem 5.3.1. (spectral decomposition theorem). *Let P be a constrictive Markov operator. Then there is an integer r, two sequences of nonnegative functions $g_i \in D$ and $k_i \in L^\infty$, $i = 1, \ldots, r$, and an operator $Q: L^1 \to L^1$ such that for every $f \in L^1$, Pf may be written in the form*

$$Pf(x) = \sum_{i=1}^r \lambda_i(f)g_i(x) + Qf(x), \qquad (5.3.8)$$

where

$$\lambda_i(f) = \int_X f(x)k_i(x)\mu(dx). \qquad (5.3.9)$$

The functions g_i and operator Q have the following properties:

(1) $g_i(x)g_j(x) = 0$ for all $i \neq j$, so that functions g_i have disjoint supports;

(2) For each integer i there exists a unique integer $\alpha(i)$ such that $Pg_i = g_{\alpha(i)}$. Further $\alpha(i) \neq \alpha(j)$ for $i \neq j$ and thus operator P just serves to permute the functions g_i.

(3) $\|P^n Qf\| \to 0$ as $n \to \infty$ for every $f \in L^1$.

Remark 5.3.2. Note from representation (5.3.8) that operator Q is automatically determined if we know the function g_i and k_i, that is,

$$Qf(x) = f(x) - \sum_{i=1}^{r} \lambda_i(f)g_i(x). \quad \square$$

From representation (5.3.8) of Theorem 5.3.1 for Pf, it immediately follows that the structure of $P^{n+1}f$ is given by

$$P^{n+1}f(x) = \sum_{i=1}^{r} \lambda_i(f)g_{\alpha^n(i)}(x) + Q_n f(x), \tag{5.3.10}$$

where $Q_n = P^n Q$, and $\alpha^n(i) = \alpha(\alpha^{n-1}(i)) = \cdots$, and $\|Q_n f\| \to 0$ as $n \to \infty$. The terms under the summation in (5.3.10) are just permuted with each application of P, and since r is finite the sequence

$$\sum_{i=1}^{r} \lambda_i(f)g_{\alpha^n(i)}(x) \tag{5.3.11}$$

must be periodic with a period $\tau \leq r!$. Since $\{\alpha^n(1), \ldots \alpha^n(r)\}$ is simply a permutation of $\{1, \ldots, r\}$, there is a unique i corresponding to each $\alpha^n(i)$. Thus it is clear that summation (5.3.11) may be rewritten as

$$\sum_{i=1}^{r} \lambda_{\alpha^{-n}(i)}(f)g_i(x),$$

where $\{\alpha^{-n}(i)\}$ denotes the inverse permutation of $\{\alpha^n(i)\}$.

Rewriting the summation in this form clarifies how successive applications of operator P really work. Since the functions g_i are supported on disjoint sets, each successive application of operator P leads to a new set of scaling coefficients $\lambda_{\alpha^{-n}}(f)$ associated with each function $g_i(x)$.

A sequence $\{P^n\}$ for which formula (5.3.8) is satisfied will be called **asymptotically periodic**. Using this notion, Theorem 5.3.1 may be rephrased as follows: If P is a constrictive operator, then $\{P^n\}$ is asymptotically periodic.

It is actually rather easy to obtain an upper bound on the integer r appearing in equation (5.3.8) if we can find an upper bound function for

$P^n f$. Assume that P is a Markov operator and there exists a function $h \in L^1$ such that

$$\lim_{n \to \infty} \|(P^n f - h)^+\| = 0 \qquad \text{for } f \in D. \qquad (5.3.12)$$

Thus, by Proposition 5.3.2, P is constrictive and representation (5.3.8) is valid. Let τ be the period of sequence (5.3.11), so that, from (5.3.8) and (5.3.12), we have

$$Lf(x) \doteq \lim_{n \to \infty} P^{n\tau} f(x) = \sum_{i=1}^{r} \lambda_i(f) g_i(x) \le h(x), \quad f \in D.$$

Set $f = g_k$ so that $Lf(x) = g_k(x) \le h(x)$. By integrating over the support of g_k, bearing in mind that the supports of the g_k are disjoint, and summing from $k = 1$ to $k = r$, we have

$$\sum_{k=1}^{r} \int_{\text{supp} g_k} g_k(x) \mu(dx) \le \sum_{k=1}^{r} \int_{\text{supp} g_k} h(x) \mu(dx) \le \|h\|.$$

Since $g_k \in D$, this reduces to

$$r \le \|h\|, \qquad (5.3.13)$$

which is the desired result.

If the explicit representation (5.3.8) for Pf for a given Markov operator P is known, then it is especially easy to check for the existence of invariant measures and to determine ergodicity, mixing, or exactness, as shown in the following sections. Unfortunately, we seldom have an explicit representation for a given Markov operator, but in the remainder of this chapter we show that the mere existence of representation (5.3.8) allows us to deduce some interesting properties.

5.4   The Existence of Stationary Densities

In this section we first show that every constrictive Markov operator has a stationary density and then give an explicit representation for $P^n f$ when that stationary density is a constant. We start with a proposition.

Proposition 5.4.1. Let $(X, \mathcal{A}, \mu)$ be a measure space and $P: L^1 \to L^1$ a constrictive Markov operator. Then P has a stationary density.

Proof. Let a density f be defined by

$$f(x) = \frac{1}{r} \sum_{i=1}^{r} g_i(x), \qquad (5.4.1)$$

where r and g_i were defined in Theorem 5.3.1. Because of property (2), Theorem 5.3.1,

$$Pf(x) = \frac{1}{r} \sum_{i=1}^{r} g_{\alpha(i)}(x)$$

and thus $Pf = f$, which completes the proof. ■

Now assume that the measure μ is normalized $[\mu(X) = 1]$ and examine the consequences for the representation of $P^n f$ when we have a constant stationary density $f = 1_X$. Remember that, if P is a Frobenius–Perron operator, this is equivalent to μ being invariant.

Proposition 5.4.2. Let (X, A, μ) be a finite measure space and $P: L^1 \to L^1$ a constrictive Markov operator. If P has a constant stationary density, then the representation for $P^{n+1} f$ takes the simple form

$$P^{n+1} f(x) = \sum_{i=1}^{r} \lambda_{\alpha^{-n}(i)}(f) \bar{1}_{A_i}(x) + Q_n f(x) \qquad \text{for all } f \in L^1, \quad (5.4.2)$$

where

$$\bar{1}_{A_i}(x) = [1/\mu(A_i)] 1_{A_i}(x).$$

The sets A_i form a partition of X, that is,

$$\bigcup_i A_i = X \quad \text{and} \quad A_i \cap A_j = \emptyset \qquad \text{for } i \neq j.$$

Furthermore, $\mu(A_i) = \mu(A_j)$ whenever $j = \alpha^n(i)$ for some n.

Proof. First observe that with $f = 1_X$ and stationary, $P1_X = 1_X$ so that $P^n 1_X = 1_X$. However, if P is constrictive, then, from Theorem 5.3.1

$$P^{n+1} 1_X(x) = \sum_{i=1}^{r} \lambda_{\alpha^{-n}(i)}(1_X) g_i(x) + Q_n 1_X(x). \qquad (5.4.3)$$

From our considerations in the preceding section, we know that the summation in equation (5.4.3) is periodic. Let τ be the period of the summation portion of P^{n+1} (remember that $\tau \leq r!$) so that

$$\alpha^{-n\tau}(i) = i$$

and

$$P^{(n+1)\tau} 1_X(x) = \sum_{i=1}^{r} \lambda_i(1_X) g_i(x) + Q_{n\tau} 1_X(x).$$

Passing to the limit as $n \to \infty$ and using the stationarity of 1_X, we have

$$1_X(x) = \sum_{i=1}^{r} \lambda_i(1_X) g_i(x). \qquad (5.4.4)$$

However, since functions g_i are supported on disjoint sets, therefore, from (5.4.4), we must have each g_i constant or, more specifically,

$$g_i(x) \equiv [1/\lambda_i(1_X)]1_{A_i}(x),$$

where $A_i \subset X$ denotes the support of g_i, that is, the set of all x such that $g_i(x) \neq 0$. From (5.4.4) it also follows that $\cup_i A_i = X$.

Apply operator P^n to equation (5.4.4) to give

$$P^n 1_X(x) \equiv 1_X(x) = \sum_{i=1}^{r} \lambda_i(1_X)g_{\alpha^n(i)}(x),$$

and, by the same reasoning employed earlier, we have

$$g_{\alpha^n(i)}(x) \equiv 1/\lambda_i(1_X) \qquad \text{for all } x \in A_{\alpha_i}.$$

Thus, the functions $g_i(x)$ and $g_{\alpha^n(i)}$ must be equal to the same constant. And, since the functions $g_i(x)$ are densities, we must have

$$\int_{A_i} g_i(x)\mu(dx) = 1 = \mu(A_i)/\lambda_i(1_X).$$

Thus $\mu(A_i) = \lambda_i(1_X)$ and

$$g_i(x) = [1/\mu(A_i)]1_{A_i}(x). \tag{5.4.5}$$

Moreover, $\mu(A_{\alpha^n(i)}) = \mu(A_i)$ for all n. ∎

5.5   Ergodicity, Mixing, and Exactness

We now turn our attention to the determination of ergodicity, mixing, and exactness for operators P that can be written in the form of equation (5.3.8). We assume throughout that $\mu(X) = 1$ and that $P1_X = 1_X$. We further note that a permutation $\{\alpha(1), \ldots, \alpha(r)\}$ of the set $\{1, \ldots, r\}$ (see Theorem 5.3.1) for which there is no invariant subset is called a **cycle** or **cyclical permutation**.

Ergodicity

Theorem 5.5.1. *Let $(X, \mathcal{A}, \mu)$ be a normalized measure space and $P: L^1 \to L^1$ a constrictive Markov operator. Then P is ergodic if and only if the permutation $\{\alpha(1), \ldots, \alpha(r)\}$ of the sequence $\{1, \ldots, r\}$ is cyclical.*

Proof. We start the proof with the "if" portion. Recall from equation (5.2.1) that the average $A_n f$ is defined by

$$A_n f(x) = \frac{1}{n}\sum_{j=0}^{n-1} P^j f(x).$$

Thus, with representation (5.4.2), $A_n f$ can be written as

$$A_n f(x) = \sum_{i=1}^{r} \left[\frac{1}{n} \sum_{j=0}^{n-1} \lambda_{\alpha^{-j}(i)}(f) \right] \bar{1}_{A_i}(x) + \tilde{Q}_n f(x),$$

where the remainder $\tilde{Q}_n f$ is given by

$$\tilde{Q}_n f = \frac{1}{n} \sum_{j=0}^{n-1} Q_j f \quad Q_0 f = -\sum_{i=1}^{r} \lambda_i(f) \bar{1}_{A_i} + f.$$

Now consider the coefficients

$$\frac{1}{n} \sum_{j=0}^{n-1} \lambda_{\alpha^{-j}(i)}(f). \tag{5.5.1}$$

Since, as we showed in Section 5.4, the sequence $\{\lambda_{\alpha^{-j}(i)}\}$ is periodic in j, the summation (5.5.1) must always have a limit as $n \to \infty$. Let this limit be $\bar{\lambda}_i(f)$. Assume there are no invariant subsets of $\{1, \ldots, r\}$ under the permutation α. Then the limits $\bar{\lambda}_i(f)$ must be independent of i since every piece of the summation (5.5.1) of length r for different i consists of the same numbers but in a different order. Thus

$$\lim_{n \to \infty} A_n f = \sum_{i=1}^{r} \bar{\lambda}(f) \bar{1}_{A_i}.$$

Further, since α is cyclical, Proposition 5.4.2 implies that $\mu(A_i) = \mu(A_j) = 1/r$ for all i, j and $\bar{1}_{A_i} = r 1_{A_i}$, so that

$$\lim_{n \to \infty} A_n f = r \bar{\lambda}(f).$$

Hence, for $f \in D$, $\bar{\lambda}(f) = 1/r$, and we have proved that if the permutation $\{\alpha(1), \ldots, \alpha(r)\}$ of $\{1, \ldots, r\}$ is cyclical, then $\{P^n f\}$ is Cesaro convergent to 1 and, therefore, ergodic.

The converse is also easy to prove. Suppose P is ergodic and that $\{\alpha(i)\}$ is not a cyclical permutation. Thus $\{\alpha(i)\}$ has an invariant subset I. As an initial f take

$$f(x) = \sum_{i=1}^{r} c_i \bar{1}_{A_i}(x)$$

wherein

$$c_i = \begin{cases} c \neq 0 & \text{if } i \text{ belongs to the invariant subset } I \text{ of the} \\ & \text{permutation of } \{1, \ldots, r\}, \\ 0 & \text{otherwise.} \end{cases}$$

Then

$$\lim_{n\to\infty} A_n f = \frac{1}{r} \sum_{i=1}^{r} \bar{\lambda}_i(f) \bar{1}_{A_i},$$

where $\bar{\lambda}_i(f) \neq 0$ if i is contained in the invariant subset I, and $\bar{\lambda}_i(f) = 0$ otherwise. Thus the limit of $A_n f$ as $n \to \infty$ is not a constant function with respect to x, so that P cannot be ergodic. This is a contradiction; hence, if P is ergodic, $\{\alpha(i)\}$ must be a cyclical permutation. ∎

Mixing and Exactness

Theorem 5.5.2. *Let $(X, \mathcal{A}, \mu)$ be a normalized measure space and $P: L^1 \to L^1$ a constrictive Markov operator. If $r = 1$ in representation (5.3.8) for P, then P is exact.*

Proof. The proof is simple. Assume $r = 1$, so by (5.4.2) we have

$$P^{n+1}f(x) = \lambda(f)1_X(x) + Q_n f(x)$$

and, thus,

$$\lim_{n\to\infty} P^{n+1}f = \lambda(f)1_X.$$

In particular, when $f \in D$ then $\lambda(f) \equiv 1$ since P preserves the norm. Hence, for all $f \in D$, $\{P^n f\}$ converges strongly to 1, and P is therefore exact (and, of course, also mixing). ∎

The converse is surprising, for we can prove that P mixing implies that $r = 1$.

Theorem 5.5.3. *Again let $(X, \mathcal{A}, \mu)$ be a normalized measure space and $P: L^1 \to L^1$ a constrictive Markov operator. If P is mixing, then $r = 1$ in representation (5.3.8).*

Proof. To see this, assume P is mixing but that $r > 1$ and take an initial $f \in D$ given by

$$f(x) = c_1 1_{A_1}(x), \qquad \text{where } c_1 = 1/\mu(A_1).$$

Therefore

$$P^n f(x) = c_1 1_{A(n)}(x),$$

where $A(n) = A_{\alpha^n(1)}$. Since P was assumed to be mixing, $\{P^n f\}$ converges weakly to 1. However, note that

$$\langle P^n f, 1_{A_1} \rangle = \begin{cases} c_1 & \text{if } \alpha^n(1) = 1 \\ 0 & \text{if } \alpha^n(1) \neq 1. \end{cases}$$

Hence $\{P^n f\}$ will converge weakly to 1 only if $\alpha^n(1) = 1$ for all sufficiently large n. Since α is a cyclical permutation, r cannot be greater than 1, thus demonstrating that $r = 1$.   ■

Remark 5.5.1. It is somewhat surprising that in this case P mixing implies P exact.   □

Remark 5.5.2. Observe that, except for the remainder $Q_n f$, $P^{n+1} f$ behaves like permutations for which the notions of ergodicity, mixing, and exactness are quite simple.   □

5.6 Asymptotic Stability of $\{P^n\}$

Our considerations of ergodicity, mixing and exactness for Markov operators in the previous section were based on the assumption that we were working with a normalized measure space $(X, \mathcal{A}, \mu)$. We now turn to a more general situation and take $(X, \mathcal{A}, \mu)$ to be an arbitrary measure space. We show how Theorem 5.3.1 allows us to obtain a most interesting result concerning the asymptotic stability of $\{P^n f\}$.

We first present a generalization for Markov operators of the concept of exactness for Frobenius–Perron operators associated with a transformation.

Definition 5.6.1. Let $(X, \mathcal{A}, \mu)$ be a measure space and $P \colon L^1 \to L^1$ a Markov operator. Then $\{P^n\}$ is said to be **asymptotically stable** if there exists a unique $f_* \in D$ such that $Pf_* = f_*$ and

$$\lim_{n \to \infty} \|P^n f - f_*\| = 0 \qquad \text{for every } f \in D. \tag{5.6.1}$$

When P is a Frobenius–Perron operator, the following definition holds.

Definition 5.6.2. Let $(X, \mathcal{A}, \mu)$ be a measure space and $P \colon L^1 \to L^1$ the Frobenius–Perron operator corresponding to a nonsingular transformation $S \colon X \to X$. If $\{P^n\}$ is asymptotically stable, then the transformation S is said to be **statistically stable**.

The following theorem is a direct consequence of Theorem 5.3.1.

Theorem 5.6.1. *Let P be a constrictive Markov operator. Assume there is a set $A \subset X$ of nonzero measure, $\mu(A) > 0$, with the property that for every $f \in D$ there is an integer $n_0(f)$ such that*

$$P^n f(x) > 0 \tag{5.6.2}$$

for almost all $x \in A$ and all $n > n_0(f)$. Then $\{P^n\}$ is asymptotically stable.

Proof. Since, by assumption, P is constrictive, representation (5.3.8) is valid. We will first show that $r = 1$.

Assume $r > 1$, and choose an integer i_0 such that A is not contained in the support of g_{i_0}. Take a density $f \in D$ of the form $f(x) = g_{i_0}(x)$ and let τ be the period of the permutation α. Then we have

$$P^{n\tau} f(x) = g_{i_0}(x).$$

Clearly, $P^{n\tau} f(x)$ is not positive on the set A since A is not contained in the support of g_{i_0}. This result contradicts (5.6.2) of the theorem and, thus, we must have $r = 1$.

Since $r = 1$, equation (5.3.10) reduces to

$$P^{n+1} f(x) = \lambda(f)g(x) + Q_n f(x)$$

so

$$\lim_{n \to \infty} P^n f = \lambda(f)g.$$

If $f \in D$, then $\lim_{n \to \infty} P^n f \in D$ also; therefore, by integrating over X we have

$$1 = \lambda(f).$$

Thus, $\lim_{n \to \infty} P^n f = g$ for all $f \in D$ and $\{P^n\}$ is asymptotically stable; this finishes the proof. ∎

The disadvantage with this theorem is that it requires checking for two different criteria: (i) that P is constrictive and (ii) the existence of the set A. It is interesting that, by a slight modification of the assumption that $P^n f$ is positive on a set A, we can completely eliminate the necessity of assuming P to be constrictive. To do this, we first introduce the notion of a lower-bound function.

Definition 5.6.3. A function $h \in L^1$ is a **lower-bound function** for a Markov operator $P: L^1 \to L^1$ if

$$\lim_{n \to \infty} \|(P^n f - h)^-\| = 0 \qquad \text{for every } f \in D. \qquad (5.6.3)$$

Condition (5.6.3) may be rewritten as

$$(P^n f - h)^- = \epsilon_n,$$

where $\|\epsilon_n\| \to 0$ as $n \to \infty$ or, even more explicitly, as

$$P^n f \geq h - \epsilon_n.$$

Thus, figuratively speaking, a lower-bound function h is one such that, for every density f, successive iterates of that density by P are eventually almost everywhere above h.

It is, of course, clear that any nonpositive function is a lower-bound function, but, since $f \in D$ and thus $P^n f \in D$ and all densities are positive, a negative lower bound function is of no interest. Thus we give a second definition.

Definition 5.6.4. A lower-bound function h is called **nontrivial** if $h \geq 0$ and $\|h\| > 0$.

Having introduced the concept of nontrivial lower-bound functions, we can now state the following theorem.

Theorem 5.6.2. *Let $P: L^1 \to L^1$ be a Markov operator. $\{P^n\}$ is asymptotically stable if and only if there is a nontrivial lower bound function for P.*

Proof. The "only if" part is obvious since (5.6.1) implies (5.6.3) with $h = f_*$. The proof of the "if" part is not so direct, and will be done in two steps. We first show that

$$\lim_{n \to \infty} \|P^n(f_1 - f_2)\| = 0 \qquad (5.6.4)$$

for every $f_1, f_2 \in D$ and then proceed to construct the function f_*.

Step I. For every pair of densities $f_1, f_2 \in D$, the $\|P^n(f_1 - f_2)\|$ is a decreasing function of n. To see this, note that, since every Markov operator is contractive,

$$\|Pf\| \leq \|f\|$$

and, as a consequence,

$$\|P^{n+m}(f_1 - f_2)\| = \|P^m P^n(f_1 - f_2)\| \leq \|P^n(f_1 - f_2)\|.$$

Now set $g = f_1 - f_2$ and note that, since $f_1, f_2 \in D$,

$$c = \|g^+\| = \|g^-\| = \tfrac{1}{2}\|g\|.$$

Assume $c > 0$. We have $g = g^+ - g^-$ and

$$\|P^n g\| = c\|(P^n(g^+/c) - h) - (P^n(g^-/c) - h)\|. \qquad (5.6.5)$$

Since g^+/c and g^-/c belong to D, by equation (5.6.3), there must exist an integer n_1 such that for all $n \geq n_1$

$$\|(P^n(g^+/c) - h)^-\| \leq \tfrac{1}{4}\|h\|$$

and

$$\|(P^n(g^-/c) - h)^-\| \leq \tfrac{1}{4}\|h\|.$$

Now we wish to establish upper bounds for $\|P^n(g^+/c) - h\|$ and $\|P^n(g^-/c) - h\|$. To do this, first note that, for any pair of nonnegative real numbers a and b,

$$|a - b| = a - b + 2(a - b)^-.$$

Next write

$$\|P^n(g^+/c) - h\| = \int_X |P^n(g^+/c)(x) - h(x)|\mu(dx)$$

$$= \int_X P^n(g^+/c)(x)\mu(dx) - \int_X h(x)\mu(dx)$$

$$+ 2\int_X (P^n(g^+/c)(x) - h(x))^-\mu(dx)$$

$$= \|P^n(g^+/c)\| - \|h\| + 2\|(P^n(g^+/c) - h)^-\|$$

$$\leq 1 - \|h\| + 2 \cdot \tfrac{1}{4}\|h\| = 1 - \tfrac{1}{2}\|h\| \qquad \text{for } n \geq n_1.$$

Analogously,

$$\|P^n(g^-/c) - h\| \leq 1 - \tfrac{1}{2}\|h\| \qquad \text{for } n \geq n_1.$$

Thus equation (5.6.5) gives

$$\|P^n g\| \leq c\|P^n(g^+/c) - h\| + c\|P^n(g^-/c) - h\|$$

$$\leq c(2 - \|h\|) = \|g\| \left(1 - \tfrac{1}{2}\|h\|\right) \qquad \text{for } n \geq n_1. \qquad (5.6.6)$$

From (5.6.6), for any $f_1, f_2 \in D$, we can find an integer n_1 such that

$$\|P^{n_1}(f_1 - f_2)\| \leq \|f_1 - f_2\| \left(1 - \tfrac{1}{2}\|h\|\right).$$

By applying the same argument to the pair $P^{n_1}f_1, P^{n_1}f_2$ we may find a second integer n_2 such that

$$\|P^{n_1+n_2}(f_1 - f_2)\| \leq \|P^{n_1}(f_1 - f_2)\| \left(1 - \tfrac{1}{2}\|h\|\right)$$

$$\leq \|f_1 - f_2\| \left(1 - \tfrac{1}{2}\|h\|\right)^2.$$

After k repetitions of this procedure, we have

$$\|P^{n_1+\cdots+n_k}(f_1 - f_2)\| \leq \|f_1 - f_2\| \left(1 - \tfrac{1}{2}\|h\|\right)^k,$$

and since $\|P^n(f_1 - f_2)\|$ is a decreasing function of n, this implies (5.6.4).

Step II. To complete the proof, we construct a maximal lower-bound function for P. Thus, let

$$\rho = \sup\{\|h\|: h \text{ is a lower-bound function for } P\}.$$

Since by assumption there is a nontrivial h, we must have $0 < \rho \leq 1$. Observe that for any two lower-bound functions h_1 and h_2, the function $h = \max(h_1, h_2)$ is also a lower-bound function. To see this, note that

$$\|(P^n f - h)^-\| \leq \|(P^n f - h_1)^-\| + \|(P^n f - h_2)^-\|.$$

Choose a sequence $\{h_j\}$ of lower-bound functions such that $\|h_j\| \to \rho$. Replacing, if necessary, h_j by $\max(h_1 \ldots, h_j)$, we can construct an increasing sequence $\{h_j\}$ of lower functions, which will always have a limit (finite or infinite). This limiting function

$$h_* = \lim_{j \to \infty} h_j$$

is also a lower-bound function since

$$\|(P^n f - h_*)\| \le \|(P^n f - h_j)\| + \|h_j - h_*\|$$

and, by the Lebesgue monotone convergence theorem,

$$\|h_j - h_*\| = \int_X h_*(x)\mu(dx) - \int_X h_j(x)\mu(dx) \to 0 \qquad \text{as } j \to \infty.$$

Now the limiting function h_* is also the maximal lower function. To see this, note that for any other lower function h, the function $\max(h, h_*)$ is also a lower function and that

$$\|\max(h, h_*)\| \le \rho = \|h_*\|,$$

which implies $h \le h_*$.

Observe that, since $(Pf)^- \le Pf^-$, for every m and n $(n > m)$,

$$\|(P^n f - P^m h_*)^-\| \le \|P^m (P^{n-m} f - h_*)^-\| \le \|(P^{n-m} f - h_*)^-\|,$$

which implies that, for every m, the function $P^m h_*$ is a lower function. Thus, since h_* is the maximal lower function, $P^m h_* \le h_*$ and since P^m preserves the integral, $P^m h_* = h_*$. Thus the function $f_* = h_*/\|h_*\|$ is a density satisfying $Pf_* = f_*$.

Finally, by equation (5.6.4), we have

$$\lim_{n \to \infty} \|P^n f - f_*\| = \lim_{n \to \infty} \|P^n f - P^n f_*\| = 0 \qquad \text{for } f \in D,$$

which automatically gives equation (5.6.1). ∎

In checking for the conditions of Theorem 5.6.2 it is once again sufficient to demonstrate that (5.6.3) holds for densities f drawn from a dense set $D_0 \subset D(X)$.

Remark 5.6.1. Before continuing, it is interesting to point out the connection between Theorems 5.3.1 and 5.6.2 concerning asymptotic periodicity and asymptotic stability. Namely, from the spectral decomposition Theorem 5.3.1 we can actually shorten the proof of asymptotic stability in Theorem 5.6.2.

To show this, assume P satisfies the lower-bound function condition (5.6.3). Pick an $f \in D$ and choose a number $n_0(f)$ such that

$$\|(P^n f - h)^-\| \le \tfrac{1}{4}\|h\| \qquad \text{for } n \ge n_0(f).$$

From $|a - b| = a - b + 2(a - b)^-$ we have

$$\|(P^n f - h)^+\| \leq \|P^n f - h\| \leq \|P^n f\| - \|h\| + 2\|(P^n f - h)^-\|,$$

and since $\|P^n f\| \equiv 1$, equation (5.6.7) gives

$$\|(P^n f - h)^+\| \leq 1 - \tfrac{1}{2}\|h\| \qquad \text{for } n \geq n_0(f).$$

Thus, by Proposition 5.3.2 we know that the operator P is constrictive. Since P is constrictive it satisfies Theorem 5.3.1 and in particular we have the decomposition formula (5.3.8). Using the assumed existence of a lower-bound function, h we will show that $r = 1$ by necessity.

Assume the contrary and take $r \geq 2$. Consider two basis functions g_1 and g_2 in the decomposition (5.3.8). From $Pg_i = g_{\alpha(i)}$ we obviously have $P^{nm} g_i = g_i$ for $m = r!$ and an arbitrary n. However, from (5.6.3) it also follows that

$$P^{nm} g_i \geq h - \epsilon^i_{nm}, \qquad i = 1, 2,$$

so $g_i \geq h - \epsilon^i_{nm}$ for $i = 1, 2$. This then implies that $g_1 g_2 > 0$, which contradicts the orthogonality of the g_i required by Theorem 5.3.1. We are thus led to a contradiction and therefore must have $r = 1$. Thus (5.3.8) implies the asymptotic stability of $\{P^n\}$ with $f_* = g_1$.

Hence, by the expedient of using Theorem 5.3.1 we have been able to considerably shorten the proof of Theorem 5.6.2.

The results of Theorem 5.6.2 with respect to the uniqueness of stationary densities for asymptotically stable Markov operators may be generalized by the following observation.

Proposition 5.6.1. *Let $(X, \mathcal{A}, \mu)$ be a measure space and $P \colon L^1 \to L^1$ a Markov operator. If $\{P^n\}$ is asymptotically stable and f_* is the unique stationary density of P, then for every normalized $f \in L^1(\|f\| = 1)$ the condition*

$$Pf = f \tag{5.6.7}$$

implies that either $f = f_*$ or $f = -f_*$.

Proof. From Proposition 3.1.3, equation (5.6.7) implies that both f^+ and f^- are fixed points of P. Assume $\|f^+\| > 0$, so that $\tilde{f} = f^+/\|f^+\|$ is a density and $P\tilde{f} = \tilde{f}$. Uniqueness of f_* implies $\tilde{f} = f_*$, hence

$$f^+ = f_* \|f^+\|,$$

which must also hold for $\|f^+\| = 0$. In an analogous fashion,

$$f^- = f_* \|f^-\|$$

so that

$$f = f^+ - f^- = (\|f^+\| - \|f^-\|) f_* = \alpha f_*.$$

Since $\|f\| = \|f_*\|$, we have $|\alpha| = 1$, and the proof is complete. ∎

Before closing this section we state and prove a result that draws the connection between statistical stability and exactness when P is a Frobenius–Perron operator.

Proposition 5.6.2. Let $(X, \mathcal{A}, \mu)$ be a measure space, $S: X \to X$ a nonsingular transformation such that $S(A) \in \mathcal{A}$ for $A \in \mathcal{A}$, and P the Frobenius–Perron operator corresponding to S. If S is statistically stable and f_* is the density of the unique invariant measure, then the transformation S with the measure

$$\mu_{f_*}(A) = \int_A f_*(x)\mu(dx) \qquad for \ A \in \mathcal{A}$$

is exact.

Proof. From Theorem 4.1.1 it follows immediately that μ_{f_*} is invariant. Thus, it only remains to prove the exactness.

Assume $\mu_{f_*}(A) > 0$ and define

$$f_A(x) = [1/\mu_{f_*}(A)]f_*(x)1_A(x) \qquad for \ x \in X.$$

Clearly, $f_A \in D(X, \mathcal{A}, \mu)$ and

$$\lim_{n \to \infty} r_n = \lim_{n \to \infty} \|P^n f_A - f_*\| = 0.$$

From the definition of μ_{f_*}, we have

$$\mu_{f_*}(S^n(A)) = \int_{S^n(A)} f_*(x)\mu(dx) \geq \int_{S^n(A)} P^n f_A(x)\mu(dx) - r_n. \qquad (5.6.8)$$

By Proposition 3.2.1, we know that $P^n f_A$ is supported on $S^n(A)$, so that

$$\int_{S^n(A)} P^n f_A(x)\mu(dx) = \int_X P^n f_A(x)\mu(dx) = 1.$$

Substituting this result into (5.6.8) and taking the limit as $n \to \infty$ gives

$$\lim_{n \to \infty} \mu_{f_*}(S^n(A)) = 1;$$

hence $S: X \to X$ is exact by definition. ∎

Remark 5.6.1. In the most general case, Proposition 5.6.2 is not invertible; that is, statistical stability of S implies the existence of a unique invariant measure and exactness, but not vice versa. Lin [1971] has shown that the inverse implication is true when the initial measure μ is invariant.   □

5.7 Markov Operators Defined by a Stochastic Kernel

As a sequel to Section 5.6, we wish to develop some important consequences of Theorems 5.6.1 and 5.6.2. Let $(X, \mathcal{A}, \mu)$ be a measure space and $K: X \times X \to R$ be a measurable function that satisfies

$$0 \le K(x, y) \tag{5.7.1}$$

and

$$\int_X K(x, y)dx = 1, \qquad [dx = \mu(dx)]. \tag{5.7.2}$$

Any function K satisfying (5.7.1) and (5.7.2) is called a **stochastic kernel**. Further, we define an integral operator P by

$$Pf(x) = \int_X K(x, y)f(y)\, dy \qquad \text{for } f \in L^1. \tag{5.7.3}$$

The operator P is clearly linear and nonnegative. Since we also have

$$\int_X Pf(x)dx = \int_X dx \int_X K(x, y)f(y)\, dy$$

$$= \int_X f(y)\, dy \int_X K(x, y)\, dx = \int_X f(y)\, dy,$$

P is therefore a Markov operator. In the special case that X is a finite set and μ is a counting measure, we have a Markov chain and P is a stochastic matrix.

Now consider two Markov operators P_a and P_b and their corresponding stochastic kernels, K_a and K_b. Clearly, $P_a P_b$ is also a Markov operator, and we wish to know how its kernel is related to K_a and K_b. Thus, write

$$(P_a P_b)f(x) = P_a(P_b f)(x) \int_X K_a(x, z)(P_b f(z))\, dz$$

$$= \int_X K_a(x, z) \left\{ \int_X K_b(z, y)f(y)dy \right\} dz$$

$$= \int_X \left\{ \int_X K_a(x, z)K_b(z, y)dz \right\} f(y)\, dy.$$

Then $P_a P_b$ is also an integral operator with the kernel

$$K(x, y) = \int_X K_a(x, z)K_b(z, y)\, dz. \tag{5.7.4}$$

We denote this composed kernel K by

$$K = K_a \star K_b \tag{5.7.5}$$

and note that the composition has the properties:

(i) $K_a \star (K_b \star K_c) = (K_a \star K_b) \star K_c$ (associative law); and

(ii) Any kernel formed by the composition of stochastic kernels is stochastic.

However, in general kernels K_A and K_b do not commute, that is, $K_a \star K_b \neq K_b \star K_a$. Note that the foregoing operation of composition definition is just a generalization of matrix multiplication.

Now we are in a position to show that Theorem 5.6.2 can be applied to operators P defined by stochastic kernels and, in fact, gives a simple sufficient condition for the asymptotic stability of $\{P^n\}$.

Corollary 5.7.1. *Let $(X, \mathcal{A}, \mu)$ be a measure space, $K: X \times X \to R$ a stochastic kernel, that is, K satisfies (5.7.1) and (5.7.2), and P the corresponding Markov operator defined by (5.7.3). Denote by K_n the kernel corresponding to P^n. If, for some m,*

$$\int_X \inf_y K_m(x, y)\, dx > 0, \tag{5.7.6}$$

then $\{P^n\}$ is asymptotically stable.

Proof. By the definition of K_n, for every $f \in D(X)$ we have

$$P^n f(x) = \int_X K_n(x, y) f(y)\, dy.$$

Furthermore, from the associative property of the composition of kernels,

$$K_{n+m}(x, y) = \int_X K_m(x, z) K_n(z, y)\, dz,$$

so that

$$P^{n+m} f(x) = \int_X K_{n+m}(x, y) f(y)\, dy$$
$$= \int_X \left\{ \int_X K_m(x, z) K_n(z, y)\, dz \right\} f(y)\, dy.$$

If we set

$$h(x) = \inf_y K_m(x, y),$$

then

$$P^{n+m} f(x) \geq h(x) \int_X \left\{ \int_X K_n(z, y) dz \right\} f(y)\, dy$$
$$= h(x) \int_X f(y)\, dy$$

since K_n is a stochastic kernel. Furthermore, since $f \in D(X)$,

$$\int_X f(y)\, dy = 1,$$

and, therefore,

$$P^{n+m} f(x) \geq h(x) \qquad \text{for } n \geq 1, \ f \in D(X).$$

Thus

$$(P^n f - h)^- = 0 \qquad \text{for } n \geq m + 1,$$

which implies that (5.6.3) holds, and we have finished the proof. ∎

In the case that X is a finite set and K is a stochastic matrix, this result is equivalent to one originally obtained by Markov.

Although condition (5.7.6) on the kernel is quite simple, it is seldom satisfied when $K(x, y)$ is defined on an unbounded space. For example, in Section 8.9 we discuss the evolution of densities under the operation of a Markov operator defined by the kernel [cf. equation (8.9.6)]

$$K(x, y) = \begin{cases} -e^y Ei(-y), & 0 < x \leq y \\ -e^y Ei(-x), & 0 < y < x, \end{cases} \qquad (5.7.7)$$

where

$$-Ei(-x) \equiv \int_x^\infty (e^{-y}/y)\, dy, \qquad x > 0,$$

is the exponential integral. In this case

$$\inf_y K(x, y) = 0 \qquad \text{for all } x > 0,$$

and the same holds for all of its iterates $K_m(x, y)$. A similar problem occurs with the kernel

$$K(x, y) = g(ax + by),$$

where $b \neq 0$ and g is an integrable function defined on R or even on R^+ (cf. Example 5.7.2).

In these and other cases where condition (5.7.6) is not satisfied, an alternative approach, reminiscent of the stability methods developed by Liapunov, offers a way to examine the asymptotic properties of iterates of densities by Markov operators.

Let G be an unbounded measurable subset of a d-dimensional Euclidian space R^d, $G \subset R^d$, and $K: G \times G \to R$ a measurable stochastic kernel. We will call any measurable nonnegative function $V: G \to R$ satisfying

$$\lim_{|x| \to \infty} V(x) = \infty \qquad (5.7.8)$$

a Liapunov function.

Next, we introduce the **Chebyshev inequality** through the following proposition.

Proposition 5.7.1. *Let $(X, \mathcal{A}, \mu)$ be a measure space, $V: X \to R$ an arbitrary nonnegative measurable function, and for all $f \in D$ set*

$$E(V|f) = \int_X V(x)f(x)\mu(dx).$$

If

$$G_a = \{x: V(x) < a\},$$

then

$$\int_{G_a} f(x)\mu(dx) \geq 1 - E(V|f)/a \qquad (5.7.9)$$

(the Chebyshev inequality).

Proof. The proof is easy. Clearly,

$$E(V)|f) \geq \int_{X \backslash G_a} V(x)f(x)\mu(dx) \geq a \int_{X \backslash G_a} f(x)\mu(dx)$$
$$\geq a\left\{ 1 - \int_{G_a} f(x)\mu(dx) \right\}.$$

Thus the Chebyshev inequality is proved.   ∎

With the lower-bound Theorem 5.6.2 and the Chebyshev inequality, it is possible to prove the following theorem.

Theorem 5.7.1. *Let $K: G \times G \to R$ be a stochastic kernel and P defined by (5.7.3), with G replacing X, be the corresponding Markov operator. If the kernel $K(x, y)$ satisfies*

$$\int_G \inf_{|y|<r} K(x, y)\, dx > 0 \qquad \text{for every } r > 0 \qquad (5.7.10)$$

and has a Liapunov function $V: G \to R$ such that

$$\int_G V(x)Pf(x)\, dx \leq \alpha \int_0^\infty V(X)f(x)\, dx + \beta \qquad 0 \leq \alpha < 1, \beta \geq 0$$
$$(5.7.11)$$

for $f \in D$, then $\{P^n\}$ is asymptotically stable.

Remark 5.7.1. Before giving the proof, we note that sometimes instead of verifying inequality (5.7.11) it is sufficient to check the simpler condition

$$\int_G K(x, y)V(x)\, dx \leq \alpha V(y) + \beta, \qquad (5.7.11a)$$

since (5.7.11a) implies (5.7.11). To see this, note that from (5.7.11a)

$$\int_G V(x)Pf(x)\,dx = \int_G \int_X V(x)K(x,y)f(y)\,dx\,dy$$
$$\leq \int_X [\alpha V(y) + \beta]f(y)\,dy = \alpha \int_X V(y)f(y)\,dy + \beta.$$

Proof. First define the function

$$E_n(V|f) \equiv \int_G V(x)P^n f(x)\,dx \qquad (5.7.12)$$

that can be thought of as the expected value of $V(x)$ with respect to the density $P^n f(x)$. From (5.7.11) we have directly

$$E_n(V|f) \leq \alpha E_{n-1}(V|f) + \beta. \qquad (5.7.13)$$

By an induction argument, it is easy to show that from this equation we obtain

$$E_n(V|f) \leq [\beta/(1-\alpha)] + a^n E_0(V|f).$$

Even though $E_0(V|f)$ is clearly dependent on our initial choice of f, it is equally clear that, for every f such that

$$E_0(V|f) < \infty, \qquad (5.7.14)$$

there is some integer $n_0 = n_0(f)$ such that

$$E_n(V|f) \leq [\beta/(1-\alpha)] + 1 \qquad \text{for all } n \geq n_0. \qquad (5.7.15)$$

Now let

$$G_a = \{x \in G : V(x) < a\}$$

so that from the Chebyshev inequality we have

$$\int_{G_a} P^n f(x)\,dx \geq 1 - \frac{E_n(V|f)}{a}. \qquad (5.7.16)$$

Further, set

$$a > 1 + [\beta/(1-\alpha)],$$

then

$$\frac{E_n(V|f)}{a} \leq \frac{1}{a}\left(1 + \frac{\beta}{1-\alpha}\right) < 1 \qquad \text{for } n \geq n_0$$

and thus (5.7.16) becomes

$$\int_{G_a} P^n f(x)\,dx \geq 1 - \frac{1}{a}\left(1 + \frac{\beta}{1-\alpha}\right) \doteq \epsilon > 0 \qquad \text{for } n \geq n_0. \qquad (5.7.17)$$

Since $V(x) \to \infty$ as $|x| \to \infty$ there is an $r > 0$ such that $V(x) > a$ for $|x| > r$. Thus the set G_a is entirely contained in the ball $|x| \le r$, and we may write

$$P^{n+1} f(x) = \int_G K(x,y) P^n f(y) \, dy \ge \int_{G_a} K(x,y) P^n f(y) \, dy$$

$$\ge \inf_{y \in G_a} K(x,y) \int_{G_a} P^n f(y) \, dy$$

$$\ge \inf_{|y| \le r} K(x,y) \int_{G_a} P^n f(y) \, dy$$

$$\ge \epsilon \inf_{|y| \le r} K(x,y) \tag{5.7.18}$$

for all $n \ge n_0$.

By setting

$$h(x) = \epsilon \inf_{|y| \le r} K(x,y)$$

in inequality (5.7.18) we have, by assumption (5.7.10), that

$$\|h\| > 0.$$

Finally, because of the continuity of V, the set $D_0 \subset D$ of all f such that (5.7.14) is satisfied is dense in D. Thus all the conditions of Theorem 5.6.2 are satisfied. ∎

Another important property of Markov operators defined by a stochastic kernel is that they may generate an asymptotically periodic sequence $\{P^n\}$ for every $f \in D$. This may happen if condition (5.7.10) on the kernel is replaced by a different one.

Theorem 5.7.2. *Let $K: G \times G \to R$ be a stochastic kernel and P be the corresponding Markov operator. Assume that there is a nonnegative $\lambda < 1$ such that for every bounded $B \subset G$ there is a $\delta = \delta(B) > 0$ for which*

$$\int_E K(x,y) \, dx \le \lambda \qquad \text{for } \mu(E) < \delta, \quad y \in B, \quad E \subset B. \tag{5.7.19}$$

Assume further there exists a Liapunov function $V: G \to R$ such that (5.7.11) holds. Then P is constrictive. Consequently, for every $f \in L^1$ the sequence $\{P^n\}$ is asymptotically periodic.

Proof. Again consider $E_n(V|f)$ defined by (5.7.12). Using condition (5.7.11) we once more obtain inequality (5.7.15). Thus by the Chebyshev inequality, with G_a defined as in the proof of Theorem 5.7.1,

$$\int_{G \setminus G_a} P^n f(x) \, dx = 1 - \int_{G_a} P^n f(x) \, dx \le \frac{E_n(V|f)}{a}$$

$$\le \frac{1}{a} \left(1 + \frac{\beta}{1-\alpha} \right) \qquad \text{for } n \ge n_0(f).$$

Set $\epsilon = \frac{1}{3}(1-\lambda)$. Choosing a sufficiently large a that satisfies

$$a \geq \frac{1}{\epsilon}\left(1 + \frac{\beta}{1-\alpha}\right),$$

we have

$$\int_{G\backslash G_a} P^n f(x)dx \leq \epsilon \qquad \text{for } n \geq n_0(f). \tag{5.7.20}$$

Consequently, from (5.7.19) we have

$$\int_{(G\backslash G_a)\cup E} P^n f(x)\,dx \leq \int_{G\backslash G_a} P^n f(x)\,dx + \int_E P^n f(x)\,dx$$

$$\leq \epsilon + \int_G P^{n-1}f(y)\,dy \int_E K(x,y)\,dx$$

$$\leq \epsilon + \int_{G\backslash G_a} P^{n-1}f(y)\,dy \int_G K(x,y)\,dx$$

$$+ \int_{G_a} P^{n-1}f(y)dy \int_E K(x,y)\,dx.$$

Using (5.7.19) and (5.7.20) applied to $B = G_a$ we finally have

$$\int_{(G\backslash G_a)\cup E} P^n f(x)\,dx \leq 2\epsilon + \lambda \int_{G_a} P^{n-1}f(y)\,dy$$

$$\leq 2\epsilon + \lambda = 1 - \epsilon \qquad \text{for } n \geq n_0(f) + 1.$$

Thus, inequality (5.3.2) in Definition 5.3.2 of constrictiveness is satisfied. A simple application of Theorem 5.3.1 completes the proof. ∎

Before passing to some examples of the application of Theorem 5.7.1 and 5.7.2, we give two simple results concerning the eventual behavior of $\{P^n\}$ when P is a Markov operator defined by a stochastic kernel.

Theorem 5.7.3. *If there exists an integer m and a $g \in L^1$ such that*

$$K_m(x,y) \leq g(x),$$

where $K_m(x,y)$ is the mth iterate of a stochastic kernel, then the sequence $\{P^n\}$ with P defined by (5.7.3) is asymptotically periodic.

Proof. Since $K_m(x,y) \leq g(x)$ we have

$$P^n f(x) = \int_X K_m(x,y)P^{n-m}f(y)\,dy \leq g(x) \qquad \text{for } n \geq m.$$

Set $h = g$ and take $\lambda = 0$ so by Proposition 5.3.2 the sequence $\{P^n\}$ is asymptotically periodic. ∎

A slight restriction on $K_m(x,y)$ in Theorem 5.7.3 leads to a different result, as given in the next result.

Theorem 5.7.4. *If there exists an integer m and a $g \in L^1$ such that*

$$K_m(x,y) \le g(x),$$

where $K_m(x,y)$ is the mth iterate of a stochastic kernel, and there is a set $A \subset X$ with $\mu(A) > 0$ such that

$$0 < K_m(x,y) \qquad for\, x \in A,\ y \in X,$$

then the sequence $\{P^n\}$ is asymptotically stable.

Proof. The proof is a trivial consequence of the constrictiveness of P from Theorem 5.7.3, the assumptions, and Theorem 5.6.1.

Example 5.7.1. To see the power of Theorem 5.7.1, we first consider the case where the kernel $K(x,y)$ is given by the exponential integrals in equation (5.7.7). It is easy to show that $-e^y(-Ei(y))$ is decreasing and consequently

$$\inf_{0 \le y \le r} K(x,y) \ge \min\{-Ei(-x), -e^r Ei(-r)\} > 0.$$

Furthermore, taking $V(x) = x$, we have, after integration,

$$\int_0^\infty xK(x,y)\,dx = \tfrac{1}{2}(1+y).$$

Therefore it is clear that $V(x) = x$ is a Liapunov function for this system when $\alpha = \beta = \tfrac{1}{2}$. Also, observe that with $f(x) = \exp(-x)$, we have

$$Pf(x) = \int_0^\infty K(x,y)e^{-y}dy = e^{-x}.$$

Thus, the limiting density attained by repeated application of the Markov operator P is $f_*(x) = \exp(-x)$. $\square$

Example 5.7.2. As a second example, let $g: R \to R$ be a continuous positive function satisfying

$$\int_{-\infty}^\infty g(x)dx = 1 \quad and \quad m_1 = \int_{-\infty}^\infty |x|g(x)dx < \infty.$$

Further, let a stochastic kernel be defined by

$$K(x,y) = |a|g(ax+by), \qquad |a| > |b|, b \ne 0$$

and consider the corresponding Markov operator

$$Pf(x) = \int_{-\infty}^{\infty} K(x,y)f(y)\, dy.$$

Let $V(x) = |x|$, so that we have

$$\int_{-\infty}^{\infty} K(x,y)V(x)\, dx = |a| \int_{-\infty}^{\infty} |x|g(ax+by)\, dx = \int_{-\infty}^{\infty} g(s) \left| \frac{s-by}{a} \right|\, ds$$

$$\leq \int_{-\infty}^{\infty} g(s) \left| \frac{s}{a} \right| ds + \int_{-\infty}^{\infty} g(s) \left| \frac{by}{a} \right| ds = \frac{m_1}{|a|} + \left| \frac{by}{a} \right|.$$

Thus, when $\alpha = |b/a|$ and $\beta = m_1/|a|$, it is clear that $V(x)$ satisfies condition (5.7.11) and hence Theorem 5.7.1 is satisfied.

As will become evident in Section 10.5, in this example Pf has the following interesting probabilistic interpretation. If ξ and η are two independent random variables with densities $f(x)$ and $g(x)$, respectively, then

$$Pf(x) = |a| \int_{-\infty}^{\infty} g(ax+by)f(y)\, dy, \qquad \text{with } a = \frac{1}{c_2} \text{ and } b = -\frac{c_1}{c_2},$$

is the density of the random variable $(c_1\xi + c_2\eta)$ [cf. equation (10.1.8)]. □

Example 5.7.3. As a final example of the applicability of the results of this section, we consider a simple model for the cell cycle [Lasota and Mackey, 1984]. First, it is assumed that there exists an intracellular substance (mitogen), necessary for mitosis and that the rate of change of mitogen is governed by

$$\frac{dm}{dt} = g(m), \qquad m(0) = r$$

with solution $m(r,t)$. The rate g is a C^1 function on $[0,\infty)$ and $g(x) > 0$ for $x > 0$. Second, it is assumed that the probability of mitosis in the interval $[t, t+\Delta t]$ is given by $\phi(m(t))\Delta t + o(\Delta t)$, where ϕ is a nonnegative function such that $q(x) = \phi(x)/g(x)$ is locally integrable (that is, integrable on bounded sets $[0, c]$) and satisfies

$$\lim_{x\to\infty} Q(x) = \infty, \qquad \text{where } Q(x) = \int_0^x q(y)\, dy. \qquad (5.7.21)$$

Finally, it is assumed that at mitosis each daughter cell receives exactly one-half of the mitogen present in the mother cell.

Under these assumptions it can be shown that for a distribution $f_{n-1}(x)$ of mitogen in the $(n-1)$st generation of a large population of cells, the mitogen distribution in the following generation is given by

$$f_n(x) = \int_0^{\infty} K(x,r)f_{n-1}(r)\, dr,$$

where

$$K(x,r) = \begin{cases} 0 & x \in [0, \tfrac{1}{2}r) \\ 2q(2x) \exp\left[-\int_r^{2x} q(y)dy\right] & x \in [\tfrac{1}{2}r, \infty). \end{cases} \quad (5.7.22)$$

It is straightforward to show that $K(x,r)$ satisfies (5.7.1) and (5.7.2) and is, thus, a stochastic kernel. Hence the operator $P: L^1(R^+) \to L^1(R^+)$ defined by

$$Pf(x) = \int_0^\infty K(x,r)f(r)\,dr \quad (5.7.23)$$

is a Markov operator. To show that there is a unique stationary density $f_* \in D$ to which $\{P^n f\}$ converges strongly, we use Theorem 5.7.1 under the assumption that

$$\lim_{x \to \infty} \inf[Q(2x) - Q(x)] > 1. \quad (5.7.24)$$

First we consider the integral

$$I = \int_0^\infty u(Q(2x))Pf(x)\,dx, \quad (5.7.25)$$

where u is a continuous nonnegative function. Using equations (5.7.21) through (5.7.23) we can rewrite (5.7.25) as follows:

$$\begin{aligned} I &= 2\int_0^\infty u(Q(2x))q(2x)\,dx \int_0^{2x} \exp[Q(y) - Q(2x)]f(y)\,dy \\ &= \int_0^\infty u(Q(z))q(z)\,dz \int_0^z \exp[Q(y) - Q(z)]f(y)\,dy \\ &= \int_0^\infty f(y)\,dy \int_y^\infty u(Q(z))\exp[Q(y) - Q(z)]q(z)\,dz. \end{aligned}$$

Setting $Q(z) - Q(y) = x$ so $q(z)dz = dx$ we finally obtain the useful equality

$$\int_0^\infty u(Q(2x))Pf(x)\,dx = \int_0^\infty f(y)\,dy \int_0^\infty u(x + Q(y))e^{-x}dx. \quad (5.7.26)$$

Note in particular from (5.7.26) that for $u(z) \equiv 1$ we have

$$\int_0^\infty Pf(x)\,dx = \int_0^\infty f(y)\,dy,$$

which also proves that P is a Markov operator.

Now take $u(x) = e^{\epsilon x}$ with $0 < \epsilon \le 1$, and $V(x) = u(Q(2x))$. From (5.7.26) it therefore follows that

$$\begin{aligned} \int_0^\infty V(x)Pf(x)\,dx &= \int_0^\infty f(y)e^{\epsilon Q(y)}dy \int_0^\infty e^{-(1-\epsilon)x}dx \\ &= \frac{1}{1-\epsilon}\int_0^\infty f(y)e^{\epsilon Q(y)}dy. \end{aligned} \quad (5.7.27)$$

Now pick a $\rho > 1$ and $x_0 \geq 0$ such that

$$Q(2y) - Q(y) \geq \rho \qquad \text{for } y \geq x_0.$$

Then we can write (5.7.27) as

$$\int_0^\infty V(x) Pf(x) dx \leq \frac{1}{1-\epsilon} \int_0^{x_0} f(y) e^{\epsilon Q(y)} dy + \frac{e^{-\epsilon\rho}}{1-\epsilon} \int_{x_0}^\infty f(y) e^{\epsilon Q(2y)} dy$$

$$\leq \frac{1}{1-\epsilon} e^{\epsilon Q(x_0)} + \frac{e^{-\epsilon\rho}}{1-\epsilon} \int_0^\infty V(y) f(y) dy.$$

For the function

$$\alpha(\epsilon) = \frac{e^{-\epsilon\rho}}{1-\epsilon}$$

we have $\alpha(0) = 1$ and $\alpha'(0) = 1 - \rho < 0$. Thus for some $\epsilon > 0$ we have $\alpha(\epsilon) < 1$. Take such an ϵ set

$$\alpha = \alpha(\epsilon), \qquad \beta = \frac{1}{1-\epsilon} e^{\epsilon Q(x_0)}$$

With these values of α and β we have shown that the operator P defined by (5.7.22)–(5.7.23) satisfies inequality (5.7.11) of Theorem 5.7.1 under the assumption of (5.7.23). It only remains to be shown that K satisfies (5.7.10).

Let $r_0 \geq 0$ be an arbitrary finite real number. Consider $K(x, r)$ for $0 \leq r \leq r_0$ and $x \geq \frac{1}{2}r$. Then

$$K(x, r) = 2q(2x) \exp\left[-\int_r^{2x} q(y) dy\right]$$

$$\geq 2q(2x) \exp\left[-\int_0^{2x} q(y) dy\right] \qquad \text{for } 0 \leq r \leq r_0, x \geq \frac{1}{2}r$$

and, as a consequence,

$$\inf_{0 \leq r \leq r_0} K(x, r) \geq h(x) = \begin{cases} 0 & \text{for } x < \frac{1}{2}r_0 \\ 2q(2x) \exp\left[-\int_0^{2x} q(y) dy\right] & \text{for } x \geq \frac{1}{2}r_0. \end{cases}$$

Further,

$$\int_0^\infty h(x) dx = \int_{r_0/2}^\infty 2q(2x) \exp\left[-\int_0^{2x} q(y) dy\right] dx$$

$$= \exp\left[-\int_0^{r_0} q(y) dy\right] > 0;$$

hence $K(x, r)$ satisfies (5.7.10). Thus, in this simple model for cell division, we know that there is a globally asymptotically stable distribution of mitogen. Generalizations of this model have appeared in the work of Tyson and Hannsgen [1986], Tyrcha [1988], and Lasota, Mackey, and Tyrcha [1992].
□

5.8   Conditions for the Existence of Lower-Bound Functions

The consequences of the theorems of this chapter for the Frobenius–Perron operator are so far-reaching that an entire theory of invariant measures for a large class of transformations on the interval [0,1], and even on manifolds, may be constructed. This forms the subject of Chapter 6. In this last section, we develop some simple criteria for the existence of lower-bound functions that will be of use in our specific examples of the next chapter.

Our first criteria for the existence of a lower bound function will be formulated in the special case when $X = (a, b)$ is an interval on the real line [(a, b) bounded or not] with the usual Borel measure. We will use some standard notions from the theory of differential inequalities [Szarski, 1967]. A function $f: (a, b) \to R$ is called **lower semicontinuous** if

$$\liminf_{\delta \to 0} f(x + \delta) \geq f(x) \qquad \text{for } x \in (a, b).$$

It is **left lower semicontinuous** if

$$\liminf_{\substack{\delta \to 0 \\ \delta > 0}} f(x - \delta) \geq f(x) \qquad \text{for } x \in (a, b).$$

For any function $f: (a, b) \to R$, we define its **right lower derivative** by setting

$$\frac{d_+ f(x)}{dx} = \liminf_{\substack{\delta \to 0 \\ \delta > 0}} \frac{1}{\delta}[f(x + \delta) - f(x)] \qquad \text{for } x \in (a, b).$$

It is well known that every left lower semicontinuous function $f: (a, b) \to R$, satisfying

$$\frac{d_+ f(x)}{dx} \leq 0 \qquad \text{for } x \in (a, b),$$

is nonincreasing on (a, b). (The same is true for functions defined on a half-closed interval $[a, b)$.)

For every $f \in D_0$ that is a dense subset of D (Definition 5.6.5) write the trajectory $P^n f$ as

$$P^n f = f_n \qquad \text{for } n \geq n_0(f). \tag{5.8.1}$$

Then we have the following proposition.

Proposition 5.8.1. *Let $P: L^1((a, b)) \to L^1((a, b))$ be a Markov operator. Assume that there exists a nonnegative function $g \in L^1((a, b))$ and a constant $k \geq 0$ such that for every $f \in D_0$ the function f_n in (5.8.1) are left lower semicontinuous and satisfy the following conditions:*

$$f_n(x) \leq g(x) \qquad \text{a.e. in } (a, b) \tag{5.8.2}$$

$$\frac{d_+ f_n(x)}{dx} \le k f_n(x) \qquad \text{for all } x \in (a,b). \tag{5.8.3}$$

Then there exists an interval $\Delta \subset (a,b)$ and an $\epsilon > 0$ such that $h = \epsilon 1_\Delta$ is a lower function for P^n.

Proof. Let $x_0 < x_1 < x_2$ be chosen in (a,b) such that

$$\int_a^{x_1} g(x)\,dx < \tfrac{1}{4} \quad \text{and} \quad \int_{x_2}^b g(x)\,dx < \tfrac{1}{4}. \tag{5.8.4}$$

Set

$$\epsilon = \min\{x_1 - x_0, M(x_2 - x_0)^{-1}\}, \qquad M = \tfrac{1}{4}\exp[-k(x_2 - x_0)].$$

Since $\|P^n f\| = 1$, condition (5.8.1) implies

$$\int_a^b f_n(x)\,dx = 1. \tag{5.8.5}$$

Now we are going to show that $h = \epsilon 1_{(x_0,x_1)}$ is a lower function. Suppose it is not. Then there is $n' \ge n_0$ and $y \in (x_0, x_1)$ such that $f_{n'}(y) < h(y) = \epsilon$. By integrating inequality (5.8.3), we obtain

$$f_{n'}(x) \le f_{n'}(y)e^{k(x-y)} \le \epsilon/4M \qquad \text{for } x \in [y, x_2]. \tag{5.8.6}$$

Furthermore, since $f_{n'} \le g$, we have

$$\int_a^b f_{n'}(x)\,dx \le \int_a^{x_1} g(x)\,dx + \int_{x_1}^{x_2} g(x)\,dx + \int_{x_2}^b g(x)\,dx.$$

Finally, by applying inequalities (5.8.4) and (5.8.6), we obtain

$$\int_a^b f_{n'}(x)\,dx \le \tfrac{1}{4} + (x_2 - y)(\epsilon/4M) + \tfrac{1}{4} \le \tfrac{3}{4},$$

which contradicts equation (5.8.5). ∎

Remark 5.8.1. In the proof of Proposition 5.8.1, the left lower semicontinuity of f_n and inequality (5.8.3) were only used to obtain the evaluation

$$f_n(x) \le f_n(y)e^{k(x-y)} \qquad \text{for } x \ge y.$$

Therefore Proposition 5.8.1 remains true under this condition; for example, it is true if all f_n are nonincreasing. □

It is obvious that in Proposition 5.8.1 we can replace (5.8.3) by $d_- f_n/dx \ge -k f_n$ and assume f_n right lower continuous (or assume f_n nondecreasing;

cf. Remark 5.8.1). In the case of a bounded interval, we may omit condition (5.8.2) and replace (5.8.3) by a two-sided inequality. This observation is summarized as follows.

Proposition 5.8.2. *Let (a, b) denote a bounded interval and let $P: L^1((a, b)) \rightarrow L^1((a, b))$ be a Markov operator. Assume that for each $f \in D_0$ the functions f_n in (5.8.1) are differentiable and satisfy the inequality*

$$\left| \frac{df_n(x)}{dx} \right| \leq k f_n(x), \qquad \text{for all } x \in (a, b), \qquad (5.8.7)$$

where $k \geq 0$ is a constant independent of f. Then there exists an $\epsilon > 0$ such that $h = \epsilon 1_{(a,b)}$ is a lower-bound function.

Proof. As in the preceding proof, we have equation (5.8.5). Set

$$\epsilon = [1/2(b - a)]e^{-k(b-a)}.$$

Now it is easy to show that $f_n \geq h$ for $n \geq n_0$. If not, then $f_{n'}(y) < \epsilon$ for some $y \in (a, b)$ and $n' \geq n_0$. Consequently, by (5.8.7),

$$f_{n'}(x) \leq f_{n'}(y)e^{k|x-y|} \leq [1/2(b - a)].$$

This evidently contradicts (5.8.5). The inequality $f_n \geq h$ completes the proof. ∎

5.9 Sweeping

Until now we have considered the situation in which the sequence $\{P^n\}$ either converges to a unique density (asymptotic stability) or approaches a set spanned by a finite number of densities (asymptotic periodicity) for every initial density f. In this section we consider quite a different property in which the densities are dispersed under the action of a Markov operator P. We call this new behavior **sweeping**, and introduce the concept through two definitions and several examples.

Our first definition is as follows.

Definition 5.9.1. Let $(X, \mathcal{A}, \mu)$ be a measure space and $\mathcal{A}_* \subset \mathcal{A}$ be a subfamily of the family of measurable sets. Also let $P: L^1(X) \rightarrow L^1(X)$ be a Markov operator. Then $\{P^n\}$ is said to be **sweeping** with respect to $\mathcal{A}_*$ if

$$\lim_{n \to \infty} \int_A P^n f(x) \mu(dx) = 0 \qquad \text{for every } f \in D \text{ and } A \in \mathcal{A}_*. \qquad (5.9.1)$$

Since every element $f \in L^1$ can be written as a linear combination of two densities

$$f = \alpha_1 f_1 + \alpha_2 f_2, \qquad \text{for } f_i \in D,$$

for a sweeping operator P, condition (5.9.1) also holds for $f \in L^1$.

In particular examples, it is sufficient to verify condition (5.9.1) for $f \in D_0$, where D_0 is an arbitrary dense subset of D. That this is so follows immediately from the inequality

$$\int_A P^n f(x)\mu(dx) \le \int_A P^n f_0(x)\mu(dx) + \|f - f_0\|, \quad \text{for } f \in D, f_0 \in D_0,$$
(5.9.2)

and the fact that both terms on the right-hand side of (5.9.2) can be made arbitrarily small.

Example 5.9.1. Let $X = R$ and μ be the standard Borel measure. Further, let

$$Pf(x) = f(x - r), \quad \text{for } f \in D$$

so

$$P^n f(x) = f(x - nr), \quad \text{for } f \in D. \tag{5.9.3}$$

With $r > 0$ the sequence $\{P^n\}$ is sweeping with respect to the family of intervals

$$A_0 = \{(-\infty, c]: c \in R\}.$$

To prove this, note that for every $f \in D$ with compact support we have

$$\int_{-\infty}^c P^n f(x)\, dx = \int_{-\infty}^c f(x - nr)\, dx = \int_{-\infty}^{c-nr} f(y)\, dy.$$

Thus the integral on the right-hand side will eventually become zero since

$$(-\infty, c - nr] \cap \operatorname{supp} f = \emptyset$$

for sufficiently large n. In an analogous fashion we can also prove that for $r < 0$ the sequence $\{P^n\}$, where P is given by (5.9.3), is sweeping with respect to the family of intervals

$$A_1 = \{[c, \infty): c \in R\}. \qquad \square$$

Example 5.9.2. Again take $X = R$ and μ to be the Borel measure. Further, let P be an integral operator with Gaussian kernel

$$Pf(x) = \frac{1}{\sqrt{2\pi\sigma^2}} \int_{-\infty}^{\infty} \exp\left[-\frac{(x-y)^2}{2\sigma^2}\right] f(y)\, dy.$$

It is easy to show (see also Example 7.4.1 and Remark 7.9.1) that

$$P^n f(x) = \frac{1}{\sqrt{2\pi\sigma^2 n}} \int_{-\infty}^{\infty} \exp\left[\frac{(x-y)^2}{2\sigma^2 n}\right] f(y)\, dy, \tag{5.9.4}$$

and as a consequence

$$P^n f(x) \le \frac{1}{\sqrt{2\pi\sigma^2 n}}.$$

Thus the sequence $\{P^n\}$ defined by (5.9.4) is sweeping with respect to the family of bounded intervals

$$\mathcal{A}_2 = \{[a, b]: -\infty < a < b < \infty\}$$

since

$$\int_a^b P^n f(x)\, dx \le \frac{b - a}{\sqrt{2\pi\sigma^2 n}} \to 0 \qquad \text{as } n \to \infty. \qquad \square$$

These two examples motivate a more restricted version of the general Definition 5.9.1 of sweeping appropriate to the situation where $X \subset R$ is an interval.

Definition 5.9.2. Let $X \subset R$ be an interval (bounded or not) with endpoints α, β and let $P: L^1(X) \to L^1(X)$ be a Markov operator. We say the following:

a. $\{P^n\}$ is **sweeping to** β if it is sweeping with respect to the family of intervals

$$\mathcal{A}_0 = \{(\alpha, c]: c < \beta\}; \tag{5.9.5}$$

b. $\{P^n\}$ is **sweeping to** α if it is sweeping with respect to the family of intervals

$$\mathcal{A}_1 = \{[c, \beta): \alpha < c\}; \tag{5.9.6}$$

c. $\{P^n\}$ is **central sweeping** if it is sweeping with respect to the family of closed intervals

$$\mathcal{A}_2 = \{[a, b]: \alpha < a < b < \beta\}. \tag{5.9.7}$$

In Examples 5.9.1 and 5.9.2 the sweeping was almost self-evident from the structure of the operator P. However, this is often not the case, and we are going to present a sufficient condition often useful for proving sweeping. We start with a definition reminiscent of the definition of the Liapunov function.

Let $(X, \mathcal{A}, \mu)$ be a measure space and let a subfamily $\mathcal{A}_* \subset \mathcal{A}$ be given. A bounded Borel measurable function $V: X \to R$ is called a **Bielecki function** if it is nonnegative and if

$$\inf_{x \in A} V(x) > 0 \qquad \text{for } A \in \mathcal{A}_*.$$

For example, if $X = [\alpha, \beta)$ and $\mathcal{A}_* = \mathcal{A}_0$ from Definition 5.9.2a, then every continuous and strictly positive function $V: [\alpha, \beta) \to R$ is a Bielecki function since the infimum of a positive continuous function on a closed bounded (compact) interval is always positive.

Having the concept of Bielecki functions it is straightforward to verify the following proposition.

Proposition 5.9.1. *Let $(X, \mathcal{A}, \mu)$ be a measure space and let $\mathcal{A}_* \subset \mathcal{A}$ be fixed. Further let $P \colon L^1(X) \to L^1(X)$ be a Markov operator for which there exists a Bielecki function $V \colon X \to R$ and a constant $\gamma < 1$ such that*

$$\int_X V(x) P f(x) \mu(dx) \leq \gamma \int_X V(x) f(x) \mu(dx) \qquad \text{for } f \in D.$$

Then $\{P^n\}$ is sweeping.

Proof. Fix an $f \in D$ and $A \in \mathcal{A}_*$. Then

$$\int_A P^n f(x) \mu(dx) \leq \frac{1}{\inf_{x \in A} V(x)} \int_A V(x) P^n f(x) \mu(dx)$$

$$\leq \frac{\gamma^n}{\inf_{x \in A} V(x)} \int_A V(x) f(x) \mu(dx).$$

Since $\gamma < 1$ by assumption, the right-hand side converges to zero as $n \to \infty$ and condition (5.9.1) holds. Thus the proof is complete. ∎

Despite the fact that the proof was relatively easy, Proposition 5.9.1 can be extremely useful in proving sweeping as the next example demonstrates.

Example 5.9.3. We once again turn to the cell cycle model of Example 5.7.3 defined by the Markov operator (5.7.23) on $L^1(R^+)$ with kernel (5.7.22) constrained by condition (5.7.21). There we showed that condition (5.7.24), that is,

$$\liminf_{x \to \infty} [Q(2x) - Q(x)] > 1$$

was sufficient to guarantee the asymptotic stability of $\{P^n\}$. In this example we will show that for the same system (5.7.22)–(5.7.23) the condition

$$\limsup_{x \to \infty} [Q(2x) - Q(x)] < 1 \tag{5.9.8}$$

implies that $\{P^n\}$ is sweeping to $+\infty$.

We start by choosing $0 < \epsilon < 1$, an $x \geq x_0$, and $\rho < 1$ such that $Q(2x) - Q(x) \leq \rho < 1$ for $x \geq x_0$. Define

$$u(z) = \begin{cases} e^{-\epsilon z_0} & \text{for } z < z_0 \\ e^{-\epsilon z} & \text{for } z \geq z_0, \end{cases}$$

where $z_0 \equiv Q(2x_0)$, and set $V(x) = u(Q(2x))$. From (5.7.26) we have

$$\int_0^\infty V(x) P f(x) \, dx = \int_0^\infty f(y) \, dy \int_0^\infty u(x + Q(y)) e^{-x} dx,$$

or

$$\int_0^\infty V(x)Pf(x)\,dx = \int_0^\infty V(y)f(y)W(y)\,dy \qquad \text{for } f \in D, \qquad (5.9.9)$$

where

$$W(y) = \frac{1}{V(y)} \int_0^\infty u(x + Q(y))e^{-x}dx. \qquad (5.9.10)$$

We will evaluate W as given by (5.9.10) separately for $y \le x_0$ and for $y \ge x_0$.

When $y < x_0$ observe that u is a nonincreasing function and that $V(y) = e^{-\epsilon z_0}$. Thus

$$W(y) \le e^{\epsilon z_0} \int_0^\infty u(x)e^{-x}dx = e^{\epsilon z_0} \left\{ \int_0^{z_0} u(x)e^{-x}dx + \int_{z_0}^\infty u(x)e^{-x}dx \right\}$$

$$= \int_0^{z_0} e^{-x}dx + \int_{z_0}^\infty e^{\epsilon(x-z_0)-x}dx$$

$$= 1 - e^{-z_0}\left[\frac{\epsilon}{1+\epsilon}\right] \equiv \alpha_1(\epsilon),$$

and it is evident that $\alpha_1(\epsilon) < 1$ for all $\epsilon > 0$.

When $y \ge x_0$ we have $V(y) = e^{-\epsilon Q(2y)}$. Furthermore $u(x) \le e^{-\epsilon x}$ for all x, so

$$W(y) \le \int_0^\infty \exp\{-x - \epsilon[x + Q(y) - Q(2y)]\}\,dx.$$

Since, by assumption, $Q(2y) - Q(y) \le \rho$ this can also rewritten as

$$W(y) \le \int_0^\infty e^{-\epsilon(x-\rho)-x}dx = \frac{e^{\epsilon\rho}}{1+\epsilon} \equiv \alpha_2(\epsilon).$$

It is clear that $\alpha_2(0) = 1$ and that $\alpha_2'(0) = \rho - 1 < 0$. Thus, there must be an $\epsilon > 0$ such that $\alpha_2(\epsilon) < 1$.

Chose an ϵ such that $\alpha(\epsilon) < 1$ and define $\bar{\alpha} = \min(\alpha_1(\epsilon), \alpha_2(\epsilon))$. Then $W(y) \le \bar{\alpha} < 1$ for all $y \ge 0$ and from (5.9.9) we have

$$\int_0^\infty V(x)Pf(x)\,dx \le \bar{\alpha} \int_0^\infty V(x)f(x)\,dx \qquad \text{for all } f \in D.$$

Thus by proposition 5.9.1 we have shown that the cell cycle model defined by equations (5.7.21)–(5.7.23) is characterized by a sweeping Markov operator when (5.9.8) holds.

5.10   The Foguel Alternative and Sweeping

From Example 5.9.3 it is clear that the demonstration of sweeping is neither necessarily straightforward nor trivial and may, in fact, require a rather strong effort. In this section we present a sufficient condition for sweeping that is sometimes especially helpful in the study of integral Markov operators with stochastic kernels.

Let $(X, \mathcal{A}, \mu)$ be a measure space and $P: L^1(x) \to L^1(X)$ be the operator

$$Pf(x) = \int_X K(x, y) f(y) \mu(dy) \tag{5.10.1}$$

where K is a stochastic kernel and thus satisfies conditions (5.7.1) and (5.7.2). We have already shown in Section 5.7 that P is a Markov operator and hence defined for all $f \in L^1$.

However, the right-hand side of (5.10.1) is well defined for every measurable $f \geq 0$ even though it may, of course, be infinite for some x. With this observation we make the following definitions.

Definition 5.10.1. Let $P: L^1 \to L^1$ be the integral Markov operator (5.10.1) and let $f: X \to R$ be a measurable and nonnegative function. We say that f is **subinvariant** if

$$Pf(x) \leq f(x) \qquad \text{for } x \in X \text{ a.e.}$$

Definition 5.10.2. Let a subfamily $\mathcal{A}_* \subset \mathcal{A}$ be fixed. We say that $\mathcal{A}_*$ is **regular** if there is a sequence of sets $A_n \in \mathcal{A}_*, n = 0, 1 \ldots$, such that

$$\bigcup_{n=0}^{\infty} A_n = X. \tag{5.10.2}$$

Definition 5.10.3. A nonnegative measurable function $f: X \to R$ is **locally integrable** if

$$\int_A f(x) \mu(dx) < \infty \qquad \text{for } A \in \mathcal{A}_*.$$

With these definitions we state the following result which will be referred to as the **Foguel alternative**.

Theorem 5.10.1. Let $(X, \mathcal{A}, \mu)$ be a measure space and $\mathcal{A}_* \in \mathcal{A}$ a regular family. Assume that $P: L^1 \to L^1$ is an integral operator with a stochastic kernel. If P has a locally integrable and positive ($f > 0$ a.e.) subinvariant function f, then either P has an invariant density or $\{P^n\}$ is sweeping.

In the statement of this theorem, there are two implications:

(1) if $\{P^n\}$ is not sweeping, then P has an invariant density; and

(2) if $\{P^n\}$ is sweeping, then P has no invariant density.

Only the first part is hard to prove, and the second part can be demonstrated using condition (5.10.2).

To prove the second implication, suppose that $\{P^n\}$ is sweeping and that $f_* = Pf_*$ is an invariant density. Further define

$$B_k = \bigcup_{i=1}^{k} A_i.$$

Then, according to (5.10.2),

$$\lim_{k \to \infty} \int_{B_k} f_*(x)\mu(dx) = \int_X f_*(x)\mu(dx) = 1,$$

and in particular for some fixed k

$$\int_{B_k} f_*(x)\mu(dx) > \tfrac{1}{2} \tag{5.10.3}$$

On the other hand, since $f_* = Pf_*$,

$$\int_{B_k} f_*(x)\mu(dx) \le \sum_{i=1}^{k} \int_{A_i} f_*(x)\mu(dx) = \sum_{i=1}^{k} \int_{A_i} P^n f_*(x)\mu(dx).$$

Since $\{P^n\}$ is sweeping by assumption, the right-hand side of this relation converges to zero. This, however, contradicts (5.10.3) and we thus conclude that $\{P^n\}$ is not sweeping.

Remark 5.10.1. This theorem was proved by Komorowski and Tyrcha [1989] and the assumptions concerning the regular family $\mathcal{A}_*$ simplified by Malczak [1992]. Similar theorems when $\mathcal{A}_*$ is the family of all measurable subsets have been proved by several authors; see Foguel [1966] and Lin [1971].

Example 5.10.1. Let $X = R^+$, and consider the integral operator

$$Pf(x) = \int_x^\infty \psi\left(\frac{x}{y}\right) f(y)\frac{dy}{y} \qquad \text{for } x \ge 0, \tag{5.10.4}$$

where $\psi\colon [0,1] \to R$ is a given integrable function such that

$$\psi(z) \ge 0 \quad \text{and} \quad \int_0^1 \psi(z)dz = 1. \tag{5.10.5}$$

The operator (5.10.4) appears on the right-hand side of the Chandrasekhar–Münch equation describing the fluctuations in the brightness of the Milky Way. This equation will be discussed in Examples 7.9.2 and 11.10.2. Here we are going to study the properties of the operator (5.10.4) alone.

Let $V: R^+ \to R$ be a nonnegative measurable function. For $f \in D$ we have

$$\int_0^\infty V(x)Pf(x)\,dx = \int_0^\infty V(x)\,dx \int_x^\infty \psi\left(\frac{x}{y}\right) f(y)\frac{dy}{y}$$
$$= \int_0^\infty f(y)\,dy \int_0^y \psi\left(\frac{x}{y}\right) V(x)\frac{dx}{y}.$$

Substituting $x/y = z$ this becomes

$$\int_0^\infty V(x)Pf(x)\,dx = \int_0^\infty f(y)dy \int_0^1 \psi(z)V(zy)\,dz. \qquad (5.10.6)$$

This equality with $V(x) \equiv 1$ gives

$$\int_0^\infty Pf(x)\,dx = \int_0^\infty f(y)\,dy \int_0^1 \psi(z)\,dz = \int_0^\infty f(y)\,dy$$

which, together with the nonnegativity of ψ, implies that (5.10.4) defines a Markov operator.

Now set $f_\beta(x) = x^{-\beta}$ in (5.10.4). Then

$$Pf_\beta(x) = \int_x^\infty \psi\left(\frac{x}{y}\right)\frac{dy}{y^{\beta+1}} = \frac{1}{x^\beta}\int_0^1 \psi(z)z^{\beta-1}dz. \qquad (5.10.7)$$

For $\beta \geq 1$ we have $\psi(z)z^{\beta-1} \leq \psi(z)$ and, as a consequence,

$$Pf_\beta(x) \leq f_\beta(x) \qquad \text{for } x \geq 0.$$

Thus, by Theorem 5.10.1, the operator P defined by (5.10.4) is either sweeping to zero or has an invariant density.

It is easy to exclude the possibility that P has an invariant density. Suppose that there is an invariant density f_*. Then the equality (5.10.6) gives

$$\int_0^\infty V(y)f_*(y)\,dy = \int_0^\infty f_*(y)\,dy \int_0^1 \psi(z)V(zy)\,dz,$$

or

$$\int_0^\infty f_*(y)\,dy \int_0^1 \psi(z)[V(y) - V(zy)]\,dz = 0. \qquad (5.10.8)$$

Now take $V:[0,\infty) \to R$ to be positive, bounded, and strictly increasing [e.g., $V(z) = z/(1+z)$].Then

$$V(y) - V(zy) > 0 \qquad \text{for } y > 0,\ 0 \leq z \leq 1,$$

and the integral

$$I(y) = \int_0^1 \psi(z)[V(y) - V(zy)]\,dz$$

is strictly positive for $y > 0$. Consequently, the product $f_*(y)I(y)$ is a non-negative and nonvanishing function. This shows that the equality (5.10.8) is not satisfied, and thus there is no invariant density for P.

Thus, for every ψ satisfying (5.10.5) the operator P given by equation (5.10.4) is sweeping. This is both interesting and surprising since we will show in Section 11.10 that the stochastic semigroup generated by the Chandrasekhar–Münch equation is asymptotically stable! □

The alternative formulated in Theorem 5.10.1 does not specify the behavior of the sequence $\{P^n\}$ in the case when an invariant density exists. We now formulate a stronger form of the Foguel alternative, first introducing the notion of an expanding operator.

Definition 5.10.4. Let $(X, \mathcal{A}, \mu)$ be a measure space and $P: L^1 \to L^1$ be a Markov operator. We say that P is **expanding** if

$$\lim_{n \to \infty} \mu(A - \text{supp} P^n f) = 0 \qquad \text{for } f \in D \text{ and } \mu(A) < \infty. \tag{5.10.9}$$

The simplest example of an expanding operator is an integral operator with a strictly positive stochastic kernel. In fact, from equation (5.7.3) with $K(x, y) > 0$ it follows that $P^n f(x) > 0$ for all $x \in X$ and $n \geq 1$. In this case, $\text{supp } P^n f = X$ and condition (5.10.9) is automatically satisfied.

A more sophisticated example of an expanding operator is given by

$$Pf(x) = \int_a^{\lambda(x)} K(x, y) f(y) \, dy, \tag{5.10.10}$$

where $K(x, y)$ is a measurable kernel satisfying

$$K(x, y) > 0 \qquad \text{for } a < y < \lambda(x), \ a < x, \tag{5.10.11}$$

and $\lambda: [a, \infty) \to [a, \infty)$ is a continuous strictly increasing function such that

$$\lambda(x) > x \qquad \text{for } a < x. \tag{5.10.12}$$

A straightforward calculation shows that P is a Markov operator on $L^1([a, \infty))$ when

$$\int_{\lambda^{-1}(y)}^{\infty} K(x, y) \, dx = 1 \qquad \text{for } y > a. \tag{5.10.13}$$

We also have the following.

Proposition 5.10.1. *If K and λ satisfy conditions (5.10.11)–(5.10.13), then the Markov operator $P: L^1([a, \infty)) \to L^1([a, \infty))$ defined by (5.10.10) is expanding.*

Proof. Let $f \in D$ be given and let

$$x_0 = \text{ess inf}\{x: f(x) > 0\}.$$

This means that x_0 is the largest possible real number satisfying

$$\mu(\text{supp}\, f \cap [0, x_0]) = 0.$$

Further, let $x_1 = \lambda^{-1}(x_0)$. It is evident from the defining equation (5.10.10) that $Pf(x) > 0$ for $\lambda((x)) > x_0$ or $x > x_1$. Define $x_n = \lambda^{-n}(x_0)$. By an induction argument it is easy to verify that $P^n f(x) > 0$ for $x > x_n$. Thus, for an arbitrary measurable set $A \subset [a, \infty)$ we have

$$\mu(A - \text{supp}\, P^n f) \le x_n - a. \tag{5.10.14}$$

The sequence $\{x_n\}$ is bounded from below ($x_n \ge a$). It is also decreasing since $x_n = \lambda^{-1}(x_{n-1}) \le x_{n-1}$. Thus $\{x_n\}$ is convergent to a number $x_* \ge a$. Since $\lambda(x_n) = x_{n-1}$, in the limit as $n \to \infty$ we have $\lambda(x_*) = x_*$. From inequality (5.10.12) it follows that $x_* = a$, which according to (5.10.14) shows that P is expanding. ∎

For expanding operators, the Foguel alternative can be formulated as follows.

Theorem 5.10.2. *Let $(X, \mathcal{A}, \mu)$ be a measure space and $\mathcal{A}_* \subset \mathcal{A}$ be a regular family of measurable sets. Assume that $P: L^1(X) \to L^1(X)$ is an expanding integral operator with a stochastic kernel. If P has a locally integrable positive ($f > 0$ a.e.) subinvariant function, then either $\{P^n\}$ is asymptotically stable or it is sweeping.*

The proof can be found in Malczak (1992). Theorem 5.10.2 can be derived from a new criterion for asymptotic stability given by Baron and Lasota (1993). See Exercise 5.8.

Example 5.10.2. We return to the modeling of the cell cycle (see Example 5.7.3) by considering the following model proposed by Tyson and Hannsgen (1986).

They assume that the probability of cell division depends on cell size m, so cell size plays the role of the mitogen considered in Example 5.7.3. It is further assumed that during the lifetime of the cell, growth proceeds exponentially, that is,

$$\frac{dm}{dt} = km.$$

When the size is smaller than a given value, which for simplicity is denoted by 1, the cell cannot divide. When the size is larger than 1, the cell must traverse two phases A and B. The end of phase B coincides with cell division. The duration of phase B is constant and is denoted by T_B. The length T_A of phase A is a random variable with the exponential distribution

$$\text{prob}(T_A \ge t) = e^{-pt}.$$

At cell division the two daughter cells have sizes exactly one-half that of the mother cell.

Using these assumptions it can be shown that the process of the replication of size may be described by the equation

$$f_{n+1}(x) = Pf_n(x) = \int_\sigma^{x/\sigma} K(x,r)f_n(r)\,dr, \qquad (5.10.15)$$

where f_n is the density function of the distribution of the initial size in the nth generation of cells, and the kernel K is given by

$$K(x,r) = \begin{cases} \left(\dfrac{p}{k\sigma}\right)\left(\dfrac{x}{\sigma}\right)^{-1-(p/k)} & \text{for } \sigma \le r < 1 \\ \left(\dfrac{p}{k\sigma}\right)\left(\dfrac{x}{\sigma}\right)^{-1-(p/k)} r^{(p/k)} & \text{for } 1 \le r \le \dfrac{x}{\sigma}. \end{cases} \qquad (5.10.16)$$

It is assumed that $\sigma = \frac{1}{2}e^{kT_B} < 1$. A straightforward calculation shows that P given by formulas (5.10.15) and (5.10.16) is a Markov operator on the space $L^1([\sigma, \infty))$.

Following Tyson and Hannsgen, we are looking for an invariant density of the form $f_*(x) = cx^{-1-\gamma}$. From the equation $f_* = Pf_*$ we obtain

$$x^{-1-\gamma} = \left(\frac{p}{k\sigma}\right)\left(\frac{x}{\sigma}\right)^{-1-(p/k)} \left\{ \int_\sigma^1 r^{-1-\gamma}dr + \int_1^{x/\sigma} r^{(p/k)-1-\gamma}dr \right\},$$

or

$$x^{-1-\gamma} = \frac{\left(\frac{p}{k\sigma}\right)\left(\frac{x}{\sigma}\right)^{-1-\gamma}}{\left(\frac{p}{k}\right)-\gamma} + \left(\frac{p}{k\sigma}\right)\left(\frac{x}{\sigma}\right)^{-1-(p/k)}\left[\frac{\sigma^{-\gamma}-1}{\gamma} - \frac{1}{\left(\frac{p}{k}\right)-\gamma}\right].$$

The above condition is satisfied when γ is a solution of the transcendental equation

$$\sigma^\gamma + \left(\frac{k}{p}\right)\gamma = 1. \qquad (5.10.17)$$

The left-hand side of this equation is equal to 1 for $\gamma = 0$ and tends to ∞ as $\gamma \to \infty$. Thus in order to have a positive solution of (5.10.17) it is sufficient to assume that

$$\frac{d}{d\gamma}\left[\sigma^\gamma + \left(\frac{k}{p}\right)\gamma\right] < 0 \qquad \text{for } \gamma = 0,$$

which is equivalent to

$$\frac{k}{p} < -\ln\sigma. \qquad (5.10.18)$$

Thus, for k, p, and σ satisfying (5.10.18), there exists $\gamma > 0$ for which the function $f_*(x) = cx^{-1-\gamma}$ is invariant with respect to P. It can be normalized on the interval $[\sigma, \infty)$, namely, for $c = \gamma\sigma^{-\gamma}$,

$$\int_\sigma^\infty f_*(x)\,dx = \gamma\sigma^{-1}\int_0^\infty x^{-1-\gamma}dx = 1.$$

Now we can apply Theorem 5.10.2. The function f_* is simultaneously a positive locally integrable (with respect to the family of compact subsets of $[\sigma, \infty)$) invariant function and an invariant density. Further, according to Proposition 5.10.1, the operator (5.10.15) is expanding since the kernel $K(x, r)$ given by equation (5.10.16) and $\lambda(x) = x/\sigma$ with $\sigma < 1$ satisfy conditions (5.10.11) and (5.10.12). Therefore, according to Theorem 5.10.2 the sequence $\{P^n\}$ is asymptotically stable. $\quad\square$

Remark 5.10.2. The asymptotic stability of $\{P^n\}$ under condition (5.10.18) was predicted by Tyson and Hannsgen (1986) and proved by Tyrcha (1989) using Liapunov function techniques (see Exercise 5.7). It should be noted that in our approach using the Foguel alternative the major effort was expended in finding the invariant density f_* since, once f_* was available, the asymptotic stability followed automatically from the properties of Markov operators with advanced arguments.

Exercises

5.1. Let $(X, \mathcal{A}, \mu)$ be a finite measure space and let $1 \le p_1 \le p_2 \le \infty$. Show that every set $\mathcal{F} \in L^{p_2}$ strongly precompact in L^{p_2} is also strongly precompact in L^{p_1}. Is the same true for weak precompactness?

5.2. Let $X = R^+$ with the standard Borel measure. Consider the four families of functions:

(1) $f_a(x) = ae^{-ax}$, for $a \ge 1$;

(2) $f_a(x) = ae^{-ax}$, for $0 \le a \le 1$;

(3) $f_a(x) = e^{-x} \sin ax$, for $a \ge 1$;

(4) $f_a(x) = e^{-x} \sin ax$, for $0 \le a \le 1$.

Which of these families is weakly and/or strongly precompact in $L^1(R^+)$?

5.3. Consider the measure space $(X, \mathcal{A}, \mu)$ described in Exercise 2.1. Let $g \in l^1 = L^1(X, \mathcal{A}, \mu)$ be nonnegative and let

$$\mathcal{F} = \{f \in l^1 : |f| \le g\}.$$

Show that $\mathcal{F}$ is strongly precompact in l^1.

5.4. Generalize the previous result and show that every weakly precompact set $\mathcal{F} \subset l^1$ is also strongly precompact in l^1.

5.5. Let $X \subset R$ be an interval and let $K: X \times X \to R$ be a continuous function. Assume that

$$Pf(x) = \int_X K(x, y) f(y) \, dy$$

is a Markov operator. Prove that K is a stochastic kernel. Try to generalize this result to the case when $(X, \mathcal{A}, \mu)$ is an arbitrary measure space and $K: X \times X \to R$ is measurable on the product space $X \times X$.

5.6. Let $X = R^+$ and let

$$Pf(x) = \int_0^x K(x, y) f(y) \, dy$$

be a Markov operator with a stochastic kernel; that is, K is measurable and

$$K(x, y) \geq 0, \qquad \int_y^\infty K(x, y) \, dx = 1.$$

Prove that if K is bounded ($K(x, y) \leq M$), then $\{P^n\}$ is sweeping to $+\infty$. Try to generalize this result of the case when K is unbounded.

5.7. Let $X = R$ and let

$$Pf(x) = (1 + e^{-x}) f(x - e^{-x}).$$

Prove that $\{P^n\}$ is sweeping with respect to the family

$$\mathcal{A}_* = \{(-\infty, c) : c \in R\},$$

but it is not sweeping with respect to

$$A_{\text{fin}} = \{A \in \mathcal{B}(R) : m(A) < \infty\},$$

where m denotes the standard Borel measure (Komorowski and Tyrcha 1989).

5.8. Let $(X, \mathcal{A}, \mu)$ be a measure space and $P: L^1 \to L^1$ be a Markov operator. We say that P **overlaps supports** if, for every $f, g \in D$ there is an integer $n_0(f, g)$ such that

$$\mu(\text{supp } P^{n_0} f \cap \text{supp } P^{n_0} g) > 0.$$

It can be proved (Baron and Lasota, 1993) that $\{P^n\}$ is asymptotically stable for every integral operator with stochastic kernel which overlaps supports and has an invariant positive (a.e.) density. Using this result and Theorem 5.10.1, prove Theorem 5.10.2.

6

The Behavior of Transformations on Intervals and Manifolds

This chapter is devoted to a series of examples of transformations on intervals and manifolds whose asymptotic behavior can be explored through the use of the material developed in Chapter 5. Although results are often stated in terms of the asymptotic stability of $\{P^n\}$, where P is a Frobenius–Perron operator corresponding to a transformation S, remember that, according to Proposition 5.6.2, S is exact when $\{P^n\}$ is asymptotically stable and S is measure preserving.

In applying the results of Chapter 5, in several examples we will have occasion to calculate the variation of a function. Thus the first section presents an exposition of the properties of functions of bounded variation.

6.1 Functions of Bounded Variation

There are a number of descriptors of the "average" behavior of a function $f\colon [a, b] \to R$. Two of the most common are the **mean value** of f,

$$m(f) = \frac{1}{b-a} \int_a^b f(x)\,dx,$$

and its **variance**, $D^2(f) = m((f - m(f))^2)$. However, these are not always satisfactory. Consider, for example, the sequence of functions $\{f_n\}$ with $f_n(x) = \sin 2n\pi x$, $n = 1, 2, \ldots$. They have the same mean value on $[0, 1]$, namely, $m(f_n) = 0$ and the same variance $D^2(f_n) = \frac{1}{2}$; but they behave quite differently for $n \gg 1$ than they do for $n = 1$. To describe these kinds of differences in the behavior of functions, it is useful to introduce the variation of a function (sometimes called the total variation).

Let f be a real-valued function defined on an interval $\Delta \subset R$ and let $[a, b]$ be a subinterval of Δ. Consider a partition of $[a, b]$ given by

$$a = x_0 < x_1 < \cdots < x_n = b \qquad (6.1.1)$$

and write

$$s_n(f) = \sum_{i=1}^{n} |f(x_i) - f(x_{i-1})|. \qquad (6.1.2)$$

If all possible sums $s_n(f)$, corresponding to all subdivisions of $[a, b]$, are bounded by a number that does not depend on the subdivision, f is said to be of **bounded variation** on $[a, b]$. Further, the smallest number c such that $s_n \leq c$ for all s_n is called the **variation** of f on $[a, b]$ and is denoted by $\bigvee_a^b f$. Notationally this is written as

$$\bigvee_a^b f = \sup s_n(f), \qquad (6.1.3)$$

where the supremum is taken over all possible partitions of the form (6.1.1).

Consider a simple example. Assume that f is a monotonic function, either decreasing or increasing. Then

$$|f(x_i) - f(x_{i-1})| = \theta[f(x_i) - f(x_{i-1})],$$

where

$$\theta = \begin{cases} 1 & \text{for } f \text{ increasing} \\ -1 & \text{for } f \text{ decreasing} \end{cases}$$

and, consequently,

$$s_n(f) = \theta \sum_{i=1}^{n} [f(x_i) - f(x_{i-1})]$$
$$= \theta[f(x_n) - f(x_0)] = |f(b) - f(a)|.$$

Thus, any function that is defined and monotonic on a closed interval is of bounded variation. It is interesting (the proof is not difficult) that any function f of bounded variation can be written in the form $f = f_1 + f_2$, where f_1 is increasing and f_2 is decreasing.

Variation of the Sum

Let f and g be of bounded variation on $[a, b]$. Then

$$|f(x_i) + g(x_i) - [f(x_{i-1}) + g(x_{i-1})]| \leq |f(x_i) - f(x_{i-1})| + |g(x_i) - g(x_{i-1})|,$$

and, consequently,

$$s_n(f+g) \leq s_n(f) + s_n(g) \leq \bigvee_a^b f + \bigvee_a^b g.$$

Thus $(f+g)$ is of bounded variation and

$$\bigvee_a^b (f+g) \leq \bigvee_a^b f + \bigvee_a^b g.$$

If $f_1, \ldots, f_n$ are of bounded variation on $[a, b]$, then by an induction argument

(V1) $\displaystyle \bigvee_a^b (f_1 + \cdots + f_n) \leq \bigvee_a^b f_1 + \cdots + \bigvee_a^b f_n$ (6.1.4)

follows immediately.

Variation on the Union of Intervals

Assume that $a < b < c$ and that the function f is of bounded variation on $[a, b]$ as well as on $[b, c]$. Consider a partition of the intervals $[a, b]$ and $[b, c]$,

$$a = x_0 < x_1 < \cdots < x_n = b = y_0 < y_1 < \cdots < y_m = c \qquad (6.1.5)$$

and the corresponding sums

$$\underset{[a,b]}{s_n}(f) = \sum_{i=1}^n |f(x_i) - f(x_{i-1})|,$$

$$\underset{[b,c]}{s_m}(f) = \sum_{i=1}^m |f(y_i) - f(y_{i-1})|.$$

It is evident that the partitions (6.1.5) jointly give a partition of $[a, c]$. Therefore,

$$\underset{[a,b]}{s_n}(f) + \underset{[b,c]}{s_m}(f) = \underset{[a,c]}{s_{n+m}}(f) \qquad (6.1.6)$$

where the right-hand side of equation (6.1.6) denotes the sum corresponding to the variation of f over $[a, c]$. Observe that (6.1.6) holds only for partitions of $[a, c]$ that contain the point b. However, any additional point in the sum s_n can only increase s_n, but, since we are interested in the supremum, this is irrelevant. From equation (6.1.6) it follows that

$$\bigvee_a^b f + \bigvee_b^c f = \bigvee_a^c f.$$

Again by an induction argument the last formula may be generalized to

$$\text{(V2)} \qquad \bigvee_{a_0}^{a_1} f + \cdots + \bigvee_{a_{n-1}}^{a_n} f = \bigvee_{a_0}^{a_n} f, \qquad \text{(6.1.7)}$$

where $a_0 < a_1 < \cdots < a_n$ and f is of bounded variation on $[a_{i-1}, a_i]$, $i = 1, \ldots, n$.

Variation of the Composition of Functions

Now let $g \colon [\alpha, \beta] \to [a, b]$ be monotonically increasing or decreasing on the interval $[\alpha, \beta]$ and let $f \colon [a, b] \to R$ be given. Then the composition $f \circ g$ is well defined and, for any partition of $[\alpha, \beta]$,

$$\alpha = \sigma_0 < \sigma_1 < \cdots < \sigma_n = \beta; \qquad \text{(6.1.8)}$$

the corresponding sum is

$$s_n(f \circ g) = \sum_{i=1}^{n} |f(g(\sigma_i)) - f(g(\sigma_{i-1}))|.$$

Observe that, due to the monotonicity of g, the points $g(\sigma_i)$ define a partition of $[a, b]$. Thus, $s_n(f \circ g)$ is a particular sum for the variation of f and, therefore,

$$s_n(f \circ g) \le \bigvee_a^b f$$

for any partition (6.1.8). Consequently,

$$\text{(V3)} \qquad \bigvee_{\alpha}^{\beta} f \circ g \le \bigvee_a^b f. \qquad \text{(6.1.9)}$$

Variation of the Product

Let f be of bounded variation on $[a, b]$ and let g be C^1 on $[a, b]$. To evaluate the variation of the product $f(x)g(x)$, $x \in [a, b]$, start from the well-known **Abel equality**,

$$\sum_{i=1}^{n} |a_i b_i - a_{i-1} b_{i-1}| = \sum_{i=1}^{n} |b_i(a_i - a_{i-1}) + a_{i-1}(b_i - b_{i-1})|.$$

Applying this equality to the sum [substituting $a_i = f(x_i)$ and $b_i = g(x_i)$]

$$s_n(fg) = \sum_{i=1}^{n} |f(x_i)g(x_i) - f(x_{i-1})g(x_{i-1})|,$$

the immediate result is

$$s_n(fg) \le \sum_{i=1}^{n} \{|g(x_i)|\,|f(x_i) - f(x_{i-1})| + |f(x_{i-1})|\,|g(x_i) - g(x_{i-1})|\}.$$

Now, by applying the mean value theorem, we have

$$s_n(fg) \le (\sup|g|)s_n(f) + \sum_{i=1}^{n} |f(x_{i-1})g'(\tilde{x}_i)|(x_i - x_{i-1})$$

$$\le (\sup|g|) \bigvee_a^b f + \sum_{i=1}^{n} |f(x_{i-1})g'(\tilde{x}_i)|(x_i - x_{i-1}).$$

with $\tilde{x}_i \in (x_{i-1}, x_i)$. Observe that the last term is simply an approximating sum for the Riemann integral of the product $|f(x)g'(x)|$. Thus the function $f(x)g(x)$ is of bounded variation and

$$(V4) \qquad \bigvee_a^b fg \le (\sup|g|) \bigvee_a^b f + \int_a^b |f(x)g'(x)|\,dx. \qquad (6.1.10)$$

Taking in particular $f \equiv 1$,

$$(V4') \qquad \bigvee_a^b g \le \int_a^b |g'(x)|\,dx. \qquad (6.1.11)$$

However, in this case, the left- and right-hand sides are strictly equal since $s_n(g)$ is a Riemann sum for the integral of g'.

Yorke Inequality

Now let f be defined on $[0,1]$ and be of bounded variation on $[a,b] \subset [0,1]$. We want to evaluate the variation of the product of f and the characteristic function $1_{[a,b]}$. Without any loss of generality, assume that the partitions of the interval $[0,1]$ will always contain the points a and b. Then

$$\underset{[0,1]}{s_n}(f1_{[a,b]}) \le \underset{[a,b]}{s_n}(f) + |f(a)| + |f(b)|.$$

Let c be an arbitrary point in $[a,b]$. Then, from the preceding inequality,

$$\underset{[0,1]}{s_n}(f1_{[a,b]}) \le \underset{[a,b]}{s_n}(f) + |f(b) - f(c)| + |f(c) - f(a)| + 2|f(c)|$$

$$\le 2\bigvee_a^b f + 2|f(c)|.$$

It is always possible to choose the point c such that

$$|f(c)| \leq \frac{1}{b-a} \int_a^b |f(x)|\,dx$$

so that

$$s_n \left(f1_{[a,b]}\right) \leq 2 \bigvee_{[0,1]}^b f + \frac{2}{b-a} \int_a^b |f(x)|\,dx,$$

which gives

(V5) $$\bigvee_0^1 f1_{[a,b]} \leq 2 \bigvee_a^b f + \frac{2}{b-a} \int_a^b |f(x)|\,dx. \qquad (6.1.12)$$

6.2 Piecewise Monotonic Mappings

Two of the most important results responsible for stimulating interest in transformations on intervals of the real line were obtained by Rényi (1957) and by Rochlin (1964). Both were considering two classes of mappings, namely,

$$S(x) = \tau(x) \qquad (\text{mod } 1), 0 \leq x \leq 1, \qquad (6.2.1)$$

where $\tau \colon [0,1] \to [0,\infty)$ is a C^2 function such that $\inf_x \tau' > 1$, $\tau(0) = 0$, and $\tau(1)$ is an integer; and the **Rényi transformation**

$$S(x) = rx \qquad (\text{mod } 1), 0 \leq x \leq 1, \qquad (6.2.2)$$

where $r > 1$, is a real constant. (The r-adic transformation considered earlier is clearly a special case of the Rényi transformation.) Using a number-theoretic argument, Rényi was able to prove the existence of a unique invariant measure for such transformations. Rochlin was able to prove that the Rényi transformations on a measure space with the Rényi measure were, in fact, exact.

In this section we unify and generalize the results of Rényi and Rochlin through the use of Theorem 5.6.2.

Consider a mapping $S \colon [0,1] \to [0,1]$ that satisfies the following four properties:

(2i) There is a partition $0 = a_0 < a_1 < \cdots < a_r = 1$ of $[0,1]$ such that for each integer $i = 1, \ldots, r$ the restriction of S to the interval $[a_{i-1}, a_i)$ is a C^2 function;

(2ii) $S(a_{i-1}) = 0$ for $i = 1, \ldots, r$;

(2iii) There is a $\lambda > 1$ such that $S'(x) \geq \lambda$ for $0 \leq x < 1$ [$S'(a_i)$ and $S''(a_i)$ denote the right derivatives]; and

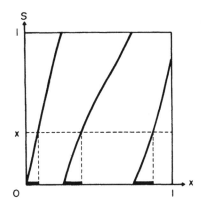

FIGURE 6.2.1. Function $S(x) = 3x + \frac{1}{4}\sin(7x/4)$ (mod 1) as an example of a transformation on $[0, 1]$ satisfying the conditions (2i)–(2iv). In this case $r = 3$, and the counterimage of the set $[0, x]$ consists of the union of the three intervals indicated as heavy lines along the x-axis.

(2iv) There is a real finite constant c such that

$$- S''(x)/[S'(x)]^2 \le c, \qquad 0 \le x < 1. \tag{6.2.3}$$

An example of a mapping satisfying these conditions is shown in Figure 6.2.1.

Then we may state the following theorem.

Theorem 6.2.1. *If* $S: [0, 1] \to [0, 1]$ *satisfies the foregoing conditions* (2i)–(2iv) *and* P *is the Frobenius–Perron operator associated with* S, *then* $\{P^n\}$ *is asymptotically stable.*

Proof. We first derive an explicit expression for the Frobenius–Perron operator. Note that, for any $x \in [0, 1)$,

$$S^{-1}([0, x]) = \bigcup_{i=1}^{r} [a_{i-1}, g_i(x)],$$

where

$$g_i(x) = \begin{cases} S_{(i)}^{-1}(x) & 0 \le x < b_i \\ a_i & b_i \le x < 1 \end{cases}$$

and $S_{(i)}(x)$ denotes the restriction of S to the interval $[a_{i-1}, a_i)$, whereas

$$b_i = \lim_{x \to a_i} S_{(i)}(x).$$

Thus, by the definition of the Frobenius–Perron operator,

$$Pf(x) = \frac{d}{dx} \int_{S^{-1}([0,x])} f(u)\,du$$

$$= \frac{d}{dx} \sum_{i=1}^{r} \int_{a_{i-1}}^{g_i(x)} f(u)\,du$$

or

$$Pf(x) = \sum_{i=1}^{r} g_i'(x) f(g_i(x)). \qquad (6.2.4)$$

If $b_i < 1$, then $g_i'(b_i)$ denotes the right derivative. Thus, $g_i'(x) = 0$ for $b_i \leq x < 1$, and all the g_i' are lower left semicontinuous.

Now let D_0 denote the subset of $D([0,1])$ consisting of all functions f that, on the interval $[0,1)$, are bounded, lower left semicontinuous, and satisfy the inequality

$$f_+'(x) = \frac{d_+ f(x)}{dx} \leq k_f f(x), \qquad \text{for } 0 \leq x < 1, \qquad (6.2.5)$$

where k_f is a constant that depends on f. For any $f \in D_0$, the function Pf as calculated from equation (6.2.4) will be bounded and lower left semicontinuous.

For every $f \in D_0$, differentiation of expression (6.2.4) for the Frobenius–Perron operator gives

$$(Pf)_+' = \sum_{i=1}^{r}(g_i')_+' (f \circ g_i) + \sum_{i=1}^{r}(g_i')^2 (f_+' \circ g_i).$$

By using the inverse function theorem, we have

$$g_i' \leq \sup(1/S') \leq 1/\lambda$$

and

$$(g_i')_+'/g_i' \leq \sup(-S''/[S']^2) \leq c$$

so, as a consequence,

$$(Pf)_+' \leq c \sum_{i=1}^{r} g_i'(f \circ g_i) + \frac{1}{\lambda} \sum_{i=1}^{r} g_i'(f_+' \circ g_i).$$

Using inequality (6.2.5), this expression may be further simplified to

$$(Pf)_+' \leq \left[c + \frac{k_f}{\lambda}\right] Pf.$$

Set $f_n = P^n f$. An induction argument shows that

$$(f_n)_+' \leq \left[\frac{c\lambda}{\lambda - 1} + \frac{k_f}{\lambda^n}\right] f_n.$$

Choose a real $k > c\lambda/(\lambda - 1)$. Then

$$(f_n)'_+ \leq k f_n \tag{6.2.6}$$

for n sufficiently large, say $n \geq n_0(f)$, and thus condition (5.8.3) of Proposition 5.8.1 is satisfied.

We now show that the f_n are bounded and hence satisfy condition (5.8.2) of Proposition 5.8.1. First note that from equation (6.2.4) we may write

$$f_{n+1}(x) = \sum_{i=1}^{r} g_i'(x) f_n(g_i(x)).$$

Thus, since $g_i' \leq 1/\lambda$ and $S(a_{i-1}) = 0$ for $i = 1, \ldots, r$,

$$f_{n+1}(0) \leq \frac{1}{\lambda} f_n(0) + \frac{1}{\lambda} \sum_{i=2}^{r} f_n(a_{i-1}). \tag{6.2.7}$$

From (6.2.6) it follows that

$$f_n(a_i) \leq f_n(x) e^k, \qquad \text{for } x \leq a_i,$$

so that

$$1 \geq \int_0^{a_i} f_n(x)\, dx \geq e^{-k} f_n(a_i) a_i, \qquad \text{for } i = 1, \ldots, r.$$

Thus $f_n(a_i) \leq e^k/a_i$, and from (6.2.7) we have

$$f_{n+1}(0) \leq (1/\lambda) f_n(0) + L/\lambda, \qquad \text{for } n \geq n_0(f),$$

where

$$L = \sum_{i=2}^{r} e^k/a_{i-1}.$$

Again, using a simple induction argument, it follows that

$$f_n(0) \leq (1/\lambda^{n-n_0}) f_{n_0}(0) + L/(\lambda - 1), \qquad \text{for } n \geq n_0(f),$$

so

$$f_n(0) \leq 1 + [L/(\lambda - 1)]$$

for sufficiently large n, say $n \geq n_1(f)$. By using this relation in conjunction with the differential inequality (6.2.6), we therefore obtain

$$f_n(x) \leq \{1 + [L/(\lambda - 1)]\} e^k, \qquad \text{for } 0 \leq x < 1, n \geq n_1. \tag{6.2.8}$$

Thus, by inequalities (6.2.6) and (6.2.8), all the conditions of Proposition 5.8.1 are satisfied and $\{P^n\}$ is asymptotically stable by Theorem 5.6.2. ∎

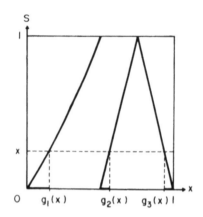

FIGURE 6.2.2. A piecewise monotonic transformation satisfying the conditions of Theorem 6.2.2.

Theorem 6.2.1 is valid only for mappings that are monotonically increasing on each subinterval $[a_{i-1}, a_i)$ of the partition of $[0, 1]$. However, by modification of some of the foregoing properties (2i)–(2iv), we may also prove another theorem valid for transformations that are either monotonically increasing or decreasing on the subintervals of the partition. The disadvantage is that the mapping must be onto for every $[a_{i-1}, a_i)$.

We now consider a mapping $S: [0, 1] \to [0, 1]$ that satisfies a condition slightly different from property (2i):

(2i)' There is a partition $0 = a_0 < a_1 < \cdots < a_r = 1$ of $[0, 1]$ such that for each integer $i = 1, \ldots, r$ the restriction of S to the interval (a_{i-1}, a_i) is a C^2 function; as well as

(2ii)' $S((a_{i-1}, a_i)) = (0, 1)$, that is, S is onto;

(2iii)' There is a $\lambda > 1$ such that $|S'(x)| \geq \lambda$, for $x \neq a_i$, $i = 0, \ldots, r$; and

(2iv)' There is a real finite constant c such that

$$|S''(x)|/[S'(x)]^2 \leq c, \qquad \text{for } x \neq a_i, i = 0, \ldots, r. \tag{6.2.9}$$

(See Figure 6.2.2 for an example.)

Then we have the following theorem.

Theorem 6.2.2. *If $S: [0, 1] \to [0, 1]$ satisfies the preceding conditions (2i)'–(2iv)' and P is the Frobenius–Perron operator associated with S, then $\{P^n\}$ is asymptotically stable.*

Proof. The proof proceeds much as for Theorem 6.2.1. Using the same notation as before, it is easy to show that for $x \in [0,1]$,

$$S^{-1}((0,x)) = \bigcup_j (a_{j-1}, g_j(x)) + \bigcup_k (g_k(x), a_k),$$

where $g_i = S_{(i)}^{-1}$, $S_{(i)}$ is as before, and the first union is over all intervals in which $S_{(i)}$ is an increasing function of x whereas the second is over intervals in which $S_{(i)}$ is decreasing. Thus

$$Pf(x) = \sum_j g_j'(x) f(g_j(x)) - \sum_k g_k'(x) f(g_k(x))$$

or, with $\sigma_i(x) = |g_i'(x)|$,

$$Pf(x) = \sum_{i=1}^r \sigma_i(x) f(g_i(x)). \tag{6.2.10}$$

Let $D_0 \subset D$ be the set of all bounded continuously differentiable densities such that

$$|f'(x)| \le k_f f(x), \qquad \text{for } 0 < x < 1, \tag{6.2.11}$$

where the constant k_f depends on f. For every $f \in D_0$, differentiating equation (6.2.10) gives

$$(Pf)' = \sum_{i=1}^r \sigma_i'(f \circ g_i) + \sum_{i=1}^r \sigma_i g_i'(f \circ g_i).$$

Exactly as in the proof of Theorem 6.2.1 we have

$$\sigma_i \le \sup(1/|S'|) \le 1/\lambda$$

and

$$|\sigma_i'|/|g_i'| \le \sup |S''|/[S']^2 \le c.$$

These two inequalities, in combination with (6.2.11), allow us to evaluate $(Pf)'$ as

$$|(Pf)'| \le \left[c + \frac{k_f}{\lambda} \right] Pf.$$

Set $f_n = P^n f$ and use an induction argument to show that

$$|f_n'| \le \left[\frac{c\lambda}{\lambda - 1} + \frac{k_f}{\lambda^n} \right] f_n.$$

Again we may always pick a $k > c\lambda/(\lambda - 1)$ such that

$$|f_n'| \le k f_n \tag{6.2.12}$$

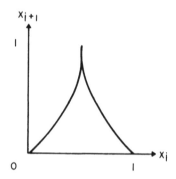

FIGURE 6.2.3. Successive maxima in the variable $x(t)$ from the Lorenz equations are labeled x_i, and one maximum is plotted against the previous (x_{i+1} vs. x_i) after rescaling so that all $x_i \in [0, 1]$.

for sufficiently large n [say $n \geq n_0(f)$], and thus Proposition 5.8.2 is satisfied. ∎

Example 6.2.1. When $\sigma = 10$, $b = 8/3$, and $r = 28$, then all three variables x, y, and z in the Lorenz [1963] equations,

$$\frac{dx}{dt} = yz - bx, \quad \frac{dy}{dt} = -xz + rz - y, \quad \frac{dz}{dt} = \sigma(y - z),$$

show very complicated dynamics. If we label successive maxima in $x(t)$ as x_i ($i = 0, 1, \ldots$), plot each maximum against the previous maximum (i.e., x_{i+1} vs. x_i), and scale the results so that the x_i are contained in the interval $[0, 1]$, then the numerical computations show that the points (x_i, x_{i+1}) are located approximately on the graph of a one-dimensional mapping, as shown in Figure 6.2.3.

As an approximation to this mapping of one maximum to the next, we can consider the transformation

$$S(x) = \begin{cases} \dfrac{(2-a)x}{1-ax} & \text{for } x \in \left[0, \tfrac{1}{2}\right] \\[2ex] \dfrac{(2-a)(1-x)}{1-a(1-x)} & \text{for } x \in \left(\tfrac{1}{2}, 1\right], \end{cases} \tag{6.2.13}$$

where $a = 1 - \varepsilon$, shown in Figure 6.2.4 for $\varepsilon = 0.01$. Clearly, $S(0) = S(1) = 0$, $S\left(\tfrac{1}{2}\right) = 1$, and, since $S'(x) = (2-a)/(1-ax)^2$, we will always have $|S'(x)| > 1$ for $x \in \left[0, \tfrac{1}{2}\right)$ if $\varepsilon > 0$. Finally, since $S''(x) = 2a(2-a)/(1-ax)^3$, $|S''(x)|$ is always bounded above. For $x \in \left(\tfrac{1}{2}, 1\right]$ the calculations are similar. Thus the transformation (6.2.13) satisfies all the requirements of Theorem 6.2.2: $\{P^n\}$ is asymptotically stable and S is exact. □

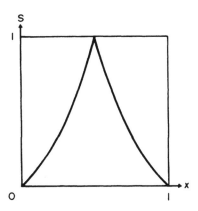

FIGURE 6.2.4. The transformation $S(x)$ given by equation (6.2.13) with $\varepsilon = 0.01$ as an approximation to the data of Figure 6.2.3.

Remark 6.2.1. The condition that $|S'(x)| > 1$ in Theorem 6.2.2 is essential for S to be exact. We could easily demonstrate this by using (6.2.13) with $\varepsilon = 0$, thus making $|S'(0)| = |S'(1)| = 1$. However, even if $|S'(x)| = 1$ for only one point $x \in [0, 1]$, it is sufficient to destroy the exactness, as can be demonstrated by the transformation

$$S(x) = \begin{cases} x/(1-x) & \text{for } x \in \left[0, \frac{1}{2}\right] \\ 2x - 1 & \text{for } x \in \left(\frac{1}{2}, 1\right], \end{cases} \tag{6.2.14}$$

which we originally considered in Section 1.3 (paradox of the weak repellor). Now, the condition $|S'(x)| > 1$ is violated only at the single point $x = 0$, and, for any $f \in L^1$, the sequence $\{P^n f\}$ converges to zero on $(0, 1]$. Thus, the only solution to the equation $Pf = f$ is the trivial solution $f \equiv 0$, and therefore there is no measure invariant under S.

This is quite difficult to prove. First write the Frobenius–Perron operator corresponding to S as

$$Pf(x) = \frac{1}{(1+x)^2} f\left(\frac{x}{1+x}\right) + \frac{1}{2} f\left(\frac{1}{2} + \frac{x}{2}\right). \tag{6.2.15}$$

Set $q_n(x) = x f_n(x)$, where $f_n = P^n f_0$, and pick the initial density to be $f_0 \equiv 1$. Thus $q_0(x) \equiv x$, and from (6.2.15) we have the recursive formula,

$$q_{n+1}(x) = \frac{1}{1+x} q_n\left(\frac{x}{1+x}\right) + \frac{x}{1+x} q_n\left(\frac{1}{2} + \frac{x}{2}\right). \tag{6.2.16}$$

Proceeding inductively, it is easy to prove that $q_n'(x) \geq 0$ for all n, so that the functions $q_n(x)$ are all positive and increasing. From equation (6.2.16) we have

$$q_{n+1}(1) \leq \tfrac{1}{2}q_n\left(\tfrac{1}{2}\right) + \tfrac{1}{2}q_n(1) \leq q_n(1),$$

which shows that

$$\lim_{n \to \infty} q_n(1) = c_0$$

exists. Write $z_0 = 1$ and $z_{k+1} = z_k/(1 + z_k)$. Then from (6.2.16) we have

$$q_{n+1}(z_k) = \frac{1}{1 + z_k} q_n(z_{k+1}) + \frac{z_k}{1 + z_k} q_n \left(\frac{1}{2} + \frac{z_k}{2}\right).$$

Take k to be fixed and assume that $\lim_{n\to\infty} q_n(x) = c_0$ for $z_k \leq x \leq 1$ (which is certainly true for $k = 0$). Since $z_k \leq \tfrac{1}{2} + \tfrac{1}{2}z_k$, taking the limit as $n \to \infty$, we have

$$c_0 = \frac{1}{1 + z_k} \lim_{n \to \infty} q_n(z_{k+1}) + \frac{z_k}{1 + z_k} c_0,$$

so $\lim_{n\to\infty} q_n(z_{k+1}) = c_0$. Since the functions $q_n(x)$ are increasing, we know that $\lim_{n\to\infty} q_n(x) = c_0$ for all $x \in [z_{k+1}, 1]$. By induction it follows that $\lim_{n\to\infty} q_n(x) = c_0$ in any interval $[z_k, 1]$ and, since $\lim_{k\to\infty} z_k = 0$, we have $\lim_{n\to\infty} q_n(x) = c_0$ for all $x \in (0, 1]$. Thus

$$\lim_{n \to \infty} f_n(x) = c_0/x.$$

Actually, the limit c_0 is zero; to show this, assume $c_0 \neq 0$. Then there must exist some $\varepsilon > 0$ such that

$$\lim_{n \to \infty} \int_\varepsilon^1 f_n(x)\, dx = \int_\varepsilon^1 (c_0/x)\, dx > 1.$$

However, this is impossible since $\|f_n\| = 1$ for every n. By induction, each of the functions $f_n(x)$ is decreasing, so the convergence of $f_n(x)$ to zero is uniform on any interval $[\varepsilon, 1]$ where $\varepsilon > 0$.

Now, let f be an arbitrary function, and write $f = f^+ - f^-$. Given $\delta > 0$, consider a constant h such that

$$\int_0^1 (f^- - h)^+ dx + \int_0^1 (f^+ - h)^+ dx \leq \delta.$$

Thus, since $|P^n f| \leq P^n |f| = P^n f^+ + P^n f^-$, we have

$$\int_\varepsilon^1 |P^n f|\, dx \leq \int_\varepsilon^1 P^n f^+\, dx + \int_\varepsilon^1 P^n f^-\, dx$$

$$= 2\int_\varepsilon^1 P^n h\, dx + \int_\varepsilon^1 P^n(f^+ - h)\, dx + \int_\varepsilon^1 P^n(f^- - h)\, dx$$

$$\leq 2h \int_\varepsilon^1 P^n 1\, dx + \delta$$

and, since $\{P^n 1\}$ converges uniformly to zero on $[\varepsilon, 1]$, we have

$$\lim_{n \to \infty} \int_{\varepsilon}^{1} |P^n f| \, dx = 0 \qquad \text{for } \varepsilon > 0.$$

Hence the sequence $\{P^n f\}$ converges to zero in $L^1([\varepsilon, 1])$ norm for every $\varepsilon > 0$ and equation $Pf = f$ cannot have a solution $f \in L^1$ except $f \equiv 0$.
$\square$

6.3 Piecewise Convex Transformations with a Strong Repellor

Although the theorems of the preceding section were moderately easy to prove using the techniques of Chapter 5, the conditions that transformation S must satisfy are highly restrictive. Thus, in specific cases of interest, it may often not be the case that $S'(x) > 1$ or $|S'(x)| > 1$, or that condition (6.2.3) or (6.2.9) is obeyed.

However, for a class of convex transformations, it is known that $\{P^n\}$ is asymptotically stable. Consider $S: [0, 1] \to [0, 1]$ having the following properties:

(3i) There is a partition $0 = a_0 < a_1 < \cdots < a_r = 1$ of $[0, 1]$ such that for each integer $i = 1, \ldots, r$ the restriction of S to $[a_{i-1}, a_i)$ is a C^2 function;

(3ii) $S'(x) > 0$ and $S''(x) \geq 0$ for all $x \in [0, 1)$, $[S'(a_i)$ and $S''(a_i)$ are right derivatives];

(3iii) For each integer $i = 1, \ldots, r$, $S(a_{i-1}) = 0$; and

(3iv) $S'(0) > 1$.

An example of a mapping satisfying these criteria is shown in Figure 6.3.1.

Remark 6.3.1. Property (3iv) implies that point $x = 0$ is a **strong repellor** (see also Section 1.3 and Remark 6.2.1), that is, trajectory $\{S(x_0), S^2(x_0), \ldots\}$, starting from a point $x_0 \in (0, a_1)$, will eventually leave $[0, a_1)$. To see this, note that as long as $S^n(x_0) \in [0, a_1)$ there is a $\xi \in (0, a_1)$ such that

$$S^n(x_0) = S(S^{n-1}(x_0)) - S(0)$$
$$= S'(\xi) S^{n-1}(x_0) \geq \lambda S^{n-1}(x_0),$$

where $\lambda = S'(0)$. By an induction argument, $S^n(x_0) \geq \lambda^n x_0$ and, since $\lambda > 1$, $S^n(x_0)$ must eventually exceed a_1. After leaving the interval $[0, a_1)$

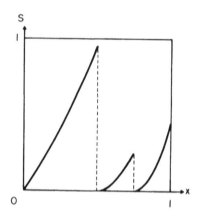

FIGURE 6.3.1. An example of a piecewise convex transformation satisfying the conditions of Theorem 6.3.1.

the trajectory will, in general, exhibit very complicated behavior. If at some point it returns to $[0, a_1)$, then it will, again, eventually leave $[0, a_1)$.   □
    With these comments in mind, we can state the following theorem.

Theorem 6.3.1. *Let $S: [0, 1] \to [0, 1]$ be a transformation satisfying the foregoing conditions* (3i)–(3iv), *and let P be the Frobenius–Perron operator associated with S. Then $\{P^n\}$ is asymptotically stable.*

Proof. The complete proof of this theorem, which may be found in Lasota and Yorke [1982], is long and requires some technical details we have not introduced. Rather than give the full proof, here we show only that $\{P^n f\}$ is bounded above, thus implying that there is a measure invariant under S.
    We first derive the Frobenius–Perron operator. For any $x \in [0, 1]$ we have

$$S^{-1}([0, x]) = \bigcup_{i=1}^{r} [a_{i-1}, g_i(x)],$$

where

$$g_i(x) = \begin{cases} S_{(i)}^{-1}(x) & \text{for } x \in S([a_{i-1}, a_i)) \\ a_i & \text{for } x \in [0, 1] \setminus S([a_{i-1}, a_i)) \end{cases}$$

and, as before, $S_{(i)}$ denotes the restriction of S to the interval $[a_{i-1}, a_i)$. Thus, as in Section 6.2, we obtain

$$Pf(x) = \sum_{i=1}^{r} g_i'(x) f(g_i(x)). \tag{6.3.1}$$

Even though equations (6.2.4) and (6.3.1) appear to be identical, the functions g_i have different properties. For instance, by using the inverse function theorem, we have

$$g_i' = 1/S' > 0 \quad \text{and} \quad g_i'' = -S''/[S']^2 \leq 0.$$

Thus, since $g_i' > 0$ we know that g_i is an increasing function of x, whereas g_i' is a decreasing function of x since $g_i'' \leq 0$.

Let $f \in D([0,1])$ be a decreasing density, that is, $x \leq y$ implies $f(x) \geq f(y)$. Then, by our previous observations, $f(g_i(x))$ is a decreasing function of x as is $g_i'(x)f(g_i(x))$. Since Pf, as given by (6.3.1), is the sum of decreasing functions, Pf is a decreasing function of x and, by induction, so is $P^n f$.

Observe further that, for any decreasing density $f \in D([0,1])$, we have

$$1 \geq \int_0^x f(u) \, du \geq \int_0^x f(x) \, du = xf(x),$$

so that, for any decreasing density,

$$f(x) \leq 1/x, \qquad x \in (0,1].$$

Hence, for $i \geq 2$, we must have

$$g_i'(x)f(g_i(x)) \leq g_i'(0)f(g_i(0))$$
$$\leq \frac{g_i'(0)}{g_i(0)} = \frac{g_i'(0)}{a_{i-1}}, \qquad i = 2, \dots, r.$$

This formula is not applicable when $i = 1$ since $a_0 = 0$. However, we do have

$$g_1'(x)f(g_1(x)) \leq g_1'(0)f(0).$$

Combining these two results with equation (6.3.1) for P, we can write

$$Pf(x) \leq g_1'(0)f(0) + \sum_{i=2}^r g_i'(0)/a_{i-1}.$$

Set

$$S'(0) = 1/g_1'(0) = \lambda > 1$$

and

$$\sum_{i=2}^r g_i'(0)/a_{i-1} = M$$

so

$$Pf(x) \leq (1/\lambda)f(0) + M.$$

Proceeding inductively, we therefore have

$$P^n f(x) \leq (1/\lambda^n)f(0) + \lambda M/(\lambda - 1)$$
$$\leq f(0) + \lambda M/(\lambda - 1).$$

Thus, for decreasing $f \in D([0,1])$, since $f(0) < \infty$ the sequence $\{P^n f\}$ is bounded above by a constant. From Corollary 5.2.1 we therefore know that there is a density, $f_* \in D$ such that $Pf_* = f_*$, and by Theorem 4.1.1 the measure μ_{f_*} is invariant. ∎

Example 6.3.1. In the experimental study of fluid flow it is commonly observed that for Reynolds numbers R less than a certain value, R_L, strictly laminar flow occurs; for Reynolds numbers greater than another value, R_T, continuously turbulent flow occurs. For Reynolds numbers satisfying $R_L < R < R_T$, a transitional type behavior (**intermittency**) is found. Intermittency is characterized by alternating periods of laminar and turbulent flow, each of a variable and apparently unpredictable length.

Intermittency is also observed in mathematical models of fluid flow, for example, the Lorenz equations [Manneville and Pomeau, 1979]. Manneville [1980] argues that, in the parameter ranges where intermittency occurs in the Lorenz equations, the model behavior can be approximated by the transformation $S: [0,1] \to [0,1]$ given by

$$S(x) = (1 + \varepsilon)x + (1 - \varepsilon)x^2 \quad (\text{mod } 1) \qquad (6.3.2)$$

with $\varepsilon > 0$, where x corresponds to a normalized fluid velocity. This transformation clearly satisfies all of the properties of Theorem 6.2.1 for $0 < \varepsilon < 2$ and is thus exact.

The utility of equation (6.3.2) in the study of intermittency stems from the fact that $x = 0$ is a strong repellor. From Remark 6.3.1 is it clear that any transformation S satisfying conditions (3i)–(3iv) will serve equally well in this approach to the intermittency problem. Exactly this point of view has been adopted by Procaccia and Schuster [1983] in their heuristic treatment of noise spectra in dynamical systems. □

6.4 Asymptotically Periodic Transformations

In order to prove the asymptotic stability of $\{P^n\}$ in the two preceding sections, we were forced to consider transformations S with very special properties. Thus, for every subinterval of the partition of $[0,1]$, we used either $S((a_{i-1}, a_i)) = (0,1)$ or $S(a_{i-1}) = 0$. Eliminating either or both of these requirements may well lead to the loss of asymptotic stability of $\{P^n\}$, as is illustrated in the following example.

Let $S: [0,1] \to [0,1]$ be defined by

$$S(x) = \begin{cases} 2x & \text{for } x \in [0, \frac{1}{4}) \\ 2x - \frac{1}{2} & \text{for } x \in [\frac{1}{4}, \frac{3}{4}) \\ 2x - 1 & \text{for } x \in [\frac{3}{4}, 1], \end{cases}$$

as shown in Figure 6.4.1. Examination of the figure shows that the Borel measure is invariant since $S^{-1}([0, x])$ always consists of two intervals whose

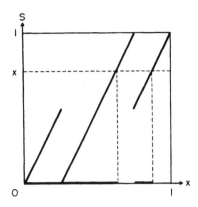

FIGURE 6.4.1. An example showing that piecewise monotonic transformation that is not onto might not even be ergodic. (See the text for details.)

union has measure x. However, S is obviously not exact and, indeed, is not even ergodic since $S^{-1}([0, \frac{1}{2}]) = [0, \frac{1}{2}]$ and $S^{-1}([\frac{1}{2}, 1]) = [\frac{1}{2}, 1]$. S that is restricted to either $[0, \frac{1}{2}]$ or $[\frac{1}{2}, 1]$ behaves like the dyadic transformation.

The loss of asymptotic stability by $\{P^n\}$ may, under certain circumstances, be replaced by the asymptotic periodicity of $\{P^n\}$. To see this, consider a mapping $S: [0, 1] \to [0, 1]$ satisfying the following three conditions:

(4i) There is a partition $0 = a_0 < a_1 < \cdots < a_r = 1$ of $[0, 1]$ such that for each integer $i = 1, \ldots, r$ the restriction of S to (a_{i-1}, a_i) is a C^2 function;

(4ii) $|S'(x)| \geq \lambda > 1, \qquad x \neq a_i, i = 1, \ldots, r;$ $\hfill$ (6.4.1)

(4iii) There is a real constant c such that

$$\frac{|S''(x)|}{[S'(x)]^2} \leq c < \infty, \qquad x \neq a_i, i = 0, \ldots, r. \qquad (6.4.2)$$

An example of a transformation satisfying these conditions is shown in Figure 6.4.2.

We now state the following theorem.

Theorem 6.4.1. *Let $S: [0, 1] \to [0, 1]$ satisfy conditions (4i)–(4iii) and let P be the Frobenius–Perron operator associated with S. Then, for all $f \in D$, $\{P^n f\}$ is asymptotically periodic.*

Proof. We first construct the Frobenius–Perron operator corresponding to

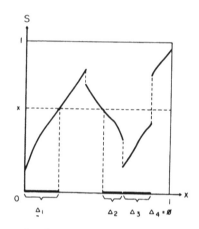

FIGURE 6.4.2. An example of a transformation on $[0,1]$ satisfying the conditions of Theorem 6.4.1.

S. For any $x \in [0,1]$, we have

$$S^{-1}((0,x)) = \bigcup_{i=1}^{r} \Delta_i(x)$$

where, for $I_i = S((a_{i-1}, a_i))$,

$$\Delta_i(x) = \begin{cases} (a_{i-1}, g_i(x)) & x \in I_i, g_i' > 0 \\ (g_i(x), a_i) & x \in I_i, g_i' < 0 \\ \emptyset \text{ or } (a_{i-1}, a_i) & x \notin I_i, \end{cases}$$

and, as before, $g_i = S_{(i)}^{-1}$ and $S_{(i)}$ denotes the restriction of S to (a_{i-1}, a_i). Therefore,

$$Pf(x) = \frac{d}{dx} \int_{S^{-1}([0,x])} f(u)\, du$$

$$= \sum_{i=1}^{r} \frac{d}{dx} \int_{\Delta_i(x)} f(u)\, du, \qquad (6.4.3)$$

where

$$\frac{d}{dx} \int_{\Delta_i(x)} f(u)\, du = \begin{cases} g_i'(x) f(g_i(x)), & x \in I_i, g_i' > 0 \\ -g_i'(x) f(g_i(x)), & x \in I_i, g_i' < 0 \\ 0 & x \notin I_i. \end{cases} \qquad (6.4.4)$$

The right-hand side of equation (6.4.3) is not defined on the set of end points of the intervals I_i, $S(a_{i-1})$, and $S(a_i)$. However, this set is finite and thus of measure zero. Since a function representing Pf that is an element of L^1 is defined up to a set of measure zero we neglect these end points.

Equation (6.4.4) may be rewritten as

$$\frac{d}{dx} \int_{\Delta_i(x)} f(u)\, du = \sigma_i(x) f(g_i(x)) 1_{I_i}(x),$$

where $\sigma_i(x) = |g_i'(x)|$ and $1_{I_i}(x)$ is the characteristic function of the interval I_i. Thus (6.4.3) may be written as

$$Pf(x) = \sum_{i=1}^{r} \sigma_i(x) f(g_i(x)) 1_{I_i}(x). \tag{6.4.5}$$

Equation (6.4.5) for the Frobenius–Perron operator is made more complicated than those in Sections 6.2 and 6.3 by the presence of the characteristic functions $1_{I_i}(x)$. The effect of these is such that even when a completely smooth initial function $f \in L^1$ is chosen, Pf and all subsequent iterates of f may be discontinuous. As a consequence we do not have simple criteria, such as decreasing functions, to examine the behavior of $P^n f$. Thus we must examine the variation of $P^n f$.

We start by examining the variation of Pf as given by equation (6.4.5). Let a function $f \in D$ be of bounded variation on $[0,1]$. From property (V1) of Section 6.1, the Yorke inequality (V5), and equation (6.4.5),

$$\bigvee_0^1 Pf(x) \le \sum_{i=1}^{r} \bigvee_0^1 [\sigma_i(x) f(g_i(x)) 1_{I_i}(x)]$$

$$\le 2 \sum_{i=1}^{r} \bigvee_{I_i} [\sigma_i(x) f(g_i(x))]$$

$$+ \sum_{i=1}^{r} \frac{2}{|I_i|} \int_{I_i} \sigma_i(x) f(g_i(x))\, dx. \tag{6.4.6}$$

Further, by property (V4),

$$\bigvee_{I_i} [\sigma_i(x) f(g_i(x))] \le \sup \sigma_i(x) \bigvee_{I_i} f(g_i(x)) + \int_{I_i} |\sigma_i'(x)| f(g_i(x))\, dx.$$

Because, from the inverse function theorem, we have $\sigma_i \le 1/\lambda$ and $|\sigma_i'| \le c\sigma_i$, the preceding inequality becomes

$$\bigvee_{I_i} [\sigma_i(x) f(g_i(x))] \le \frac{1}{\lambda} \bigvee_{I_i} f(g_i(x)) + c \int_{I_i} \sigma_i(x) f(g_i(x))\, dx,$$

and, thus, (6.4.6) becomes

$$\bigvee_0^1 Pf(x) \le \frac{2}{\lambda} \sum_{i=1}^r \bigvee_{I_i} f(g_i(x))$$

$$+ 2 \sum_{i=1}^r \left[c + \frac{1}{|I_i|} \right] \int_{I_i} \sigma_i(x) f(g_i(x)) \, dx. \qquad (6.4.7)$$

Define a new variable $y = g_i(x)$ for the integral in (6.4.7) and use property (V3) for the first term to give

$$\bigvee_0^1 Pf(x) \le \frac{2}{\lambda} \sum_{i=1}^r \bigvee_{a_i-1}^{a_i} f + 2 \sum_{i=1}^r \left[c + \frac{1}{|I_i|} \right] \int_{a_i-1}^{a_i} f(y) \, dy.$$

Set $L = \max_i 2(c + 1/|I_i|)$ and use property (V2) to rewrite this last inequality as

$$\bigvee_0^1 Pf(x) \le \frac{2}{\lambda} \bigvee_0^1 f + L \int_0^1 f(y) \, dy = \frac{2}{\lambda} \bigvee_0^1 f + L \qquad (6.4.8)$$

since $f \in D([0,1])$.

By using an induction argument with inequality (6.4.8), we have

$$\bigvee_0^1 P^n f \le \left(\frac{2}{\lambda} \right)^n \bigvee_0^1 f + L \sum_{j=0}^{n-1} \left(\frac{2}{\lambda} \right)^j. \qquad (6.4.9)$$

Thus, if $\lambda > 2$, then

$$\bigvee_0^1 P^n f \le \left(\frac{2}{\lambda} \right)^n \bigvee_0^1 f + \frac{\lambda L}{\lambda - 2}$$

and, therefore, for every $f \in D$ of bounded variation,

$$\lim_{n \to \infty} \sup \bigvee_0^1 P^n f < K, \qquad (6.4.10)$$

where $K > \lambda L/(\lambda - 2)$ is independent of f.

Now let the set $\mathcal{F}$ be defined by

$$\mathcal{F} = \left\{ g \in D : \bigvee_0^1 g \le K \right\}.$$

From (6.4.10) it follows that $P^n f \in \mathcal{F}$ for a large enough n and, thus, $\{P^n f\}$ converges to $\mathcal{F}$ in the sense that $\lim_{n \to \infty} \inf_{P^n f \in \mathcal{F}} \| P^n f - g \| = 0$. We want to show that $\mathcal{F}$ is weakly precompact. From the definition of the variation, it is clear that, for any positive function g defined on $[0,1]$,

$$g(x) - g(y) \le \bigvee_0^1 g$$

for all $x, y \in [0, 1]$. Since $g \in D$, there is some $y \in [0, 1]$ such that $g(y) \leq 1$ and, thus,

$$g(x) \leq K + 1.$$

Hence, by criterion 1 of Section 5.1, $\mathcal{F}$ is weakly precompact. (Actually, it is strongly precompact, but we will not use this fact.) Since $\mathcal{F}$ is weakly precompact, then P is constrictive by Proposition 5.3.1. Finally, by Theorem 5.3.1, $\{P^n f\}$ is asymptotically periodic and the theorem is proved when $\lambda > 2$.

To see that the theorem is also true for $\lambda > 1$, consider another transformation $\tilde{S} \colon [0, 1] \to [0, 1]$ defined by

$$\tilde{S}(x) = S \circ \overset{q}{\cdots} \circ S(x) = S^q(x). \tag{6.4.11}$$

Let q be the smallest integer such that $\lambda^q > 2$ and set $\tilde{\lambda} = \lambda^q$. It is easy to see that $\tilde{S}$ satisfies conditions (4i)–(4ii). By the chain rule,

$$|\tilde{S}'(x)| \geq (\inf |S'(x)|)^q \geq \lambda^q = \tilde{\lambda} > 2.$$

Thus, by the preceding part of the proof, $\{\tilde{P}^n\}$ satisfies

$$\lim_{n \to \infty} \sup \bigvee_0^1 \tilde{P}^n f < \tilde{K}$$

for every $f \in D$ of bounded variation, where the constant $\tilde{K}$ is independent of f. Write an integer n in the form $n = mq + s$, where the remainder s satisfies $0 \leq s < q$. Take m sufficiently large, $m > m_0$, so that

$$\bigvee_0^1 \tilde{P}^m f \leq \tilde{K}, \qquad m > m_0.$$

Now, using inequality (6.4.9), we have

$$\bigvee_0^1 P^n f = \bigvee_0^1 P^s(\tilde{P}^m f) \leq \left(\frac{2}{\lambda}\right)^s \bigvee_0^1 \tilde{P}^m f + L \sum_{j=0}^{q-1} \left(\frac{2}{\lambda}\right)^j$$

$$\leq \tilde{K} \sup_{0 \leq j \leq q-1} \left(\frac{2}{\lambda}\right)^j + L \sum_{j=0}^{q-1} \left(\frac{2}{\lambda}\right)^j, \qquad n \geq (m_0 + 1)q.$$

Thus, for n sufficiently large, the variation of $P^n f$ is bounded by a constant independent of f and the proof proceeds as before. ∎

Remark 6.4.1. From the results of Kosjakin and Sandler [1972] or Li and Yorke [1978a], it follows that transformations S satisfying the assumptions of Theorem 6.4.1 are ergodic if $r = 2$. □

Example 6.4.1. In this example we consider one of the simplest heuristic models for the effects of periodic modulation of an autonomous oscillator [Glass and Mackey, 1979].

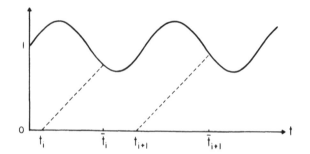

FIGURE 6.4.3. The periodic threshold $\theta(t)$ is shown as a solid curved line, and the activity $x(t)$ as dashed lines. (See Example 6.4.1 for further details.)

Consider a system (see Figure 6.4.3) whose activity $x(t)$ increases linearly from a starting time t_i until it reaches a periodic threshold $\theta(t)$ at time $\bar{t}_i$:

$$x(\bar{t}_i) = \theta(\bar{t}_i). \tag{6.4.12}$$

We take

$$x(t) = \lambda(t - t_i) \quad \text{and} \quad \theta(t) = 1 + \phi(t),$$

where ϕ is a continuous periodic function with period 1 whose amplitude satisfies

$$1 \geq \sup \phi(t) = -\inf \phi(t) = K \geq 0.$$

When the activity reaches threshold it instantaneously resets to zero, and the process begins anew at the starting time,

$$t_{i+1} = \bar{t}_i + \gamma^{-1} x(\bar{t}_i). \tag{6.4.13}$$

In (6.4.13), $\bar{t}_i$ is an implicit function of t_i given by (6.4.12) or by

$$\lambda(\bar{t}_i - t_i) = 1 + \phi(\bar{t}_i). \tag{6.4.14}$$

Equation (6.4.14) has exactly one smallest solution $\bar{t}_i \geq t_i$ for every $t_i \in R$. We wish to examine the behavior of the starting times t_i. Set

$$F(t_i) = \bar{t}_i(t_i) + \gamma^{-1} x(\bar{t}_i(t_i))$$

so that the transformation

$$S(t) = F(t) \quad \pmod{1} \tag{6.4.15}$$

gives the connection between successive starting times.

Many authors have considered the specific cases of $\phi(t) = K \sin 2\pi t$, $\gamma^{-1} = 0$, so $\bar{t}_i = t_{i+1}$ and, thus, t_{i+1} is given implicitly by

$$\lambda(t_{i+1} - t_i) = 1 + K \sin 2\pi t_{i+1}.$$

Here, to illustrate the application of the material of this and previous sections, we restrict ourselves to the simpler situation in which $\phi(t)$ is a piecewise linear function of t and θ given by

$$\theta(t) = \begin{cases} 4Kt + 1 - K & t \in \left[0, \frac{1}{2}\right] \\ 4K(1 - t) + 1 - K & t \in \left(\frac{1}{2}, 1\right]. \end{cases}$$

The calculation of $F(t)$ depends on the sign of $\lambda - 4K$. For example, if $\lambda > 4K$, a simple computation shows that

$$F(t) = \begin{cases} \frac{1+\alpha}{1-\beta}t + \left(\frac{1}{\lambda} + \frac{1}{\gamma}\right)\frac{1-K}{1-\beta} & t \in \left[-a, \frac{1}{2}(1 - \beta) - a\right] \\ \frac{1-\alpha}{1+\beta}t + \left(\frac{1}{\lambda} + \frac{1}{\gamma}\right)\frac{1+3K}{1+\beta} & t \in \left(\frac{1}{2}(1 - \beta) - a, 1 - a\right], \end{cases} \tag{6.4.16}$$

where $\alpha = 4K/\gamma$, $\beta = 4K/\lambda$, and $a = (1 - K)/\lambda$.

Since $0 \leq \beta < 1$, it is clear that $F'(t) > 1$ for all $t \in [-a, \frac{1}{2}(1 - \beta) - a]$. However, if $(1 - \alpha)/(1 + \beta) < -1$, then $|S'(t)| > 1$ for all t and $\{P^n\}$ is asymptotically periodic by Theorem 6.4.1. Should it happen in this case that S is onto for every subinterval of the partition, then $\{P^n\}$ is asymptotically stable by Theorem 6.2.2.

Despite the obvious simplifications in such models they have enjoyed great popularity in neurobiology: the "integrate and fire" model [Knight, 1972a,b]; in respiratory physiology, the "inspiratory off switch" model [Petrillo and Glass, 1984]; in cardiac electrophysiology, the "circle model" [Guevara and Glass, 1982]; and in cell biology, the "mitogen" model [Kauffman, 1974; Tyson and Sachsenmaier, 1978].   □

Example 6.4.2. An interesting problem arises in the rotary drilling of rocks. Usually the drilling tool is in the form of a toothed cone (mass M and radius R) that rotates on the surface of the rock with tangential velocity u. At rest the tool exerts a pressure Q on the rock. In practice it is found that, for sufficiently large tool velocities, after each impact of a tooth with the rock the tool rebounds before the next blow. The energy of each impact, and thus the efficiency of the cutting process, is a function of the angle at which the impact occurs.

Let x be the normalized impact angle that is in the interval $[0, 1]$. Lasota and Rusek [1974] have shown that the next impact angle is given by the transformation $S: [0, 1] \to [0, 1]$ defined by

$$S(x) = x + aq(x) - \sqrt{[aq(x)]^2 + 2axq(x) - aq(x)[1 + q(x)]} \pmod{1}, \tag{6.4.17}$$

where

$$q(x) = 1 + \text{int}[(1 - 2x)/(\alpha - 1)];$$

$\text{int}(y)$ denotes the integer part of y, namely, the largest integer smaller than or equal to y, and

$$\alpha = F/(F - 1),$$

where
$$F = Mu^2/QR$$

is Freude's number, the ratio of the kinetic and potential energies.

The Freude number F contains all of the important parameters characterizing this process. It is moderately straightforward to show that with $\tilde{S} = S \circ S$, $|\tilde{S}'(x)| > 1$ if $F > 2$. However, the transformation (6.4.17) is not generally onto, so that by Theorem 6.4.1 the most that we can say is that for $F > 2$, if P is the Frobenius–Perron operator corresponding to S then $\{P^n\}$ is asymptotically periodic. However, it seems natural to expect that $\{P^n\}$ is in fact asymptotically stable. This prediction is supported experimentally, because, once $u > (2QR/M)^{1/2}$, there is a transition from smooth cutting to extremely irregular behavior (chattering) of the tool. □

Example 6.4.3. Kitano, Yabuzaki, and Ogawa [1983] experimentally examined the dynamics of a simple, nonlinear, acoustic feedback system with a time delay. A voltage x, the output of an operational amplifier with response time γ^{-1}, is fed to a speaker. The resulting acoustic signal is picked up by a microphone after a delay τ (due to the finite propogation velocity of sound waves), passed through a full-wave rectifier, and then fed back to the input of the operational amplifier.

Kitano and co-workers have shown that the dynamics of this system are described by the delay-differential equation

$$\gamma^{-1}\dot{x}(t) = -x(t) + \mu F(x(t-\tau)), \tag{6.4.18}$$

where

$$F(x) = -\left|x + \tfrac{1}{2}\right| + \tfrac{1}{2} \tag{6.4.19}$$

is the output of the full-wave rectifier with an input x, and μ is the circuit loop gain.

In a series of experiments, Kitano et al. found that increasing the loop gain μ above 1 resulted in very complicated dynamics in x, whose exact nature depends on the value of $\gamma\tau$. To understand these behaviors they considered the one-dimensional difference equation,

$$x_{n+1} = \mu F(x_n),$$

derived from expressions (6.4.18) and (6.4.19) as $\gamma^{-1} \to 0$. In our notation this is equivalent to the map $T: [-\mu/(\mu-1), \mu/2] \to [-\mu/(\mu-1), \mu/2]$, defined by

$$T(x) = \begin{cases} \mu(1+x) & \text{for } x \in \left[-\frac{\mu}{\mu-1}, -\frac{1}{2}\right] \\ -\mu x & \text{for } x \in \left(-\frac{1}{2}, \frac{\mu}{2}\right]. \end{cases} \tag{6.4.20}$$

for $1 < \mu \le 2$. Make the change of variables

$$x \to -\frac{\mu}{\mu - 1} + x'\frac{\mu}{2}\frac{\mu + 1}{\mu - 1}$$

so that (6.4.20) is equivalent to the transformation $S: [0, 1] \to [0, 1]$, defined by

$$S(x') = \begin{cases} \mu x' & \text{for } x' \in [0, 1/\mu] \\ 2 - \mu x' & \text{for } x' \in (1/\mu, 1]. \end{cases} \tag{6.4.21}$$

For $1 < \mu \le 2$, the transformation S defined by (6.4.21) satisfies all the conditions of Theorem 6.4.1, and S is thus asymptotically periodic. If $\mu = 2$, then, by Theorem 6.2.2, S is statistically stable. Furthermore, from Remark 6.4.1 it follows that S is ergodic for $1 < \mu < 2$ and will, therefore, exhibit disordered dynamical behavior. This is in agreement with the experimental results.   □

Remark 6.4.2. As we have observed in the example of Figure 6.4.1, piecewise monotonic transformations satisfying properties (4i)–(4iii) may not have a unique invariant measure. If the transformation is ergodic, and the invariant measure is thus unique by Theorem 4.2.2, then the invariant measure has many interesting properties. For example, in this case Kowalski [1976] has shown that the invariant measure is continuously dependent on the transformation.   □

6.5   Change of Variables

In the three preceding sections, we have examined transformations $S: [0, 1] \to [0, 1]$ with very restrictive conditions on the derivatives $S'(x)$ and $S''(x)$. However, most transformations do not satisfy these conditions. A good example is the quadratic transformation,

$$S(x) = 4x(1 - x), \qquad \text{for } x \in [0, 1].$$

For this transformation, $S'(x) = 4 - 8x$, and $|S'(x)| < 1$ for $x \in \left(\frac{3}{8}, \frac{5}{8}\right)$. Furthermore, $|S''(x)/[S'(x)]^2| = \frac{1}{2}(1 - 2x)^{-2}$, which is clearly not bounded at $x = \frac{1}{2}$. However, iteration of any initial density on $[0, 1]$ indicates that the iterates rapidly approach the same density (Figure 1.2.2), leading one to suspect that, for the quadratic transformation, $\{P^n\}$ is asymptotically stable.

In this section we show how, by a change of variables, we can sometimes utilize the results of the previous sections to prove asymptotic stability. The idea is originally due to Ruelle [1977] and Pianigiani [1983].

Theorem 6.5.1. *Let $S: [0, 1] \to [0, 1]$ be a transformation satisfying properties (2i)' and (2ii)' of Section 6.2, and P_S be the Frobenius–Perron operator* corresponding to S. If there exists an a.e. positive C^1 function $\phi \in L^1([0,1])$ such that, for some real λ and c,

$$p(x) \equiv \frac{|S'(x)|\phi(S(x))}{\phi(x)} \geq \lambda > 1, \qquad 0 < x < 1 \qquad (6.5.1)$$

and

$$\left| \frac{1}{\phi(x)} \frac{d}{dx} \left(\frac{1}{p(x)} \right) \right| \leq c < \infty, \qquad 0 < x < 1, \qquad (6.5.2)$$

then $\{P_S^n\}$ is asymptotically stable.

Proof. Set

$$g(x) = \frac{1}{\|\phi\|} \int_0^x \phi(u)\, du, \qquad \text{for } x \in [0,1], \qquad (6.5.3)$$

and consider a new transformation T defined by

$$T(x) = g(S(g^{-1}(x))) \qquad (6.5.4)$$

with associated Frobenius–Perron operator P_T. From (6.5.4), $T(g(x)) = g(S(x))$, so

$$\frac{dT}{dg} \frac{dg}{dx} = \frac{dg}{dS} \frac{dS}{dx}$$

or

$$T'(g(x)) = \frac{g'(S(x))S'(x)}{g'(x)}.$$

Using (6.5.3) this may be rewritten as

$$T'(g(x)) = \frac{S'(x)\phi(S(x))}{\phi(x)}.$$

Hence, by (6.5.1), we have $|T'(g)| \geq \lambda > 1$. Further, by comparing this equation with (6.5.1), we see that $p(x) = |T'(g(x))|$. It follows that

$$\frac{1}{\phi(x)} \frac{d}{dx} \left(\frac{1}{p(x)} \right) = \frac{1}{\phi(x)} \frac{d}{dx} \left(\frac{1}{|T'(g)|} \right) = \pm \frac{T''(g)}{[T'(g)]^2 \|\phi\|},$$

so that, from inequality (6.5.2),

$$\left| \frac{T''(g)}{[T'(g)]^2} \right| \leq c\|\phi\| < \infty.$$

Thus the new transformation T satisfies all the conditions of Theorem 6.2.2, and $\{P_T^n\}$ is asymptotically stable as is $\{P_S^n\}$ by (6.5.14).   ■

Example 6.5.1. Consider the quadratic transformation $S(x) = 4x(1-x)$ with $x \in [0,1]$ and set

$$\phi(x) = \frac{1}{\pi\sqrt{x(1-x)}}. \qquad (6.5.5)$$

Using equations (6.5.3) and (6.5.5), it is easy to verify that all the conditions of Theorem 6.5.1 are satisfied in this case and, thus, for the quadratic transformation, $\{P^n\}$ is asymptotically stable.

Note that with ϕ as given by (6.5.5), the associated function g, as defined by (6.5.3), is given by

$$g(x) = \frac{1}{\pi} \int_0^x \frac{du}{\sqrt{u(1-u)}} = \frac{1}{2} - \frac{1}{\pi} \sin^{-1}(1-2x), \qquad (6.5.6)$$

and thus

$$g^{-1}(x) = \tfrac{1}{2} - \tfrac{1}{2}\cos(\pi x). \qquad (6.5.7)$$

Hence, when $S(x) = 4x(1-x)$, the transformation $T: [0,1] \to [0,1]$, defined by

$$T(x) = g \circ S \circ g^{-1}(x), \qquad (6.5.8)$$

is easily shown to be

$$T(x) = \begin{cases} 2x & \text{for } x \in \left[0, \tfrac{1}{2}\right) \\ 2(1-x) & \text{for } x \in \left[\tfrac{1}{2}, 1\right]. \end{cases} \qquad (6.5.9)$$

[The transformation defined by (6.5.9) is often referred to as the **tent map** or **hat map**.] The Frobenius–Perron operator, P_T, corresponding to T is given by

$$P_T f(x) = \tfrac{1}{2} f\left(\tfrac{1}{2}x\right) + \tfrac{1}{2} f\left(1 - \tfrac{1}{2}x\right),$$

and, by Theorem 6.2.2, $\{P_T^n\}$ is asymptotically stable. Furthermore, it is clear that $f_* \equiv 1$ is the unique stationary density of P_T, so T is, in fact, exact by Theorem 4.4.1. Reversing the foregoing procedure by constructing a transformation $S = g^{-1} \circ T \circ g$ from T given by (6.5.9) and from g, g^{-1} given by equations (6.5.6) and (6.5.7) yields the transformation $S(x) = 4x(1-x)$. From this $\{P_S^n\}$ is asymptotically stable, and ϕ, given by (6.5.5), is the stationary density of P_S. $\square$

These comments illustrate the construction of a statistically stable transformation S with a *given* stationary density from an exact transformation T. Clearly, the use of a different exact transformation T_1 will yield a different statistically stable transformation S_1, but one that has the same stationary density as S. Thus we are led to the next theorem.

Theorem 6.5.2. *Let* $T: (0,1) \to (0,1)$ *be a measurable, nonsingular transformation and let* $\phi \in D((a,b))$, *with* a *and* b *finite or not, be a given positive density, that is,* $\phi > 0$ *a.e. Let a second transformation* $S: (a,b) \to (a,b)$ *be given by* $S = g^{-1} \circ T \circ g$, *where*

$$g(x) = \int_a^x \phi(y)\, dy, \qquad a < x < b. \qquad (6.5.10)$$

Then T is exact if and only if S is statistically stable and ϕ is the density of the measure invariant with respect to S.

Proof. Let P_T and P_S be Frobenius–Perron operators corresponding to the transformations T and S, respectively. We start with the derivation of the relation between P_T and P_S. By the definition of P_S, we have

$$\int_a^y P_S f(x)\, dx = \int_{S^{-1}((a,y))} f(x)\, dx, \qquad \text{for } f \in L^1((a,b)),$$

where $S^{-1}((a,y)) = g^{-1}(T^{-1}(g(a), g(y)))$. Set $x = g^{-1}(z)$ and use equation (6.5.10) to change the variables so that the last integral may be rewritten to give

$$\int_a^y P_S f(x)\, dx = \int_{T^{-1}(g(a),g(y))} f(g^{-1}(z)) \frac{dz}{\phi(g^{-1}(z))}.$$

Defining

$$P_g f(x) = \frac{f(g^{-1}(x))}{\phi(g^{-1}(x))}, \qquad \text{for } f \in L^1((a,b)), \tag{6.5.11}$$

we have

$$\int_a^y P_S f(x)\, dx = \int_{T^{-1}(g(a),g(y))} P_g f(z)\, dz = \int_{g(a)}^{g(y)} P_T P_g f(z)\, dz. \tag{6.5.12}$$

Setting

$$P_g^{-1} \tilde{f}(x) = \tilde{f}(g(x))\phi(x), \qquad \text{for } \tilde{f} \in L^1((0,1)), \tag{6.5.13}$$

and substituting $z = g(x)$ in the last integral in (6.5.12) yields

$$\int_a^y P_S f(x)\, dx = \int_a^y P_g^{-1} P_T P_g f(x)\, dx.$$

Thus P_S and P_T are related by

$$P_S f = P_g^{-1} P_T P_g f, \qquad \text{for } f \in L^1((a,b)). \tag{6.5.14}$$

By integrating equation (6.5.11) over the entire space, we have

$$\|P_g f\|_{L^1(0,1)} = \|f\|_{L^1(a,b)}, \qquad \text{for } f \in L^1((a,b)).$$

Further, P_g^{-1}, as given by (6.5.13), is the inverse operator to P_g, and integration of (6.5.13) gives

$$\|P_g^{-1} \tilde{f}\|_{L^1(a,b)} = \|\tilde{f}\|_{L^1(0,1)}, \qquad \text{for } \tilde{f} \in L^1((0,1)). \tag{6.5.15}$$

If T is measure preserving, we have $P_T 1 = 1$. Furthermore, from the definition of P_g in (6.5.11), we have $P_g \phi = 1$. As a consequence

$$P_S \phi = P_g^{-1} P_T P_g \phi = P_g^{-1} P_T 1 = P_g^{-1} 1 = \phi,$$

which shows that ϕ is the density of the measure invariant with respect to S. Analogously from $P_S \phi = \phi$, it follows that $P_T 1 = 1$. By using an induction argument with equation (6.5.14), we obtain

$$P_S^n f = P_g^{-1} P_T^n P_g f, \qquad \text{for } f \in L^1((a,b)).$$

This, in conjunction with (6.5.15) and the equality $P_g \phi = 1$ gives

$$\|P_S^n f - \phi\|_{L^1(a,b)} = \|P_g^{-1} P_T^n P_g f - P_g^{-1} P_g \phi\|_{L^1(a,b)}$$
$$= \|P_T^n P_g f - P_g \phi\|_{L^1(0,1)} = \|P_T^n P_g f - 1\|_{L^1(0,1)} \qquad (6.5.16)$$

By substituting

$$f = P_g^{-1} \tilde{f}, \qquad \text{for } \tilde{f} \in L^1((0,1)),$$

into (6.5.16), we have

$$\|P_S^n P_g^{-1} \tilde{f} - \phi\|_{L^1(a,b)} = \|P_T^n \tilde{f} - 1\|_{L^1(0,1)}. \qquad (6.5.17)$$

Thus, from equations (6.5.16) and (6.5.17), it follows that the strong convergence of $\{P_S^n f\}$ to ϕ for $f \in D((a,b))$ is equivalent to the strong convergence of $\{P_T^n \tilde{f}\}$ to 1 for $\tilde{f} \in D((0,1))$. ∎

Example 6.5.2. Let T be the hat transformation of (6.5.9) and pick $\phi(x) = k \exp(-kx)$ for $0 < x < \infty$, which is the density distribution function for the lifetime of an atom with disintegration constant $k > 0$. Then it is straightforward to show that the transformation $S = g^{-1} \circ T \circ g$ is given by

$$S(x) = \ln \left\{ \frac{1}{|1 - 2e^{-kx}|^{1/k}} \right\}.$$

The Frobenius–Perron operator associated with S is given by

$$P_S f(x) = \frac{e^{-kx}}{1 + e^{-kx}} f\left(\frac{1}{k} \ln \frac{2}{1 + e^{-kx}} \right) + \frac{e^{-kx}}{1 - e^{-kx}} f\left(\frac{1}{k} \ln \frac{2}{1 - e^{-kx}} \right).$$

By Theorem 6.5.2, $\{P_S^n\}$ is asymptotically stable with the stationary density $\phi(x) = k \exp(-kx)$. □

Example 6.5.3. As a second example, consider the Chebyshev polynomials $S_m : (-2, 2) \to (-2, 2)$,

$$S_m(x) = 2 \cos[m \cos^{-1}(x/2)], \qquad m = 0, 1, 2, \ldots .$$

Define

$$g(x) = \frac{1}{\pi} \int_{-2}^{x} \frac{du}{\sqrt{4 - u^2}}$$

corresponding to the density

$$\phi(x) = \frac{1}{\pi\sqrt{4 - x^2}}. \qquad (6.5.18)$$

The Chebyshev polynomials satisfy $S_{m+1}(x) = xS_m(x) - S_{m-1}(x)$ with $S_0(x) = 2$ and $S_1(x) = x$. It is straightforward, but tedious, to show that the transformation $T_m = g \circ S_m \circ g^{-1}$ is given by

$$T_m(x) = \begin{cases} m\left(x - \frac{2n}{m}\right) & \text{for } x \in \left[\frac{2n}{m}, \frac{2n+1}{m}\right) \\ m\left(\frac{2n+2}{m} - x\right) & \text{for } x \in \left[\frac{2n+1}{m}, \frac{2n+2}{m}\right), \end{cases} \qquad (6.5.19)$$

where $n = 0, 1, \ldots, [(m-1)/2]$, and $[y]$ denotes the integer part of y. For $m \geq 2$, by Theorem 6.2.2, $\{P_{T_m}^n\}$ is asymptotically stable. An explicit computation is easy and shows that $f_* \equiv 1$ is the stationary density of P_{T_m}. Thus T_m is exact. Hence, by Theorem 6.5.2, the Chebyshev polynomials S_m are statistically stable for $m \geq 2$ with a stationary density given by equation (6.5.18). This may also be proved more directly as shown by Adler and Rivlin [1964].

This example is of interest from several standpoints. First, it illustrates in a concrete way the nonuniqueness of statistically stable transformations (S_m) with the same stationary density derived from different exact transformations (T_m). Second, it should be noted that the transformation $\tilde{S}_m : (0, 1) \to (0, 1)$, given by

$$\tilde{S}_m(x) = -\tfrac{1}{4}S_m(4x - 2) + \tfrac{1}{2}$$

when $m = 2$, is just the familiar parabola, $\tilde{S}_2(x) = 4x(1 - x)$. Finally, we note in passing that cubic maps equivalent to S_3 have arisen in a study of a simple genetic model involving one locus and two alleles [May, 1980] and have also been studied in their own right by Rogers and Whitley [1983]. $\square$

Example 6.5.4. As a further illustration of the power of Theorem 6.5.2, we consider an example drawn from quantum mechanics. Consider a particle of mass m free to move in the x direction and subjected to a restoring force, $-kx$. This is equivalent to the particle being placed in a potential $V(x) = kx^2/2$. The standard solution to this quantized harmonic oscillator problem is [Schiff, 1955]

$$u_n(x) = \left[\frac{\alpha}{\sqrt{\pi}\, 2^n n!}\right]^{1/2} H_n(\alpha x)e^{-(1/2)\alpha^2 x^2}, \qquad \text{for } n = 0, 1, \ldots,$$

where

$$\alpha^4 = mk/\hbar^2$$

($\hbar$ is Planck's constant) and $H_n(y)$ denotes the nth-order Hermite polynomial, defined recursively by

$$H_n(y) = (-1)^n e^{y^2} \frac{d^n}{dy^n} (e^{-y^2})$$

$[H_0(y) = 1, H_1(y) = 2y, H_2(y) = 4y^2 - 2, \ldots]$. In accord with the usual interpretation of quantum mechanics, the associated densities are given by $\phi_n(x) = [u_n(x)]^2$, or

$$\phi_n(x) = \frac{\alpha}{\sqrt{\pi} \, 2^n n!} H_n^2(\alpha x) e^{-\alpha^2 x^2}, \qquad \text{for } n = 0, 1, \ldots,$$

and the g_n are

$$g_n(x) = \frac{\alpha}{\sqrt{\pi} \, 2^n n!} \int_{-\infty}^{x} H_n^2(\alpha y) e^{-\alpha^2 y^2} \, dy, \qquad \text{for } n = 0, 1, \ldots.$$

Then for any exact transformation T, the transformations $S_n(x) = g_n^{-1} \circ T \circ g_n(x)$ have the requisite stationary densities ϕ. $\square$

To close this section we note that the following result is a direct extension of Theorem 6.5.2.

Corollary 6.5.1. Let $S: (a, b) \to (a, b)$, with a and b finite or not, be a statistically stable transformation with a stationary density $\phi \in D((a, b))$ and let $\tilde{\phi} \in D((\alpha, \beta))$ be given, with α and β also finite or not. Further, set

$$g(x) = \int_a^x \phi(y) \, dy \quad and \quad \tilde{g}(x) = \int_\alpha^x \tilde{\phi}(y) \, dy.$$

Then the transformation $\tilde{S}: (\alpha, \beta) \to (\alpha, \beta)$, defined by

$$\tilde{S} = \tilde{g}^{-1} \circ g \circ S \circ g^{-1} \circ \tilde{g},$$

is statistically stable with stationary density $\tilde{\phi}$.

Proof. First set $T: (0, 1) \to (0, 1)$ equal to $T = g \circ S \circ g^{-1}$. This is equivalent to $S = g^{-1} \circ T \circ g$ and, by Theorem 6.5.2, T is exact. Again, using Theorem 6.5.2 with the exactness of T, we have that $\tilde{S} = \tilde{g}^{-1} \circ T \circ \tilde{g}$ is statistically stable. ∎

Remark 6.5.1. Nonlinear transformations with a specified stationary density can be used as pseudorandom number generators. For details see Li and Yorke [1978]. $\square$

6.6   Transformations of the Real Line

All of the transformations considered in previous sections were defined on the interval $[0, 1]$. The particular choice of the interval $[0, 1]$ is not restrictive since, given $S: [a, b] \to [a, b]$, we can always consider $T(x) = Q^{-1}(S(Q(x)))$, $T: [0, 1] \to [0, 1]$, where $Q(x) = a + (b-a)x$. All of the asymptotic properties of S are the same as those of T.

However, if S maps the whole real line (or half-line) into itself, no linear change of variables is available to reduce this problem to an equivalent transformation on a finite interval. Further, transformations on the real line may have some anomalous properties. For example, the requirement that $|S'(x)| \geq \lambda > 1$ for $S: R \to R$ is not sufficient for the asymptotic stability of $\{P^n\}$. This is amply illustrated by the specific example $S(x) = 2x$, which was considered in Section 1.3.

There are, however, transformations on the real line for which the asymptotic stability of $\{P^n\}$ can be demonstrated; one example is $S(x) = \beta \tan(\gamma x + \delta)$, $|\beta\gamma| > 1$. This section will treat a class of such transformations.

Assume the transformation $S: R \to R$ satisfies the following conditions:

(6i)   There is a partition $\cdots a_{-2} < a_{-1} < a_0 < a_1 < a_2 \cdots$ of the real line such that, for every integer $i = 0, \pm 1, \pm 2, \ldots$, the restriction $S_{(i)}$ of S to the interval (a_{i-1}, a_i) is a C^2 function;

(6ii) $S((a_{i-1}, a_i)) = R$;

(6iii)   There is a constant $\lambda > 1$ such that $|S'(x)| \geq \lambda$ for $x \neq a_i$, $i = 0, \pm 1, \pm 2, \ldots$;

(6iv)   There is a constant $L \geq 0$ and a function $q \in L^1(R)$ such that

$$a_i - a_{i-1} \leq L, \qquad |g_i'(x)| \leq q(x)(a_i - a_{i-1}) \qquad (6.6.1)$$

where $g_i = S_{(i)}^{-1}$, for $i = 0, \pm 1, \ldots$; and

(6v)   There is a real constant c such that

$$\frac{|S''(x)|}{[S'(x)]^2} \leq c, \qquad \text{for } x \neq a_i, \ i = 0, \pm 1, \ldots. \qquad (6.6.2)$$

Then the following theorem summarizes results of Kemperman [1975], Schweiger [1978], Jabłoński and Lasota [1981], and Bugiel [1982].

Theorem 6.6.1. *If $S: R \to R$ satisfies conditions (6i)–(6v) and P is the associated Frobenius–Perron operator, then $\{P^n\}$ is asymptotically stable.*

Proof. We first calculate the Frobenius–Perron operator. To do this note that

$$S^{-1}((-\infty, x)) = \bigcup_j (a_{j-1}, g_j(x)) + \bigcup_k (g_k(x), a_k),$$

where the first union is over intervals in which g_i is an increasing function of x, and the second is for intervals in which g_i is decreasing. Thus

$$Pf(x) = \frac{d}{dx} \int_{S^{-1}((-\infty, x))} f(u)\, du$$

$$= \frac{d}{dx} \sum_j \int_{a_{j-1}}^{g_j(x)} f(u)\, du + \frac{d}{dx} \sum_k \int_{g_k(x)}^{a_k} f(u)\, du$$

or

$$Pf(x) = \sum_{i=-\infty}^{\infty} \sigma_i(x) f(g_i(x)), \qquad (6.6.3)$$

where $\sigma_i(x) = |g_i'(x)|$.

Having an expression for $Pf(x)$ we now calculate the variation of Pf to show that the sequence $f_n = P^n f$ satisfies assumptions (5.8.2) and (5.8.3) of Proposition 5.8.1. Denote by $D_0 \subset D(R)$ the set of all densities of bounded variation on R that are positive, continuously differentiable, and satisfy

$$|f'(x)| \le k_f f(x), \qquad \text{for } x \in R, \qquad (6.6.4)$$

where the constant k_f depends on f. Now

$$\bigvee_{-\infty}^{\infty} Pf(x) \le \sum_{i=-\infty}^{\infty} \bigvee_{-\infty}^{\infty} \sigma_i(x) f(g_i(x))$$

$$\le \sum_{i=-\infty}^{\infty} \left\{ \frac{1}{\lambda} \bigvee_{-\infty}^{\infty} f(g_i(x)) + \int_{-\infty}^{\infty} |\sigma_i'(x)| f(g_i(x))\, dx \right\}.$$

Using $|\sigma_i'(x)| \le c\sigma_i(x)$, which follows from inequality (6.6.2), and making the change of variables $y = g_i(x)$, we have

$$\bigvee_{-\infty}^{\infty} Pf(x) \le \sum_{i=-\infty}^{\infty} \left\{ \frac{1}{\lambda} \bigvee_{a_{i-1}}^{a_i} f(y) + c \int_{a_{i-1}}^{a_i} f(y)\, dy \right\}$$

$$\le \frac{1}{\lambda} \bigvee_{-\infty}^{\infty} f(y) + c.$$

By an induction argument, we obtain

$$\bigvee_{-\infty}^{\infty} P^n f(x) \le \frac{1}{\lambda^n} \bigvee_{-\infty}^{\infty} f(y) + \frac{\lambda c}{\lambda - 1}.$$

Since $\lambda > 1$, then, for real $\alpha > \lambda c/(\lambda - 1)$, there must exist a sufficiently large n, say $n > n_0(f)$, such that

$$\bigvee_{-\infty}^{\infty} P^n f(x) \leq \alpha. \tag{6.6.5}$$

Now we are in a position to evaluate $P^n f$. From inequalities (6.6.1) and (6.6.3), we have

$$Pf(x) \leq q(x) \sum_{i=-\infty}^{\infty} f(g_i(x))(a_i - a_{i-1}). \tag{6.6.6}$$

For every interval (a_{i-1}, a_i) pick a $z_i \in (a_{i-1}, a_i)$ such that

$$(a_i - a_{i-1})f(z_i) \leq \int_{a_{i-1}}^{a_i} f(x)\,dx, \qquad \text{for } i = 0, \pm 1, \ldots .$$

Thus, from (6.6.1) and (6.6.6), we obtain

$$Pf(x) \leq q(x) \sum_{i=-\infty}^{\infty} \left\{ L|f(g_i(x)) - f(z_i)| + \int_{a_{i-1}}^{a_i} f(x)\,dx \right\}$$

$$\leq Lq(x) \bigvee_{-\infty}^{\infty} f(x) + q(x) \int_{-\infty}^{\infty} f(x)\,dx$$

$$= q(x) \left\{ L \bigvee_{-\infty}^{\infty} f(x) + 1 \right\}.$$

By substituting $P^{n-1}f$ instead of f in this expression and using (6.6.5), we have

$$P^n f(x) \leq q(x)(\alpha L + 1), \qquad x \in R, n > n_0(f) + 1. \tag{6.6.7}$$

Thus the sequence of functions $f_n = P^n f$ satisfies conditions (5.8.2) of Proposition 5.8.1.

Now, differentiating equation (6.6.3) and using $|\sigma_i'| \leq c\sigma_i$, $\sigma_i < 1/\lambda$, and $|f'| \leq k_f f$ gives

$$\frac{(Pf)'}{Pf} \leq \frac{\sum_i |\sigma_i'|(f \circ g_i)}{\sum_i \sigma_i(f_i \circ g_i)} + \frac{\sum_i (\sigma_i)^2 |f' \circ g_i|}{\sum_i \sigma_i(f_i \circ g_i)}$$

$$\leq c + \frac{1}{\lambda} \frac{\sum_i \sigma_i k_f(f \circ g)}{\sum_i \sigma_i(f \circ g_i)} \leq c + \frac{1}{\lambda} k_f,$$

and, by induction,

$$\frac{|(P^n f)'|}{P^n f} \leq \frac{c\lambda}{\lambda - 1} + \frac{1}{\lambda^n} k_f.$$

Pick a constant $K > \lambda c/(\lambda - 1)$ so that, since $\lambda > 1$, for n sufficiently large $(n > n_1(f))$, we have

$$|(P^n f)'| \le K P^n f. \tag{6.6.8}$$

Thus the iterates $f_n = P^n f$ satisfy condition (5.8.3) of Proposition 5.8.1. Therefore, by Proposition 5.8.1, $P^n f$ has a nontrivial lower-bound function, and thus, by Theorem 5.6.2, $\{P^n\}$ is asymptotically stable. $\blacksquare$

Remark 6.6.1. Observe that in the special case where S is periodic (in x) with period $L = a_i - a_{i-1}$, condition (6iv) is automatically satisfied. In fact, in this case $g_i'(x) = g_0'(x)$ so, by setting $q = |g_0'|/L$, we obtain inequality (6.6.1) and, moreover,

$$\|q\| = \frac{1}{L} \int_{-\infty}^{\infty} |g_i'(x)| dx = \frac{1}{L} \left| \int_{-\infty}^{\infty} g_i'(x) dx \right| = |[g_i(x)]_{-\infty}^{\infty}| = \frac{L}{L} = 1,$$

showing that $q \in L^1$. The remaining conditions simply generalize the properties of the transformation $S(x) = \beta \tan(\gamma x + \delta)$ with $|\beta\gamma| > 1$. $\square$

Example 6.6.1. It is easy to show that the Frobenius–Perron operator P associated with $S(x) = \beta \tan(\gamma x + \delta)$, $|\beta\gamma| > 1$, is asymptotically stable. We have

$$S'(x) = \frac{\beta\gamma}{\cos^2(\gamma x + \delta)}$$

hence $|S'(x)| \ge \beta\gamma$. Further

$$-\frac{S''(x)}{[S'(x)]^2} = -\frac{1}{\beta} \sin[2(\gamma x + \delta)]$$

so that

$$\left| \frac{S''(x)}{[S'(x)]^2} \right| \le \frac{1}{|\beta|}. \quad \square$$

6.7 Manifolds

The last goal of this chapter is to show how the techniques described in Chapter 5 may be used to study the behavior of transformations in higher-dimensional spaces. The simplest, and probably most striking, use of the Frobenius–Perron operator in d-dimensional spaces is for expanding mappings on manifolds. To illustrate this, the results of Krzyżewski and Szlenk (1969), which may be considered as a generalization of the results of Rényi presented in Section 6.2, are developed in detail in Section 6.8. However, in this section we preface these results by presenting some basic concepts from the theory of manifolds, which will be helpful for understanding the geometrical ideas related to the Krzyżewski–Szlenk results. This elementary description of manifolds is by no means an exhaustive treatment of differential geometry.

First consider the paraboloid $z = x^2 + y^2$. This paraboloid is embedded in three-dimensional space, even though it is a two-dimensional object. If the paraboloid is the state space of a system, then, to study this system, each point on the paraboloid must be described by precisely two numbers. Thus, any point m on the paraboloid with coordinates $(x, y, x^2 + y^2)$ is simply described by its x, y-coordinates. This two-dimensional system of coordinates may be described in a more abstract way as follows. Denote by M the graph of the paraboloid, that is,

$$M = \{(x, y, z): z = x^2 + y^2\},$$

and, as a consequence, there is a one-to-one transformation $\phi: M \to R^2$ described by $\phi(x, y, z) = (x, y)$ for $(x, y, z) \in M$. Of course, other coordinate systems on M are possible, that is, another one-to-one mapping, $\phi^*: M \to R^2$, but ϕ is probably the simplest one.

Now let M be the unit sphere,

$$M = \{(x, y, z): x^2 + y^2 + z^2 = 1\}.$$

In this example it is impossible to find a single smooth invertible function $\phi: M \to R^2$. However, six functions $\phi_i: M \to R^2$ may be defined as follows:

$$
\begin{aligned}
\phi_1(x, y, z) &= (x, y), &&\text{for } z > 0; \\
\phi_2(x, y, z) &= (x, y), &&\text{for } z < 0; \\
\phi_3(x, y, z) &= (x, z), &&\text{for } y > 0; \\
\phi_4(x, y, z) &= (x, z), &&\text{for } y < 0; \\
\phi_5(x, y, z) &= (y, z), &&\text{for } x > 0; \\
\phi_6(x, y, z) &= (y, z), &&\text{for } x < 0.
\end{aligned}
$$

Each of these functions ϕ_i maps a hemisphere of M onto an open unit disk. This coordinate system has the property that for any $m \in M$ there is an open hemisphere that contains m and on each of these hemispheres one ϕ_i is defined.

In the same spirit, we give a general definition of a smooth manifold.

Definition 6.7.1. A smooth d-dimensional manifold consists of a topological Hausdorff space M and a system $\{\phi_i\}$ of local coordinates satisfying the following properties:

(a) Each function ϕ_i is defined and continuous on an open subset $W_i \subset M$ and maps it onto an open subset $U_i = \phi_i(W_i)$ of R^d. The inverse functions ϕ_i^{-1} exist and are continuous (i.e., ϕ_i is a homeomorphism of W_i onto U_i);

(b) For each $m \in M$ there is a W_i such that $m \in W_i$, that is, $M = \bigcup_i W_i$;

(c) If the intersection $W_i \cap W_j$ is nonempty, then the mapping $\phi_i \circ \phi_j^{-1}$, which is defined on $\phi_j(W_i \cap W_j) \subset R^d$ and having values in R^d, is a C^∞ mapping.

(Note that a topological space is called a **Hausdorff space** if every two distinct points have nonintersecting neighborhoods.)

Any map ϕ_i gives a coordinate system of a part of M, namely, W_i. A local coordinate of a point $m \in W_i$ is $\phi_i(m)$. Having a coordinate system, we may now define what we mean by a C^k function on M. We say that $f: M \to R$ is of **class** C^k if for each $\phi_i: W_i \to U_i$ the composed mapping $f \circ \phi_i^{-1}$ is of class C^k on U_i.

Next consider the gradient of a function defined on the manifold. For $f: R^d \to R$ the **gradient** of f at a point $x \in R^d$ is simply the vector (sequence of real numbers),

$$\operatorname{grad} f(x) = \left(\frac{\partial f(x)}{\partial x_1}, \ldots, \frac{\partial f(x)}{\partial x_d} \right).$$

For $f: M \to R$ of class C^1, the gradient of f at a point $m \in M$ can be calculated in local coordinates as follows:

$$\operatorname{grad} f(m) = (D_{x_1}(m)f, \ldots, D_{x_d}(m)f), \qquad (6.7.1a)$$

where

$$D_{x_i}(m)f = \frac{\partial}{\partial x_i} \left[f(\phi^{-1}(x)) \right]_{x=\phi(m)}. \qquad (6.7.1b)$$

Thus the gradient is again a sequence of real numbers that depends on the choice of the local coordinates.

The most important notion from the theory of manifolds is that of tangent vectors and tangent spaces. A continuous mapping $\gamma: [a, b] \to M$ represents an arc on M with the end points $\gamma(a)$ and $\gamma(b)$. We say that γ starts from $m = \gamma(a)$. The arc γ is C^k if, for any coordinate system ϕ, the composed function $\phi \circ \gamma$ is of class C^k. The **tangent vector** to γ at a point $m = \gamma(a)$ in a coordinate system ϕ is defined by

$$\frac{d}{dt} \left[\phi(\gamma(t)) \right]_{t=a} = (\xi^1, \ldots, \xi^d), \qquad (6.7.2)$$

where, again, the numbers $\xi^1, \ldots, \xi^d$ depend on the choice of the coordinate system ϕ. Of course, γ must be at least of class C^1. Two arcs γ_1 and γ_2 starting from m are called **equivalent** if they produce the same coordinates, that is,

$$\frac{d}{dt} \left[\phi(\gamma_1(t)) \right]_{t=a_1} = \frac{d}{dt} \left[\phi(\gamma_2(t)) \right]_{t=a_2}, \qquad (6.7.3)$$

where $\gamma_1(a_1) = \gamma_2(a_2) = m$. Observe that, if (6.7.3) holds in a given system of coordinates ϕ, then it holds in any other coordinate system. The class of all equivalent arcs produces the same sequence (6.7.3) for any given system of coordinates. Such a class represents the tangent vector. Tangent vectors are denoted by the Greek letters ξ and η.

Assume that a tangent vector ξ in a coordinate system ϕ has components $\xi_1, \ldots, \xi^d$. What are the components in another coordinate system ψ? Now,

$$\frac{d}{dt}[\psi(\gamma(t))] = \frac{d}{dt}[H(\phi(\gamma(t)))],$$

where $H = \psi \circ \phi^{-1}$ and, therefore, setting $d(\psi \circ \gamma)/dt = (\eta^1, \ldots, \eta^d)$,

$$\eta^i = \sum_{j=1}^{d} \frac{\partial H_i}{\partial x_j} \xi^j. \tag{6.7.4}$$

Equation (6.7.4) shows the transformations of the tangent vector coordinates under the change of coordinate system. Thus from an abstract (tensor analysis) point of view the tangent vector at a point m is nothing but a sequence of numbers in each coordinate system given in such a way that these numbers satisfy condition (6.7.4) when we pass from one coordinate system to another. From this description it is clear that the tangent vectors at m form a linear space, the **tangent space**, which we denote by T_m.

Now consider a transformation F from a d-dimensional manifold M into a d-dimensional manifold N, $F: M \to N$. The transformation F is said to be of class C^k if, for any two coordinate systems ϕ on M and ψ on N, the composed function $\psi \circ F \circ \phi^{-1}$ is of class C^k, or its domain is empty. Let ξ be a tangent vector at m, represented by a C^1 arc $\gamma: [a, b] \to M$ starting from m. Then $F \circ \gamma$ is an arc starting from $F(m)$, and it is of class C^1 if F is of class C^1. The tangent vector to $F \circ \gamma$ in a coordinate system ψ is given by

$$\frac{d}{dt}[\psi \circ F \circ \gamma]_{t=a} = (\eta^1, \ldots, \eta^d).$$

Setting $\sigma = \psi \circ F \circ \phi^{-1}$, where ϕ is a coordinate system on M,

$$\eta^i = \sum_{j=1}^{d} \frac{\partial \sigma_i}{\partial x_j} \xi^j \tag{6.7.5}$$

results. Equation (6.7.5) gives the linear transformation of a tangent vector ξ at m to a tangent vector η at $F(m)$ without explicit reference to the arc γ. This transformation is called the **differential** of F at a point m and is denoted by $dF(m)$, and thus symbolically

$$\eta = dF(m)\xi.$$

Note that the differential of F is represented in any two coordinate systems, ϕ on M and ψ on N, by the matrix

$$\left(\frac{\partial \sigma_i}{\partial x_j}\right), \qquad i, j = 1, \ldots, d.$$

The same matrix appears in the formula for the gradient of the composed function: If $F: M \to N$ and $f: N \to R$ are C^1 functions, then the differentiation of $(f \circ F) \circ \phi^{-1} = (f \circ \psi^{-1}) \circ (\psi \circ F \circ \phi^{-1})$ gives

$$\text{grad}(f \circ F)(m) = (D_{x_1}(m)(f \circ F), \ldots, D_{x_d}(m)(f \circ F)),$$

where

$$D_{x_i}(m)(f \circ F) = \sum_{j=1}^{d} \frac{\partial}{\partial x_j}\left[f(\psi^{-1}(x))\right]_{x=\psi(F(m))} \frac{\partial \sigma_j}{\partial x_i}.$$

This last formula may be written more compactly as

$$\text{grad}((f \circ F)(m)) = (\text{grad } f)(dF(m)).$$

Observe that now $dF(m)$ appears on the right-hand side of the vector.

Finally observe the relationship between tangent vectors and gradients. Let $f: M \to R$ be of class C^1 and let $\gamma: [a, b] \to M$ start from m. Consider the composed function $f \circ \gamma: [a, b] \to R$ that is also of class C^1. Using the local system of coordinates,

$$f \circ \gamma = (f \circ \phi^{-1}) \circ (\phi \circ \gamma),$$

and, consequently,

$$\left.\frac{d(f \circ \gamma)}{dt}\right|_{t=a} = \sum_{i=1}^{d} [D_{x_i}(\gamma(a))f]\xi^i. \tag{6.7.6}$$

Observe that the numbers $D_{x_i}f$ and ξ^i depend on ϕ even though the left-hand side of (6.7.6) does not. Equation (6.7.6) may be more compactly written as

$$\left.\frac{d(f \circ \gamma)}{dt}\right|_{t=a} = [\text{grad } f(\gamma(a))]\gamma'. \tag{6.7.7}$$

In order to construct a calculus on manifolds, concepts such as the length of a tangent vector, the norm of a gradient, and the area of Borel subsets of M are necessary. The most effective way of introducing these is via the Riemannian metric. Generally speaking the Riemannian metric is a scalar product on T_m. This means that, for any two vectors $\xi_1, \xi_2 \in T_m$, there corresponds a real number denoted by $\langle \xi_1, \xi_2 \rangle$. However, the coordinates

$$(\xi_1^1, \ldots, \xi_1^d), (\xi_2^1, \ldots, \xi_2^d)$$

depend on the coordinate system ϕ. Thus the rule that allows $\langle \xi_1, \xi_2 \rangle$ to be calculated given (ξ_1^i), (ξ_2^i) must also depend on ϕ. These facts are summarized in the following definition.

Definition 6.7.2. A **Riemannian metric** on the manifold M is a system of functions

$$g_{ij}^{\phi}(m): M \to R, \quad i, j = 1, \ldots, d,$$

such that

(a) For any choice of local coordinates $\phi: W \to U$ the functions $g_{ij}^{\phi}(\phi^{-1}(x))$ are defined and C^{∞} for $x \in U$.

(b) For each $m \in M$ the quadratic form

$$\sum_{i,j=1}^{d} g_{ij}^{\phi} \lambda^i \lambda^j$$

is symmetric and positive definite (i.e., $g_{ij}^{\phi} = g_{ji}^{\phi}$, and the value of this sum is positive except if all $\lambda^i = 0$).

(c) For every $\xi_1, \xi_2 \in T_m$ the **scalar product**

$$\langle \xi_1, \xi_2 \rangle = \sum_{i,j=1}^{d} g_{ij}^{\phi}(m) \xi_1^i \xi_2^j \tag{6.7.8}$$

does not depend on ϕ.

The last condition (6.7.8) looks somewhat mysterious, but it simply means that

$$\sum_{k,l} g_{kl}^{\psi}(m) \eta_1^k \eta_2^l = \sum_{i,j} g_{ij}^{\phi}(m) \xi_1^i \xi_2^j$$

where η_1, η_2 are calculated from ξ_1, ξ_2 by equation (6.7.4). Thus

$$\sum_{k,l} g_{kl}^{\psi}(m) \sum_{i,j} \frac{\partial H_k}{\partial x_i} \frac{\partial H_l}{\partial x_j} \xi_1^i \xi_2^j = \sum_{i,j} g_{ij}^{\phi}(m) \xi_1^i \xi_2^j,$$

which implies that

$$g_{ij}^{\phi}(m) = \sum_{k,l} g_{kl}^{\psi}(m) \frac{\partial H_k}{\partial x_i} \frac{\partial H_l}{\partial x_j}. \tag{6.7.9}$$

Now having introduced the scalar product, the **norm** of $\xi \in T_m$ is defined by $\|\xi\| = (\langle \xi, \xi \rangle)^{1/2}$. If a C^1 arc $\gamma: [a,b] \to M$ is given, it defines, at each point $m = \gamma(t_0)$, the tangent vector $\gamma'(t_0)$. Thus the length of an arc γ is just

$$l(\gamma) = \int_a^b \|\gamma'(t)\| \, dt. \tag{6.7.10}$$

This equation may be used for any arc γ that is continuous and piecewise C^1. If a manifold M is such that any two points $m_0, m_1 \in M$ can be joined by a continuous piecewise C^1 arc, it is said to be **connected**. On connected manifolds the **distance** between points is given by

$$\rho(m_0, m_1) = \inf l(\gamma),$$

where the inf is taken over all possible arcs joining m_0 and m_1. With this distance, M becomes a metric space.

From equation (6.7.7) it is easy to define the **length** of grad f at a point m. It is, by definition,

$$|\text{grad } f(m)| = \sup \left| \frac{d}{dt}[f(\gamma(t))]_{t=a} \right|,$$

where the sup is taken over all possible arcs $\gamma: [a, b] \to M$ with $\gamma(a) = m$ and $\|\gamma'(a)\| = 1$. From this definition, it follows that, for an arbitrary C^1 arc γ and C^1 function f,

$$\left| \frac{d}{dt}[f(\gamma(t))] \right| \leq |\text{grad } f(\gamma(t))| \cdot \|\gamma'(t)\|. \qquad (6.7.11)$$

Analogously, for a C^1 mapping $F: M \to N$ we introduce the norm of the differential $dF(m)$ by

$$|dF(m)| = \sup \|dF(m)\xi\|,$$

where the supremum is taken over all $\xi \in T_m$ such that $\|\xi\| = 1$. Using this notion it can be verified that

$$\|dF(m)\xi\| \leq |dF(m)| \cdot \|\xi\|, \qquad m \in M, \xi \in T_m,$$

and

$$|\text{grad}(f \circ F)(m)| \leq |(\text{grad } f)(F(m))| \cdot |dF(m)|, \qquad m \in M,$$

where $f: N \to R$ is a C^1 function. Differentiation on manifolds satisfies some other properties analogous to those on R^d. We have, for example,

$$|\text{grad}(fg)| \leq |f| \cdot |\text{grad } g| + |g| \cdot |\text{grad } f|, \qquad (fg)(m) = f(m)g(m)$$

and

$$\left| \text{grad} \sum_i f_i \right| \leq \sum_i |\text{grad } f_i|,$$

where f, g, and f_i are C^1 functions.

To introduce the measure on M associated with the Riemannian metric, it is first necessary to define the **unit volume function** $V_\phi(m)$. Consider the Riemannian form $g_{ij}^\phi(m)$ corresponding to a coordinate system $\phi: W \to U$. We can find d normalized vectors

$$\begin{pmatrix} \xi_1^1 \\ \vdots \\ \xi_1^d \end{pmatrix} \quad \overset{d}{\cdots} \quad \begin{pmatrix} \xi_d^1 \\ \vdots \\ \xi_d^d \end{pmatrix} \qquad (6.7.12)$$

orthogonal with respect to this form, that is,

$$\langle \xi_k, \xi_l \rangle = \sum_{i,j=1}^{d} g_{ij}^{\phi}(m) \xi_k^i \xi_l^j = \delta_{kl},$$

where $\delta_{kl} = 1$ if $k = l$, and $\delta_{kl} = 0$, $k \neq l$, is the Kronecker delta. Write

$$V_\phi(m) = \left| \det \begin{vmatrix} \xi_1^1 & \cdots & \xi_d^1 \\ \cdots & \cdots & \cdots \\ \xi_1^d & \cdots & \xi_d^d \end{vmatrix} \right|.$$

This same procedure can be carried out algebraically by setting $V_\phi(m) = |\det(g_{ij}^{\phi}(m))|^{-1/2}$.

Function $V_\phi(m)$ has a simple heuristic interpretation. Vectors $\xi_1, \ldots, \xi_d$, which correspond to the components (6.7.12), are orthogonal and normalized. Thus the volume spanned by them should be equal to 1. The volume spanned by the representation of $\xi_1, \ldots, \xi_d$ in a local coordinate system is $V_\phi(m)$. Thus $V_\phi(m)$ gives the volume in local coordinates corresponding to a unit volume in M. By using this interpretation we define the **measure** μ of a Borel set $B \subset W$ as

$$\mu(B) = \int_{\phi(B)} \frac{dx}{V_\phi(\phi^{-1}(x))}. \tag{6.7.13}$$

The idea leading to this definition is obvious, as the elementary volume $dx = dx_1 \cdots dx_d$ in R^d corresponds to the volume $V_\phi(\phi^{-1}(x))dx_1 \cdots dx_d$ in M. Thus, in order to reproduce the "original" volume in M, we must divide dx by V_ϕ. It can be shown that $\mu(B)$ defined by (6.7.13) does not depend on the choice of ϕ, which is quite obvious from the heuristic interpretation of $V_\phi(m)$.

Analogous considerations lead to the definition of the **determinant of the differential** of a C^1 transformation F from a d-dimensional manifold M into a d-dimensional manifold N. Take a point $m \in M$ and define

$$|\det dF(m)| = \left| \frac{d\sigma}{dx} \right| \frac{V_\phi(m)}{V_\psi(F(m))},$$

where $|d\sigma/dx|$ denotes the absolute value of the determinant of the $d \times d$ matrix

$$\left(\frac{\partial \sigma_i}{\partial x_j} \right), \qquad i, j = 1, \ldots, d.$$

It can be shown that this definition does not depend on the choice of coordinate systems ϕ and ψ in M and N, respectively. Note also that the determinant per se is not defined, but only its absolute value. This is because our manifolds M, N are not assumed to be oriented.

The following calculation will justify our definition of $|\det dF(m)|$. Let B be a small set on M, and $F(B)$ its image on N. What is the ratio $\mu(F(B))/\mu(B)$? From equation (6.7.13),

$$\frac{\mu(F(B))}{\mu(B)} = \int_{\psi(F(B))} \frac{dy}{V(\psi^{-1}(y))} \bigg/ \int_{\phi(B)} \frac{dx}{V_\phi(\phi^{-1}(x))}.$$

Setting $\sigma = \psi \circ F \circ \phi^{-1}$ and substituting $y = \sigma(x)$,

$$\frac{\mu(F(B))}{\mu(B)} = \int_{\phi(B)} \left|\frac{d\sigma}{dx}\right| \frac{dx}{V_\psi(F(\phi^{-1}(x)))} \bigg/ \int_{\phi(B)} \frac{dx}{V_\phi(\phi^{-1}(x))}$$

results. Thus, for small B containing a point m, we have approximately

$$\frac{\mu(F(B))}{\mu(B)} \simeq \left|\frac{d\sigma}{dx}\right| \frac{V_\phi(m)}{V_\psi F(m)} = |\det(dF(m))|. \tag{6.7.14}$$

6.8 Expanding Mappings on Manifolds

With the background material of the preceding section, we now turn to an examination of the asymptotic behavior of expanding mappings on manifolds.

We assume that M is a finite-dimensional compact connected smooth (C^∞) manifold with a Riemannian metric. As we have seen in Section 6.7, this metric induces the natural (Borel) measure μ and distance ρ on M. We use $|f'(m)|$ to denote the length of the gradient of f at point $m \in M$.

Before starting and proving our main result, we give a sufficient condition for the existence of a lower-bound function in the same spirit as contained in Propositions 5.8.1 and 5.8.2. We use the notation of Section 5.8.

Proposition 6.8.1. Let $P: L^1(M) \to L^1(M)$ be a Markov operator and if we assume that there is a set D_0, dense in D, so that for every $f \in D_0$ the trajectory

$$P^n f = f_n, \qquad \text{for } n \geq n_0(f), \tag{6.8.1}$$

is such that the functions f_n are C^1 and satisfy

$$|f_n'(m)| \leq k f_n(m), \qquad \text{for } m \in M, \tag{6.8.2}$$

where $k \geq 0$ is a constant independent of f, then there exists $\varepsilon > 0$ such that $h = \varepsilon 1_M$ is a lower-bound function for P.

Proof. The proof of this proposition proceeds much as for Proposition 5.8.2. As before, $\|f_n\| = 1$. Set

$$\varepsilon = [1/2\mu(M)]e^{-kr},$$

where

$$r = \sup_{m_0, m_1 \in M} \rho(m_0, m_1).$$

Let $\gamma(t)$, $a \leq t \leq b$, be a piecewise smooth arc joining points $m_0 = \gamma(a)$ and $m_1 = \gamma(b)$. Differentiation of $f_n \circ \gamma$ gives [see inequality (6.7.11)]

$$\left| \frac{d[f_n(\gamma(t))]}{dt} \right| \leq |f'_n(\gamma(t))| \cdot \|\gamma'(t)\|$$
$$\leq k\|\gamma'(t)\| f_n(\gamma(t))$$

so that

$$f_n(m_1) \leq f_n(m_0) \exp\left\{ k \int_a^b \|\gamma'(s)\|\, ds \right\}.$$

Since γ was an arbitrary arc, this gives

$$f_n(m_1) \leq f_n(m_0) e^{k\rho(m_0, m_1)} \leq f_n(m_0) e^{kr}.$$

Now suppose that $h = \varepsilon 1_M$ is not a lower-bound function for P. This means that there must be some $n' > n_0$ and $m_0 \in M$ such that $f_{n'}(m_0) < \varepsilon$. Therefore,

$$f_{n'}(m_1) \leq \varepsilon e^{kr} = (1/2\mu(M)), \qquad \text{for } m_1 \in M,$$

which contradicts $\|f_n\| = 1$ for all $n > n_0(f)$. Thus we must have $f_n \geq h = \varepsilon 1_M$ for $n > n_0$. ∎

Next we turn to a definition of an expanding mapping on a manifold.

Definition 6.8.1. Let M be a finite-dimensional compact connected smooth (C^∞) manifold with Riemannian metric and let μ be the corresponding Borel measure. A C^1 mapping $S: M \to M$ is called **expanding** if there exists a constant $\lambda > 1$ such that the differential $dS(m)$ satisfies

$$\|dS(m)\xi\| \geq \lambda\|\xi\| \tag{6.8.3}$$

at each $m \in M$ for each tangent vector $\xi \in T_m$.

With this definition, Krzyżewski and Szlenk [1969] and Krzyżewski [1977] demonstrate the existence of a unique absolutely continuous normalized measure invariant under S and establish many of its properties. Most of these results are contained in the next theorem.

Theorem 6.8.1. Let $S: M \to M$ be an expanding mapping of class C^2, and P the Frobenius–Perron operator corresponding to S. Then $\{P^n\}$ is asymptotically stable.

Proof. From equation (6.7.5) with $F = S$, since S is expanding, $\eta \neq 0$ for any $\xi \neq 0$, and, thus, the matrix $(\partial\sigma_i/\partial x_j)$ must be nonsingular for every $m \in M$.

In local coordinates the transformation S has the form

$$x \to \phi(S(\phi^{-1}(x))) = \sigma(x)$$

and consequently is locally invertible. Therefore, for any point $m \in M$ the counterimage $S^{-1}(m)$ consists of isolated points, and, since M is compact, the number of these points is finite. Denote the counterimages of m by $m_1, \ldots, m_k$. Because S is locally invertible there exists a neighborhood W of m and neighborhoods W_i of m_i such that S restricted to W_i is a one to one mapping from W_i onto W. Denote the inverse mapping of S on W_i by g_i. We have $S \circ g_i = I_{w_i}$, where I_{w_i} is the identity mapping on W_i and, consequently, $(dS) \circ (dg_i)$ is the identity mapping on the tangent vector space. From this, in conjunction with (6.8.3), it immediately follows that

$$\|(dg_i)\xi\| \leq (1/\lambda)\|\xi\|. \tag{6.8.4}$$

Now take a set $B \subset W$, so

$$S^{-1}(B) = \bigcup_{i=1}^{k} g_i(B),$$

and, by the definition of the Frobenius–Perron operator,

$$\int_B Pf(m)\mu(dm) = \int_{S^{-1}(B)} f(m)\mu(dm) = \sum_{i=1}^{k} \int_{g_i(B)} f(m)\mu(dm).$$

This may be rewritten as

$$\frac{1}{\mu(B)} \int_B Pf(m)\mu(dm) = \sum_{i=1}^{k} \frac{\mu(g_i(B))}{\mu(B)} \cdot \frac{1}{\mu(g_i(B))} \int_{g_i(B)} f(m)\mu(dm).$$

If B shrinks to m, then $g_i(B)$ shrinks to $g_i(m)$,

$$\frac{1}{\mu(B)} \int_B Pf(m)\mu(dm) \to Pf(m) \quad \text{a.e.}$$

and

$$\frac{1}{\mu(g_i(B))} \int_{g_i(B)} f(m)\mu(dm) \to f(g_i(m)) \quad \text{a.e.}, \qquad i = 1, \ldots, k.$$

Moreover, by (6.7.14),

$$\frac{\mu(g_i(B))}{\mu(B)} \to |\det(dg_i(m))|.$$

Thus, by combining all the preceding expressions, we have

$$Pf(m) = \sum_{i=1}^{k} |\det(dg_i(m))| f(g_i(m)), \tag{6.8.5}$$

which is quite similar to the result in equation (6.2.10).

Now let $D_0 \subset D(M)$ be the set of all strictly positive C^1 densities. For $f \in D_0$, differentiation of $Pf(m)$ as given by (6.8.5) yields

$$
\begin{aligned}
\frac{|(Pf)'|}{Pf} &= \frac{\sum_{i=1}^{k} |(J_i(f \circ g_i))'|}{\sum_{i=1}^{k} J_i(f \circ g_i)} \\
&\leq \frac{\sum_{i=1}^{k} |J_i'|(f \circ g_i)}{\sum_{i=1}^{k} J_i(f \circ g_i)} + \frac{\sum_{i=1}^{k} J_i|f' \circ g_i|\,|dg_i|}{\sum_{i=1}^{k} J_i(f \circ g_i)} \\
&\leq \max_i \frac{|J_i'|}{J_i} + \max_i \frac{|f' \circ g_i|\,|dg_i|}{(f \circ g_i)},
\end{aligned}
$$

where $J_i = |\det dg_i(m)|$. From equation (6.8.4), it follows that $|dg_i| \leq 1/\lambda$, so that

$$
\sup \frac{|(Pf)'|}{Pf} \leq c + \frac{1}{\lambda} \sup \frac{|f'|}{f},
$$

where

$$
c = \sup_{i,m} \frac{|J_i'(m)|}{J_i(m)}.
$$

Thus, by induction, for $n = 1, 2, \ldots$, we have

$$
\sup \frac{|(P^n f)'|}{P^n f} \leq \frac{c\lambda}{\lambda - 1} + \frac{1}{\lambda^n} \sup \frac{|f'|}{f}.
$$

Choose a real $K > \lambda c/(\lambda - 1)$, then

$$
\sup \frac{|(P^n f)'|}{P^n f} \leq K \tag{6.8.6}
$$

for n sufficiently large, say $n > n_0(f)$. A straightforward application of Proposition 6.8.1 and Theorem 5.6.2 finishes the proof. ∎

Example 6.8.1. Let M be the two-dimensional torus, namely, the Cartesian product of two unit circles:

$$
M = \{(m_1, m_2) : m_1 = e^{ix_1}, m_2 = e^{ix_2}, x_1 x_2 \in R\}.
$$

M is evidently a Riemannian manifold, and the inverse functions to

$$
m_1 = e^{ix_1} \quad \text{and} \quad m_2 = e^{ix_2} \tag{6.8.7}
$$

define the local coordinate system. In these local coordinates the Riemannian metric is given by $g_{jk} = \delta_{jk}$, the Kronecker delta, and defines a Borel measure μ identical with that obtained from the product of the Borel measures on the circle.

We define a mapping $S : M \to M$ that, in local coordinates, has the form

$$
S(x_1, x_2) = (3x_1 + x_2, x_1 + 3x_2) \quad (\bmod\ 2\pi). \tag{6.8.8}
$$

Thus S maps each point (m_1, m_2) given by (6.8.7) to the point $(\tilde{m}_1, \tilde{m}_2)$, where

$$\tilde{m}_1 = \exp[i(3x_1 + x_2)] \quad \text{and} \quad \tilde{m}_2 = \exp[i(x_1 + 3x_2)].$$

We want to show that S is an expanding mapping.

From (6.8.8) we see that $dS(m)$ maps the vector $\xi = (\xi^1, \xi^2)$ into the vector $(3\xi^1 + \xi^2, \xi^1 + 3\xi^2)$. Also, since $g_{jk} = \delta_{jk}$, hence $\langle \xi^1, \xi^2 \rangle = (\xi^1)^2 + (\xi^2)^2$ from (6.7.8). Thus

$$\begin{aligned}
\|dS(m)\xi\|^2 &= (3\xi^1 + \xi^2)^2 + (\xi^1 + 3\xi^2)^2 \\
&= 4[(\xi^1)^2 + (\xi^2)^2] + 6(\xi^1 + \xi^2)^2 \\
&\geq 4\|\xi\|^2,
\end{aligned}$$

and we see that inequality (6.8.3) is satisfied with $\lambda = 2$, therefore S is an expanding mapping. Further, if P is the Frobenius–Perron operator corresponding to S, then, by Theorem 6.8.1, $\{P^n\}$ is asymptotically stable. It is also possible to show that S is measure preserving, so by Proposition 5.6.2 this transformation is exact. This proves our earlier assertion in Section 4.3. □

Exercises

6.1. Let $(X, \mathcal{A}, \mu)$ be a measure space and let $S: X \to X$ be a nonsingular transformation. Fix an integer $k \geq 1$. Prove that S^k is statistically stable if and only if S is statistically stable.

6.2. Consider the transformation $S: [0, 1] \to [0, 1]$ defined by

$$S(x) = \begin{cases} 2x & 0 \leq x < \frac{1}{2} \\ \frac{1-x}{x} & \frac{1}{2} \leq x \leq 1. \end{cases}$$

Using the result proved in Exercise 6.1 show that S is statistically stable.

6.3. Consider the transformation $S: [0, 1] \to [0, 1]$ of the form

$$S(x) = cx(1 - x^2).$$

Fix c at the value for which S maps $[0, 1]$ onto $[0, 1]$ and, using the change of variables formulas (Theorem 6.5.1), show that for this value of c the transformations S is statistically stable.

6.4. Consider the transformations $S: [0, 1] \to [0, 1]$ of the form

$$S(x) = cx^2(1 - x).$$

Again fix c to be that value at which S maps $[0, 1]$ onto $[0, 1]$ and use a change of variables and the Birkhoff individual ergodic theorem to show that

$$\lim_{n \to \infty} S^n(x) = 0, \quad \text{for } x \in [0, 1] \text{ a.e.}$$

Observe that S has periodic points of period 3 and thus is chaotic in the sense of Šarkovskiĭ and Li–Yorke [Jama, 1989; Li and Yorke, 1975].

6.5. Consider the transformation $S: [0, l] \rightarrow [0, l]$ defined by

$$S(x) = x \left[-a + \frac{b}{1+x} \right],$$

where $l = (b/a) - 1 > 0$. Find $b = b_0$ such that S maps $[0, l]$ onto $[0, l]$ and using the change of variables formulas prove that S is statistically stable.

6.6. Consider the "tent" transformation $S: [0, 1] \rightarrow [0, 1]$ of the form

$$S(x) = \begin{cases} cx & 0 \leq x < \frac{1}{2} \\ c(1 - x) & \frac{1}{2} \leq x \leq 1 \end{cases} .$$

where $1 < c \leq 2$. Let P be the corresponding Frobenius–Perron operator. Determine the values of parameter c for which $\{P^n\}$ is asymptotically periodic but not asymptotically stable ($r \geq 2$ in formula (5.3.8)).

6.7. Let M be the two-dimensional torus described in Example 6.8.1. Consider the transformation $S: M \rightarrow M$ which in local coordinates has the form

$$S(x_1, x_2) = (a_{11}x_1 + a_{12}x_2, a_{21}x_1 + a_{22}x_2).$$

Assume that the coefficients a_{ij} are positive integers and find conditions concerning the matrix (a_{ij}) which imply the exactness of S.

6.8. Using TRAJ, BIFUR, DENTRAJ, and DENITER numerically study the behavior of transformations defined in Exercises 6.5 and 6.6 for $a < b \leq b_0$, and $0 < c < 2$, respectively.

7
Continuous Time Systems: An Introduction

In previous chapters we concentrated on discrete time systems because they offer a convenient way of introducing many concepts and techniques of importance in the study of irregular behaviors in model systems. Now we turn to a study of continuous time systems.

Continuous and discrete systems differ in several important and interesting ways, which we will touch on throughout the remainder of this book. For example, in a continuous time system, complicated irregular behaviors are possible only if the dimension of the phase space of the system is three or greater. As we have seen, this is in sharp contrast to discrete time processes that can have extremely complicated dynamics in only one dimension. Further, continuous time processes in a finite-dimensional phase space are in general invertible, which immediately implies that exactness is a property that will not occur for these systems (recall that noninvertibility is a necessary condition for exactness). However, systems in an infinite-dimensional phase space, namely, time delay equations and some partial differential equations, are generally not invertible and, thus, may display exactness.

This chapter is devoted to an introduction of the concept of continuous time systems, an extension of many properties developed previously for discrete time systems, and the development of tools and techniques specifically designed for studying continuous time systems.

7.1   Two Examples of Continuous Time Systems

Here a continuous time process in a phase space X is given by a family of mappings

$$S_t: X \to X, \qquad t \geq 0.$$

As illustrated in Figure 7.1.1, the value $S_t(x^0)$ is the position of the system at a time t that started from an initial point $x^0 \in X$ at time $t = 0$. We consider only those processes in which the dynamical law S does not explicitly depend on time so that the property

$$S_t(S_{t'}(x)) = S_{t+t'}(x) \tag{7.1.1}$$

holds. This simply means that the dynamics governing the evolution of the system are the same on the intervals $[0, t']$ and $[t, t + t']$.

Example 7.1.1. A well-known example of a continuous time process is given by an autonomous d-dimensional system of ordinary differential equations

$$\frac{dx}{dt} = F(x) \tag{7.1.2}$$

where $x = (x_1, \ldots, x_d)$ and $F: R^d \to R^d$ is sufficiently smooth to ensure the existence and uniqueness of solutions, such as F is C^1 and satisfies $|F(x)| \leq \alpha + \beta|x|$, with α and β finite. In this case, $S_t(x^0)$ is the solution of (7.1.2) with the initial condition

$$x(0) = x^0. \tag{7.1.3}$$

In this example time t need not be restricted to $t \geq 0$, and the system can also be studied for $t \leq 0$. As we will see in Section 7.8, this is a commonly encountered situation for problems in finite dimensional phase spaces.   □

Example 7.1.2. Consider the delay-differential equation

$$\frac{dx(t)}{dt} = F(x(t), x(t - 1)) \tag{7.1.4}$$

with the initial condition

$$x(\tau) = x^0(\tau), \qquad \text{for } \tau \in [-1, 0). \tag{7.1.5}$$

For rather simple restrictions on F (namely, that F is C^1 and $|F(x, y)| \leq \alpha(y) + \beta(y)|x|$, where α and β are arbitrary continuous functions of y), there is a unique solution to (7.1.4) with (7.1.5) [see Hale, 1977].

Let X be the space of all functions $[-1, 0] \to R$ with the usual uniform convergence topology. Given $x^0 \in X$ and the solution x of (7.1.4) and (7.1.5), we may define

$$S_t x^0(\tau) = x(t + \tau), \qquad \text{for } \tau \in [-1, 0]. \tag{7.1.6}$$

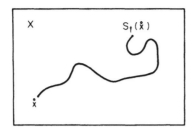

FIGURE 7.1.1. The trajectory of a continuous time process in the phase space X. At time $t = 0$ the system is at x^0, and at time t it is at $S_t(x^0)$.

If $x(t)$ is the solution of (7.1.4)–(7.1.5), then, since F is not an explicit function of t, $x(t + a)$, $a \geq 0$, is also a solution to the problem. Using this fact it is easy to verify that transformation (7.1.6) satisfies property (7.1.1), although it is impossible to define $S_t x(\tau)$ for $t < 0$. □

A very important difference exists between these two examples with respect to their invertibility. Thus, although the solution to the system of ordinary differential equations in Example 7.1.1 may be studied for $t \leq 0$, in general no solution exists for the differential-delay equation of Example 7.1.2 when $t < 0$. This lack of invertibility is generally the case for delay-differential equations and, indeed, for many continuous time systems whose phase space X is not finite dimensional (e.g., some partial differential equations).

7.2 Dynamical and Semidynamical Systems

It is possible to establish many results for continuous time processes in a phase space X endowed with no other property than a measure μ, as was done in earlier chapters for discrete time processes. However it is simpler to consider continuous time processes in a measure space that is also equipped with a topology. Thus, from this point on, let X be a topological Hausdorff space and $\mathcal{A}$ the σ-algebra of Borel sets, that is, the smallest σ-algebra that contains all open, and thus closed, subsets of X.

Dynamical Systems

Definition 7.2.1. A **dynamical system** $\{S_t\}_{t \in R}$ on X is a family of transformations $S_t \colon X \to X$, $t \in R$, satisfying

(a) $S_0(x) = x$ for all $x \in X$;

(b) $S_t(S_{t'}(x)) = S_{t+t'}(x)$ for all $x \in X$, with $t, t' \in R$; and

(c) The mapping $(t, x) \to S_t(x)$ from $X \times R$ into X is continuous.

Remark 7.2.1. System (7.1.2) of ordinary differential equations, introduced in the preceding section, is clearly an example of a dynamical system. □

Remark 7.2.2. It is clear from the group property of Definition 7.2.1 that

$$S_t(S_{-t}(x)) = x \quad \text{and} \quad S_{-t}(S_t(x)) = x \qquad \text{for all } t \in R.$$

Thus, for all $t_0 \in R$, any transformation S_{t_0} of a dynamical system $\{S_t\}_{t \in R}$ is invertible. □

In applied problems the space X is customarily called the **phase space** of the dynamical system $\{S_t\}_{t \in R}$, whereas, for every fixed $x^0 \in X$, the function $S_t(x^0)$, considered as a function of t, is called a **trajectory** of the system. The trajectories of a dynamical system $\{S_t\}_{t \in R}$ in its phase space X are of only three possible types, as shown in Figure 7.2.1a,b,c for $X = R^2$. First (Figure 7.2.1a), the trajectory can be a stationary point x^0 such that

$$S_t(x^0) = x^0 \qquad \text{for all } t \in R.$$

Second, as shown in Figure 7.2.1b, the trajectory can be periodic with period $\omega > 0$, that is,

$$S_{t+\omega}(x^0) = S_t(x^0) \qquad \text{for all } t \in R.$$

Finally, the trajectory can be nonintersecting (see Figure 7.2.1c), by which we mean that

$$S_{t_1}(x^0) \neq S_{t_2}(x^0) \qquad \text{for all } t_1 \neq t_2, \qquad \text{with } t_1, t_2 \in R.$$

It is straightforward to show that the trajectory of a dynamical system cannot be of the intersecting nonperiodic form shown in Figure 7.2.1d. To demonstrate this, assume the contrary, that, for a given $x^0 \in X$, we have

$$S_{t_1}(x^0) = S_{t_2}(x^0) \qquad t_2 > t_1.$$

By applying S_{t-t_1} to both sides of this equation, we have

$$S_{t-t_1}(S_{t_1}(x^0)) = S_{t-t_1}(S_{t_2}(x^0)).$$

By the group property (b) of Definition 7.2.1, we also have

$$S_{t-t_1}(S_{t_1}(x^0)) = S_t(x^0)$$

and

$$S_{t-t_1}(S_{t_2}(x^0)) = S_{t+(t_2-t_1)}(x^0).$$

Hence, with $\omega = (t_2 - t_1)$, our assumption leads to

$$S_t(x^0) = S_{t+\omega}(x^0),$$

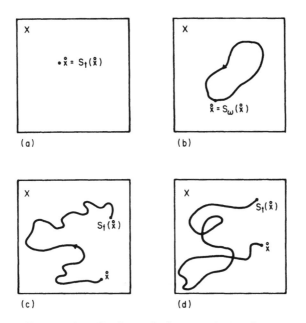

FIGURE 7.2.1. Trajectories of a dynamical system in its phase space X. In (a) the trajectory is a stationary point, whereas in (b) the trajectory is a periodic orbit. Trajectory (c) is of the nonintersecting type. The intersecting trajectory shown in (d) is not possible in a dynamical system.

implying that the only possible intersecting trajectories of a dynamical system are periodic.

However, it is often the case that the evolution in time of data is observed to be of the intersecting nonperiodic type. For example, the two-dimensional projection of the trajectory of a three-dimensional system might easily be of this type. The projection of a trajectory of a dynamical system is called the trace of the system. The following is a more precise definition.

Definition 7.2.2. Let X and Y be two topological Hausdorff spaces, $\phi: Y \to X$ a given continuous function, and $S_t: Y \to Y$ a given dynamical system on Y. A function $g: R \to X$ is called the **trace** of the dynamical system $\{S_t\}_{t \in R}$ if there is a $y \in Y$ such that

$$g(t) = \phi(S_t(y)) \qquad \text{for all } t \in R.$$

From our precise definition of the trace of a dynamical system, the following obvious question arises: Given an observed continuous function in a space X that is intersecting and nonperiodic, when is this function the trace of a dynamical system $\{S_t\}_{t\in R}$ operating in some higher-dimensional phase space Y? The answer is as surprising as it is simple, but it turns out that *all continuous functions in X are traces of a single dynamical system!* This is stated more formally in the following theorem.

Theorem 7.2.1. *Let X be an arbitrary topological Hausdorff space. Then there is another topological Hausdorff space Y, a dynamical system $\{S_t\}_{t\in R}$ operating in Y, and a continuous function $\phi\colon Y \to X$ such that every continuous function $g\colon R \to X$ is the trace of $\{S_t\}_{t\in R}$ that is, for every g there is a $y \in Y$ such that*

$$g(t) = \phi(S_t(y)) \qquad for\ all\ t \in R.$$

Proof. Let Y be the space of all continuous functions from R into X (note that the elements of space Y are functions, not points). Let a dynamical system $\{S_{t'}\}_{t'\in R}, S_{t'}\colon Y \to Y$, operating on Y, be a simple shift so that starting from a given $y \in Y$ we have, after the operation of $S_{t'}$, a new function $y(t + t')$. This may be represented by a diagram,

$$y(t) \xrightarrow{\ S_{t'}\ } y(t+t'),$$

or, more formally,

$$S_{t'}(y)(t) = y(t+t').$$

Define a projection $\phi\colon Y \to X$ by

$$\phi(y) = y(0),$$

then projection ϕ is just the evaluation of y at point $t = 0$. Let $g\colon R \to X$ be an arbitrary continuous function so that, by our definitions,

$$S_{t'}(g)(t) = g(t+t')$$

and

$$\phi(S_{t'}(g)) = S_{t'}g(0) = g(t'),$$

showing that g is the trace of a trajectory of the dynamical system $\{S_{t'}\}_{t'\in R}$ operating in Y, namely, a trajectory starting from the initial point g. Further, Y will be a topological Hausdorff space, and $(t', y) \to S_{t'}(y)$ a continuous mapping if we equip the function space Y with the topology of uniform convergence on compact intervals. This, coupled with the trivial observation that $S_{t'+t''}(y) = S_{t'}(S_{t''}(y))$ and $S_0(y) = y$, completes the proof of the theorem. ■

Remark 7.2.3. Note that the proof of this theorem rests on the identification of the functions on X as the objects on which the new dynamical system $\{S_{t'}\}_{t'\in R}$ operates. □

Semidynamical Systems

Definition 7.2.3. A semidynamical system $\{S_t\}_{t \geq 0}$ on X is a family of transformations $S_t \colon X \to X$, $t \in R^+$, satisfying

(a) $S_0(x) = x$ for all $x \in X$;

(b) $S_t(S_{t'}(x)) = S_{t+t'}(x)$ for all $x \in X$, with $t, t' \in R^+$;

(c) The mapping $(t, x) \to S_t(x)$ from $X \times R^+$ into X is continuous.

Remark 7.2.4. The only difference between dynamical and semidynamical systems is contained in the group property [compare conditions (b) of Definitions 7.2.1 and 7.2.3]. The consequence of this difference is most important, however, because semidynamical systems, in contrast to dynamical systems, are not invertible. It is this property that makes the study of semidynamical systems so important for applications. Henceforth, we will confine our attention to semidynamical systems. □

Remark 7.2.5. An examination of the proof of Theorem 7.2.1 shows that it is also true for semidynamical systems. □

Remark 7.2.6. On occasion a family of transformations $\{S_t\}_{t \geq 0}$ satisfying properties (a) and (b) will be called a **semigroup of transformations**. This is because property (b) in Definition 7.2.3 ensures that transformations S_t form an Abelian semigroup in which the group operation is the composition of two functions. Thus a semidynamical system is a **continuous semigroup**. □

Remark 7.2.7. The area of topological dynamics examines the behavior of semidynamical systems from a topological perspective. Here, however, since we are primarily interested in highly irregular behaviors, our main tools will be measures on X. □

7.3 Invariance, Ergodicity, Mixing, and Exactness in Semidynamical Systems

Invariance and the Individual Ergodic Theorem

From the continuity property (c) of Definition 7.2.3, all our transformations S_t are measurable, that is, for all $A \in \mathcal{A}$,

$$S_t^{-1}(A) \in \mathcal{A}$$

where, as usual, $S_t^{-1}(A)$ denotes the counterimage of A, namely, the set of all points x such that $S_t(x) \in A$. Thus we can state the following definition.

Definition 7.3.1. A measure μ is called **invariant** under a family $\{S_t\}$ of measurable transformations $S_t\colon X \to X$ if

$$\mu(S_t^{-1}(A)) = \mu(A) \qquad \text{for all } A \in \mathcal{A}. \tag{7.3.1}$$

As for discrete time processes, we will say interchangeably either that a measure is invariant under $\{S_t\}$ or that transformations $\{S_t\}$ are **measure preserving** when equation (7.3.1) holds.

Given a finite invariant measure μ, we can formulate a continuous time analog of Theorem 4.2.3, which is also known as the Birkhoff individual ergodic theorem.

Theorem 7.3.1. *Let μ be a finite invariant measure with respect to the semidynamical system $\{S_t\}_{t\geq0}$, and let $f\colon X \to R$ be an arbitrary integrable function. Then the limit*

$$f^*(x) = \lim_{T\to\infty} \frac{1}{T} \int_0^T f(S_t(x))\,dt \tag{7.3.2}$$

exists for all $x \in X$ except perhaps for a set of measure zero.

Proof. This theorem may be rather easily demonstrated using the corresponding discrete time result, Theorem 4.2.3, if we assume, in addition, that for almost all $x \in X$ the integrand $f(S_t(x))$ is a bounded measurable function of t.

Set

$$g(x) = \int_0^1 f(S_t(x))\,dt$$

and assume at first that T is an integer, $T = n$. Note also that the group property (b) of semidynamical systems implies that

$$f(S_t(x)) = f(S_{t-k}(S_k(x))).$$

Then the integral on the right-hand side of (7.3.2) may be written as

$$\frac{1}{T} \int_0^T f(S_t(x))\,dt = \frac{1}{n} \int_0^n f(S_t(x))\,dt$$

$$= \frac{1}{n} \sum_{k=0}^{n-1} \int_k^{k+1} f(S_t(x))\,dt$$

$$= \frac{1}{n} \sum_{k=0}^{n-1} \int_k^{k+1} f(S_{t-k}(S_k(x)))\,dt$$

$$= \frac{1}{n} \sum_{k=0}^{n-1} \int_0^1 f(S_{t'}(S_k(x))) \, dt'$$

$$= \frac{1}{n} \sum_{k=0}^{n-1} g(S_k(x)).$$

However, $S_k = S_1 \circ S_{k-1} = S_1 \circ \overset{k}{\cdots} \circ S_1 = S_1^k$, so that

$$\lim_{n \to \infty} \frac{1}{n} \int_0^n f(S_t(x)) \, dt = \lim_{n \to \infty} \frac{1}{n} \sum_{k=0}^{n-1} g(S_1^k(x)),$$

and the right-hand side exists by Theorem 4.2.3. Call this limit $f^*(x)$.

If T is not an integer, let n be the largest integer such that $n < T$. Then we may write

$$\frac{1}{T} \int_0^T f(S_t(x)) \, dt = \frac{n}{T} \cdot \frac{1}{n} \int_0^n f(S_t(x)) \, dt + \frac{1}{T} \int_n^T f(S_t(x)) \, dt.$$

As $T \to \infty$, the first term on the right-hand side converges to $f^*(x)$, as we have shown previously, whereas the second term converges to zero since $f(S_t(x))$ is bounded. ∎

As in the discrete time case, the limit $f^*(x)$ satisfies two conditions:

(C1) $f^*(S_t(x)) = f^*(x)$, a.e. in x for every $t \geq 0$, $\qquad$ (7.3.3)

and

(C2) $\int_X f^*(x) \, dx = \int_X f(x) \, dx.$ $\qquad$ (7.3.4)

Ergodicity and Mixing

We now develop the notions of ergodicity and mixing for semidynamical systems. Exact semidynamical systems are considered in the next section.

Under the action of a semidynamical system $\{S_t\}_{t \geq 0}$, a set $A \in \mathcal{A}$ is called **invariant** if

$$S_t^{-1}(A) = A \qquad \text{for } t \geq 0. \qquad (7.3.5)$$

Again we require that for every $t \geq 0$ the equality (7.3.5) is satisfied modulo zero (see Remark 3.1.3). By using this notion of invariant sets, we can define ergodicity for semidynamical systems.

Definition 7.3.2. A semidynamical system $\{S_t\}_{t \geq 0}$, consisting of nonsingular transformations $S_t: X \to X$ is **ergodic** if every invariant set $A \in \mathcal{A}$

is such that either $\mu(A) = 0$ or $\mu(X \setminus A) = 0$. (Recall that a set A for which $\mu(A) = 0$ or $\mu(X \setminus A) = 0$ is called **trivial**.)

Example 7.3.1. Again we consider the example of rotation on the unit circle, originally introduced in Example 4.2.2. Now $X = [0, 2\pi)$ and

$$S_t(x) = x + \omega t \quad (\text{mod } 2\pi). \tag{7.3.6}$$

S_t is measure preserving (with respect to the natural Borel measure on the circle) and, for $\omega \neq 0$, it is also ergodic. To see this, first pick $t = t_0$ such that $\omega t_0 / 2\pi$ is irrational. Then the transformation $S_{t_0}: X \to X$ is ergodic, as was shown in Example 4.4.1. Since S_{t_0} is ergodic for at least one t_0, every (invariant) set A that satisfies $S_{t_0}^{-1}(A) = A$ must be trivial by Definition 4.2.1. Thus, any set A that satisfies (7.3.5) must likewise be trivial, and the semidynamical system $\{S_t\}_{t \geq 0}$ with S_t given by (7.3.6) is ergodic. $\square$

Remark 7.3.1. It is interesting to note that, for any t_0 commensurate with $2\pi/\omega$ (e.g., $t_0 = \pi/\omega$), the transformation S_{t_0} is not ergodic. This curious result illustrates a very general property of semidynamical systems: For a given ergodic semidynamical system $\{S_t\}_{t \geq 0}$, there might be a specific t_0 for which S_{t_0} is not ergodic. However, if at least one S_{t_0} is ergodic, then the entire semidynamical system $\{S_t\}_{t \geq 0}$ is ergodic. $\square$

We now turn our attention to mixing in semidynamical systems, starting with the following definition.

Definition 7.3.3. A semidynamical system $\{S_t\}_{t \geq 0}$ on a measure space $(X, \mathcal{A}, \mu)$ with a normalized invariant measure μ is **mixing** if

$$\lim_{t \to \infty} \mu(A \cap S_t^{-1}(B)) = \mu(A)\mu(B) \quad \text{for all } A, B \in \mathcal{A}. \tag{7.3.7}$$

Thus, in continuous time systems, the interpretation of mixing is the same as for discrete time systems. For example, consider all points x in the set $A \cap S_t^{-1}(B)$, that is, points x such that $x \in A$ and $S_t(x) \in B$. From (7.3.7), for large t the measure of these points is just $\mu(A)\mu(B)$, which means that the fraction of points starting in A that eventually are in B is given by the product of the measures of A and B in the phase space X.

By Definition 7.3.3 the semidynamical system $\{S_t\}_{t \geq 0}$, consisting of rotation on the unit circle given by (7.3.6), is evidently not mixing. This is because, given any two nontrivial disjoint sets, $A, B \in \mathcal{A}$, the left-hand side of (7.3.7) is always zero for $\omega t = 2\pi n$ (n an integer), whereas $\mu(A)\mu(B) \neq 0$. A continuous time system that is mixing is illustrated in Example 7.7.2.

Remark 7.3.2. The concepts of ergodicity and mixing are also applicable to dynamical systems. In this case, condition (7.3.7) can be replaced by

$$\lim_{t \to \infty} \mu(A \cap S_t(B)) = \mu(A)\mu(B) \tag{7.3.8}$$

since

$$\mu(A \cap S_t(B)) = \mu(S_t^{-1}(A \cap S_t(B))) = \mu(S_t^{-1}(A) \cap B). \quad \square$$

Exactness

Definition 7.3.4. Let $(X, \mathcal{A}, \mu)$ be a normalized measure space. A measure-preserving semidynamical system $\{S_t\}_{t \geq 0}$ such that $S_t(A) \in \mathcal{A}$ for $A \in \mathcal{A}$ is **exact** if

$$\lim_{t \to \infty} \mu(S_t(A)) = 1 \qquad \text{for all } A \in \mathcal{A}, \mu(A) > 0. \qquad (7.3.9)$$

Example 11.1.1 illustrates exactness for a continuous time semidynamical system.

Remark 7.3.3. As in discrete time systems, exactness of $\{S_t\}_{t \geq 0}$ implies that $\{S_t\}_{t \geq 0}$ is mixing. $\square$

Remark 7.3.4. Due to their invertibility, dynamical systems cannot be exact. This is easily seen, since $\mu(S_{-t}(S_t(A))) = \mu(A)$ and, thus, the limit in (7.3.9) is $\mu(A)$ and not 1, for all $A \in \mathcal{A}$. If the system is nontrivial and contains a set A such that $0 < \mu(A) < 1$, then, of course, condition (7.3.9) is not satisfied. $\square$

7.4 Semigroups of the Frobenius–Perron and Koopman Operators

As we have seen in the discrete time case, many properties of dynamical systems are more easily studied by examining ensembles of trajectories rather than single trajectories. This is primarily because the ensemble approach leads to semigroups of linear operators, and, hence, the techniques of linear functional analysis may be applied to a study of their properties.

Since, for any fixed t in a semidynamical system $\{S_t\}_{t \geq 0}$, the transformation S_t is measurable, we can adopt the discrete time definitions of the Frobenius–Perron and Koopman operators directly for the continuous time case.

Frobenius–Perron Operator

Assume that a measure μ on X is given and that all transformations S_t of a semidynamical system $\{S_t\}_{t \geq 0}$ are nonsingular, that is,

$$\mu(S_t^{-1}(A)) = 0 \qquad \text{for each } A \in \mathcal{A} \text{ such that } \mu(A) = 0.$$

Then, analogously to (3.2.2), the condition

$$\int_A P_t f(x)\mu(dx) = \int_{S_t^{-1}(A)} f(x)\mu(dx) \qquad \text{for } A \in \mathcal{A} \qquad (7.4.1)$$

for each fixed $t \geq 0$ uniquely defines the Frobenius–Perron operator P_t: $L^1(X) \to L^1(X)$, corresponding to the transformation S_t.

It is easy to show, with the aid of (7.4.1), that P_t has the following properties:

(FP1) $P_t(\lambda_1 f_1 + \lambda_2 f_2) = \lambda_1 P_t f_1 + \lambda_2 P_t f_2,$ for all $f_1, f_2 \in L^1$,

$\lambda_1, \lambda_2 \in R;$ (7.4.2)

(FP2) $P_t f \geq 0,$ if $f \geq 0;$ (7.4.3)

(FP3) $\int_X P_t f(x)\mu(dx) = \int_X f(x)\mu(dx),$ for all $f \in L^1.$ (7.4.4)

Thus, for every fixed t, the operator P_t: $L^1(X) \to L^1(X)$ is a **Markov operator**.

The entire family of Frobenius–Perron operators P_t: $L^1(X) \to L^1(X)$ satisfies some properties similar to (a) and (b) of Definition 7.2.3. To see this, first note that since $S_{t+t'} = S_t \circ S_{t'}$, then $S_{t+t'}^{-1} = S_{t'}^{-1}(S_t^{-1})$ and, thus,

$$\int_A P_{t+t'} f(x)\mu(dx) = \int_{S_{t+t'}^{-1}(A)} f(x)\mu(dx) = \int_{S_{t'}^{-1}(S_t^{-1}(A))} f(x)\mu(dx)$$

$$= \int_{S_t^{-1}(A)} P_{t'} f(x)\mu(dx)$$

$$= \int_A P_t(P_{t'} f(x))\mu(dx).$$

This implies that

$$P_{t+t'} f = P_t(P_{t'} f) \qquad \text{for all } f \in L^1(X), t, t' \geq 0 \qquad (7.4.5)$$

and, thus, P_t satisfies a group property analogous to (b) of Definition 7.2.3. Further, since $S_0(x) = x$, we have $S_0^{-1}(A) = A$ and, consequently,

$$\int_A P_0 f(x)\mu(dx) = \int_{S_0^{-1}(A)} f(x)\mu(dx) = \int_A f(x)\mu(dx)$$

implying that

$$P_0 f = f \qquad \text{for all } f \in L^1(X). \qquad (7.4.6)$$

Hence P_t satisfies properties (a) and (b) of the definition of a semidynamical system.

The properties of P_t in (7.4.2)–(7.4.6) are important enough to warrant the following definition.

Definition 7.4.1. Let $(X, \mathcal{A}, \mu)$ be a measure space. A family of operators P_t: $L^1(X) \to L^1(X)$, $t \geq 0$, satisfying properties (7.4.2)–(7.4.6) is called a **stochastic semigroup**. Further, if, for every $f \in L^1$ and $t_0 \geq 0$,

$$\lim_{t \to t_0} \|P_t f - P_{t_0} f\| = 0,$$

then this semigroup is called **continuous**.

A very important and useful property of stochastic semigroups is that

$$\|P_t f_1 - P_t f_2\| \leq \|f_1 - f_2\| \qquad \text{for } f_1, f_2 \in L^1, \tag{7.4.7}$$

and, thus, from the group property (7.4.5), the function $t \to \|P_t f_1 - P_t f_2\|$ is a nonincreasing function of t. This is simply shown by

$$\|P_{t+t'} f_1 - P_{t+t'} f_2\| = \|P_{t'}(P_t f_1 - P_t f_2)\| \leq \|P_t f_1 - P_t f_2\|,$$

which follows from (7.4.7). By using this property, we may now proceed to prove a continuous time analog of Theorem 5.6.2.

Theorem 7.4.1. Let $\{P_t\}_{t \geq 0}$ be a semigroup of Markov operators, not necessarily continuous. Assume that there is an $h \in L^1$, $h(x) \geq 0$, $\|h\| > 0$ such that

$$\lim_{t \to \infty} \|(P_t f - h)^-\| = 0 \qquad \text{for every } f \in D. \tag{7.4.8}$$

Then there is a unique density f_* such that $P_t f_* = f_*$ for all $t \geq 0$. Furthermore,

$$\lim_{t \to \infty} P_t f = f_* \qquad \text{for every } f \in D. \tag{7.4.9}$$

Proof. Take any $t_0 > 0$ and define $P = P_{t_0}$ so that $P_{nt_0} = P^n$. Then, from (7.4.8)

$$\lim_{n \to \infty} \|(P^n f - h)^-\| = 0 \qquad \text{for each } f \in D.$$

Thus, by Theorem 5.6.2, there is a unique $f_* \in D$ such that $Pf_* = f_*$ and

$$\lim_{n \to \infty} P^n f = f_* \qquad \text{for every } f \in D.$$

Having shown that $P_t f_* = f_*$ for the set $\{t_0, 2t_0, \ldots\}$, we now turn to a demonstration that $P_t f_* = f_*$ for all t. Pick a particular time t', set $f_1 = P_{t'} f_*$, and note that $f_* = P^n f_* = P_{nt_0} f_*$. Therefore,

$$\begin{aligned}
\|P_{t'} f_* - f_*\| &= \|P_{t'}(P_{nt_0} f_*) - f_*\| \\
&= \|P_{nt_0}(P_{t'} f_*) - f_*\| \\
&= \|P^n(P_{t'} f_*) - f_*\| \\
&= \|P^n f_1 - f_*\|. \tag{7.4.10}
\end{aligned}$$

Thus, since,

$$\lim_{n \to \infty} \|P^n f_1 - f_*\| = 0$$

and the left-hand side of (7.4.10) is independent of n, we must have $\|P_{t'} f_* - f_*\| = 0$ so $P_{t'} f_* = f_*$. Since t' is arbitrary, we have $P_t f_* = f_*$ for all $t \geq 0$.

Finally, to show (7.4.9), pick a function $f \in D$ so that $\|P_t f - f_*\| = \|P_t f - P_t f_*\|$ is a nonincreasing function. Pick a subsequence $t_n = n t_0$. We know from before that $\lim_{n \to \infty} \|P_{t_n} f - f_*\| = 0$. Thus we have a nonincreasing function that converges to zero on a subsequence and, hence

$$\lim_{t \to \infty} \|P_t f - f_*\| = 0. \quad \blacksquare$$

Remark 7.4.1. The proof of this theorem illustrates a very important property of stochastic semigroups: namely, a stochastic semigroup $\{P_t\}_{t \geq 0}$ is called **asymptotically stable** if there exists a unique $f_* \in D$ such that $P f_* = f_*$ and if condition (7.4.9) holds for every $f \in D$.   □

Remark 7.4.2. From the above definition, it immediately follows that the asymptotic stability of a semigroup $\{P_t\}_{t \geq 0}$ implies the asymptotic stability of the sequence $\{P_{t_0}^n\}$ for arbitrary $t_0 > 0$. The proof of Theorem 7.4.1 shows that the converse holds, that is, if for some $t_0 > 0$ the sequence $\{P_{t_0}^n\}$ is asymptotically stable, then the semigroup $\{P_t\}_{t \geq 0}$ is also asymptotically stable.   □

Stochastic semigroups that are not semigroups of Frobenius–Perron operators can arise, as illustrated by the following example.

Example 7.4.1. Let $X = R$, $f \in L^1(X)$, and define $P_t: L^1(X) \to L^1(X)$ by

$$P_t f(x) = \int_{-\infty}^{\infty} K(t, x, y) f(y)\, dy, \qquad P_0 f(x) = f(x), \qquad (7.4.11)$$

where

$$K(t, x, y) = \frac{1}{\sqrt{2\pi\sigma^2 t}} \exp\left[-\frac{(x - y)^2}{2\sigma^2 t}\right]. \qquad (7.4.12)$$

It may be easily shown that the kernel $K(t, x, y)$ satisfies:

(a) $K(t, x, y) \geq 0$;

(b) $\displaystyle\int_{-\infty}^{\infty} K(t, x, y)\, dx = 1$; and

(c) $K(t + t', x, y) = \displaystyle\int_{-\infty}^{\infty} K(t, x, z) K(t', z, y)\, dz.$

From these properties it follows that P_t defined by (7.4.11) forms a continuous stochastic semigroup. The demonstration that $\{P_t\}_{t\geq 0}$ defined by (7.4.11) and (7.4.12) is not a semigroup of Frobenius–Perron operators is postponed to Remark 7.10.2.

That (7.4.11) and (7.4.12) look familiar should come as no surprise as the function $u(t,x) = P_t f(x)$ is the solution to the heat equation

$$\frac{\partial u}{\partial t} = \frac{\sigma^2}{2}\frac{\partial^2 u}{\partial x^2} \qquad \text{for } t > 0, x \in R \qquad (7.4.13)$$

with the initial condition

$$u(0,x) = f(x) \qquad \text{for } x \in R. \quad \square \qquad (7.4.14)$$

The Koopman Operator

Again let $\{S_t\}_{t\geq 0}$ be a semigroup of nonsingular transformations S_t in our topological Hausdorff space X with Borel σ-algebra $\mathcal{A}$ and measure μ. Recall that the S_t are nonsingular if, and only if, for every $A \in \mathcal{A}$ such that $\mu(A) = 0$, $\mu(S_t^{-1}(A)) = 0$. Further, let $f \in L^\infty(X)$. Then the function $U_t f$, defined by

$$U_t f(x) = f(S_t(x)), \qquad (7.4.15)$$

is again a function in $L^\infty(X)$. Equation (7.4.15) defines, for every $t \geq 0$, the Koopman operator associated with the transformation S_t. The family of operators $\{U_t\}_{t\geq 0}$, defined by (7.4.15), satisfies all the properties of the discrete time Koopman operator introduced in Section 3.3.

It is also straightforward to show that $\{U_t\}_{t\geq 0}$ is a semigroup. To check this, first note from the defining formula (7.4.15) that

$$U_{t+t'}f(x) = f(S_{t+t'}(x)) = f(S_t(S_{t'}(x)))$$
$$= U_t(U_{t'}f(x)),$$

which implies

$$U_{t+t'}f \equiv U_t(U_{t'}f) \qquad \text{for all } f \in L^\infty.$$

Furthermore, $U_0 f(x) = f(S_0(x)) = f(x)$, or

$$U_0 f \equiv f \qquad \text{for all } f \in L^\infty,$$

so that $\{U_t\}_{t\geq 0}$ is a semigroup.

Finally, the Koopman operator is adjoint to the Frobenius–Perron operator, or

$$\langle P_t f, g\rangle = \langle f, U_t g\rangle \qquad \text{for all } f \in L^1(X), g \in L^\infty(X) \text{ and } t \geq 0. \quad (7.4.16)$$

The family of Koopman operators is, in general, not a stochastic semigroup because U_t does not map L^1 into itself (though it does map L^∞ into itself) and satisfies the inequality

$$\text{ess sup}\,|U_t f| \leq \text{ess sup}\,|f|$$

FIGURE 7.4.1. Plots of $f(x)$ and $T_t f(x) = f(x - ct)$, for $c > 0$.

instead of preserving the norm. In order to have a common notion for families of operators such as $\{P_t\}$ and $\{U_t\}$, we introduce the following definition.

Definition 7.4.2. Let $L = L^p$, $1 \leq p \leq \infty$. A family $\{T_t\}_{t \geq 0}$ of operators, $T_t \colon L \to L$, defined for $t \geq 0$, is called a **semigroup of contracting linear operators** (or a **semigroup of contractions**) if T_t satisfies the following conditions:

(a) $T_t(\lambda_1 f_1 + \lambda_2 f_2) = \lambda_1 T_t f_1 + \lambda_2 T_t f_2$, for $f_1, f_2 \in L$, $\lambda_1, \lambda_2 \in R$;

(b) $\|T_t f\|_L \leq \|f\|_L$ for $f \in L$;

(c) $T_0 f = f$, for $f \in L$; and

(d) $T_{t+t'} f = T_t(T_{t'} f)$, for $f \in L$.

Moreover, if

$$\lim_{t \to t_0} \|T_t f - T_{t_0} f\|_L = 0, \qquad \text{for } f \in L, t_0 \geq 0,$$

then this semigroup is called **continuous**.

Example 7.4.2. Consider the family of operators $\{T_t\}_{t \geq 0}$ defined by (see Figure 7.4.1)
$$T_t f = f(x - ct) \qquad \text{for } x \in R, t \geq 0. \tag{7.4.17}$$
These operators map $L = L^p(R)$, $1 \leq p \leq \infty$, into itself, satisfy properties (a)–(d) of Definition 7.4.2., and form a semigroup of contractions.

To see that property (b) holds for T_t, use the "change of variables" formula,

$$\|T_t f\|_{L^p}^p = \int_{-\infty}^{\infty} |f(x - ct)|^p \, dx = \int_{-\infty}^{\infty} |f(y)|^p \, dy = \|f\|_{L^p}^p$$

when $p < \infty$, and the obvious equality,

$$\|T_t f\|_{L^\infty} = \operatorname*{ess\,sup}_x |f(x - ct)| = \operatorname*{ess\,sup}_x |f(x)| = \|f\|_{L^\infty}$$

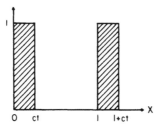

FIGURE 7.4.2. Function $|1_{(1,1+ct]}(x) - 1_{[0,ct)}(x)|$ versus x.

when $p = \infty$. The remaining properties (a), (c), and (d) follow immediately from the definition of T_t in equation (7.4.17).

Finally, we note that if $p = 1$ then this semigroup of contractions is continuous. To see this, first use

$$\|T_t f - T_{t_0} f\|_{L^1} = \int_{-\infty}^{\infty} |f(x - ct) - f(x - ct_0)| \, dx$$
$$= \int_{-\infty}^{\infty} |f(y) - f(y - c(t_0 - t))| \, dy$$

and note that the right-hand side converges to zero by Corollary 5.1.1. A slightly more complicated calculation shows that $\{T_t\}_{t \geq 0}$ is a continuous semigroup of contractions for every $1 \leq p < \infty$. However, in L^∞ the semigroup $\{T_t\}_{t \geq 0}$ given by (7.4.17) is not continuous except in the trivial case when $c = 0$. This may be easily shown by setting $f = 1_{[0,1]}$. We then have

$$T_t f(x) = 1_{[0,1]}(x - ct) = 1_{[ct,ct+1]}(x)$$

and, as a consequence,

$$\|T_t f - f\|_{L^\infty} = \text{ess} \sup_x |1_{(1,1+ct]}(x) - 1_{[0,ct)}(x)| = 1$$

for $0 < ct < 1$. Thus $\|T_t f - f\|_{L^\infty}$ does not converge to zero as $t \to 0$. This may be simply interpreted as shown in Figure 7.4.2 where the hatched areas corresponding to the function $|1_{(1,1_ct]} - 1_{[0,ct)}|$ disappear as $t \to 0$ but the heights do not. $\square$

7.5   Infinitesimal Operators

The problems associated with the study of continuous time processes are more difficult than those encountered in discrete time systems. This is partially due to concerns over continuity of processes with respect to time. Also, equivalent formulations of discrete and continuous time properties

may appear more complicated in the continuous case because of the use
of integrals rather than summations, for example, in the Birkhoff ergodic
theorem. However, there is one great advantage in the study of continuous
time problems over discrete time dynamics, and this is the existence of a
new tool—the infinitesimal operator.

In the case of a semidynamical system $\{S_t\}_{t\geq 0}$ arising from a system of
ordinary differential equations (7.1.2), the infinitesimal operator is simply
the function $F(x)$. This connection between the infinitesimal operator and
$F(x)$ stems from the formula

$$\lim_{t\to 0} \frac{x(t) - x(0)}{t} = F(x^0),$$

where $x(t)$ is the solution of (7.1.2) with the initial condition (7.1.3). This
can be rewritten in terms of the transformations S_t as

$$\lim_{t\to 0} \frac{S_t(x^0) - x^0}{t} = F(x^0).$$

This relation offers some insight into how the infinitesimal operator may
be defined for semigroups of contractions in general, and for semigroups of
the Frobenius–Perron and Koopman operators in particular.

Definition 7.5.1. Let $L = L^p$, $1 \leq p \leq \infty$, and $\{T_t\}_{t\geq 0}$ be a semigroup of
contractions. We define by $\mathcal{D}(A)$ the set of all $f \in L$ such that the limit

$$Af = \lim_{t\to 0} \frac{T_t f - f}{t} \tag{7.5.1}$$

exists, where the limit is considered in the sense of strong convergence (cf.
Definition 2.3.3). Thus (7.5.1) is equivalent to

$$\lim_{t\to 0} \left\| Af - \frac{T_t f - f}{t} \right\|_L = 0.$$

Operator $A: \mathcal{D}(A) \to L$ is called the **infinitesimal operator**. It is evident
that the subspace $\mathcal{D}(A)$ is linear or that

$$\lambda_1 f_1 + \lambda_2 f_2 \in \mathcal{D}(A) \qquad \text{for all } f_1, f_2 \in \mathcal{D}(A), \text{ and } \lambda_1, \lambda_2 \in R.$$

Furthermore, operator $A: \mathcal{D}(A) \to L$ is linear or

$$A(\lambda_1 f_1 + \lambda_2 f_2) = \lambda_1 A f_1 + \lambda_2 A f_2 \qquad \text{for all } f_1, f_2 \in \mathcal{D}(A) \text{ and } \lambda_1, \lambda_2 \in R.$$

In general, the domain $\mathcal{D}(A)$ of operator A is not the entire space L.

Before deriving the infinitesimal operators for the Frobenius–Perron and
Koopman semigroups, we consider the following example.

Example 7.5.1. Let $X = R$ and $L = L^p(R)$, $1 \leq p < \infty$. Consider a
semigroup $\{T_t\}_{t\geq 0}$ on L defined, as in Example 7.4.2, by

$$T_t f(x) = f(x - ct)$$

cf. Figure 7.4.1). By the mean value theorem, if f is C^1 on R, then

$$\frac{f(x - ct) - f(x)}{t} = -cf'(x - \theta ct),$$

where $|\theta| \leq 1$ and $f' \equiv df/dx$. Thus, if f' is bounded and uniformly continuous on R, then

$$Af = \lim_{t \to 0} \frac{T_t f - f}{t} = -cf',$$

and the limit is uniform on R and consequently strong in L^∞. Further, if f (and thus f') has **compact support** (zero outside a bounded interval), then the limit is strong in every L^p, $1 \leq p \leq \infty$. Thus, all such f belong to $\mathcal{D}(A)$ and for them A is just differentiation with respect to x and multiplication by $-c$. □

In studying infinitesimal operators and many other problems of analysis, functions that are equal to zero outside a compact set play an important role. It is customary to call such mappings **functions with compact support**. This notion does not coincide very well with our definition of support given by equation (3.1.8) even though it is commonly accepted. Thus, we will also use this traditional terminology, hoping that it will not lead to confusion or misunderstanding.

Having introduced the notion of infinitesimal operators, and illustrated their calculation in Example 7.5.1, we now wish to state a theorem that makes explicit the relation among semigroups of contractions, infinitesimal operators, and differential equations.

First, however, we must define the strong derivative of a function with values in $L = L^p$. Given a function $u: \Delta \to L$, where $\Delta \subset R$, and a point $t_0 \in \Delta$, we define the **strong derivative** $u'(t_0)$ by

$$u'(t_0) = \lim_{t \to t_0} \frac{u(t) - u(t_0)}{t - t_0},$$

where the limit is considered in the sense of strong convergence. This definition is equivalent to

$$\lim_{t \to t_0} \left\| \frac{u(t) - u(t_0)}{t - t_0} - u'(t_0) \right\|_L = 0. \tag{7.5.2}$$

By using this concept, we can see that the value of the infinitesimal operator for $f \in \mathcal{D}(A)$, Af, is simply the derivative of the function $u(t) = T_t f$ at $t = 0$. The following theorem gives a more sophisticated relation between the strong derivative and the infinitesimal operator.

Theorem 7.5.1. Let $\{T_t\}_{t \geq 0}$ be a continuous semigroup of contractions acting on L, and $A: \mathcal{D}(A) \to L$ the corresponding infinitesimal operator. Further, let $u(t) = T_t f$ for fixed $f \in \mathcal{D}(A)$. Then $u(t)$ satisfies the following properties:

(1) $u(t) \in \mathcal{D}(A)$ for $t \geq 0$;

(2) $u'(t)$ exists for $t \geq 0$; and

(3) $u(t)$ satisfies the differential equation

$$u'(t) = Au(t) \qquad \text{for } t \geq 0 \tag{7.5.3}$$

and the initial condition

$$u(0) = f. \tag{7.5.4}$$

Proof. For $t = 0$, properties (1)–(3) are satisfied by assumption. Thus we may concentrate on $t > 0$. Let $t_0 > 0$ be fixed. By the definition of $u(t)$, we have

$$\frac{u(t) - u(t_0)}{t - t_0} = \frac{T_t f - T_{t_0} f}{t - t_0}.$$

Noting that $T_t = T_{t-t_0} T_{t_0}$ for $t > t_0$ this differential quotient may be rewritten as

$$\frac{u(t) - u(t_0)}{t - t_0} = T_{t_0} \left(\frac{T_{t-t_0} f - f}{t - t_0} \right) \qquad \text{for } t > t_0. \tag{7.5.5}$$

Because $f \in \mathcal{D}(A)$, the limit of

$$\frac{T_{t-t_0} f - f}{t - t_0}$$

exists as $t \to t_0$ and gives Af. Thus the limit of (7.5.5) as $t \to t_0$ also exists and is equal to $T_{t_0} Af$. In an analogous fashion, if $t < t_0$, we have $T_{t_0} = T_t T_{t_0-t}$ and, as a consequence,

$$\frac{u(t) - u(t_0)}{t - t_0} = T_t \left(\frac{T_{t_0-t} f - f}{t_0 - t} \right) \qquad \text{for } t < t_0 \tag{7.5.6}$$

and

$$\left\| \frac{u(t) - u(t_0)}{t - t_0} - T_{t_0} Af \right\|_L \leq \left\| T_t \left(\frac{T_{t_0-t} f - f}{t_0 - t} - Af \right) \right\|_L$$
$$+ \| T_t Af - T_{t_0} Af \|_L \leq \left\| \frac{T_{t_0-t} f - f}{t_0 - t} - Af \right\|_L$$
$$+ \| T_t Af - T_{t_0} Af \|_L.$$

Again, since $T_t Af$ converges to $T_{t_0} Af$ as $t \to t_0$, the limit of (7.5.6) exists as $t \to 0$ and is equal to $T_{t_0} Af$. Thus the existence of the derivative $u'(t_0)$ is proved.

Now we can rewrite equation (7.5.5) in the form

$$\frac{u(t) - u(t_0)}{t - t_0} = \frac{T_{t-t_0}(T_{t_0} f) - (T_{t_0} f)}{t - t_0} \qquad \text{for } t > t_0.$$

Since the limit of the differential quotient on the left-hand side exists as $t \to t_0$, the limit on the right-hand side also exists as $t \to t_0$, and we obtain

$$u'(t_0) = AT_{t_0}f,$$

which proves that $T_{t_0}f \in \mathcal{D}(A)$ and that $u'(t_0) = Au(t_0)$.   ■

Remark 7.5.1. The main property of the set $\mathcal{D}(A)$ that follows directly from Theorem 7.5.1 is that, for $f \in \mathcal{D}(A)$, the function $u(t) = T_t f$ is a solution of equations (7.5.3) and (7.5.4). Moreover, the solution can be proved to be unique. Unfortunately, in general $\mathcal{D}(A)$ is not the entire space L, although it can be proved that, for continuous semigroups of contractions, $\mathcal{D}(A)$ is dense in L.   □

In Theorem 7.5.1, the notion of a function $u: [0, \infty) \to L$, where L is again a space of functions, may seem strange. In fact, u actually represents a function of two variables, t and x, since, for each $t \geq 0$, $u(t) \in L^p$. Thus we frequently write $u(t)(x) = u(t, x)$, and equation (7.5.3) is to be interpreted as an equation in two variables.

Applying this theorem to the semigroup considered in Examples 7.4.2 and 7.5.1 with $L = L^p$, $1 \leq p < \infty$, it is clear that this semigroup satisfies equation (7.5.3), where

$$u(t, x) = T_t f(x) = f(x - ct)$$

and

$$Af = -c\frac{df}{dx}, \qquad f \in \mathcal{D}(A).$$

These relations can, in turn, be interpreted as meaning that $u(t, x)$ satisfies the first-order partial differentiation equation

$$\frac{\partial u}{\partial t} + c\frac{\partial u}{\partial x} = 0 \tag{7.5.7}$$

with the initial condition

$$u(0, x) = f(x).$$

Remark 7.5.2. It is important to stress the large difference in the two interpretations of this problem as embodied in equations (7.5.3) and (7.5.7). From the point of view of (7.5.7), $u(t, x)$ is thought of as a *function of isolated coordinates* t and x that evolve independently and whose derivatives $\partial u / \partial t$ and $\partial u / \partial x$ are evaluated at specific points in the (t, x)-plane. However, in the semigroup approach that leads to (7.5.3), we are considering the evolution in time of a *family of functions*, and the derivative $du(t)/dt$ is to be thought of as taken over an entire ensemble of points. This is made somewhat clearer when we take into account that $u(t) = T_t f$ has a time derivative $u'(t_0)$ at a point t_0 if (7.5.2) is satisfied, that is,

$$\lim_{t \to t_0} \int_{-\infty}^{\infty} \left| \frac{u(t)(x) - u(t_0)(x)}{t - t_0} - u'(t_0)(x) \right|^p dx = 0$$

Moreover, $u(t)(x)$ and $u'(t)(x)$ with fixed t are defined as functions of x up to a set of measure zero.   □

7.6   Infinitesimal Operators for Semigroups Generated by Systems of Ordinary Differential Equations

We now turn to an explicit calculation of the infinitesimal operators for the semigroups $\{P_t\}_{t\geq 0}$ and $\{U_t\}_{t\geq 0}$ generated by a d-dimensional system of ordinary differential equations

$$\frac{dx}{dt} = F(x) \tag{7.6.1a}$$

or

$$\frac{dx_i}{dt} = F_i(x), \qquad i = 1,\ldots,d, \tag{7.6.1b}$$

where $x = (x_1,\ldots,x_d)$.

The semigroup of transformations $\{S_t\}_{t\geq 0}$ corresponding to equations (7.6.1) is defined by the formula

$$S_t(x^0) = x(t), \tag{7.6.2}$$

where $x(t)$ is the solution of (7.6.1) corresponding to the initial condition

$$x(0) = x^0. \tag{7.6.3}$$

We will assume that the F_i have continuous derivatives $\partial F_i/\partial x_j$, $i,j = 1,\ldots,d$, and that for every $x^0 \in R^d$ the solution $x(t)$ exists for all $t \in R$. This guarantees that (7.6.2) actually defines a group of transformations. Because of a well-known theorem on the continuous dependence of solutions of differential equations on the initial condition, $\{S_t\}_{t\geq 0}$ is a dynamical system (see Example 7.1.1).

As the derivative of the infinitesimal operator A_K for the Koopman operator is simpler, we start from there. By definition we have

$$U_t f(x^0) = f(S_t(x^0)).$$

Therefore

$$\frac{U_t f(x^0) - f(x^0)}{t} = \frac{f(S_t(x^0)) - f(x^0)}{t} = \frac{f(x(t)) - f(x^0)}{t},$$

so that, if f is continuously differentiable with compact support, then by the mean value theorem

$$\frac{U_t f(x^0) - f(x^0)}{t} = \sum_{i=1}^{d} f_{x_i}(x(\theta t))x_i'(\theta t) = \sum_{i=1}^{d} f_{x_i}(x(\theta t))F_i(x(\theta t)),$$

where $0 < \theta < 1$. Now by using equation (7.6.2), we obtain

$$\frac{U_t f(x^0) - f(x^0)}{t} = \sum_{i=1}^{d} f_{x_i}(S_{\theta t}(x^0)) F_i(S_{\theta t}(x^0)). \tag{7.6.4}$$

Since the derivatives f_{x_i} have compact support

$$\lim_{t \to 0} f_{x_i}(S_{\theta t}(x^0)) F_i(S_{\theta t}(x^0)) = f_{x_i}(x^0) F_i(x^0)$$

uniformly for all x^0. Thus (7.6.4) has a strong limit in L^∞, and the infinitesimal operator A_K is given by

$$A_K f(x) = \sum_{i=1}^{d} \frac{\partial f}{\partial x_i} F_i(x). \tag{7.6.5}$$

Observe that equation (7.6.5) was derived only for functions f with some special properties, namely, continuously differentiable f with compact support. These functions do not form a dense set in L^∞, which is not surprising since it can be proved that the semigroup $\{U_t\}_{t \geq 0}$ is not, in general, continuous in L^∞. It does become continuous in a subspace of L^∞ consisting of all continuous functions with compact support (see Remark 7.6.2).

Hence, if f is continuously differentiable with compact support, then by Theorem 7.5.1 for such f the function

$$u(t, x) = U_t f(x)$$

satisfies the first-order partial differential equation (7.5.3). From (7.6.5) it may be written as

$$\frac{\partial u}{\partial t} - \sum_{i=1}^{d} F_i(x) \frac{\partial u}{\partial x_i} = 0. \tag{7.6.6}$$

Remark 7.6.1. It should be noted that the same equation can be immediately derived for $u(t, x) = f(S_t(x))$ by differentiating the equality $u(t, S_{-t}(x)) = f(x)$ with respect to t. In this case f may be an arbitrary continuously differentiable function, not necessarily having compact support. However, in this case (7.6.6) is satisfied locally at every point (t, x) and is not an evolution equation in L^∞ (cf. Remark 7.5.2). $\square$

We now turn to a derivation of the infinitesimal operator for the semigroup of Frobenius–Perron operators generated by the semigroup of (7.6.1a). This is difficult to do if we start from the formal definition of the Frobenius–Perron operator, that is,

$$\int_A P_t f(x) \mu(dx) = \int_{S_t^{-1}(A)} f(x) \mu(dx) \qquad \text{for } A \in \mathcal{A}.$$

However, the derivation is straightforward if we start from the fact that the Frobenius–Perron and Koopman operators are adjoint, that is,

$$\langle P_t f, g \rangle = \langle f, U_t g \rangle, \qquad \text{for } f \in L^1, g \in L^\infty. \tag{7.6.7}$$

Subtract $\langle f, g \rangle$ from both sides of (7.6.7) to give

$$\langle P_t f - f, g \rangle = \langle f, U_t g - g \rangle$$

or, after division on both sides by t,

$$\langle (P_t f - f)/t, g \rangle = \langle f, (U_t g - g)/t \rangle. \tag{7.6.8}$$

Now let $f \in \mathcal{D}(A_{FP})$ and $g \in \mathcal{D}(A_K)$, where A_{FP} and A_K denote, respectively, the infinitesimal operators for the semigroups of Frobenius–Perron and Koopman operators. Take the limit as $t \to 0$ in (7.6.8) to obtain

$$\langle A_{FP} f, g \rangle = \langle f, A_K g \rangle. \tag{7.6.9}$$

However, from equation (7.6.5) the right-hand side of (7.6.9) can be written as

$$\left\langle f, \sum_{i=1}^{d} F_i \frac{\partial g}{\partial x_i} \right\rangle,$$

provided g is a continuously differentiable function with compact support. If we write out this scalar product explicitly and note that $X = R^d$ and $d\mu = dx_1 \cdots dx_d = dx$, we obtain

$$\left\langle f, \sum_{i=1}^{d} F_i \frac{\partial g}{\partial x_i} \right\rangle = \int_{R^d} f \sum_{i=1}^{d} F_i \frac{\partial g}{\partial x_i} \, dx$$

$$= \sum_{i=1}^{d} \int_{R^d} \left\{ \frac{\partial (f F_i g)}{\partial x_i} - g \frac{\partial (f F_i)}{\partial x_i} \right\} \, dx$$

for $f \in \mathcal{D}(A_{FP})$, which is also continuously differentiable. Since g has compact support,

$$\sum_{i=1}^{d} \int_{R^d} \frac{\partial (f F_i g)}{\partial x_i} \, dx = 0$$

and thus

$$\left\langle f, \sum_{i=1}^{d} F_i \frac{\partial g}{\partial x_i} \right\rangle = - \sum_{i=1}^{d} \int_{R^d} g \frac{\partial (f F_i)}{\partial x_i} \, dx$$

$$= \left\langle - \sum_{i=1}^{d} \frac{\partial (f F_i)}{\partial x_i}, g \right\rangle,$$

which is a d-dimensional version of the "integration by parts" formula. From this and equation (7.6.9), we finally obtain

$$\langle A_{FP}f, g\rangle = \left\langle -\sum_{i=1}^{d} \frac{\partial(fF_i)}{\partial x_i}, g\right\rangle.$$

This formula holds for every continuously differentiable $f \in \mathcal{D}(A_{FP})$ and for every continuously differentiable g with compact support. Such a function g is automatically contained in $\mathcal{D}(A_K)$.

Therefore

$$A_{FP}f = -\sum_{i=1}^{d} \frac{\partial(fF_i)}{\partial x_i} \tag{7.6.10}$$

for continuously differentiable $f \in \mathcal{D}(A_{FP})$. Again, by using Theorem 7.5.1, we conclude that the function

$$u(t, x) = P_t f(x)$$

satisfies the partial differential equation (continuity equation)

$$\frac{\partial u}{\partial t} + \sum_{i=1}^{d} \frac{\partial(uF_i)}{\partial x_i} = 0. \tag{7.6.11}$$

Example 7.6.1. As a special case of the system (7.6.1) of ordinary differential equations, let $d = 2n$ and consider a **Hamiltonian system** whose dynamics are governed by the canonical equations of motion (Hamilton's equations)

$$\frac{dq_i}{dt} = \frac{\partial H}{\partial p_i} \quad \text{and} \quad \frac{dp_i}{dt} = -\frac{\partial H}{\partial q_i}, \qquad i = 1, \ldots, n, \tag{7.6.12}$$

where $H(p, q)$ is the system **Hamiltonian**. In systems of this type, q and p are referred to as the generalized position and momenta, respectively, whereas H is called the energy. Equation (7.6.11) for Hamiltonian systems takes the form

$$\frac{\partial u}{\partial t} + \sum_{i=1}^{n} \frac{\partial u}{\partial q_i}\frac{\partial H}{\partial p_i} - \frac{\partial u}{\partial p_i}\frac{\partial H}{\partial q_i} = 0,$$

which is often written as

$$\frac{\partial u}{\partial t} + [u, H] = 0,$$

where $[u, H]$ is the **Poisson bracket** of u with H. For Hamiltonian systems, the change with time of an arbitrary function g of the variables $q_1, \ldots, q_n$, $p_1, \ldots, p_n$ is given by

$$\frac{dg}{dt} = \sum_{i=1}^{n} \frac{\partial g}{\partial q_i}\frac{\partial H}{\partial p_i} - \frac{\partial g}{\partial p_i}\frac{\partial H}{\partial q_i} = [g, H].$$

In particular, if we take g to be a function of the energy H, then

$$\frac{dg}{dt} = \frac{dg}{dH}\frac{dH}{dt} = \frac{dg}{dH}[H, H] \equiv 0$$

since $[H, H] \equiv 0$. Thus any function of the generalized energy H is a **constant of the motion.** $\square$

Remark 7.6.2. The semigroup of Frobenius–Perron operators $\{P_t\}_{t\geq 0}$ corresponding to the system $\{S_t\}_{t\geq 0}$ generated by equation (7.6.1) is continuous.

To show this note that, since S_t is invertible ($S_t^{-1} = S_{-t}$), by Corollary 3.2.1 we have

$$P_t f(x) = f(S_{-t}(x))J_{-t}(x), \qquad (7.6.13)$$

where J_{-t} is the Jacobian of the transformation S_{-t}. Thus, for every continuous f with compact support,

$$\lim_{t\to t_0} f(S_{-t}(x))J_{-t}(x) = f(S_{-t_0}(x))J_{-t_0}(x)$$

uniformly with respect to x. This implies that

$$\lim_{t\to t_0} \|P_t f - P_{t_0} f\| = \lim_{t\to t_0} \int_{R^d} |P_t f(x) - P_{t_0} f(x)|\, dx = 0$$

since the integrals are, in actuality, over a bounded set. Because continuous functions with compact support form a dense subset of L^1, this completes the proof that $\{P_t\}_{t\geq 0}$ is continuous.

Much the same argument holds for the semigroup $\{U_t\}_{t\geq 0}$ if we restrict ourselves to continuous functions with compact support. In this case, from the relation

$$U_t f(x) = f(S_t(x)),$$

it immediately follows that $U_t f$ is uniformly convergent to $U_{t_0} f$ as $t \to t_0$. For this class of functions the proof of Theorem 7.5.1 can be repeated, thus showing that equation (7.5.3) is true for $f \in \mathcal{D}(A_K)$. $\square$

In the whole space L^∞, it may certainly be the case that $\{U_t\}_{t\geq 0}$ is not a continuous semigroup. As an example, consider the differential equation

$$\frac{dx}{dt} = -c$$

whose corresponding dynamical system is $S_t x = x - ct$. Thus the semigroup $\{U_t\}_{t\geq 0}$ is given by $U_t f(x) = f(x - ct)$. As we know from Example 7.4.2, when $c \neq 0$, this semigroup is certainly not continuous in L^∞.

The continuity of $\{P_t\}_{t\geq 0}$ is very important since it proves that the set $\mathcal{D}(A_{FP})$ is dense in L^1. Using equation (7.6.13) it may also be shown that $\mathcal{D}(A_{FP})$ contains all f with compact support, that have continuous first- and second-order derivatives.

7.7 Applications of the Semigroups of the Frobenius–Perron and Koopman Operators

After developing the concept of the semigroups of the Frobenius–Perron operators in Section 7.4 and introducing the general notion of an infinitesimal operator in Section 7.5 and of infinitesimal operators for semigroups generated by a system of ordinary differential equations in Section 7.6, we are now in a position to examine the utility and applications of these semigroups to questions concerning the existence of invariant measures and ergodicity. This material forms the core of this and the following section.

Theorem 7.7.1. *Let $(X, \mathcal{A}, \mu)$ be a measure space, and $S_t: X \to X$ a family of nonsingular transformations. Also let $P_t: L^1 \to L^1$ be the Frobenius–Perron operator corresponding to $\{S_t\}_{t \geq 0}$. Then the measure*

$$\mu_f(A) = \int_A f(x)\mu(dx)$$

is invariant with respect to $\{S_t\}_{t \geq 0}$ if and only if $P_t f = f$ for all $t \geq 0$.

Proof. The proof is trivial, since the invariance of μ_f implies

$$\mu_f(A) = \mu_f(S_t^{-1}(A)) \qquad \text{for } A \in \mathcal{A},$$

which, with the definition of P_t, implies $P_t f = f$. The converse is equally easy to prove. ∎

Now assume that μ_f is invariant. Since by the preceding theorem we know that $P_t f = f$, and

$$A_{FP}f = \lim_{t \to 0} \frac{P_t f - f}{t},$$

then $A_{FP}f = 0$. Thus the condition $A_{FP}f = 0$ is necessary for μ_f to be invariant. To demonstrate that $A_{FP}f = 0$ is also sufficient for μ_f to be invariant is not so easy, since we must pass from the infinitesimal operator to the semigroup. To deal with this very general and difficult problem, we must examine the way in which semigroups are constructed from their infinitesimal operators. This construction is very elegantly demonstrated by the Hille–Yosida theorem, which is described in Section 7.8.

Analogously to the way in which the semigroup of the Frobenius–Perron operator is employed in studying invariant measures of a semidynamical system $\{S_t\}_{t \geq 0}$, the semigroup of the Koopman operator can be used to study the ergodicity of $\{S_t\}_{t \geq 0}$.

We start by stating the following theorem.

Theorem 7.7.2. *A semidynamical system $\{S_t\}_{t \geq 0}$, with nonsingular transformations $S_t: X \to X$, is ergodic if and only if the fixed points of $\{U_t\}_{t \geq 0}$ are constant functions.*

Proof. The proof is quite similar to that of Theorem 4.2.1. First note that if $\{S_t\}_{t\geq 0}$ is not ergodic then there is an invariant nontrivial subset $C \subset X$, that is,

$$S_t^{-1}(C) = C \qquad \text{for } t \geq 0.$$

By setting $f = 1_C$, we have

$$U_t f = 1_C \circ S_t = 1_{S_t^{-1}(C)} = 1_C = f.$$

Since C is not a trivial set, f is not a constant function (cf. Theorem 4.2.1). Thus, if $\{S_t\}_{t\geq 0}$ is not ergodic, then there is a nonconstant fixed point of $\{U_t\}_{t\geq 0}$.

Conversely, assume there exists a nonconstant fixed point f of $\{U_t\}_{t\geq 0}$. Then it is possible to find a number r such that the set

$$C = \{x \colon f(x) < r\}$$

is nontrivial (cf. Figure 4.2.1). Since, for each $t \geq 0$,

$$S_t^{-1}(C) = \{x \colon S_t(x) \in C\} = \{x \colon f(S_t(x)) < r\}$$
$$= \{x \colon U_t f < r\} = \{x \colon f(x) < r\} = C,$$

subset C is invariant, implying that $\{S_t\}_{t\geq 0}$ is not ergodic. ■

Proceeding further with an examination of the infinitesimal operator generated by the Koopman operator, note that the condition $U_t f = f$, $t \geq 0$, implies that

$$A_K f = \lim_{t\to 0} \frac{U_t f - f}{t} = 0.$$

Thus, if the only solutions of $A_K f = 0$ are constant, then the semidynamical system $\{S_t\}_{t\geq 0}$ must be ergodic.

Example 7.7.1. In this example we consider the ergodic motion of a point on a d-dimensional torus, which is a generalization of the rotation of the circle treated in Example 7.3.1. We first note that the unit circle S^1 is a circle of radius 1, or

$$S^1 = \{m \colon m = e^{ix}, x \in R\}.$$

Formally, the d-**dimensional torus** T^d is defined as the Cartesian product of d unit circles S^1, that is,

$$T^d = S^1 \times \overset{d}{\cdots} \times S^1$$
$$= \{(m_1, \ldots, m_d) \colon m_k = e^{ix_k}, x_k \in R, k = 1, \ldots, d\}$$

(cf. Example 6.8.1 where we introduced the two-dimensional torus). T^d is clearly a d-dimensional Riemannian manifold, and the functions $m_k = e^{ix_k}$,

$k = 1, \ldots, d$, give a one to one correspondence between points on the torus T^d and points on the Cartesian product

$$[0, 2\pi) \times \overset{d}{\cdots} \times [0, 2\pi). \tag{7.7.1}$$

The x_k have an important geometrical interpretation since they are lengths on S^1. The natural Borel measure on S^1 is generated by these arc lengths and, by Fubini's theorem, these measures, in turn, generate a Borel measure on T^d. Thus, from a measure theoretic point of view, we identify T^d with the Cartesian product (7.7.1), and the measure μ on T^d with the Borel measure on R^d. We have, in fact, used exactly this identification in the intuitively simpler cases $d = 1$ (r-adic transformation; see Example 4.1.1 and Remark 4.1.2) and $d = 2$ (Anosov diffeomorphism; see Example 4.1.4 and Remark 4.1.6). The disadvantage of this identification is that curves that are continuous on the torus may not be continuous on the Cartesian product (7.7.1).

Thus we consider a dynamical system $\{S_t\}_{t \in R}$ that, in the coordinate system $\{x_k\}$, is defined by

$$S_t(x_1, \ldots, x_d) = (x_1 + \omega_1 t, \ldots, x_d + \omega_d t) \qquad (\text{mod } 2\pi).$$

We call this system **rotation on the torus** with angular velocities $\omega_1, \ldots, \omega_d$. Since $\det(dS_t(x)/dx) = 1$, the transformation S_t preserves the measure. We will prove that $\{S_t\}_{t \in R}$ is ergodic if and only if the angular velocities $\omega_1, \ldots, \omega_d$ are linearly independent over the ring of integers. This linear independence means that the only integers $k_1, \ldots, k_d$ satisfying

$$k_1 \omega_1 + \cdots + k_d \omega_d = 0 \tag{7.7.2}$$

are $k_1 = \cdots = k_d = 0$.

To prove this, we will use Theorem 7.7.2. Choose $f \in L^2(T^d)$ and assume $U_t f = f$ for $t \in R$, where $U_t f = f \circ S_t$ is the group of Koopman operators corresponding to S_t. Write f as a Fourier series

$$f(x_1, \ldots, x_d) = \sum a_{k_1 \cdots k_d} \exp[i(k_1 x_1 + \cdots + k_d x_d)],$$

where the summation is taken over all possible integers $k_1, \ldots, k_d$. Substitution of this series into the identity $f(x) = f(S_t(x))$ yields

$$\sum a_{k_1 \cdots k_d} \exp[i(k_1 x_1 + \cdots + k_d x_d)]$$
$$= \sum a_{k_1 \cdots k_d} \exp[it(\omega_1 k_1 + \cdots + \omega_d k_d)]$$
$$\exp[i(k_1 x_1 + \cdots + k_d x_d)].$$

As a consequence we must have

$$a_{k_1 \cdots k_d} = a_{k_1 \cdots k_d} \exp[it(\omega_1 k_1 + \cdots + \omega_d k_d)] \qquad \text{for } t \in R \tag{7.7.3}$$

and all sequences $k_1, \ldots, k_d$. Equation (7.7.3) will be satisfied either when $a_{k_1 \cdots k_d} = 0$ or when (7.7.2) holds. If $\omega_1, \ldots, \omega_d$ are linearly independent, then the only Fourier coefficient that can be different from zero is $a_{0 \cdots 0}$. In this case, then, $f(x) = a_{0 \cdots 0}$ is constant and the ergodicity of $\{S_t\}_{t \in R}$ is proved.

Conversely, if the $\omega_1, \ldots, \omega_d$ are not linearly independent, and condition (7.7.2) is thus satisfied for a nontrivial sequence $k_1, \ldots, k_d$, then (7.7.3) holds for $a_{k_1 \cdots k_d} = 1$. In this case the nonconstant function

$$f(x) = \exp[i(k_1 x_1 + \cdots + k_d x_d)]$$

satisfies $f(x) = f(S_t(x))$ and $\{S_t\}_{t \in R}$ is not ergodic.   □

Remark 7.7.1. The reason why rotation on the torus is so important stems from its frequent occurrence in applied problems. As a simple example, consider a system of d independent and autonomous oscillators

$$\frac{dp_k}{dt} + \omega_k^2 q_k = 0, \quad \frac{dq_k}{dt} = p_k, \quad k = 1, \ldots, d, \tag{7.7.4}$$

where $q_1, \ldots, q_d$ are the positions of the oscillators and $p_1, \ldots, p_d$ are their corresponding velocities. For this system the total energy of each oscillator is given by

$$E_k = \tfrac{1}{2} p_k^2 + \tfrac{1}{2} \omega_k^2 q_k^2, \quad k = 1, \ldots, d,$$

and it is clear that the E_k are constants of the motion. Assuming that $E_1, \ldots, E_d$ are given and positive, equations (7.7.4) may be solved to give

$$p_k(t) = A_k \omega_k \cos(\omega_k t + \alpha_k), \quad q_k(t) = A_k \sin(\omega_k t + \alpha_k),$$

where $A_k = \sqrt{2 E_k}/\omega_k$ and the α_k are determined, modulo 2π, by the initial conditions of the system. Set $\tilde{p}_k = p_k / A_k \omega_k$ and $\tilde{q}_k = q_k / A_k$ so that the vector $(\tilde{p}(t), \tilde{q}(t))$ describes the position of a point on a d-dimensional torus moving with the angular velocities $\omega_1, \ldots, \omega_d$. Thus, for fixed and positive $E_1, \ldots, E_d$, all possible trajectories of the system (7.7.4) are described by the group $\{S_t\}_{t \in R}$ of the rotation on the torus.

At first it might appear that the set of oscillators described by (7.7.4) is a very special mechanical system. Such is not the case, as equations (7.7.4) are approximations to a very general situation. We present an argument below that supports this claim.

Consider a Hamiltonian system

$$\frac{dq_k}{dt} = \frac{\partial H}{\partial p_k}, \quad \frac{dp_k}{dt} = -\frac{\partial H}{\partial q_k} \quad k = 1, \ldots, d. \tag{7.7.5}$$

Typically the energy H has the form

$$H(p, q) = \tfrac{1}{2} \sum_{j,k} a_{jk}(q) p_j p_k + V(q),$$

where the first term represents the kinetic energy and V is a potential function. Because the first term in H is associated with the kinetic energy, the quadratic form $\sum_{j,k} a_{jk}(q)$ is symmetric and positive definite. Further, if q^0 is a stable equilibrium point, then

$$\left.\frac{\partial V}{\partial q_k}\right|_{q=q^0} = 0 \qquad k = 1, \ldots, d$$

and the quadratic form,

$$\sum_{j,k} \frac{\partial^2 V}{\partial q_j \partial q_k},$$

is also positive definite (we neglect some special cases in which it might be semidefinite). Further, we assume that $H(0, q^0) = V(q^0) = 0$ since the potential is only defined up to an additive constant. Thus, developing H in a Taylor series in the neighborhood of $(0, q^0)$, and neglecting terms of order three and higher, we obtain

$$H(p, q) = \tfrac{1}{2} \sum_{j,k} a_{jk} p_j p_k + \tfrac{1}{2} \sum_{j,k} b_{jk}(q_j - q_j^0)(q_k - q_k^0) \qquad (7.7.6)$$

where $a_{jk} = a_{jk}(q^0)$ and $b_{jk} = (\partial^2 V/\partial q_j \partial q_k)|_{q^0}$. Both the quadratic forms $\sum_{j,k} a_{jk}$ and $\sum_{j,k} b_{jk}$ are symmetric and positive definite. With approximation (7.7.6), the original Hamiltonian equations (7.7.5) may be rewritten as

$$\frac{d(q_k - q_k^0)}{dt} = \sum_j a_{jk} p_j, \qquad \frac{dp_k}{dt} = -\sum_j b_{jk}(q_j - q_j^0), \qquad (7.7.7)$$

where the variables p_k and $q_k - q_k^0$ denote, respectively, the deviation of the system from the equilibrium point $(0, q^0)$.

Since matrices $A = (a_{jk})$ and $B = (b_{jk})$ are symmetric and positive definite, there exists a nonsingular matrix C such that (Gantmacher, 1959)

$$CBC^T = \begin{pmatrix} \lambda^1 & \cdots & 0 \\ \vdots & & \vdots \\ 0 & \cdots & \lambda_d \end{pmatrix} \quad \text{and} \quad CA^{-1}C^T = \begin{pmatrix} 1 & \cdots & 0 \\ \vdots & & \vdots \\ 0 & \cdots & 1 \end{pmatrix}$$

with positive elements λ_i on the diagonal. By introducing new variables $q - q^0 = C^T \bar{q}$ and $p = C^{-1}\bar{p}$ into equations (7.7.7), we obtain

$$\frac{d\bar{q}_k}{dt} = \bar{p}_k, \qquad \frac{d\bar{p}_k}{dt} = -\lambda_k \bar{q}_k. \qquad (7.7.8)$$

This new system is completely equivalent to our system (7.7.4) of independent oscillators with angular velocities $\omega_k^2 = \lambda_k$.

Finally we note that, although our approximation shows the correspondence between rotation on the torus and Hamiltonian systems, the terms we neglected in our expansion of H might play a very important role in modifying the eventual asymptotic behavior of a Hamiltonian system.   □

Remark 7.7.2. Note that the statement and proof of Theorem 7.7.2 are virtually identical with the corresponding discrete time result given in Theorem 4.2.1. Indeed, necessary and sufficient conditions for ergodicity, mixing, and exactness using the Frobenius–Perron operator, identical to those in Theorem 4.4.1, can be stated by replacing n by t. Analogously, conditions for ergodicity and mixing in continuous time systems using the Koopman operator can be obtained from Proposition 4.4.1 by setting $n = t$. Since all of these conditions are completely equivalent we will not rewrite them for continuous time systems.   □

Example 7.7.2. To illustrate the property of mixing in a continuous time system we consider a model for an ideal gas in R^3 adapted from Cornfeld, Fomin, and Sinai [1982]. However, our proof of the mixing property is based on a different technique. At any given moment of time the state of this system is described by the set of pairs

$$y = \{(x_i, v_i)\}, \qquad x_i \in R^3, \; v_i \in R^3,$$

where x_i denotes the position, and v_i, the velocity of a particle. We emphasize that y is a set of pairs and not a sequence of pairs, which means that the coordinate pairs (x_i, v_i) are not taken in any specific order. Physically this means that the particles are not distinguishable. It is further assumed that the gas is sufficiently dilute, both in spatial position and in velocity, so that the only states that must be considered are such that in every bounded set $B \subset R^6$ there is, at most, a finite number of pairs (x_i, v_i).

The collection of all possible states of this gas will be denoted by Y, and we assume that the motion of each particle at the gas is governed by a group of transformations $S_t \colon Y \to Y$ given by

$$S_t(y) = \{(x_i + v_i t, v_i)\} \qquad \text{for } y = \{(x_i, v_i)\},$$

or, more compactly, by $S_t(y) = \{s_t(x_i, v_i)\}$, where $\{s_t\}_{t \in R}$ is the family of transformations in R^6 such that

$$s_t(x, v) = (x + vt, v).$$

Thus particles move with a constant speed and do not interact. The surprising result, proved below, is that this system is mixing.

To study the asymptotic properties of $\{S_t\}_{t \in R}$, we must define a σ-algebra and a measure on Y. We do this by first introducing a special measure on R^6, which is the phase space for the motion of a single particle. Let g be a density on R^3. As usual, the measure associated with g is

$$m_g(A) = \int_A g(v) \, dv$$

for every Borel set $A \subset R^3$, and the measure m in $R^6 = R^3 \times R^3$ is defined as the product of the usual Borel measure and m_g, that is,

$$m(A_1 \times A_2) = \int_{A_1} dx \int_{A_2} g(v)\,dv, \qquad A_1, A_2 \subset R^3.$$

From a physical point of view this definition of the measure simply reflects the fact that the particle positions are uniformly distributed in R^3, whereas the velocities are distributed with a given density g, for instance, the Maxwellian $g(v) = c \exp(-|v|^2)$.

With these comments we now proceed to define a σ-algebra and a measure on Y. Let $B_1, \ldots, B_n$ be a given sequence of bounded Borel subsets of R^6 for an arbitrary n, and $k_1, \ldots, k_n$ be a given sequence of integers. We use $C(B_1, \ldots, B_n; k_1, \ldots, k_n)$ to denote the set of all $y = \{(x_i, v_i)\}$ such that the number of elements (x_i, v_i) that belong to B_j is equal to k_j, that is,

$$C(B_1, \ldots, B_n; k_1, \ldots, k_n) = \{y \in Y\colon {}^{\#}(y \cap B_1) = k_1, \ldots, {}^{\#}(y \cap B_n) = k_n\},$$
(7.7.9)

where ${}^{\#}Z$ denotes the number of elements of the set Z. Sets of the form (7.7.9) are called **cylinders**. If the sets $B_1, \ldots, B_n$ are disjoint, then the cylinder is said to be **proper**. For every proper cylinder, we define

$$\mu(C(B_1, \ldots, B_n; k_1, \ldots, k_n))$$
$$= \frac{[m(B_1)]^{k_1} \cdots [m(B_n)]^{k_n}}{k_1! \cdots k_n!} \exp\left[-\sum_{i=1}^{n} m(B_i)\right]. \quad (7.7.10)$$

From (7.7.10) it follows immediately that

$$\mu(C(B_1, \ldots, B_n; k_1, \ldots, k_n)) = \mu(C(B_1; k_1)) \cdots \mu(C(B_n; k_n)) \quad (7.7.11)$$

whenever the sets $B_1, \ldots, B_n$ are mutually disjoint.

It is also easy to calculate the measure of $C(B_1, B_2; k_1, k_2)$ when B_1 and B_2 are not disjoint by writing C as the union of proper cylinders. Thus, y belongs to $C(B_1, B_2; k_1, k_2)$ if, for some $r \leq \min(k_1, k_2)$, the set $B_1^0 = B_1 \setminus B_2$ contains $k_1 - r$ particles, $B^0 = B_1 \cap B_2$ contains r particles, and $B_2^0 = B_2 \setminus B_1$ has $k_2 - r$ particles. As a consequence,

$$C(B_1, B_2; k_1, k_2) = \bigcup_{r=0}^{k} [C(B_1^0; k_1 - r) \cap C(B^0; r) \cap C(B_2^0; k_2 - r)],$$

where $k = \min(k_1, k_2)$, and, thus,

$$\mu(C(B_1, B_2; k_1, k_2)) = \sum_{r=0}^{k} \frac{[m(B_1^0)]^{k_1 - r}[m(B^0)]^r[m(B_2^0)]^{k_2 - r}}{(k_1 - r)! r! (k_2 - r)!}$$
$$\cdot \exp[-m(B_1^0) - m(B^0) - m(B_2^0)] \quad (7.7.12)$$

By employing arguments of this type we can calculate the measure μ of any cylinder. However, the formulas for arbitrary cylinders are much more complicated as it is necessary to sum these various contributions first with respect to $q = \binom{n}{2}$ parameters $r_1, \ldots, r_q$, corresponding to all possible intersections $B_i \cap B_j$, $i \neq j$, then with respect to $\binom{n}{3}$ parameters corresponding to all possible intersections $B_i \cap B_j \cap B_l$, $i \neq j \neq l$, and so forth.

With respect to the σ-algebra, we define $\mathcal{A}$ to be the smallest σ-algebra that contains all the cylinders or, equivalently, all proper cylinders. Using standard results from measure theory, it is possible to prove that μ given by (7.7.10) for proper cylinders can be uniquely extended to a measure on $\mathcal{A}$ and that the characteristic functions of proper cylinders

$$1_{C(B_1, \ldots, B_n; k_1, \ldots k_n)}$$

form a linearly dense subset of $L^2(Y, \mathcal{A}, \mu)$. We omit the proof of these facts as they are quite technical in nature and, instead, turn to consider the asymptotic properties of system $\{S_t\}_{t \in R}$ on the phase space Y.

First we note that the measure μ is normalized. To show this, take an arbitrary bounded Borel set B. Then

$$Y = \bigcup_{k=0}^{\infty} C(B; k)$$

since every y belongs to one of the cylinders $C(B; k)$, namely, the one for which ${}^{\#}(y \cap B) = k$. As the cylinders $C(B; k)$, $k = 0, 1, \ldots$, are mutually disjoint, we have

$$\mu(Y) = \sum_{k=0}^{\infty} \mu(C(B; k)) = \sum_{k=0}^{\infty} \frac{[m(B)]^k}{k!} e^{-m(B)} = 1.$$

Second, the measure μ is invariant with respect to $\{S_t\}_{t \in R}$. To show this, note that for every cylinder

$$S_t(C(B_1, \ldots, B_n; k_1, \ldots, k_n)) = C(s_t(B_1), \ldots, s_t(B_n); k_1, \ldots, k_n).$$

It is clear that $(x, v) \in s_t(B_j)$ if and only if $(\bar{x}, \bar{v}) \in B_j$, where $\bar{x} = x - vt$, $\bar{v} = v$, and, as a consequence,

$$m(s_t(B_j)) = \iint_{s_t(B_j)} g(v) \, dx \, dv = \iint_{B_j} g(\bar{v}) \, d\bar{x} \, d\bar{v} = m(B_j).$$

From this equality, $m(s_t(B_j)) = m(B_j)$ and, from equation (7.7.10), we, therefore, have

$$\mu(S_t(C(B_1,\ldots,B_n;k_1,\ldots,k_n))) = \mu(C(B_1,\ldots,B_n;k_1,\ldots,k_n))$$

for every proper cylinder. Writing $\mu_t(E) = \mu(S_t(E))$ for $E \in \mathcal{A}$, we define for every fixed t a measure μ_t, on $\mathcal{A}$ that is identical with μ for proper cylinders. Since μ is uniquely determined by its values on cylinders, we must have $\mu_t(E) = \mu(E)$ for all $E \in \mathcal{A}$, and thus the invariance of μ with respect to S_t is proved.

With these results in hand, we now prove that the dynamical system $\{S_t\}_{t \in R}$ is mixing. Since the characteristic functions of proper cylinders are linearly dense in $L^2(Y, \mathcal{A}, \mu)$, by Remark 7.7.2 it is sufficient to verify the condition

$$\lim_{t \to \infty} \langle U_t 1_{C_1}, 1_{C_2} \rangle = \langle 1_{C_1}, 1 \rangle \langle 1, 1_{C_2} \rangle \tag{7.7.13}$$

for every two proper cylinders C_1 and C_2. Since

$$U_t 1_{C_1}(y) = 1_{C_1}(S_t(y)) = 1_{S_{-t}(C_1)}(y)$$

and $\langle 1_{C_j}, 1 \rangle = \mu(C_j)$, condition (7.7.13) is equivalent to

$$\lim_{t \to \infty} \mu(S_{-t}(C_1) \cap C_2) = \mu(C_1)\mu(C_2). \tag{7.7.14}$$

We will verify that (7.7.14) holds only in the simplest case when each of the cylinders C_j is determined by only one bounded Borel set. Thus we assume

$$C_j = C(B_j; k_j), \qquad j = 1, 2. \tag{7.7.15}$$

(This is not an essential simplification, since the argument proceeds in exactly the same way for arbitrary proper cylinders. However, in the general case the formulas are so complicated that the simple geometrical ideas behind the calculations are obscured.) When the C_j are given by (7.7.15), the right-hand side of equation (7.7.14) may be easily calculated by (7.7.10). Thus

$$\mu(C_1)\mu(C_2) = \frac{[m(B_1)]^{k_1}[m(B_2)]^{k_2}}{k_1!k_2!} \exp[-m(B_1) - m(B_2)]. \tag{7.7.16}$$

To compute the left-hand side of equation (7.7.14), observe that

$$S_{-t}(C_1) = C(s_{-t}(B_1); k_1)$$

so

$$\mu(S_{-t}(C_1) \cap C_2) = \mu(C(s_{-t}(B_1); k_1) \cap C(B_2; k_2))$$
$$= \mu(C(s_{-t}(B_1), B_2; k_1, k_2)). \tag{7.7.17}$$

With (7.7.12) we have

$$\mu(S_{-t}(C_1) \cap C_2) = \sum_{r=0}^{k} \frac{[m(B_1^t)]^{k_1-r}[m(B^t)]^r[m(B_2^t)]^{k_2-r}}{(k_1-r)!r!(k_2-r)!}$$
$$\cdot \exp[-m(B_1^t) - m(B^t) - m(B_2^t)], \qquad (7.7.18)$$

where $B_1^t = s_{-t}(B_1) \setminus B_2$, $B^t = s_{-t}(B_1) \cap B_2$, and $B_2^t = B_2 \setminus s_{-t}(B_1)$. From our definition of m, we have

$$m(B^t) = \iint\limits_{s_{-t}(B_1) \cap B_2} g(v)\, dx\, dv = \iint\limits_{B_2} 1_{s_{-t}(B_1)}(x,v)g(v)\, dx\, dv$$

$$= \iint\limits_{B_2} 1_{B_1}(x + vt, v)g(v)\, dx\, dv.$$

Since B_1 and B_2 are bounded, $1_{B_1}(x + vt, v) = 0$ for almost every point $(x, v) \in B_2$ if t is sufficiently large (except for some points at which $v = 0$). Thus, by the Lebesgue dominated convergence theorem,

$$\lim_{t \to \infty} m(B^t) = 0. \qquad (7.7.19)$$

Furthermore, since $B_2^t = B_2 \setminus B^t$, it follows that

$$\lim_{t \to \infty} m(B_2^t) = \lim_{t \to \infty} [(m(B_2) - m(B^t))] = m(B_2). \qquad (7.7.20)$$

Finally, since $B_1^t = s_{-t}(B_1) \setminus B^t$ and s_t is measure preserving,

$$m(B_1^t) = m(s_{-t}(B_1)) - m(B^t) = m(B_1) - m(B^t),$$

and

$$\lim_{t \to \infty} m(B_1^t) = m(B_1). \qquad (7.7.21)$$

Passing to the limit in equation (7.7.18) and using (7.7.19) through (7.7.21) gives

$$\lim_{t \to \infty} \mu(S_{-t}(C_1) \cap C_2) = \left\{ \frac{[m(B_1)]^{k_1}[m(B_2)]^{k_2}}{k_1!k_2!} \right\} \exp[-m(B_1) - m(B_2)],$$

which, together with (7.7.16), proves (7.7.14).

From this proof, it should be clear that mixing in this model is a consequence of the following two facts. The first is that, for disjoint B_1 and B_2 and given k_1 and k_2, the events consisting of B_1 containing k_1 particles and B_2 containing k_2 particles are independent [this follows from equation (7.7.11)]. Second, for every two bounded Borel sets B_1 and B_2, the sets $s_{-t}(B_1)$ and B_2 are "almost" disjoint for large t. Taken together these produce the surprising result that mixing can appear in a system without particle interaction.   □

Example 7.7.3. The preceding example gave a continuous time, dynamical system that was mixing. The phase space of this system was infinite dimensional. This fact is not essential. There is a large class of finite dimensional, mixing, dynamical systems that play an important role in classical mechanics. In this example we briefly describe these systems. An exhaustive treatment requires highly specialized techniques from differential geometry and cannot be given within the measure-theoretic framework that we have adopted. All necessary information can be found in the books by Arnold and Avez [1968], by Abraham and Marsden [1978], and articles by Anosov [1967] and by Smale [1967].

Let M be a compact connected smooth Riemannian manifold. Having M, we define the sphere bundle Σ as the set of all pairs (m, ξ), where m is an arbitrary point of M and ξ is a unit tangent vector starting at m. This definition can be written as

$$\Sigma = \{(m, \xi): m \in M, \xi \in T_m, \|\xi\| = 1\}.$$

It can be proved that Σ, with an appropriately defined metric, is also a Riemannian manifold. Thus a measure μ_Σ is automatically given on Σ. In a physical interpretation, M is the configuration space of a system that moves with constant speed and Σ is its phase space. To describe precisely the dynamical system that corresponds to this interpretation we need only the concept of geodesics. Let $\gamma: R \to M$ be a C^1 curve. This curve is called a **geodesic** if for every point $m_0 = \gamma(t_0)$ there is an $\varepsilon > 0$ such that for every $m_1 = \gamma(t_1)$, with $|t_1 - t_0| \le \varepsilon$, the length of the arc γ between the points m_0 and m_1 is equal to the distance between m_0 and m_1. It can be proved that, for every $(m, \xi) \in \Sigma$, there exists exactly one geodesic satisfying

$$\gamma(0) = m, \quad \gamma'(0) = \xi, \quad \|\gamma'(t)\| = 1 \quad \text{for } t \in R. \tag{7.7.22}$$

We define a dynamical system $\{S_t\}_{t \in R}$ on Σ by setting

$$S_t(m, \xi) = (\gamma(t), \gamma'(t)) \qquad \text{for } t \in R,$$

where the geodesic γ satisfies (7.7.22). This system is called a **geodesic flow**.

In the case dim $M = 2$, the geodesic flow has an especially simple interpretation: It describes the motion of a point that moves on the surface M in the absence of external forces and without friction. The motion described by the geodesic flow looks quite specific but, in fact, it represents a rather general situation. If M is the configuration space of a mechanical system with the typical Hamiltonian function (see Remark 7.7.1),

$$H(q, p) = \tfrac{1}{2} \sum_{j,k} a_{jk}(q) p_j p_k + V(q),$$

then it is possible to change the Riemannian metric on M in such a way that trajectories of the system become geodesics.

The behavior of the geodesic flow depends on the geometrical properties of the manifold M and most of all on its curvature. In the simplest case, dim $M = 2$, the curvature K is a scalar function and has a clear geometrical interpretation. In order to define K at a point $m \in M$, we consider, in a neighborhood W of m, a triangle made by three geodesics. We denote the angles of that triangle by α_1, α_2, α_3, and its area by σ. Then

$$K(m) = \lim[(\alpha_1 + \alpha_2 + \alpha_3 - \pi)/\sigma],$$

where the limit is taken over a sequence of neighborhoods that shrinks to the point m. In the general case, dim $M > 2$, the curvature must be defined separately for every two-dimensional section of a neighborhood of the point m. (Thus, in this case, the curvature becomes a tensor.) When the curvature of M is negative, the behavior of the geodesic flow is quite specific and highly chaotic. Such flows have been studied since the beginning of the century, starting with Hadamard [1898]. Results were first obtained for manifolds with constant negative curvature and then finally completed by Anosov [1967]. It follows that the geodesic flow on a compact, connected, smooth Riemannian manifold with negative curvature is mixing and even a K-flow (a continuous time analog of K-automorphism). This fact has some profound consequences for the foundations of classical statistical mechanics. A heuristic geometrical argument of Arnold [1963] shows that the Boltzmann–Gibbs model of a dilute gas (ideal balls with elastic collisions) may be considered as a geodesic flow on a manifold with negative curvature. Thus, such a system is not only ergodic but also mixing. A sophisticated proof of the ergodicity and mixing of the Boltzmann–Gibbs model has been given by Sinai [1963, 1970].   □

7.8   The Hille–Yosida Theorem and Its Consequences

Theorem 7.8.1 (Hille–Yosida). *Let $A: \mathcal{D}(A) \to L$ be a linear operator, where $\mathcal{D}(A) \subset L$ is a linear subspace of L. In order for A to be an infinitesimal operator for a continuous semigroup of contractions, it is necessary and sufficient that the following three conditions are satisfied:*

(a) *$\mathcal{D}(A)$ is dense in L, that is, every point in L is a strong limit of a sequence of points from $\mathcal{D}(A)$;*

(b) *For each $f \in L$ there exists a unique solution $g \in \mathcal{D}(A)$ of the resolvent equation*

$$\lambda g - Ag = f; \tag{7.8.1}$$

(c) *For every $g \in \mathcal{D}(A)$ and $\lambda > 0$,*

$$\|\lambda g - Ag\|_L \geq \lambda \|g\|_L. \tag{7.8.2}$$

Further, if A satisfies (a)–(c), then the semigroup corresponding to A is unique and is given by

$$T_t f = \lim_{\lambda \to \infty} e^{tA_\lambda} f, \qquad f \in L, \tag{7.8.3}$$

*where $A_\lambda = \lambda A R_\lambda$ and $R_\lambda f = g$ (the **resolvent operator**) is the unique solution of $\lambda g - Ag = f$.*

Consult Dynkin [1965] or Dunford and Schwartz [1957] for the proof.

Operator $A_\lambda = \lambda A R_\lambda$ can be written in several alternative forms, each of which is useful in different situations. Thus, after substitution of $g = R_\lambda f$ into (7.8.1), we have

$$\lambda R_\lambda f - A R_\lambda f = f \qquad \text{for } f \in L. \tag{7.8.4}$$

By applying the operator R_λ to both sides of (7.8.1) and using $g = R_\lambda f$, we also obtain

$$\lambda R_\lambda g - R_\lambda A g = g \qquad \text{for } g \in \mathcal{D}(A). \tag{7.8.5}$$

Equations (7.8.4) and (7.8.5) immediately give

$$R_\lambda A f = A R_\lambda f \qquad \text{for } f \in \mathcal{D}(A). \tag{7.8.6}$$

Equation (7.8.4) also gives

$$A R_\lambda f = (\lambda R_\lambda - I) f \qquad \text{for } f \in L, \tag{7.8.7}$$

where I is the identity operator ($If \equiv f$ for all f). Thus we have three possible representations for A_λ: the original definition,

$$A_\lambda = \lambda A R_\lambda; \tag{7.8.8}$$

or, from (7.8.7),

$$A_\lambda = \lambda(\lambda R_\lambda - I); \tag{7.8.9}$$

and, finally, from (7.8.6),

$$A_\lambda = \lambda R_\lambda A. \tag{7.8.10}$$

The representations in (7.8.8) and (7.8.9) hold in the entire space L, whereas (7.8.10) holds in $\mathcal{D}(A)$.

From conditions (b) and (c) of the Hille–Yosida theorem, using $g = R_\lambda f$, it follows that

$$\|f\|_L \geq \lambda \|R_\lambda f\|_L. \tag{7.8.11}$$

Consequently, using (7.8.9),

$$\|A_\lambda f\|_L = \|\lambda^2 R_\lambda f - \lambda f\|_L \leq \|\lambda^2 R_\lambda f\|_L + \|\lambda f\|_L \leq 2\lambda\|f\|_L,$$

so that the operator $\exp(tA_\lambda)$ can be interpreted as the series

$$e^{tA_\lambda} f = \sum_{n=0}^{\infty} \frac{t^n}{n!} A_\lambda^n f, \tag{7.8.12}$$

which is strongly convergent.

In addition to demonstrating the existence of a semigroup $\{T_t\}_{t\geq 0}$ corresponding to a given operator A, the Hille–Yosida theorem also allows us to determine some properties of $\{T_t\}_{t\geq 0}$.

One very interesting corollary is the following. Suppose we have an operator $A: \mathcal{D}(A) \to L$ (remembering that $L = L^p$) that satisfies conditions (a)–(c) of the Hille–Yosida theorem, and such that the solution $g = R_\lambda f$ of equation (7.8.1) has the property that $R_\lambda f \geq 0$ for $f \geq 0$. Then, as we will show next, $T_t f \geq 0$ for every $f \geq 0$.

To see this, note that from (7.8.9) we have

$$e^{tA_\lambda} f = e^{-t\lambda}\left(e^{t\lambda^2 R_\lambda} f\right), \tag{7.8.13}$$

where

$$e^{t\lambda^2 R_\lambda} f = \sum_{n=0}^{\infty} \frac{t^n \lambda^n}{n!} (\lambda R_\lambda)^n f. \tag{7.8.14}$$

Further, for any $f \geq 0$, $R_\lambda f \geq 0$ and, by induction, $R_\lambda^n f \geq 0$. Thus, from (7.8.14), since $\lambda > 0$ and $t \geq 0$, $\exp(t\lambda^2 R_\lambda)f \geq 0$ and so, from (7.8.13), $\exp(tA_\lambda)f \geq 0$. Finally, from (7.8.3), we have $T_t f \geq 0$ since it is the limit of nonnegative functions.

Now suppose that $L = L^1$ and that the operator λR_λ preserves the integral, that is,

$$\lambda \int_X R_\lambda f(x)\mu(dx) = \int_X f(x)\mu(dx) \qquad \text{for all } f \in L^1, \lambda > 0. \tag{7.8.15}$$

We will show that these properties imply that

$$\int_X T_t f(x)\mu(dx) = \int_X f(x)\mu(dx), \qquad f \geq 0, t \geq 0.$$

This is straightforward. Since (7.8.14) is strongly convergent, and using equation (7.8.15), we obtain

$$\int_X e^{t\lambda^2 R_\lambda} f(x)\mu(dx) = \sum_{n=0}^{\infty} \frac{t^n \lambda^n}{n!} \int_X (\lambda R_\lambda)^n f(x)\mu(dx)$$

$$= \sum_{n=0}^{\infty} \frac{t^n \lambda^n}{n!} \int_X f(x)\mu(dx)$$

$$= e^{t\lambda} \int_X f(x)\mu(dx). \tag{7.8.16}$$

Now,

$$\int_X T_t f(x)\mu(dx) = \lim_{\lambda\to\infty} \int_X e^{tA_\lambda} f(x)\mu(dx)$$

$$= \lim_{\lambda\to\infty} \int_X e^{-t\lambda}\big(e^{t\lambda^2 R_\lambda} f(x)\big)\mu(dx) = \int_X f(x)\mu(dx)$$

by the use of equation (7.8.16), and the claim is demonstrated.

These two results may be summarized in the following corollary.

Corollary 7.8.1. *Let $A\colon \mathcal{D}(A) \to L^1$ be an operator satisfying conditions (a)–(c) of the Hille–Yosida theorem. If the solution $g = R_\lambda f$ of (7.8.1) is such that λR_λ is a Markov operator, then $\{T_t\}_{t\geq 0}$ generated by A is a continuous semigroup of Markov operators.*

In fact, in this corollary only conditions (a) and (b) of the Hille–Yosida theorem need be checked, as condition (c) is automatically satisfied for any Markov operator.

To see this, set $f = \lambda g - Ag$ and write inequality (7.8.2) in the form

$$\|f\| \geq \|\lambda R_\lambda f\|.$$

This is always satisfied if λR_λ is a Markov operator, as we have shown in Section 3.1 [cf. inequality (3.1.6)].

The Hille–Yosida theorem has several other important applications. The first is that it provides an immediate and simple way to demonstrate that $A_{FP} f = 0$ is a sufficient condition that μ_f is an invariant measure.

Thus, $Af = 0$ implies, from (7.8.10), that $A_\lambda f = 0$ and from (7.8.12)

$$e^{tA_\lambda} f = f.$$

This, combined with (7.8.3), gives

$$T_t f = f \qquad \text{for all } t \geq 0.$$

Thus, in the special case $A_{FP} f = 0$ this implies that $P_t f = f$ and thus μ_f is invariant.

By combining this result with that of Section 7.7, we obtain the following theorem.

Theorem 7.8.2. *Let $\{S_t\}_{t\geq 0}$ be a semidynamical system such that the corresponding semigroup of Frobenius–Perron operators is continuous. Under this condition, an absolutely continuous measure μ_f is invariant if and only if $A_{FP} f = 0$.*

Consider the special case where A_{FP} is the infinitesimal operator for a d-dimensional system of ordinary differential equations (cf. equation 7.6.10).

Then the necessary and sufficient condition that μ_f be invariant, that is $A_{FP}f = 0$, reduces to

$$\sum_{i=1}^{d} \frac{\partial (fF_i)}{\partial x_i} = 0 \qquad (7.8.17)$$

for continuously differentiable $f \in L^1$. This result was originally obtained by Liouville using quite different techniques and is known as **Liouville's theorem**.

Remark 7.8.1. Equation (7.8.17) is also a necessary and sufficient condition for the invariance of the measure

$$\mu_f(A) = \int_A f(x)\mu(dx)$$

even if f is an arbitrary continuously differentiable function that is not necessarily integrable on R^d. This is related to the fact that operators $P_t f$ as given by (7.6.13) can also be considered for nonintegrable functions. Thus, if one wishes to determine when the Lebesgue measure

$$\mu(A) = \int_A dx_1 \ldots dx_d = \int_A dx$$

is invariant, it is necessary to substitute its density $f(x) \equiv 1$ into (7.8.17). This gives

$$\sum_{i=1}^{d} \frac{\partial F_i}{\partial x_i} = 0 \qquad (7.8.18)$$

as a necessary and sufficient condition for the invariance of the Lebesgue measure. [In many sources, equation (7.8.18) is called Liouville's equation, even though it is a special case of equation (7.8.17).]   □

Remark 7.8.2. It is quite straightforward to show that Hamiltonian systems (see Example 7.6.1) satisfy (7.8.18) since

$$\sum_{i=1}^{n} \left[\frac{\partial}{\partial q_i} \left(\frac{\partial H}{\partial p_i} \right) + \frac{\partial}{\partial p_i} \left(-\frac{\partial H}{\partial q_i} \right) \right] = 0$$

automatically, and thus they preserve the Lebesgue measure.   □

Returning now to the problem of determining the ergodicity of a semi-dynamical system $\{S_t\}_{t \geq 0}$, recall that $U_t g = g$ implies $A_K g = 0$. Using this relation and Theorem 7.7.2 we are going to prove the following theorem.

Theorem 7.8.3. Let $\{S_t\}_{t \geq 0}$ be a semidynamical system such that the corresponding semigroup $\{P_t\}$ of Frobenius–Perron operators is continuous.

Then $\{S_t\}_{t\geq 0}$ is ergodic if and only if $A_K g = 0$ has only constant solutions in L^∞.

Proof. The "if" part follows from Theorem 7.7.2. The proof of the "only if" part is more difficult since, in general, the semigroup $\{U_t\}$ is not continuous and we cannot use the Hille–Yosida theorem. Thus, assume that $A_K g = 0$ for some nonconstant g. Choose an arbitrary $f \in L^1$ and define the real-valued function ϕ by the formula

$$\phi(t) = \langle f, U_t g \rangle = \langle P_t f, g \rangle.$$

Due to the continuity of $\{P_t\}$, function ϕ is also continuous. Further, we have

$$\frac{\phi(t+h) - \phi(t)}{h} = \left\langle f, \frac{U_{t+h} g - U_t g}{h} \right\rangle$$
$$= \left\langle P_t f, \frac{U_h g - g}{h} \right\rangle \qquad \text{for } h > 0,\, t \geq 0.$$

Since $A_K g = 0$, passing to the limit as $h \to 0$, we obtain

$$\phi'(t) = \langle P_t f, A_K g \rangle = 0.$$

Function ϕ is continuous with the right-hand derivative identically equal to zero, implying that $\phi(t) = \phi(0)$ for all $t \geq 0$. Consequently,

$$\langle f, U_t g - g \rangle = \phi(t) - \phi(0) = 0 \qquad \text{for } t \geq 0.$$

Since f is arbitrary this, in turn, implies that $U_t g = g$ for $t \geq 0$, which, by Theorem 7.7.2, completes the proof. ∎

In particular, if $\{S_t\}_{t\geq 0}$ is a semigroup generated by a system of ordinary differential equations then, from equation (7.6.5), $A_K f = 0$ is equivalent to

$$\sum_{i=1}^{d} F_i(x) \frac{\partial f}{\partial x_i} = 0 \qquad (7.8.19)$$

for continuously differentiable f with compact support. However, it must be pointed out that (7.8.19) is of negligible usefulness in checking ergodicity, because the property "$A_K f = 0$ implies f constant for *all* functions in L^∞" must be checked and not just the continuously differentiable functions. This is quite different from the situation where one is using the Liouville theorem (7.8.17) to check for invariant measures. In the latter case, it is necessary to find only a *single* solution of $A_{FP} f = 0$.

Example 7.8.1. Theorem 7.8.3 allows us easily to prove that Hamiltonian systems (see Example 7.6.1) are not ergodic. To show this, note that for a

Hamiltonian system defined by equation (7.6.12), equation (7.6.5) becomes

$$A_K f = \sum_{i=1}^{n} \left[\frac{\partial f}{\partial q_i} \frac{\partial H}{\partial p_i} - \frac{\partial f}{\partial p_i} \frac{\partial H}{\partial q_i} \right] = [f, H].$$

Take $f \in L^\infty$ to be any nonconstant function of the energy H. By Example 7.6.1, we know that $A_K f \equiv 0$ since

$$[f(H), H] = \frac{\partial f}{\partial H}[H, H] = 0$$

and therefore Hamiltonian systems are not ergodic on the whole space. However, if we fix the total energy, or the energy for each degree of freedom as in Remark 7.7.1, then the system *may* become ergodic.   □

7.9 Further Applications of the Hille–Yosida Theorem

Thus far we have used the Hille–Yosida theorem to demonstrate some simple properties of semigroups that followed directly from properties of the infinitesimal operator A and the resolvent equation (7.8.1). In these cases the semigroups were given. Now we are going to show a simple application of the theorem to the problem of determining a semigroup corresponding to a given infinitesimal operator A.

Let $X = R$ and $L = L^1(R)$, and consider the infinitesimal operator

$$Af = \frac{d^2 f}{dx^2} \tag{7.9.1}$$

that can, of course, only be defined for some $f \in L^1$. Let $\mathcal{D}(A)$ be the set of all $f \in L^1$ such that $f''(x)$ exists almost everywhere, is integrable on R, and

$$f'(x) = f'(0) + \int_0^x f''(s)\, ds.$$

In other words, $\mathcal{D}(A)$ is the set of all f such that f' is absolutely continuous and f'' is integrable on R. We will show that there is a unique semigroup corresponding to the infinitesimal operator A.

The set $\mathcal{D}(A)$ is evidently dense in L^1 (even the set of C^∞ functions is dense in L^1), therefore we may concentrate on verifying properties (b) and (c) of the Hille–Yosida theorem.

The resolvent equation (7.8.1) has the form

$$\lambda g - \frac{d^2 g}{dx^2} = f, \tag{7.9.2}$$

which is a second-order ordinary differential equation in the unknown function g. Using standard arguments, the general solution of (7.9.2) may be written as

$$g(x) = C_1 e^{-\alpha x} + C_2 e^{\alpha x} + \frac{1}{2\alpha} \int_{x_0}^{x} e^{-\alpha(x-y)} f(y)\, dy - \frac{1}{2\alpha} \int_{x_1}^{x} e^{\alpha(x-y)} f(y)\, dy$$

where $\alpha = \sqrt{\lambda}$, and C_1, C_2, x_0, and x_1 are arbitrary constants. To be specific, pick $x_0 = -\infty$, $x_1 = +\infty$, and set

$$K(x - y) = (1/2\alpha)e^{-\alpha|(x-y)|}. \tag{7.9.3}$$

Then the solution of (7.9.2) can be written in the more compact form

$$g(x) = C_1 e^{-\alpha x} + C_2 e^{\alpha x} + \int_{-\infty}^{\infty} K(x-y) f(y)\, dy. \tag{7.9.4}$$

The last term on the right-hand side of (7.9.4) is an integrable function on R, since

$$\int_{-\infty}^{\infty} dx \int_{-\infty}^{\infty} K(x-y) f(y)\, dy = \int_{-\infty}^{\infty} K(x-y)\, dx \int_{-\infty}^{\infty} f(y)\, dy$$

$$= \frac{1}{\lambda} \int_{-\infty}^{\infty} f(y)\, dy. \tag{7.9.5}$$

Thus, since neither $\exp(-\alpha x)$ nor $\exp(\alpha x)$ are integrable over R, a necessary and sufficient condition for f to be integrable over R is that $C_1 = C_2 = 0$. In this case we have shown that the resolvent equation (7.9.1) has a unique solution $g \in L^1$ given by

$$g(x) = R_\lambda f(x) = \int_{-\infty}^{\infty} K(x-y) f(y)\, dy, \tag{7.9.6}$$

and thus condition (b) of the Hille–Yosida theorem is satisfied.

Combining equations (7.9.5) and (7.9.6) it follows immediately that the operator λR_λ preserves the integral. Moreover, $\lambda R_\lambda \geq 0$ if $f \geq 0$, so that λR_λ is a Markov operator. Thus condition (c) of the Hille–Yosida theorem is automatically satisfied, and we have shown that the operator d^2/dx^2 is an infinitesimal operator of a continuous semigroup $\{T_t\}_{t\geq 0}$ of Markov operators, where

$$T_t f = \lim_{\lambda \to \infty} e^{-t\lambda} \sum_{n=0}^{\infty} \frac{t^n \lambda^n}{n!} (\lambda R_\lambda)^n f \tag{7.9.7}$$

and R_λ is defined by (7.9.3) and (7.9.6).

It is interesting that the limit (7.9.7) can be calculated explicitly. To do this, denote by ϕ_f the Fourier transformation of f, that is,

$$\phi_f(\omega) = \int_{-\infty}^{\infty} e^{-i\omega x} f(x)\, dx.$$

The Fourier transformation of $K(x)$ given by equation (7.9.3) is

$$1/(\lambda + \omega^2),$$

where $\lambda = \alpha^2$. Since, by (7.9.6), $R_\lambda f$ is the convolution of the functions K and f, and it is well known that

$$\phi_{f \star g}(\omega) = \phi_f(\omega)\phi_g(\omega), \tag{7.9.8}$$

where $f \star g$ denotes the convolution of f with g, the Fourier transformation of $R_\lambda^n f$ is

$$[1/(\lambda + \omega^2)^n]\phi_f(\omega).$$

As a consequence, the Fourier transformation of the series in (7.9.7) is

$$\sum_{n=0}^{\infty} \frac{t^n \lambda^{2n}}{(\lambda + \omega_2)^n n!}\phi_f(\omega) = \exp[\lambda^2 t/(\lambda + \omega^2)]\phi_f(\omega).$$

Thus the Fourier transformation of $T_t f$ is

$$\lim_{\lambda \to \infty} \exp(-\lambda t)\exp[\lambda^2 t/(\lambda + \omega^2)]\phi_f(\omega) = \exp(-\omega^2 t)\phi_f(\omega).$$

Using the fact that $\exp(-\omega^2 t)$ is the Fourier transformation of

$$\frac{1}{\sqrt{4\pi t}}\exp(-x^2/4t)$$

and (7.9.8), we then have

$$T_t f(x) = \frac{1}{\sqrt{4\pi t}} \int_{-\infty}^{\infty} \exp[-(x-y)^2/4t]f(y)\, dy. \tag{7.9.9}$$

Hence, using the semigroup method we have shown that $u(t, x) = T_t f(x)$ is the solution of the heat equation

$$\frac{\partial u}{\partial t} = \frac{\partial^2 u}{\partial x^2}$$

with the initial condition

$$u(0, x) = f(x).$$

Remark 7.9.1. It is a direct consequence of the elementary properties of the differential quotient (see Definition 7.5.1) that if A is the infinitesimal operator corresponding to a semigroup $\{T_t\}_{t \geq 0}$, then cA is the infinitesimal operator corresponding to $\{T_{ct}\}_{t \geq 0}$. Thus, since we have proved that $A = d^2/dx^2$ is the infinitesimal operator corresponding to the semigroup $\{T_t\}_{t \geq 0}$ given by (7.9.9), we know immediately that

$$T_{\sigma^2 t/2}f(x) = \frac{1}{\sqrt{2\pi\sigma^2 t}} \int_{-\infty}^{\infty} \exp[-(x-y)^2/2\sigma^2 t]f(y)\, dy$$

has a corresponding infinitesimal operator equal to $(\sigma^2/2)(d^2/dx^2)$. (This is in perfect agreement with our observations in Example 7.4.1.) For simplicity, we have omitted the coefficient $(\sigma^2/2)$ in the foregoing calculations. □

The proof that d^2/dx^2 is an infinitesimal operator for a stochastic semigroup on R may be extended to R^d. Thus, for example, the operator

$$Af \equiv \Delta f = \sum_{i=1}^{n} \frac{\partial^2 f}{\partial x_i^2} \qquad (7.9.10)$$

on R^d may be shown to be an infinitesimal operator for a stochastic semigroup, as can

$$Af = \sum_{i,j=1}^{n} a_{ij} \frac{\partial^2 f}{\partial x_i \partial x_j}, \qquad (7.9.11)$$

where the a_{ij} are constant, or sufficiently regular functions of x, and $\sum_{i,j} a_{ij}\xi_i\xi_j$ is positive definite. The procedure for proving these assertions is similar to that for operator d^2/dx^2 on R, but requires some special results from the theory of partial differential equations and functional analysis, allowing us to extend the definitions of the differential operators (7.9.10) and (7.9.11).

Operators such as d^2/dx^2, (7.9.10), or (7.9.11) may be considered not on the whole space (R or R^d), but also on bounded subspaces. However, in this case other boundary conditions must be specified, for example,

$$Af = \frac{d^2 f}{dx^2} \qquad \text{on } L^1([a,b])$$

with

$$\left.\frac{df}{dx}\right|_a = 0 \quad \text{and} \quad \left.\frac{df}{dx}\right|_b = 0$$

is an infinitesimal operator for a stochastic semigroup. More details concerning such general elliptic operators may be found in Dynkin [1965].

Finally, we note that all semigroups that are generated by second-order differential operators are not semigroups of Frobenius–Perron operators for a semidynamical system and, thus, cannot arise from deterministic processes. This is quite contrary to the situation for first-order differential operators, as already discussed in Section 7.8.

Remark 7.9.2. Equation (7.8.3) of the Hille–Yosida theorem allows the construction of the semigroup $\{T_t\}_{t\geq 0}$ if the resolvent operator R_A is known. As it turns out, the construction of the resolvent operator when the continuous semigroup of contractions is given is even simpler. Thus it can be shown that (Dynkin, 1965)

$$R_\lambda f = \int_0^\infty e^{-\lambda t} T_t f \, dt \qquad \text{for } f \in L, \; \lambda > 0. \qquad (7.9.12)$$

In (7.9.12) the integral on the half-line $[0, \infty)$ is considered as the limit of Riemann integrals on $[0, a]$ as $a \to \infty$. This limit exists since

$$\left\| \int_0^\infty e^{-\lambda t} T_t f \, dt \right\| \leq \int_0^\infty e^{-\lambda t} \|T_t f\| \, dt \leq \frac{1}{\lambda} \|f\|.$$

It is an immediate consequence of (7.9.12) that for every stochastic semigroup $T_t: L^1 \to L^1$, the operator λR_λ is a Markov operator. To show this note first that, for $f \geq 0$, equation (7.9.12) implies $\lambda R_\lambda \geq 0$. Furthermore, for $f \geq 0$,

$$\|R_\lambda f\| = \int_X R_\lambda f(x) \, dx = \int_0^\infty e^{-\lambda t} \left\{ \int_X T_t f(x) \, dx \right\} dt$$
$$= \int_0^\infty e^{-\lambda t} \|f\| \, dt = \frac{1}{\lambda} \|f\|.$$

In addition to demonstrating that λR_λ is a Markov operator, (7.9.12) also demonstrates that the semigroup corresponding to a given resolvent R_λ is unique. To see this, choose $g \in L^\infty$ and take the scalar product of both sides of equation (7.9.12) with g. We obtain

$$\langle g, R_\lambda f \rangle = \int_0^\infty e^{-\lambda t} \langle g, T_t f \rangle \, dt \qquad \text{for } \lambda > 0,$$

which shows that $\langle g, R_\lambda f \rangle$, as a function of λ, is the Laplace transformation of $\langle g, T_t f \rangle$ with respect to t. Since the Laplace transformation is one to one, this implies that $\langle g, T_t f \rangle$ is uniquely determined by $\langle g, R_\lambda f \rangle$. Further, since $g \in L^\infty$ is arbitrary, $\{T_t f\}$ is uniquely determined by $\{R_\lambda f\}$. The same argument also shows that for a bounded continuous function $u(t)$, with values in L^1, the equality

$$R_\lambda f = \int_0^\infty e^{-\lambda t} u(t) \, dt \qquad \text{implies } u(t) = T_t f. \quad \square$$

Some of the most sophisticated applications of semigroup theory occur in treating integro-differential equations. Thus we may not only prove the existence and uniqueness of solutions to such equations, but also determine the asymptotic properties of the solutions. One of the main tools in this area is the following extension of the Hille–Yosida theorem, generally known as the **Phillips perturbation theorem.**

Theorem 7.9.1. *Let a continuous stochastic semigroup $\{T_t\}_{t \geq 0}$ and a Markov operator P be given. Further, let A be the infinitesimal operator of $\{T_t\}_{t \geq 0}$. Then there exists a unique continuous stochastic semigroup $\{P_t\}_{t \geq 0}$ for which*

$$A_0 = A + P - I$$

(I is the identity operator on L^1) is the infinitesimal operator. Furthermore, the semigroup $\{P_t\}_{t \geq 0}$ is defined by

$$P_t f = e^{-t} \sum_{n=0}^{\infty} T_n(t) f \qquad f \in L^1, \tag{7.9.13}$$

where $T_0(t) = T_t$ and

$$T_n(t) f = \int_0^t T_0(t - \tau) P T_{n-1}(\tau) f \, d\tau. \tag{7.9.14}$$

Proof. Denote by $R_\lambda(A)$ the resolvent corresponding to operator A, that is, $g = R_\lambda(A) f$ is the solution of

$$\lambda g - A g = f \qquad \text{for } f \in L^1.$$

Since $\{T_t\}_{t \geq 0}$ is a stochastic semigroup, $\lambda R_\lambda(A)$ is a Markov operator (see Remark 7.9.2). Now we observe that the resolvent equation for operator A_0,

$$\lambda g - A_0 g = f, \tag{7.9.15}$$

may be rewritten as

$$(\lambda + 1) g - A g = f + P g.$$

Thus (7.9.15) is equivalent to

$$g = R_{\lambda+1}(A) f + R_{\lambda+1}(A) P g. \tag{7.9.16}$$

From inequality (7.8.11) we have $\|R_{\lambda+1}(A) P g\| \leq \|P g\| / (\lambda + 1)$. Since P is a Markov operator, this becomes

$$\|R_{\lambda+1}(A) P g\| \leq \frac{\|g\|}{(\lambda + 1)}.$$

Thus, equation (7.9.16) has a unique solution that can be constructed by the method of successive approximations. The result is given by

$$g = R_\lambda(A_0) f = \sum_{n=0}^{\infty} [R_{\lambda+1}(A) P]^n R_{\lambda+1}(A) f, \tag{7.9.17}$$

and the existence of a solution g to (7.9.15) is proved. Further, from (7.9.17) it follows that $R_\lambda(A_0) f \geq 0$ for $f \geq 0$ and that

$$\|R_\lambda(A_0) f\| = \sum_{n=0}^{\infty} \left(\frac{1}{\lambda + 1} \right)^{n+1} \|f\| = \frac{1}{\lambda} \|f\| \qquad \text{for } f \geq 0.$$

Thus $\lambda R_\lambda(A_0)$ is a Markov operator and A_0 satisfies all of the assumptions of the Hille–Yosida theorem. Hence the infinitesimal operator A_0 generates a unique stochastic semigroup and the first part of the theorem is proved.

Now we show that this semigroup is given by equations (7.9.13) and (7.9.14). Using (7.9.14) it is easy to show by induction that

$$\|T_n(t)f\| \le (t^n/n!)\|f\|. \tag{7.9.18}$$

Thus, the series (7.9.13) is uniformly convergent, with respect to t, on bounded intervals and $P_t f$ is a continuous function of t. Now set

$$Q_{\lambda,n}f = \int_0^\infty e^{-\lambda t} T_n(t)f\, dt, \qquad n = 0, 1, \ldots,$$

so

$$Q_{\lambda,0}f = \int_0^\infty e^{-\lambda t} T_t f\, dt = R_\lambda(A)f$$

and

$$
\begin{aligned}
Q_{\lambda,n}f &= \int_0^\infty e^{-\lambda t}\left\{ \int_0^t T_0(t-\tau) P T_{n-1}(\tau)f\, d\tau \right\} dt \\
&= \int_0^\infty \left\{ \int_\tau^\infty e^{-\lambda t} T_0(t-\tau) P T_{n-1}(\tau)f\, dt \right\} d\tau \\
&= \int_0^\infty \left\{ e^{-\lambda \tau} \int_0^\infty e^{-\lambda t} T_0(t) P T_{n-1}(\tau)f\, dt \right\} d\tau \\
&= \int_0^\infty \left\{ e^{-\lambda t} T_0(t) P \int_0^\infty e^{-\lambda \tau} T_{n-1}(\tau)f\, d\tau \right\} dt \\
&= R_\lambda(A) P Q_{\lambda,n-1}f.
\end{aligned}
$$

Hence, by induction, we have

$$Q_{\lambda,n} = [R_\lambda(A)P]^n R_\lambda(A).$$

Define

$$Q_\lambda f = \int_0^\infty e^{-\lambda t} P_t f\, dt$$

and substitute equation (7.9.13) to give

$$
\begin{aligned}
Q_\lambda f &= \sum_{n=0}^\infty \int_0^\infty e^{-(\lambda+1)t} T_n(t)f\, dt = \sum_{n=0}^\infty Q_{\lambda+1,n}f \\
&= \sum_{n=0}^\infty [R_{\lambda+1}(A)P]^n R_{\lambda+1}(A)f.
\end{aligned}
$$

By comparing this result with (7.9.17), we see that $Q_\lambda = R_\lambda(A_0)$ or

$$R_\lambda(A_0)f = \int_0^\infty e^{-\lambda t} P_t f\, dt. \tag{7.9.19}$$

From (7.9.19) (see also the end of Remark 7.9.2), it follows that $\{P_t f\}_{t \geq 0}$ is the semigroup corresponding to A_0.   ∎

Example 7.9.1. Consider the integro-differential equation

$$\frac{\partial u(t,x)}{\partial t} + u(t,x) = \frac{\sigma^2}{2} \frac{\partial^2 u(t,x)}{\partial x^2} + \int_{-\infty}^{\infty} K(x,y) u(t,y)\, dy,$$

$$t > 0,\ x \in R \qquad (7.9.20)$$

with the initial condition

$$u(0,x) = \phi(x) \qquad x \in R. \qquad (7.9.21)$$

We assume that the kernel is measurable and stochastic, that is,

$$K(x,y) \geq 0 \quad \text{and} \quad \int_{-\infty}^{\infty} K(x,y)\, dx = 1.$$

To treat the initial value problem, equations (7.9.20) and (7.9.21), using semigroup theory, we rewrite it in the form

$$\frac{du}{dt} = (A + P - I)u, \qquad u(0) = \phi, \qquad (7.9.22)$$

where $A = \frac{1}{2}\sigma^2(d^2/dx^2)$ is the infinitesimal operator for the semigroup

$$T_t f(x) = \frac{1}{\sqrt{2\sigma^2 \pi t}} \int_{-\infty}^{\infty} \exp[-(x-y)^2/2\sigma^2 t] f(y)\, dy \qquad (7.9.23)$$

(see Remark 7.9.1) and

$$P f(x) = \int_{-\infty}^{\infty} K(x,y) f(y)\, dy.$$

From Theorem 7.9.1 it follows that there is a unique continuous semigroup $\{P_t\}_{t \geq 0}$ corresponding to operator $A_0 = A + P - I$, and, by Theorem 7.5.1, the function $u(t) = P_t \phi$ is the solution of (7.9.22) for every $\phi \in \mathcal{D}(A_0) = \mathcal{D}(A)$. Thus $u(t,x) = P_t \phi(x)$ can be interpreted as the generalized solution to equations (7.9.20) and (7.9.21) for every $\phi \in L^1(R)$.

This method of treating equation (7.9.20) is convenient from several points of view. First, it demonstrates the existence and uniqueness of the solution $u(t,x)$ for every $\phi \in L^1(R)$, and stochastic kernel K. Second, it shows that $P_t \phi$ is a density for $t \geq 0$ whenever ϕ is a density. Furthermore, some additional properties of the solution can be demonstrated by using the explicit representation for P_t given in Theorem 7.9.1. For this example, it follows directly from (7.9.13) and (7.9.14) that

$$P_t \phi = e^{-t} \int_0^t T_0(t-\tau) g_\tau\, d\tau + e^{-t} T_0(t)\phi,$$

where

$$g_t = \sum_{n=1}^{\infty} PT_{n-1}(t)\phi.$$

Thus, using (7.9.23) (with $T_0(t) = T_t$), we have the explicit representation

$$P_t\phi(x) = e^{-t} \int_0^t \left\{ \frac{1}{\sqrt{2\sigma^2\pi(t-\tau)}} \int_{-\infty}^{\infty} \exp[-(x-y)^2/2\sigma^2(t-\tau)] \cdot \right.$$
$$\left. \cdot g_\tau(y)\, dy \right\} d\tau + e^{-t} \frac{1}{\sqrt{2\sigma^2\pi t}} \int_{-\infty}^{\infty} \exp[-(x-y)^2/2\sigma^2 t]\phi(y)\, dy.$$

This shows directly that the function $u(t, x) = P_t\phi(x)$ is continuous and strictly positive for $t > 0$ and every $\phi \in L^1(R)$, even if ϕ and the stochastic kernel K are not continuous! Finally, we will come back to this semigroup approach in Section 11.10 and use it to demonstrate some asymptotic properties of the solution $u(t, x)$.   □

Example 7.9.2. As a second example of the applicability of the Phillips perturbation theorem, we consider the first-order integro-differential equation

$$\frac{\partial u(t, x)}{\partial t} + \frac{\partial u(t, x)}{\partial x} + u(t, x) = \int_x^{\infty} K(x, y)u(t, y)\, dy,$$
$$t > 0, \ x \geq 0 \qquad (7.9.24)$$

with

$$u(t, 0) = 0 \quad \text{and} \quad u(0, x) = \phi(x). \qquad (7.9.25)$$

Again the kernel K is assumed to be measurable and stochastic, that is,

$$K(x, y) \geq 0 \quad \text{and} \quad \int_0^y K(x, y)\, dx = 1. \qquad (7.9.26)$$

Equation (7.9.24) occurs in queuing theory and astrophysics [Bharucha-Reid, 1960]. In its astrophysical form,

$$K(x, y) = (1/y)\psi(x/y), \qquad (7.9.27)$$

and, with this specific expression for K, equation (7.9.24) is called the Chandrasekhar–Münch equation. As developed by Chandrasekhar and Münch [1952], equation (7.9.24) with K as given by (7.9.27) describes fluctuations in the brightness x of the Milky Way as a function of the extent of the system t along the line of sight. The unknown function $u(t, x)$ is the probability density of the fluctuations, and the given function ψ in (7.9.27) is related to the probability density of light transmission through interstellar gas clouds. This function satisfies

$$\psi(z) \geq 0 \quad \text{and} \quad \int_0^1 \psi(z)\, dz = 1 \qquad (7.9.28)$$

and, thus, K as given by (7.9.27) automatically satisfies (7.9.26).

To rewrite (7.9.24) as a differential equation in L^1, recall (see Example 7.5.1) that $-d/dx$ is the infinitesimal operator for the semigroup $T_t f(x) = f(x - t)$ defined on $L^1(R)$. On $L^1([0, \infty))$,

$$T_t f(x) = 1_{[0,\infty)}(x - t)f(x - t) \tag{7.9.29}$$

plays an analogous role. Proceeding much as in Example 7.5.1, a simple calculation shows that for continuously differentiable f with compact support in $[0, \infty)$ the infinitesimal operator corresponding to the semigroup in (7.9.29) is given by $Af = -df/dx$. Further, it is clear that $u(t, x) = T_t f(x)$ satisfies $u(t, 0) = 0$ for $t > 0$. Hence we may rewrite equations (7.9.24)–(7.9.25) in the form

$$\frac{du}{dt} = (A + P - I)u, \qquad u(0) = \phi, \tag{7.9.30}$$

where

$$Pf(x) = \int_x^\infty K(x, y)f(y)\, dy.$$

By Theorem 7.9.1 there is a continuous semigroup $\{P_t\}_{t \geq 0}$ corresponding to the infinitesimal operator $A + P - I$. For every $\phi \in \mathcal{D}(A)$, the function $u(t) = P_t\phi$ is a solution of (7.9.30). $\quad\square$

7.10 The Relation Between the Frobenius–Perron and Koopman Operators

The semigroup of Frobenius–Perron operators $\{P_t\}$ and the semigroup $\{U_t\}$ of Koopman operators, both generated by the same semidynamical system $\{S_t\}_{t \geq 0}$, are closely related because they are adjoint. However, each describes the behavior of the system $\{S_t\}_{t \geq 0}$ in a different fashion, and in this section we show the connection between the two.

Equation (7.4.16), $\langle P_t f, g \rangle = \langle f, U_t g \rangle$, which says that P_t and U_t are adjoint, may be written explicitly as

$$\int_X g(x)P_t f(x)\mu(dx) = \int_X f(x)g(S_t(x))\mu(dx) \qquad \text{for } f \in L^1, g \in L^\infty.$$

For some $A \subset X$ such that A and $S_t(A)$ are in $\mathcal{A}$, take $f(x) = 0$ for all $x \notin A$ and $g = 1_{X \setminus S_t(A)}$ so the preceding formula becomes

$$\int_X 1_{X \setminus S_t(A)}(x)P_t f(x)\mu(dx) = \int_X f(x)1_{X \setminus S_t(A)}(S_t(x))\mu(dx)$$

$$= \int_A f(x)1_{X \setminus S_t(A)}(S_t(x))\mu(dx).$$

The right-hand side of this equation is obviously equal to zero since $S_t(x) \notin X \setminus S_t(A)$ for $x \in A$. The left-hand side is, however, just the L^1 norm of the integrand, so that

$$\|1_{X \setminus S_t(A)} P_t f\| = 0.$$

This, in turn, implies

$$1_{X \setminus S_t(A)}(x) P_t f(x) = 0$$

or

$$P_t f(x) = 0 \qquad \text{for } x \notin S_t(A). \tag{7.10.1}$$

Thus the operator P_t "carries" the function f, supported on A, forward in time to a function supported on a subset of $S_t(A)$ (see Example 3.2.1 and Proposition 3.2.1). Figuratively speaking, we may say that the density is transformed by P_t analogously to the way in which initial points x are transformed into $S_t(x)$.

Now consider the definition of the Koopman operator,

$$U_t f(x) = f(S_t(x)).$$

Assume $f \in L^\infty$ is zero outside a set A, so we have

$$f(S_t(x)) = 0 \qquad \text{if } S_t(x) \notin A. \tag{7.10.2}$$

This, in turn, implies that

$$U_t f(x) = 0 \qquad \text{for } x \notin S_t^{-1}(A). \tag{7.10.3}$$

In contrast to P_t, therefore, U_t may be thought of as transporting the function supported on A, backward in time to a function supported on $S_t^{-1}(A)$.

These observations become even clearer when $\{S_t\}$ is a group of transformations, that is, when the group property holds for both positive and negative time,

$$S_{t+t'}(x) = S_t(S_{t'}(x)) \qquad \text{for all } t, t' \in R, x \in X,$$

and all the S_t are at least nonsingular. In this case, $S_t^{-1}(x) = S_{-t}(x)$ and (7.10.3) becomes

$$U_t f(x) = 0 \qquad \text{for } x \notin S_{-t}(A).$$

If, in addition, the group $\{S_t\}$ preserves the measure μ, we have

$$\int_A P_t f(x) \mu(dx) = \int_{S_{-t}(A)} f(x) \mu(dx) = \int_A f(S_{-t}(x)) \mu(dx),$$

which gives

$$P_t f(x) = f(S_{-t}(x))$$

or, finally,

$$P_t f(x) = U_{-t} f(x). \tag{7.10.4}$$

Equation (7.10.4) makes totally explicit our earlier comments on the forward and backward transport of densities in time by the Frobenius–Perron and Koopman operators.

Furthermore, from (7.10.4) we have directly that

$$\lim_{t \to 0}[(P_t f - f)/t] = \lim_{t \to 0}[(U_{-t} f - f)/t]$$

and, thus, for f in a dense subset of L^1,

$$A_{FP} f = -A_K f. \tag{7.10.5}$$

This relation was previously derived, although not explicitly stated, for dynamical systems generated by a system of ordinary differential equations [cf. equations (7.6.5) and (7.6.10)].

Remark 7.10.1. Equation (7.10.4) may, in addition be interpreted as saying that the operator adjoint to P_t is also its inverse. In the terminology of Hilbert spaces [and thus in $L^2(X)$] this means simply that $\{P_t\}$ is a semigroup of unitary operators. The original discovery that $\{U_t\}$, generated by a group $\{S_t\}$ of measure-preserving transformations, forms a group of unitary operators is due to Koopman [1931]. It was later used by von Neumann [1932] in his proof of the statistical ergodic theorem. □

Remark 7.10.2. Equation (7.10.1) can sometimes be used to show that a semigroup of Markov operators cannot arise from a deterministic dynamical system, which means that it is not a semigroup of Frobenius–Perron operators for any semidynamical system $\{S_t\}_{t \geq 0}$.

For example, consider the semigroup $\{P_t\}$ given by equations (7.4.11) and (7.4.12):

$$P_t f(x) = \frac{1}{\sqrt{2\pi\sigma^2 t}} \int_{-\infty}^{\infty} f(y) \exp\left[-\frac{(x-y)^2}{2\sigma^2 t}\right] dy. \tag{7.10.6}$$

Setting $f(y) = 1_{[0,1]}(y)$, it is evident that we obtain

$$P_t f(x) > 0 \qquad \text{for all } x \text{ and } t > 0.$$

However, according to (7.10.1), if $P_t f(x)$ was the Frobenius–Perron operator generated by a semidynamical system $\{S_t\}_{t \geq 0}$, then it should be zero outside a bounded interval $S_t([0,1])$. [The interval $S_t([0,1])$ is a bounded interval since a continuous function maps bounded intervals into bounded intervals.] Thus $\{P_t\}$, where $P_t f(x)$ is given by (7.10.6), does not correspond to any semidynamical system. □

7.11 Sweeping for Stochastic Semigroups

The notion of sweeping for operators as developed in Section 5.9 is easily extended to semigroups. We start with the following.

Definition 7.11.1. Let $(X, \mathcal{A}, \mu)$ be a measure space and $\mathcal{A}_* \subset \mathcal{A}$ be a given family of measurable sets. A stochastic semigroup $P_t: L^1(X) \to L^1(X)$ is called **sweeping** with respect to $\mathcal{A}_*$ if

$$\lim_{t \to \infty} \int_A P_t f(x) \mu(dx) = 0 \qquad \text{for } f \in D \text{ and } A \in \mathcal{A}_*. \qquad (7.11.1)$$

As in the discrete time case, it is easy to verify that condition (7.11.1) for a sweeping semigroup $\{P_t\}_{t \geq 0}$ also holds for every $f \in L^1(X)$. Alternately, if $D_0 \subset D$ is dense in D, then it is sufficient to verify (7.11.1) for $f \in D_0$.

In the special case that $X \subset R$ is an interval (bounded or not) with endpoints α and β, $\alpha < \beta$, we will use notions analogous to those in Definition 5.9.2. Namely, we will say that a stochastic semigroup $P_t: L^1(X) \to L^1(X)$ is **sweeping to** α, **sweeping to** β, or simply **sweeping** if it is sweeping with respect to the families A_0, A_1, or A_2 defined in equations (5.9.5)–(5.9.7), respectively.

Example 7.11.1. Let $X = R$. We consider the semigroup generated by the infinitesimal operators cd/dx and $(\sigma^2/2)d^2/dx^2$ discussed in Example 7.5.1 and Remark 7.9.1.

The operator cd/dx corresponds to the semigroup

$$P_t f(x) = f(x - ct)$$

which, for $c > 0$, is sweeping to $+\infty$ and for $c < 0$ to $-\infty$. The verification of these properties is analogous to the procedure in Example 5.9.1. Thus, for $c > 0$ we have

$$\int_{-\infty}^b P_t f(x)\, dx = \int_{-\infty}^b f(x - ct)\, dx = \int_{-\infty}^{b-ct} f(y)\, dy = 0$$

when f has compact support and t is sufficiently large. For $c < 0$ the argument is similar.

The operator $(\sigma^2/2)d^2/dx^2$ generates the semigroup

$$P_t f(x) = \frac{1}{\sqrt{2\pi\sigma^2 t}} \int_{-\infty}^{+\infty} \exp\left[-\frac{(x-y)^2}{2\sigma^2 t}\right] f(y)\, dy$$

which is evidently sweeping since, for $f \in D$,

$$\int_a^b P_t f(x)\, dx \leq \frac{b-a}{\sqrt{2\pi\sigma^2 t}} \to 0 \qquad \text{as } t \to \infty. \quad \square$$

Comparing Examples 5.9.1, 5.9.2, and 7.11.1 we observe that the sweeping property of a semigroup $\{P_t\}_{t\geq 0}$ appears simultaneously with the sweeping of the sequence $\{P_{t_0}^n\}$ for some $t_0 > 0$. This is not a coincidence. It is evident from Definitions 5.9.1 and 7.11.1 that if $\{P_t\}_{t\geq 0}$ is sweeping, then $\{P_{t_0}^n\}$ is also sweeping for an arbitrary $t_0 > 0$. The converse is more delicate, but is assured by the following result.

Theorem 7.11.1. *Let* $(X, \mathcal{A}, \mu)$ *be a measure space,* $\mathcal{A}_* \subset \mathcal{A}$ *be a given family of measurable sets, and* $P_t\colon L^1(X) \to L^1(X)$ *a continuous stochastic semigroup. If for some* $t_0 > 0$ *the sequence* $\{P_{t_0}^n\}$ *is sweeping, then the semigroup* $\{P_t\}_{t\geq 0}$ *is also sweeping.*

Proof. Fix an $\varepsilon > 0$ and $f \in D$. Since P_t is continuous there is a $\delta > 0$ such that
$$\|P_t f - f\| \leq \varepsilon \qquad \text{for } 0 \leq t \leq \delta.$$
Let
$$0 = s_0 < s_1 < \cdots < s_k = t_0$$
be a partition of the interval $[0, t_0]$ such that
$$s_i - s_{i-1} \leq \delta \qquad \text{for } i = 1, \ldots, k.$$
Define $f_i = P_{s_i} f$. Every value $t_0 \geq 0$ can be written in the form
$$t = nt_0 + s_i + \tau,$$
where n and i are integers $(n = 0, 1, \ldots; \ i = i, \ldots, k)$ and $0 \leq \tau < \delta$. Therefore,
$$P_t f = P_{t_0}^n P_{s_i} P_\tau f = P_{t_0}^n f_i + P_{t_0}^n P_{s_i} (P_\tau f - f).$$
Since $\|P_\tau f - f\| \leq \varepsilon$ and $P_{t_0}^n$ and P_{s_i} are contractive, we have
$$\|P_{t_0}^n P_{s_i} (P_\tau f - f)\| \leq \varepsilon.$$
As a consequence, for every $A \in \mathcal{A}_*$
$$\int_A P_t f(x) \mu(dx) \leq \int_A P_{t_0}^n f_i(x) \mu(dx) + \varepsilon.$$
Evidently, $n \to \infty$ as $t \to \infty$ and the integrals on the right-hand side converge to zero, thus completing the proof. ∎

The main advantage of Theorem 7.11.1 is that it allows us to obtain many corollaries concerning sweeping for semigroups from previous results for iterates of a single operator. As an example, from Theorem 7.11.1 and Proposition 5.9.1 we have the following:

Proposition 7.11.1. *Let* $(X, \mathcal{A}, \mu)$ *be a measure space, and* $\mathcal{A}_* \subset \mathcal{A}$ *be a given family of measurable sets. Furthermore, let* $P_t\colon L^1(X) \to L^1(X)$ *be a*

continuous stochastic semigroup for which there exists a Bielecki function
$V: X \to R$, *a constant* $\gamma < 1$, *and a point* $t_0 > 0$ *such that*

$$\int_X V(x) P_{t_0} f(x) \mu(dx) \leq \gamma \int_X V(x) f(x) \mu(dx) \qquad \text{for } f \in D.$$

Then the semigroup $\{P_{t_0}\}_{t \geq 0}$ *is sweeping.*

Proof. Since the operator P_{t_0} satisfies the conditions of Proposition 5.9.1, the sequence $\{P_{t_0}^n\}$ is sweeping. Theorem 7.11.1 completes the proof. ∎

More sophisticated applications of Theorem 7.11.1 will be given in the next section.

7.12   Foguel Alternative for Continuous Time Systems

We start from a question concerning the relationship between the existence of an invariant density for a stochastic semigroup $\{P_t\}_{t \geq 0}$ and for an operator P_{t_0} with a fixed t_0. Clearly, if f_* is invariant with respect to P_t so $P_t f_* = f_*$ for all $t \geq 0$, then f_* is invariant for every operator P_{t_0}. The converse is, however, unfortunately false. Rather we have the following result.

Proposition 7.12.1. *If* $P_t: L^1(X) \to L^1(X)$ *is a continuous stochastic semigroup and if* $P_{t_0} f_0 = f_0$ *for some* $t_0 > 0$ *with* $f_0 \in D$, *then*

$$f_*(x) = \frac{1}{t_0} \int_0^{t_0} P_t f_0(x) \, dt$$

is a density and satisfies $P_t f_* = f_*$ *for all* $t \geq 0$.

Proof. From the definition of f_* we have

$$\int_X f_*(x) \mu(dx) = \int_X \left[\frac{1}{t_0} \int_0^{t_0} P_t f_0(x) \, dt \right] \mu(dx)$$

$$= \frac{1}{t_0} \int_0^{t_0} \left[\int_X P_t f_0(x) \mu(dx) \right] dt$$

$$= 1.$$

Furthermore,

$$P_t f_* = \frac{1}{t_0} \int_0^{t_0} P_{s+t} f_0 \, ds = \frac{1}{t_0} \int_t^{t_0+t} P_s f_0 \, ds$$

$$= \frac{1}{t_0} \int_0^{t_0} P_s f_0 \, ds + \frac{1}{t_0} \int_{t_0}^{t_0+t} P_s f_0 \, ds - \frac{1}{t_0} \int_0^t P_s f_0 \, ds$$

$$= f_* + \frac{1}{t_0} \int_0^t (P_{s+t_0} f_0 - P_s f_0) \, ds.$$

Since $P_{t_0} f_0 = f_0$ we have $P_{s+t_0} f_0 - P_s f_0 = 0$ and the last integral vanishes, thus completing the proof. ∎

Now, using Theorems 5.9.1, 5.9.2, and 7.12.1, it is easy to establish the following alternative.

Theorem 7.12.1. *Let $(X, \mathcal{A}, \mu)$ be a measure space, and $\mathcal{A}_* \subset \mathcal{A}$ be a given regular family of measurable sets. Furthermore, let $P_t \colon L^1(X) \to L^1(X)$ be a continuous stochastic semigroup such that for some $t_0 > 0$ the operator P_{t_0} satisfies the following conditions:*

(1) P_{t_0} is an integral operator given by a stochastic kernel; and

(2) There is a locally integrable function f_* such that

$$P_{t_0} f_* \le f_* \quad and \quad f_* > 0 \quad a.e.$$

Under these conditions, the semigroup $\{P_t\}_{t \ge 0}$ either has an invariant density, or it is sweeping. If an invariant density exists and, in addition, P_{t_0} is an expanding operator, then the semigroup is asymptotically stable.

Proof. The proof is quite straightforward. Assume first that $\{P_t\}_{t \ge 0}$ is not sweeping so by Theorem 7.11.1 the sequence $\{P_{t_0}^n\}$ is also not sweeping. In this case, by Theorem 5.10.1 the operator P_{t_0} has an invariant density. Proposition 7.12.1 then implies that $\{P_t\}_{t \ge 0}$ must have an invariant density. In the particular case that P_{t_0} is also an expanding operator, it follows from Theorem 5.10.2 that $\{P_{t_0}^n\}$ is asymptotically stable. Finally, Remark 7.4.2 implies that $\{P_t\}_{t \ge 0}$ is also asymptotically stable.

In the second case that $\{P_t\}_{t \ge 0}$ is sweeping, $\{P_{t_0}^n\}$ is also, and by Theorem 5.10.1 the operator P_{t_0} does not have an invariant density. As a consequence, $\{P_t\}_{t \ge 0}$ also does not have an invariant density. ∎

Exercises

7.1. Let $A \colon L \to L$ be a linear bounded operator, that is,

$$\|A\| = \sup\{\|Af\| \colon \|f\| \le 1\} < \infty.$$

Using a comparison series prove that

(a) $e^{tA}f = \sum_{n=0}^{\infty}(t^n/n!)A^n f$ is strongly convergent in L for $t \in R$ and $f \in L$,

(b) $e^{(t_1+t_2)A}f = e^{t_1 A}e^{t_2 A}f$ for $t_1, t_2 \in R$, $f \in L$.

7.2. Again, let $A: L \to L$ be a linear bounded operator. Using the results of Exercise 7.1, prove that A is the infinitesimal operator of the semigroup

$$T_t f = e^{tA}f$$

and that $\mathcal{D}(A) = L$.

7.3. A linear operator $A: \mathcal{D}(A) \to L^1$ is called **closed** if the conditions

$$\|f_n - f\| \to 0; \quad \|Af_n - g\| \to 0; \quad f_n \in \mathcal{D}(A); \quad f, g \in L$$

imply that $f \in \mathcal{D}(A)$ and $g = Af$. Prove that the following operators are closed:

(a) The operator $Af = df/dx$ defined on the set $\mathcal{D}(A) \subset L^1$ of all absolutely continuous $f \in L^1$ such that $f' \in L^1$.

(b) The operator $Af = d^2 f/dx^2$ defined on the set $\mathcal{D}(A) \subset L^1$ of all $f \in L^1$ such that f' is absolutely continuous and $f'' \in L^1$.

7.4. Generalize the previous results and show that every operator A satisfying the conditions of the Hille–Yosida theorem is closed.

7.5. In Section 7.9 using the Hille–Yosida theorem we have proved that $A = d^2/dx^2$ generates the semigroup $\{T_t\}$ given by formula (7.9.9). Reverse the calculation and assuming that $\{T_t\}$ is defined by (7.9.9) show that $A = d^2/dx^2$ is its infinitesimal operator.

7.6. Consider the one-dimensional $(X = R)$ case of the continuity equation

$$\frac{\partial u}{\partial t} + \frac{\partial}{\partial x}(F(x)u) = 0, \qquad u(0, x) = f(x),$$

where $F: R \to R$ is a C^1 function satisfying

$$|F(x)| \leq c(1 + |x|^r) \qquad \text{for } x \in R.$$

Assuming the above inequality with $r = 1$ show that the formula $P_t f(x) = u(t, x)$ defines a stochastic semigroup on $L^1(R)$. Find counter-examples showing that for $r > 1$ the semigroup $\{P_t\}$ may not be well defined or stochastic.

7.7. Consider the semigroup $\{P_t\}$ defined in the previous exercise (with $r = 1$). Show that

(a) $\{P_t\}$ is not asymptotically stable;

(b) $\{P_t\}$ is sweeping to $+\infty$ if and only if all solutions of the equation $x' = F(x)$ satisfy $\lim_{t\to\infty} x(t) = \infty$.

7.8. Consider the integro-differential equation

$$\frac{\partial u(t,x)}{\partial t} + u(t,x) = \frac{\partial^2 u(t,x)}{\partial x^2} + \int_0^\pi K(x,y)u(x,y)\,dy \quad \text{for } t > 0,\ 0 \le x \le \pi,$$

where $K \colon [0,\pi] \times [0,\pi] \to R$ is a stochastic kernel. Using the Hille–Yosida and the Phillips perturbation theorem, show that this equation with the boundary value condition

$$u_x'(t,0) = u_x'(t,\pi) = 0$$

and the initial condition

$$u(0,x) = f(x)$$

generates the stochastic semigroup $P^t f(x) = u(t,x)$ on the space $L^1([0,\pi])$. In particular, define precisely the domain $\mathcal{D}(A)$ of $A = d^2/dx^2$ for which the conditions of the Hille–Yosida theorem are satisfied (Jama 1986).

8
Discrete Time Processes Embedded in Continuous Time Systems

In this chapter, our goal is to introduce a way in which discrete time processes may be embedded in continuous time systems without altering the phase space. To do this, we adopt a strictly probabilistic point of view, not embedding the deterministic system $S: X \to X$ in a continuous time process, but rather embedding its Frobenius–Perron operator $P: L^1(X) \to L^1(X)$ that acts on L^1 functions. The result of this embedding is an abstract form of the Boltzmann equation. This chapter requires some elementary definitions from probability theory and a knowledge of Poisson processes, which are introduced following the preliminary remarks of the next section.

8.1 The Relation Between Discrete and Continuous Time Processes

For a semidynamical system $\{S_t\}_{t \geq 0}$ on a phase space X, if we fix the time t at some value t_0, then by property (b) of Definition 7.2.3,

$$S_{nt_0}(x) = S_{t_0} \circ \overset{n}{\cdots} \circ S_{t_0}(x) = S_{t_0}^n(x) \qquad \text{for all } x \in X.$$

Thus, in this fashion, a discrete time system may be generated from any continuous time (semidynamical) system. It is possible that a study of the discrete time system may yield some partial information concerning the continuous time system from which it was derived.

Another way of obtaining a discrete time system from a continuous one

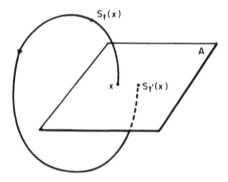

FIGURE 8.1.1. Determination of the first return (or Poincaré) map for a semi-dynamical system $\{S_t\}_{t\geq 0}$.

Also assume that we can find a closed set $A \subset X$ such that, if $x \in A$, then, for $t > 0$ sufficiently small $S_t(x) \notin A$, that is, each trajectory leaves A immediately (see Figure 8.1.1). Further, if every trajectory that starts in A eventually returns to A, that is, for every $x \in A$ there is a $t' > 0$ such that $S_{t'}(x) \in A$, then we may define a new mapping, the **first return map**. This is given by

$$\tilde{S}(x) = S_{t'}(x),$$

where t' is the smallest time $t' > 0$ such that $S_{t'}(x) \in A$. Again, by studying $\tilde{S}$, we may gain some insight into the properties of $\{S_t\}_{t\geq 0}$. This method was introduced by Poincaré, and the first return map is often called the **Poincaré map**.

Thus it is relatively straightforward to devise ways to study continuous time processes by a reduction to a discrete time system. However, given a discrete time system $S: X \to X$, it is much more difficult to embed it in a continuous time system and, indeed, such embedding is, in general, impossible. That is, given $\tilde{S}: X \to X$, generally there does not exist a $\{S_t\}_{t\geq 0}$ such that $\tilde{S}(x) = S_{t_0}(x)$ for some $t_0 \geq 0$ [see Zdun, 1977]. For example, in previous chapters we considered the quadratic transformation $S(x) = 4x(1 - x)$, $x \in [0, 1]$. It can be proved that there does not exist a semidynamical system $\{S_t\}_{t\geq 0}$ on $[0, 1]$ such that $S_{t_0}(x) = 4x(1 - x)$. Of course, it is always possible to embed a discrete time process into a continuous time system by altering the phase space in an appropriate way.

8.2   Probability Theory and Poisson Processes

Up to this point we have almost never used the word probabilistic even though we have dealt, from the outset, with normalized measures that are also measures on a probability space. In this short section, we review all of the material necessary for an understanding of Poisson processes.

The fundamental notion of probability theory is that of a **probability space** $(\Omega, \mathcal{F}, \text{prob})$, where Ω is a nonempty set called the **space of all possible elementary events**, $\mathcal{F}$ is a σ-algebra of subsets of Ω, which are called **events**, and "prob" is a normalized measure on $\mathcal{F}$. The equality

$$\text{prob}(A) = p, \qquad A \in \mathcal{F},$$

means that the probability of event A is p. From the fact that prob is a measure, it immediately follows that

$$\text{prob}\left(\bigcup_i A_i\right) = \sum_i \text{prob}(A_i), \qquad (8.2.1)$$

where the $A_i \in \mathcal{F}$ are mutually disjoint, that is, $A_i \cap A_j = \emptyset$ for all $i \neq j$. To introduce the concept of independence, we define it as follows.

Definition 8.2.1. In a sequence of events $A_1, A_2, \ldots$ (finite or not), the events are called **independent** if, for any increasing sequence of integers $k_1 < k_2 < \cdots < k_n$,

$$\text{prob}(A_{k_1} \cap A_{k_2} \cap \cdots \cap A_{k_n}) = \text{prob}(A_{k_1}) \cdots \text{prob}(A_{k_n}). \qquad (8.2.2)$$

Equation (8.2.2) just means that the probability of all the events A_{k_i} occurring is the product of the probabilities that each will occur separately.

Random variables are defined next.

Definition 8.2.2. A **random variable** ξ is a measurable transformation from Ω into R. More precisely, $\xi: \Omega \to R$ is a random variable if, for any Borel set $B \subset R$,

$$\xi^{-1}(B) = \{\omega \in \Omega : \xi(\omega) \in B\} \in \mathcal{F}.$$

This set is customarily written in the more compact notation $\{\xi \in B\}$. Thus, for any Borel set $B \subset R$, $\text{prob}\{\xi \in B\}$ is well defined.

A function $f \in D(R)$ is called the **density** of the random variable ξ if

$$\text{prob}\{\xi \in B\} = \int_B f(x) \, dx \qquad (8.2.3)$$

for any Borel set $B \subset R$.

Let $\xi_1, \xi_2, \ldots$ be a sequence of random variables. We say the ξ_i are **independent** if, for any sequence of Borel sets $B_1, B_2, \ldots$, the events

$$\{\xi_1 \in B_1\}, \{\xi_2 \in B_2\}, \ldots$$

are independent. Thus a finite sequence of independent random variables satisfies

$$\text{prob}\{\xi_1 \in B_1, \ldots, \xi_n \in B_n\} = \text{prob}\{\xi_1 \in B_1\} \cdots \text{prob}\{\xi_n \in B_n\}, \qquad (8.2.4)$$

is as follows. Again, suppose we are given a semidynamical system $\{S_t\}_{t \geq 0}$.

and the probability that all events $\{\xi_i \in B_i\}$ will occur is simply given by the product of the probabilities that each will occur separately.

We are now in a position to make the concept of a stochastic process precise with the following definition.

Definition 8.2.3. A **stochastic process** $\{\xi_t\}$ is a family of random variables that depends on a parameter t, usually called time. If t assumes only integer values, $t = 1, 2, \ldots$, then the stochastic process reduces to a sequence $\{\xi_n\}$ of random variables called a **discrete time stochastic process**. However, if t belongs to an interval (bounded or not) of R, then the stochastic process is called a **continuous time stochastic process**.

By its very definition, a stochastic process $\{\xi_t\}$ is a function of two variables, namely, time t and event ω, but this is seldom made explicit by writing $\{\xi_t(\omega)\}$. If the time is fixed, then ξ_t is simply a random variable. However, if ω is fixed, then the mapping $t \to \xi_t(\omega)$ is called the **sample path** of the stochastic process.

Two important properties that stochastic processes may have are given in the following definition.

Definition 8.2.4. A continuous time stochastic process $\{\xi_t\}_{t \geq 0}$ has **independent increments** if, for any sequence of times $t_0 < t_1 < \cdots < t_n$, the random variables

$$\xi_{t_1} - \xi_{t_0}, \xi_{t_2} - \xi_{t_1}, \ldots, \xi_{t_n} - \xi_{t_{n-1}}$$

are independent. Further, if for any t_1 and t_2 and Borel set $B \subset R$,

$$\text{prob}\{\xi_{t_2+t'} - \xi_{t_1+t'} \in B\} \tag{8.2.5}$$

does not depend on t', then the continuous time stochastic process $\{\xi_t\}$ has **stationary independent increments**.

Before giving the definition of a Poisson process, we note that a stochastic process $\{\xi_t\}$ is called a **counting process** if its sample paths are nondecreasing functions of time with integer values. Counting processes will be denoted by $\{N_t\}_{t \geq 0}$.

Definition 8.2.5. A **Poisson process** is a counting process $\{N_t\}_{t \geq 0}$ with stationary independent increments satisfying:

(a) $N_0 = 0$; \hfill (8.2.6a)

(b) $\lim_{t \to 0}(1/t) \text{prob}\{N_t \geq 2\} = 0$; \hfill (8.2.6b)

(c) The limit
$$\lambda = \lim_{t \to 0}(1/t) \text{prob}\{N_t = 1\} \tag{8.2.6c}$$

exists and is positive; and

(d) $\text{prob}\{N_t = k\}$ as functions of t are continuous.

A classic example of a Poisson process is illustrated by a radioactive substance placed in a chamber equipped with a device for detecting and counting the total number of atomic disintegrations N_t that have occurred up to a time t. The amount of the substance must be sufficiently large such that during the time of observation there is no significant decrease in the mass. This ensures that the probability (8.2.5) is independent of t'. It is an experimental observation that the number of disintegrations that occur during any given interval of time is independent of the number occurring during any other disjoint interval, thus giving stationary independent increments. Conditions (a)–(c) in Definition 8.2.5 have the following interpretations within this example: $N_0 = 0$ simply means that we start to count disintegrations from time $t = 0$. Condition (b) states that two or more disintegrations are unlikely in a short time, whereas (c) simply means that during a short time t the probability of one disintegration is proportional to t.

Also, the classical derivations of the Boltzmann equation implicitly assume that molecular collisions are a Poisson process. This fact will turn out to be important later.

It is interesting that from the properties of the Poisson process we may derive a complete description of the way the process depends on time. Thus we may derive an explicit formula for

$$p_k(t) = \text{prob}\{N_t = k\}. \tag{8.2.7}$$

This is carried out in two steps. First we derive an ordinary differential equation for $p_k(t)$, and then we solve it. In our construction it will be useful to rewrite equations (8.2.6a) through (8.2.6c) using the notation of (8.2.7):

$$p_0(0) = 1, \tag{8.2.8a}$$

$$\lim_{t \to 0} \frac{1}{t} \sum_{i=2}^{\infty} p_i(t) = 0, \tag{8.2.8b}$$

and

$$\lambda = \lim_{t \to 0}(1/t)p_1(t). \tag{8.2.8c}$$

To obtain the differential equation for $p_k(t)$, we first start with $p_0(t)$, noting that $p_0(t + h)$ may be written as

$$p_0(t + h) = \text{prob}\{N_{t+h} = 0\} = \text{prob}\{N_{t+h} - N_t + N_t - N_0 = 0\}.$$

Since N_t is not decreasing, hence $(N_{t+h} - N_t) + (N_t - N_0) = 0$ if and only if $(N_{t+h} - N_t) = 0$ and $(N_t - N_0) = 0$. Thus,

$$p_0(t + h) = \text{prob}\{(N_{t+h} - N_t) = 0 \text{ and } (N_t - N_0) = 0\}$$

$$= \text{prob}\{N_{t+h} - N_t = 0\} \text{prob}\{N_t - N_0 = 0\}$$
$$= \text{prob}\{N_h - N_0 = 0\} \text{prob}\{N_t - N_0 = 0\}$$
$$= p_0(h)p_0(t), \qquad (8.2.9)$$

where we have used the property of stationary independent increments. From (8.2.9) we may write

$$\frac{p_0(t+h) - p_0(t)}{h} = \frac{p_0(h) - 1}{h} p_0(t). \qquad (8.2.10)$$

Since $\sum_{i=0}^{\infty} p_i(t) = 1$, we have

$$\frac{p_0(h) - 1}{h} = -\frac{p_1(h)}{h} - \frac{1}{h} \sum_{i=2}^{\infty} p_i(h),$$

and, thus, by taking the limit of both sides of (8.2.10) as $h \to 0$, we obtain

$$\frac{dp_0(t)}{dt} = -\lambda p_0(t). \qquad (8.2.11)$$

The derivation of the differential equation for $p_k(t)$ proceeds in a similar fashion. Thus

$$p_k(t+h) = \text{prob}\{N_{t+h} = k\}$$
$$= \text{prob}\{N_{t+h} - N_t + N_t - N_0 = k\}$$
$$= \text{prob}\{N_t - N_0 = k \text{ and } N_{t+h} - N_t = 0\}$$
$$+ \text{prob}\{N_t - N_0 = k - 1 \text{ and } N_{t+h} - N_t = 1\}$$
$$+ \sum_{i=2}^{k} \text{prob}\{N_t - N_0 = k - i \text{ and } N_{t+h} - N_t = i\}$$
$$= p_k(t)p_0(h) + p_{k-1}(t)p_1(h) + \sum_{i=2}^{k} p_{k-i}(t)p_i(h).$$

As before, we have

$$\frac{p_k(t+h) - p_k(t)}{h} = \frac{p_0(h) - 1}{h} p_k(t) + \frac{p_1(h)}{h} p_{k-1}(t) + \frac{1}{h} \sum_{i=2}^{k} p_{k-i}(t)p_i(h),$$

and, by taking the limit as $h \to 0$, we obtain

$$\frac{dp_k(t)}{dt} = -\lambda p_k(t) + \lambda p_{k-1}(t). \qquad (8.2.12)$$

The initial conditions for $p_0(t)$ and $p_k(t)$, $k \geq 1$, are just $p_0(0) = 1$ (by definition), and this immediately gives $p_k(0) = 0$ for all $k \geq 1$. Thus, from (8.2.11), we have

$$p_0(t) = e^{-\lambda t}. \qquad (8.2.13)$$

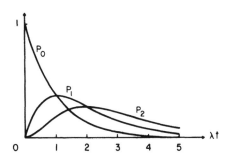

FIGURE 8.2.1. Probabilities $p_0(t)$, $p_1(t)$, $p_2(t)$ versus λt for a Poisson process.

Substituting this into (8.2.12) when $k = 1$ gives

$$\frac{dp_1(t)}{dt} = -\lambda p_1(t) + \lambda e^{-\lambda t}$$

whose solution is

$$p_1(t) = \lambda t e^{-\lambda t}.$$

Repeating this procedure for $k = 2, \ldots$ we find, by induction, that

$$p_k(t) = \frac{(\lambda t)^k}{k!} e^{-\lambda t}. \tag{8.2.14}$$

The behavior of $p_k(t)$ as a function of t is shown in Figure 8.2.1 for $k = 0$, 1, and 2. Figure 8.2.2 shows $p_k(t)$ versus k for several values of λt.

Remark 8.2.1. Note that in our derivation of equation (8.2.12) we have only used $h > 0$ and, therefore, the derivative p_k' on the left-hand side of (8.2.12) is, in fact, the right-hand derivative of p_k. However, it is known [Szarski, 1967] that, if the right-hand derivative p_k' exists and the p_k are continuous [as they are here by assumption (d) of Definition 8.2.5], then there is a unique solution to (8.2.12). Thus the functions (8.2.14) give the unique solution to the problem. □

Although the way we have introduced Poisson processes and derived the expressions for $p_k(t)$ is the most common, there are other ways in which this may be accomplished. However, all these derivations, as indicated by properties (a)–(c) of Definition 8.2.5, show that a Poisson process results if the events counted by N_t are caused by a large number of independent factors, each of which has a small probability of incrementing N_t.

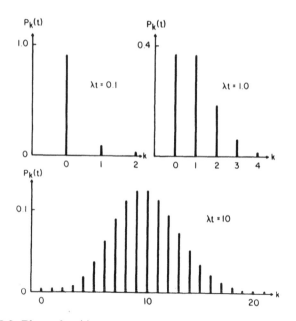

FIGURE 8.2.2. Plots of $p_k(t)$ versus k for a Poisson process with $\lambda t = 0.1$, 1.0, or 10.

8.3   Discrete Time Systems Governed by Poisson Processes

A particular sample path for a Poisson process might look like the one shown in Figure 8.3.1. In this section we develop some ideas and tools that will allow us to study the behavior of a deterministic discrete time process given by a nonsingular transformation $S: X \to X$ on a measure space $(X, \mathcal{A}, \mu)$ coupled with a Poisson process $\{N_t\}_{t \geq 0}$. The coupling is such that, even though the dynamics are deterministic, the *times* at which the transformation S operates are determined by the Poisson process. Thus we consider the situation in which each point $x \in X$ is transformed into $S^{N_t}(x)$. This may be written symbolically as

$$x \to S^{N_t}(x)$$

for times in the interval $[0, \infty)$. Specifically, we consider the following problem. Given an initial distribution of points $x \in X$, with density f, how does this distribution evolve in time? We denote the time-dependent density by $u(t, x)$ and set $u(0, x) = f(x)$.

The solution of this problem starts with a calculation of the probability

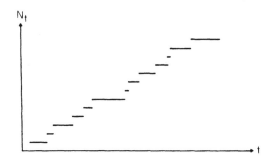

FIGURE 8.3.1. A sample path for a Poisson process.

that

$$S^{N_t}(x) \in A \tag{8.3.1}$$

for a given set $A \subset \mathcal{A}$ and time $t > 0$. This probability depends on two factors: the initial density f and the counting process $\{N_t\}_{t\geq0}$.

To be more precise, we need to calculate the measure of the set

$$\{(\omega, x) \colon S^{N_t(\omega)}(x) \in A\}. \tag{8.3.2}$$

This, in turn, requires some assumptions concerning the **product space** $\Omega \times X$ given by

$$\Omega \times X = \{(\omega, x) \colon \omega \in \Omega, x \in X\}$$

that contains all sets of the form (8.3.2). In the space $\Omega \times X$ we define (see Theorem 2.2.2) a **product measure** that, for the sets $C \times A$, $C \in \mathcal{F}$, $A \in \mathcal{A}$, is given by $\text{prob}(C)\mu_f(A)$, and we denote it by $\text{Prob}(C \times A)$ or

$$\text{Prob}(C \times A) = \text{prob}(C)\mu_f(A), \tag{8.3.3}$$

where, as usual,

$$\mu_f(A) = \int_A f(x)\mu(dx).$$

This measure is denoted by "Prob" since it is a probability measure. Equation (8.3.3) intuitively corresponds to the assumption that the initial position x and the stochastic process $\{N_t\}_{t\geq0}$ are independent.

Now we may proceed to calculate the measure of the set (8.3.2). This set may be rewritten as the union of disjoint subsets in the following way:

$$\{(\omega, x) \colon S^{N_t(\omega)}(x) \in A\} = \bigcup_{k=0}^{\infty} \{N_t(\omega) = k, S^k(x) \in A\}$$

$$= \bigcup_{k=0}^{\infty} \{N_t(\omega) = k\} \times \{S^k(x) \in A\}.$$

Thus the Prob of this set is

$$\text{Prob}\{S^{N_t} \in A\} = \sum_{k=0}^{\infty} \text{Prob}\{N_t(\omega) = k, S^k(x) \in A\}$$

$$= \sum_{k=0}^{\infty} \text{prob}\{N_t = k\}\mu_f(x \in S^{-k}(A))$$

$$= \sum_{k=0}^{\infty} p_k(t) \int_{S^{-k}(A)} f(x)\mu(dx)$$

$$= \sum_{k=0}^{\infty} p_k(t) \int_A P^k f(x)\mu(dx) \qquad (8.3.4)$$

so that

$$\text{Prob}\{S^{N_t} \in A\} = \int_A \sum_{k=0}^{\infty} p_k(t) P^k f(x)\mu(dx) \qquad \text{for } A \in \mathcal{A}, \qquad (8.3.5)$$

where, as before, P denotes the Frobenius–Perron operator associated with S, and we have assumed that $S\colon X \to X$ is nonsingular.

The integrand on the right-hand side of (8.3.5) is just the desired density, $u(t, x)$:

$$u(t, x) = \sum_{k=0}^{\infty} p_k(t) P^k f(x). \qquad (8.3.6)$$

[Note that the change in order of integration and summation in arriving at (8.3.5) is correct since $\|P^k f\| = 1$ and $\sum_{k=0}^{\infty} p_k(t) \equiv 1$. Thus the sequence on the right-hand side of (8.3.6) is strongly convergent in L^1.]

Differentiating (8.3.6) with respect to t and using (8.2.12), we have

$$\frac{\partial u(t, x)}{\partial t} = \sum_{k=0}^{\infty} \frac{dp_k(t)}{dt} P^k f(x)$$

$$= -\lambda \sum_{k=0}^{\infty} p_k(t) P^k f(x) + \lambda \sum_{k=1}^{\infty} p_{k-1}(t) P^k f(x).$$

Since the last two series are strongly convergent in L^1, the initial differentiation was proper. Thus we have

$$\frac{\partial u(t, x)}{\partial t} = -\lambda u(t, x) + \lambda \sum_{k=0}^{\infty} p_k(t) P^{k+1} f(x)$$

$$= -\lambda u(t, x) + \lambda P \sum_{k=0}^{\infty} p_k(t) P^k f(x)$$

$$= -\lambda u(t, x) + \lambda P u(t, x).$$

Therefore $u(t, x)$ satisfies the differential equation

$$\frac{\partial u(t, x)}{\partial t} = -\lambda u(t, x) + \lambda P u(t, x) \qquad (8.3.7)$$

with, from (8.3.6), the initial condition

$$u(0, x) = f(x).$$

We may always change the time scale in (8.3.7) to give

$$\frac{\partial u(t, x)}{\partial t} = -u(t, x) + P u(t, x). \qquad (8.3.8)$$

From a formal point of view, equation (8.3.7) is a generalization of the system of differential equations (8.2.11) and (8.2.12) derived for the Poisson process. Consider the special case where X is the set of nonnegative integers $\{0, 1, \ldots\}$, μ is a counting measure, and $S(x) = x + 1$. For a single point $n \geq 1$,

$$P f(n) = f(n - 1)$$

and when $n = 0$, $P f(0) = 0$. Thus, from (8.3.7), we have

$$\frac{\partial u(t, x)}{\partial t} = -\lambda u(t, n) + \lambda u(t, n - 1) \qquad n \geq 1$$

and

$$\frac{\partial u(t, 0)}{\partial t} = -\lambda u(t, 0),$$

which are identical with equations (8.2.12) and (8.2.11), respectively, except that the initial condition is more general than for the Poisson process since $u(0, n) = f(n)$.

8.4 The Linear Boltzmann Equation: An Intuitive Point of View

Our derivation in the preceding section of equation (8.3.8) for the density $u(t, x)$ was quite long as we wished to be precise and show the connection with Poisson processes. In this section we present a more intuitive derivation of the same result, using arguments similar to those often employed in statistical mechanics.

Assume that we have a hypothetical system consisting of N particles enclosed in a container, where N is a large number. Each particle may change its velocity $x = (v_1, v_2, v_2)$ from x to $S(x)$ only by colliding with the walls of the container. Our problem is to determine how the velocity distribution of particles evolves with time. Thus we must determine the function $u(t, x)$ such that

$$N \int_A u(t, x) \, dx$$

is the number of particles having, at time t, velocities in the set A.

The change in the number of particles, whose velocity is in A, between t and $t + \Delta t$ is given by

$$N \int_A u(t + \Delta t, x) \, dx - N \int_A u(t, x) \, dx. \qquad (8.4.1)$$

From our assumption, such a change can only take place through collisons with the walls of the container. Take Δt to be sufficiently small so that a negligible number of particles make two or more collisions with a wall during Δt. Thus, the number of particles striking the wall during a time Δt with velocity in A before the collision [and, therefore, having velocities in $S(A)$ after the collision] is

$$N \lambda \Delta t \int_A u(t, x) \, dx, \qquad (8.4.2)$$

where λN is the number of particles striking the walls per unit time. In this idealized, abstract example we neglect the quite important physical fact that the faster particles are striking the walls of the container more frequently than are the slower particles.

Conversely, to find the number of particles whose velocity is in A after the collision, we must calculate the number having velocities in the set $S^{-1}(A)$ before the collision. Again, assuming Δt to be sufficiently small to make the number of double collisions by single particles negligible, we have

$$N \lambda \Delta t \int_{S^{-1}(A)} u(t, x) \, dx. \qquad (8.4.3)$$

Hence the total change in the number of particles with velocity in the set A over a short time Δt is given by the difference between (8.4.3) and (8.4.2):

$$N \lambda \Delta t \int_{S^{-1}(A)} u(t, x) \, dx - N \lambda \Delta t \int_A u(t, x) \, dx. \qquad (8.4.4)$$

By combining equation (8.4.1) with equation (8.4.4), we have

$$N \int_A [u(t + \Delta t, x) - u(t, x)] dx = \lambda N \Delta t \left\{ \int_{S^{-1}(A)} u(t, x) \, dx - \int_A u(t, x) \, dx \right\},$$

and, since

$$\int_{S^{-1}(A)} u(t, x) \, dx = \int_A P u(t, x) \, dx,$$

where P is the Frobenius–Perron operator associated with S, we have

$$N \int_A [u(t + \Delta t, x) - u(t, x)] \, dx$$

$$= \lambda N \, \Delta t \int_A [-u(t, x) + Pu(t, x)] \, dx. \qquad (8.4.5)$$

Equation (8.4.5) is exact to within an error that is small compared to Δt.

By dividing through in (8.4.5) by Δt and passing to the limit $\Delta t \to 0$, we obtain

$$\int_A \frac{\partial u(t, x)}{\partial t} \, dx = \lambda \int_A [-u(t, x) + Pu(t, x)] \, dx,$$

which gives

$$\frac{\partial u(t, x)}{\partial t} = -\lambda u(t, x) + \lambda Pu(t, x).$$

Thus we have again arrived at equation (8.3.7).

In this derivation we assumed that the particle, upon striking the wall, changed its velocity from x to $S(x)$, where $S \colon X \to X$ is a point-to-point transformation. An alternative physical assumption, which is more general from a mathematical point of view, would be to assume that the change in velocity is not uniquely determined but is a probabilistic event. In other words, we might assume that collision with the walls of the container alters the distribution of particle velocities. Thus, if before the collision the particles have a velocity distribution with density g, then after collision they have a distribution with density Pg, where $P \colon L^1(X) \to L^1(X)$ is a Markov operator.

So, assume as before that $u(t, x)$ is the density of the distribution of particles having velocity x at time t, so

$$N \int_A u(t, x) \, dx$$

is the number of particles with velocities in A. Once again,

$$\lambda N \, \Delta t \int_A u(t, x) \, dx$$

is the number of particles with velocity in A that will collide with the walls in a time Δt, whereas

$$\lambda N \, \Delta t \int_A Pu(t, x) \, dx$$

is the number of particles whose velocities go into A because of collisions over a time Δt. Thus,

$$-\lambda N \, \Delta t \int_A u(t, x) \, dx + \lambda N \, \Delta t \int_A Pu(t, x) \, dx$$

is the net change, due to collisions over a time Δt in the number of particles whose velocities are in A.

Combining this result with (8.4.1), we immediately obtain the balance equation (8.4.5), which leads once again to (8.3.7). The only difference is that P is no longer a Frobenius–Perron operator corresponding to a given one-to-one deterministic transformation S, but it is an arbitrary Markov operator.

Since in our intuitive derivations of (8.3.7) presented in this section, we used arguments that are employed to derive a Boltzmann equation, we will call equation (8.3.7) a **linear abstract Boltzmann equation** corresponding to a collision (Markov) operator P. To avoid confusion with the usual Boltzmann equation, bear in mind that x corresponds to the particle velocity and not to position. Indeed, it is because we assume that the only source of change for particle velocity is collisions with the wall, that drift and external force terms do not appear in (8.3.7).

Our next goal will not be to apply equation (8.3.7) to specific physical systems. Rather, we will demonstrate the interdependence between the properties of discrete time deterministic processes, governed by $S: X \to X$ or a Markov operator, and the continuous time process, determined by (8.3.7). The next four sections are devoted to an examination of the most important properties of (8.3.7), and then in the last section we demonstrate that the Tjon–Wu representation of the Boltzmann equation is a special case of (8.3.7).

8.5   Elementary Properties of the Solutions of the Linear Boltzmann Equation

To facilitate our study of the linear Boltzmann equation (8.3.7), we will consider the solution $u(t, x)$ as a function from the positive real numbers, R^+, into L^1

$$u: R^+ \to L^1.$$

Thus, by writing (8.3.8) in the form

$$\frac{du}{dt} = (P - I)u, \tag{8.5.1}$$

where P is a given Markov operator and I is the identity operator, we may apply the Hille–Yosida theorem 7.8.1 to the study of equation (8.3.8).

All three assumptions (a)–(c) of the Hille–Yosida theorem are easily shown to be satisfied by the operator $(P - I)$ of equation (8.5.1). First, since $A = P - I$ is defined on the whole space, L^1, $\mathcal{D}(A) = L^1$ and property (a) is thus trivially satisfied.

To check property (b), rewrite the resolvent equation $\lambda f - Af = g$ using

$A = P - I$ to give

$$(\lambda + 1)f - Pf = g. \tag{8.5.2}$$

Equation (8.5.2) may be easily solved by the method of successive approximations. Starting from an arbitrary f_0, we define f_n by

$$(\lambda + 1)f_n - Pf_{n-1} = g,$$

so, as a consequence,

$$f_n = \frac{1}{(\lambda + 1)^n}P^n f_0 + \sum_{k=1}^{n}\frac{1}{(\lambda + 1)^k}P^{k-1}g. \tag{8.5.3}$$

Since $\|P^k g\| \le \|g\|$, the series in (8.5.3) is, therefore, convergent, and the unique solution f of the resolvent equation (8.5.2) is

$$f \equiv R_\lambda g = \lim_{n \to \infty} f_n = \sum_{k=1}^{\infty}\frac{1}{(\lambda + 1)^k}P^{k-1}g. \tag{8.5.4}$$

Remark 8.5.1. The method of successive approximations applied to an equation such as (8.5.2) will always result in a solution (8.5.3) that converges to a unique limit, as $n \to \infty$, when $\|P\| \le \lambda + 1$. The limiting solution given by (8.5.4) is called a **von Neumann series.**   □

To check that the linear Boltzmann equation satisfies property (c) of the Hille–Yosida theorem, integrate (8.5.4) over the entire space X to give

$$\int_X R_\lambda g(x)\mu(dx) = \sum_{k=1}^{\infty}\frac{1}{(\lambda + 1)^k}\int_X P^{k-1}g(x)\mu(dx)$$

$$= \sum_{k=1}^{\infty}\frac{1}{(\lambda + 1)^k}\int_X g(x)\mu(dx)$$

$$= \frac{1}{\lambda}\int_X g(x)\mu(dx) = \frac{1}{\lambda},$$

where we used the integral-preserving property of Markov operators in passing from the first to the second line. Thus,

$$\int_X \lambda R_\lambda g(x)\mu(dx) = 1,$$

and, since λR_λ is linear, nonnegative, and also preserves the integral, it is a Markov operator. Thus condition (c) is automatically satisfied (see Corollary 7.8.1).

Therefore, by the Hille–Yosida theorem, the linear Boltzmann equation (8.3.8) generates a continuous semigroup of Markov operators, $\{\hat{P}_t\}_{t \ge 0}$.

To determine an explicit formula for $\hat{P}_t$, we first write

$$A_\lambda f = \lambda A R_\lambda f = \lambda (P - I) R_\lambda f$$

$$= \lambda (P - I) \sum_{k=1}^{\infty} \frac{1}{(\lambda + 1)^k} P^{k-1} f$$

$$= \lambda \sum_{k=1}^{\infty} \frac{1}{(\lambda + 1)^k} P^k f - \lambda \sum_{k=1}^{\infty} \frac{1}{(\lambda + 1)^k} P^{k-1} f,$$

so

$$\lim_{\lambda \to \infty} A_\lambda f = Pf - f.$$

Thus, by the Hille–Yosida theorem and equation (7.8.3), the unique semigroup corresponding to $A = P - I$ is given by

$$\hat{P}_t f = e^{t(P-I)} f, \tag{8.5.5}$$

and the unique solution to equation (8.3.8) with the initial condition $u(0, x) = f(x)$ is

$$u(t, x) = e^{t(P-I)} f(x). \tag{8.5.6}$$

Although we have determined the solution of (8.3.8) using the Hille–Yosida theorem, precisely the same result could have been obtained by applying the method of successive approximations to equation (8.5.1). However, our derivation once again illustrates the techniques involved in using the Hille–Yosida theorem and establishes that (8.3.8) generates a continuous semigroup of Markov operators. Finally, we note that if P in equation (8.3.8) is a Frobenius–Perron operator corresponding to a nonsingular transformation S, the solution can be obtained by substituting equation (8.2.14) into equation (8.3.6).

In addition to the existence and uniqueness of the solution to (8.3.8), other properties of $\hat{P}_t$ may be demonstrated.

Property 1. From inequality (7.4.7) we know that, given $f_1, f_2 \in L^1$, the norm

$$\|\hat{P}_t f_1 - \hat{P}_t f_2\| \tag{8.5.7}$$

is a nonincreasing function of time t.

Property 2. If for some $f \in L^1$ the limit

$$f_* = \lim_{t \to \infty} \hat{P}_t f \tag{8.5.8}$$

exists, then, for the same f,

$$\lim_{t \to \infty} \hat{P}_t(Pf) = f_*. \tag{8.5.9}$$

To show this, we prove even more, namely that

$$\lim_{t \to \infty} \hat{P}_t(f - Pf) = 0 \tag{8.5.10}$$

for all $f \in L^1$. Now,

$$\hat{P}_t f = e^{t(P-I)} f = e^{-t} e^{tP} f = e^{-t} \sum_{n=0}^{\infty} \frac{t^n}{n!} P^n f, \qquad (8.5.11)$$

and

$$\hat{P}_t(Pf) = e^{-t} \sum_{n=0}^{\infty} \frac{t^n}{n!} P^{n+1} f = e^{-t} \sum_{n=1}^{\infty} \frac{t^{n-1}}{(n-1)!} P^n f.$$

Taking the norm of $\hat{P}_t f - \hat{P}_t(Pf)$, we have

$$\|\hat{P}_t f - \hat{P}_t(Pf)\| \le e^{-t} \left\| \sum_{n=1}^{\infty} \left[\frac{t^n}{n!} - \frac{t^{n-1}}{(n-1)!} \right] P^n f \right\| + e^{-t} \|f\|$$

$$\le e^{-t} \sum_{n=1}^{\infty} \left| \frac{t^n}{n!} - \frac{t^{n-1}}{(n-1)!} \right| \|f\| + e^{-t} \|f\|.$$

If $t = m$, an integer, then

$$e^{-t} \sum_{n=1}^{\infty} \left| \frac{t^n}{n!} - \frac{t^{n-1}}{(n-1)!} \right| = 2e^{-m} \left(\frac{m^m}{m!} - \frac{1}{2} \right)$$

since almost all of the terms in the series cancel. However, by Stirling's formula, $m! = m^m e^{-m} \sqrt{1\pi m} \theta_m$, where $\theta_m \to 1$ as $m \to \infty$. Thus for integer t,

$$\|\hat{P}_t f - \hat{P}_t(Pf)\|$$

converges to zero as $t \to \infty$. Since, by property 1 this quantity is a non-increasing function, then (8.5.10) is demonstrated for all $t \to \infty$. Finally, inserting (8.5.8) into (8.5.10) gives the desired result, (8.5.9).

Remark 8.5.2. Note that the sum of the coefficients of $\hat{P}_t f$ given in equation (8.5.11) is identically 1, and thus the solutions of the linear Boltzmann equation $u(t, x) = \hat{P}_t f(x)$ bear a strong correspondence to the averages $A_n f$ studied earlier in Chapter 5, with n and t playing analogous roles. $\square$

Property 3. The operators P and $\hat{P}_t$ commute, that is, $P\hat{P}_t f = \hat{P}_t Pf$ for all $f \in L^1$. This is easily demonstrated by applying P to (8.5.11):

$$P(\hat{P}_t f) = e^{-t} \sum_{n=0}^{\infty} \frac{t^n}{n!} P^{n+1} f$$

$$= e^{-t} \sum_{n=0}^{\infty} \frac{t^n}{n!} P^n(Pf) = \hat{P}_t(Pf).$$

Property 4. If for some $f \in L^1$ the limit (8.5.8), $f_* = \lim_{t \to \infty} \hat{P}_t f$, exists, then f_* is a fixed point of the Markov operator P, that is,

$$Pf_* = f_*.$$

To show this, note that if

$$f_* = \lim_{t \to \infty} \hat{P}_t f,$$

then, by (8.5.9),

$$Pf_* = \lim_{t \to \infty} P(\hat{P}_t f) = \lim_{t \to \infty} \hat{P}_t(Pf) = f_*,$$

which gives the desired result. Further, the same argument shows that, if $f_* = \lim_{n \to \infty} \hat{P}_{t_n} f$ exists for some subsequence $\{t_n\}$, then $Pf_* = f_*$.

Property 5. If $Pf_* = f_*$ for some $f_* \in L^1$, then also $\hat{P}_t f_* = f_*$. This is also easy to show. Write $Pf_* = f_*$ as

$$(P - I)f_* = 0.$$

Since $(P - I) = A$ is an infinitesimal operator, and every solution of $Af = 0$ is a fixed point of the semigroup (see Section 7.8), we have immediately that $\hat{P}_t f_* = f_*$.

8.6 Further Properties of the Linear Boltzmann Equation

As shown in the preceding section, the solutions of the linear Boltzmann equation are rather regular in their behavior, that is, the distance between any two solutions never increases. Now we will show that, under a rather mild condition, $\hat{P}_t f$ always converges to a limit.

Recall our definition of precompactness (Section 5.1) and observe that every sequence $\{f_n\}$ that is weakly precompact contains a subsequence that is weakly convergent. Analogously, if for a given f, the trajectory $\{\hat{P}_t f\}$ is weakly precompact, then there exists a sequence $\{t_n\}$ such that $\{\hat{P}_{t_n} f\}$ is weakly convergent as $t_n \to \infty$. To see this, take an arbitrary sequence of numbers $t_n' \to \infty$ and, then, applying the definition of precompactness to $\{\hat{P}_{t_n'} f\}$, choose a weakly convergent subsequence $\{\hat{P}_{t_n} f\}$ of $\{\hat{P}_{t_n'} f\}$.

Theorem 8.6.1. If the trajectory $\{\hat{P}_t f\}$ is weakly precompact, then there exists a fixed point of P.

Proof. If $\{\hat{P}_t f\}$ is weakly precompact, then there exists a sequence $\{t_n\}$ such that

$$\lim_{n \to \infty} \hat{P}_{t_n} f = f_* \qquad \text{weakly} \tag{8.6.1}$$

exists. This implies the weak convergence of

$$\lim_{n \to \infty} \hat{P}_{t_n}(Pf) = \lim_{n \to \infty} P(\hat{P}_{t_n} f) = Pf_*. \tag{8.6.2}$$

However, from (8.5.10), we have

$$\lim_{n \to \infty} \hat{P}_{t_n}(f - Pf) = 0,$$

and, thus, from equations (8.6.1) and (8.6.2), we have

$$Pf_* = f_*,$$

which establishes the claim. Note also from property 5 of $\hat{P}_t$ (Section 8.5) that this implies $\hat{P}_t f_* = f_*$. ∎

Theorem 8.6.2. *For a given $f \in L^1$, if the trajectory $\{\hat{P}_t f\}$ is weakly precompact, then $\hat{P}_t f$ strongly converges to a limit.*

Proof. From Theorem 8.6.1 we know that $\hat{P}_{t_n} f$ converges weakly to an f_* that is a fixed point of P and $\hat{P}_t$. Write $f \in L^1$ in the form

$$f = f - f_* + f_*.$$

Assume that for every $\varepsilon > 0$ the function $f - f_*$ may be written in the form

$$f - f_* = Pg - g + r, \qquad (8.6.3)$$

where $g \in L^1$ and $\|r\| \leq \varepsilon$. (We will prove in the following that this representation is possible.) By using (8.6.3), we may write

$$\hat{P}_t f = \hat{P}_t(f - f_* + f_*) = \hat{P}_t(Pg - g) + \hat{P}_t f_* + \hat{P}_t r.$$

However $\hat{P}_t f_* = f_*$ and, thus,

$$\|\hat{P}_t f - f_*\| \leq \|\hat{P}_t(Pg - g)\| + \|\hat{P}_t r\|.$$

From (8.5.10), the first term on the right-hand side approaches zero as $t \to \infty$, whereas the second term is not greater than ε. Thus

$$\|\hat{P}_t f - f_*\| \leq 2\varepsilon$$

for t sufficiently large, and, since ε is arbitrary,

$$\lim_{t \to \infty} \|\hat{P}_t f - f_*\| = 0,$$

which completes the proof if (8.6.3) is true. Suppose (8.6.3) is not true, which implies that

$$f - f_* \notin \text{closure}(P - I)L^1(X).$$

This, in turn, implies by the Hahn–Banach theorem (see Proposition 5.2.3) that there is a $g_0 \in L^\infty$ such that

$$\langle f - f_*, g_0 \rangle \neq 0 \qquad (8.6.4)$$

and
$$\langle h, g_0 \rangle = 0$$
for all $h \in \text{closure}(P - I)L^1(X)$. In particular
$$\langle (P - I)P^n f, g_0 \rangle = 0,$$
since $(P - I)P^n f \in (P - I)L^1(X)$, so
$$\langle P^{n+1} f, g_0 \rangle = \langle P^n f, g_0 \rangle$$
for $n = 0, 1, \ldots$. Thus, by induction, we have
$$\langle P^n f, g_0 \rangle = \langle f, g_0 \rangle. \tag{8.6.5}$$

Furthermore, since $e^{-t} \sum_{n=0}^{\infty} t^n / n! = 1$, we may multiply both sides of (8.6.5) by $e^{-t} t^n / n!$ and sum over n to obtain

$$\left\langle e^{-t} \sum_{n=0}^{\infty} \frac{t^n}{n!} P^n f, g_0 \right\rangle = \langle f, g_0 \rangle,$$

or
$$\langle \hat{P}_t f, g_0 \rangle = \langle f, g_0 \rangle. \tag{8.6.6}$$
Substituting $t = t_n$ and taking the limit as $t \to \infty$ in (8.6.6) gives
$$\langle f_*, g_0 \rangle = \langle f, g_0 \rangle,$$
and, thus,
$$\langle f_* - f, g_0 \rangle = 0,$$
which contradicts equation (8.6.4). Thus (8.6.3) is true. ∎

8.7   Effect of the Properties of the Markov Operator on Solutions of the Linear Boltzmann Equation

From the results of Section 8.6, some striking properties of the solutions of the linear Boltzmann equation emerge. The first of these is stated in the following corollary.

Corollary 8.7.1. *If for $f \in L^1$ there exists a $g \in L^1$ such that*
$$|\hat{P}_t f| \leq g, \qquad t \geq 0, \tag{8.7.1}$$
then the (strong) limit
$$\lim_{t \to \infty} \hat{P}_t f \tag{8.7.2}$$

exists. That is, either $\hat{P}_t f$ is not bounded by any integrable function or $\hat{P}_t f$ is strongly convergent.

Proof. Observe that $\{\hat{P}_t f\}$ is weakly precompact by our first criterion of precompactness; see Section 5.1. Thus the limit (8.7.2) exists according to Theorem 8.6.2. ∎

With this result available to us, we may go on to state and demonstrate some important corollaries that give information concerning the convergence of solutions $\hat{P}_t f$ of (8.3.8) when the operator P has various properties.

Corollary 8.7.2. *If the (Markov) operator P has a positive fixed point f_*, $f_*(x) > 0$ a.e., then the strong limit, $\lim_{t\to\infty} \hat{P}_t f$, exists for all $f \in L^1$.*

Proof. First note that when the initial function f satisfies

$$|f| \le cf_* \tag{8.7.3}$$

for some sufficiently large constant $c > 0$, we have

$$|P^n f| \le P^n (cf_*) = cP^n f_* = cf_*.$$

Multiply both sides by $e^{-t}t^n/n!$ and sum the result over n to give

$$\left| e^{-t} \sum_{n=0}^{\infty} \frac{t^n}{n!} P^n f \right| \le ce^{-t} \sum_{n=0}^{\infty} \frac{t^n}{n!} f_* = cf_*.$$

The left-hand side of this inequality is just $|\hat{P}_t f|$, so that

$$|\hat{P}_t f| \le cf_*,$$

and, since $\hat{P}_t f$ is bounded, by Corollary 8.7.1 we know that the strong limit $\lim_{t\to\infty} \hat{P}_t f$ exists.

In the more general case when the initial function f does not satisfy (8.7.3), we proceed as follows. Define a new function by

$$f_c(x) = \begin{cases} f(x) & \text{if } |f(x)| \le cf_*(x) \\ 0 & \text{if } |f(x)| > cf_*(x). \end{cases}$$

It follows from the Lebesgue dominated convergence theorem that

$$\lim_{c\to\infty} \|f_c - f\| = 0.$$

Thus, by writing $f = f_c + f - f_c$, we have

$$\hat{P}_t f = \hat{P}_t f_c + \hat{P}_t(f - f_c).$$

Since f_c satisfies
$$|f_c| \leq cf_*,$$
from (8.7.3) we know that $\{\hat{P}_t f_c\}$ converges strongly. Now take $\varepsilon > 0$. Since $\{\hat{P}_t f_c\}$ is strongly convergent, there is a $t_0 > 0$, which in general depends on c, such that

$$\|\hat{P}_{t+t'} f_c - \hat{P}_t f_c\| \leq \varepsilon \qquad \text{for } t \geq t_0, \ t' \geq 0. \tag{8.7.4}$$

Further,

$$\|\hat{P}_t f - \hat{P}_t f_c\| \leq \|f - f_c\| \leq \varepsilon \qquad \text{for } t \geq 0 \tag{8.7.5}$$

for a fixed but sufficiently large c. From equations (8.7.4) and (8.7.5) it follows that

$$\|\hat{P}_{t+t'} f - \hat{P}_t f\| \leq 3\varepsilon \qquad \text{for } t \geq t_0, \ t' \geq 0,$$

which is the Cauchy condition for $\{\hat{P}_t f\}$. Thus $\{\hat{P}_t f\}$ also converges strongly, and the proof is complete. ∎

The existence of the strong limit (8.7.2) is interesting, but from the point of view of applications we would like to know what the limit is. In the following corollary we give a sufficient condition for the existence of a unique limit to (8.7.2), noting, of course, that, since (8.7.2) is linear, uniqueness is determined only up to a multiplicative constant.

Corollary 8.7.3. *Assume that in the set of all densities $f \in D$ the equation $Pf = f$ has a unique solution f_* and $f_*(x) > 0$ a.e. Then, for any initial density, $f \in D$*

$$\lim_{t \to \infty} \hat{P}_t f = f_*, \tag{8.7.6}$$

and the convergence is strong.

Proof. The proof is straightforward. From Corollary 8.7.2 the $\lim_{t \to \infty} \hat{P}_t f$ exists and is also a nonnegative normalized function. However, by property 4 of $\hat{P}_t$ (Section 8.5), we know that this limit is a fixed point of the Markov operator P. Since, by our assumption, the fixed point is unique it must be f_*, and the proof is complete. ∎

In the special case that P is a Frobenius–Perron operator for a nonsingular transformation $S \colon X \to X$, the condition $Pf_* = f_*$ is equivalent to the fact that the measure

$$\mu_{f_*}(A) = \int_A f_*(x)\mu(dx)$$

is invariant with respect to S. Thus, in this case, from Corollary 8.7.2 the existence of an invariant measure μ_{f_*} with a density $f_*(x) > 0$ is sufficient for the existence of the strong limit (8.7.2) for the solutions of (8.3.8). Since, for ergodic transformations f_* is unique (cf. Theorem 4.2.2), these results may be summarized in the following corollary.

Corollary 8.7.4. *Suppose $S: X \to X$ is a nonsingular transformation and P is the corresponding Frobenius–Perron operator. Then with respect to the trajectories $\{\hat{P}_t f\}$ that generate the solutions of the linear Boltzmann equation (8.3.8):*

1. *If there exists an absolutely continuous invariant measure μ_{f_*} with a positive density $f_*(x) > 0$ a.e., then for every $f \in L^1$ the strong limit, $\lim_{t \to \infty} \hat{P}_t f$ exists; and*

2. *If, in addition, the transformation S is ergodic, then*

$$\lim_{t \to \infty} \hat{P}_t f = f_* \tag{8.7.7}$$

for all $f \in D$.

Now consider the more special case where $(X, \mathcal{A}, \mu)$ is a finite measure space and $S: X \to X$ is a measure-preserving transformation. Since S is measure preserving, f_* exists and is given by

$$f_*(x) = 1/\mu(X) \qquad \text{for } x \in X.$$

Thus $\lim_{t \to \infty} \hat{P}_t f$ always exists. Furthermore, this limit is unique, that is,

$$\lim_{t \to \infty} P_t f = f_* = 1/\mu(X) \tag{8.7.8}$$

if and only if S is ergodic (cf. Theorem 4.2.2).

In closing this section we would like to recall that, from Definition 4.4.1, a Markov operator $P: L^1 \to L^1$ is exact if and only if the sequence $\{P^n f\}$ has a strong limit that is a constant for every $f \in L^1$. Although the term exactness is never used in talking about the behavior of stochastic semigroups, for the situation where (8.7.8) holds, then, the behavior of the trajectory $\{\hat{P}_t f\}$ is precisely analogous to our original definition of exactness. Figuratively speaking, then, we could say that S is ergodic if and only if $\{\hat{P}_t\}_{t \geq 0}$ is exact.

8.8 Linear Boltzmann Equation with a Stochastic Kernel

In this section we consider the linear Boltzmann equation

$$\frac{\partial u(t, x)}{\partial t} + u(t, x) = Pu$$

where the Markov operator P is given by

$$Pf(x) = \int_X K(x,y)f(y)\,dy \qquad (8.8.1)$$

and $K(x,y): X \times X \to R$ is a stochastic kernel, that is,

$$K(x,y) \geq 0 \qquad (8.8.2)$$

and

$$\int_X K(x,y)\,dx = 1. \qquad (8.8.3)$$

For this particular formulation of the linear Boltzmann equation, we will show some straightforward applications of the general results presented earlier.

The simplest case occurs when we are able to evaluate the stochastic kernel from below. Thus we assume that for some integer m the function $\inf_y K_m(x,y)$ is not identically zero, so that

$$\int_X \inf_y K_m(x,y)\,dx > 0 \qquad (8.8.4)$$

(K_m is the m times iterated kernel K). In this case we will show that the strong limit

$$\lim_{t \to \infty} \hat{P}_t f = f_* \qquad (8.8.5)$$

exists for all densities $f \in D$, where f_* is the unique density that is a solution of

$$f(x) = \int_X K(x,y)f(y)\,dy. \qquad (8.8.6)$$

The proof of this is quite direct. Set

$$h(x) = \inf_y K_m(x,y).$$

By using the explicit formula (8.5.11) for the solution $\hat{P}_t f$, we have

$$\hat{P}_t f = e^{-t} \sum_{n=0}^{\infty} \frac{t^n}{n!} P^n f.$$

However, for $n \geq m$, we may write

$$P^n f(x) = \int_X K_m(x,y)P^{n-m}f(y)\,dy \geq h(x),$$

and thus the explicit solution $\hat{P}_t f$ becomes

$$\hat{P}_t f(x) \geq e^{-t} \sum_{n=0}^{m} \frac{t^n}{n!} P^n f(x) + e^{-t} \sum_{n=m+1}^{\infty} \frac{t^n}{n!} h(x)$$

$$= e^{-t} \sum_{n=0}^{m} \frac{t^n}{n!} P^n f(x) + h(x) \left[1 - e^{-t} \sum_{n=0}^{m} \frac{t^n}{n!} \right]$$

$$\geq h(x) \left[1 - e^{-t} \sum_{n=0}^{m} \frac{t^n}{n!} \right].$$

Thus we have immediately that

$$\hat{P}_t f(x) - h(x) \geq \left(-e^{-t} \sum_{n=0}^{m} \frac{t^n}{n!} \right) h(x),$$

so that

$$(\hat{P}_t f - h)^- \leq \left(e^{-t} \sum_{n=0}^{m} \frac{t^n}{n!} \right) h.$$

Since, however, $e^{-t} t^n \to 0$ as $t \to \infty$, we have

$$\lim_{t \to \infty} \|(\hat{P}_t f - h)^-\| = 0,$$

and, by Theorem 7.4.1, the strong limit f_* of (8.8.5) is unique. Properties 4 and 5 of the solution $\hat{P}_t f$, outlined in Section 8.5, tell us that f_* is the unique solution of $Pf = f$, namely, equation (8.8.6). Thus the proof is complete.

Now we assume, as before, that $K(x, y)$ is a stochastic kernel for which there is an integer m and a $g \in L^1$ such that

$$K_m(x, y) \leq g(x) \qquad \text{for } x, y \in X. \tag{8.8.7}$$

Then the strong limit

$$\lim_{t \to \infty} \hat{P}_t f \tag{8.8.8}$$

exists for all $f \in L^1$.

As before, to prove this we use the explicit series representation of $\hat{P}_t f$, noting first that, because of (8.8.7), we have, for $n \geq m$,

$$|P^n f(x)| = |P^m (P^{n-m} f(x))| \leq \int_X K_m(x, y) |P^{n-m} f(y)| \, dy$$

$$\leq g(x) \int_X |P^{n-m} f(y)| \, dy \leq g(x) \|f\|.$$

Thus we can evaluate $\hat{P}_t f$ as

$$|\hat{P}_t f| \le e^{-t} \sum_{n=0}^{m} \frac{t^n}{n!} |P^n f| + \left(e^{-t} \sum_{n=m+1}^{\infty} \frac{t^n}{n!} g \right) \|f\|$$

$$\le e^{-t} \sum_{n=0}^{m} \frac{t^n}{n!} |P^n f| + g\|f\|.$$

Further, setting

$$r = c \sum_{n=0}^{m} |P^n f|, \qquad c = \sup_{\substack{0 \le t \\ 0 \le n \le m}} e^{-t} \frac{t^n}{n!},$$

we finally obtain

$$|\hat{P}_t f| \le g\|f\| + r.$$

Evidently, $(g\|f\| + r)$ is an integrable function, and from Corollary 8.7.1 we know that the strong limit (8.8.8) exists.

Under assumption (8.8.7) we have no assurance that the strong limit (8.8.8) is unique. However, some additional properties of $K(x, y)$ may en-sure this uniqueness. For example, if X is a bounded interval of the real line or the half-line, (8.8.7) holds, and $K_m(x, y)$ is monotonically increasing or decreasing in x, then

$$\lim_{t \to \infty} \hat{P}_t f = f_* \qquad \text{for all } f \in D, \tag{8.8.9}$$

where f_* is the unique solution of (8.8.6).

To demonstrate this, note that by repeating the proof of Proposition 5.8.1 we may construct an $h(x)$, $h(x) \ge 0$, $\|h\| > 0$, such that $K_m(x, y) \ge h(x)$. Then the proof follows directly from the assertion following equation (8.8.4).

Analogously, if (8.8.7) holds and $K_m(x, y) > 0$ for $x \in A$, $y \in X$, where A is a set of positive measure, then the limit (8.8.9) exists and is unique. To prove this set $\hat{P}_m = \tilde{P}$ and observe that for $f \in D$ the operator $\tilde{P}$ satisfies

$$\tilde{P}f \le g \quad \text{and} \quad \tilde{P}f(x) > 0 \qquad \text{for } x \in A.$$

Thus by Theorem 5.6.1 the limiting function $\lim_{n \to \infty} \tilde{P}^n f$ does not depend on f for $f \in D$. Since $\tilde{P}^n = \hat{P}_{mn}$, the limit (8.8.9) is also independent of f.

It should be noted that the same result holds under even weaker condi-tions, that is, if (8.8.7) holds and for some integer k

$$\sum_{n=1}^{k} K_n(x, y) > 0 \qquad \text{for } x \in A, y \in X.$$

8.9   The Linear Tjon–Wu Equation

To illustrate the application of the results developed in this chapter we close with an example drawn from the kinetic theory of gases [see Dłotko and Lasota, 1983].

In the theory of dilute gases [Chapman and Cowling, 1960] the Boltzmann equation

$$\frac{DF(t, x, v)}{Dt} = C(F(t, x, v))$$

is studied to obtain information about the particle distribution function F that depends on time (t), position (x), and velocity (v). DF/Dt denotes the total rate of change of F due to spatial gradients and any external forces, whereas the collision operator $C(\cdot)$ determines the way in which particle collisions affect F. In the case of a spatially homogeneous gas with no external forces the Boltzmann equation reduces to

$$\frac{\partial F(t, v)}{\partial t} = C(F(t, v)). \tag{8.9.1}$$

Bobylev [1976], Krook and Wu [1977], and Tjon and Wu [1979] have shown that in some cases equation (8.9.1) may be transformed into

$$\frac{\partial u(t, x)}{\partial t} = -u(t, x) + \int_x^\infty \frac{dy}{y} \int_0^y u(t, y - z) u(t, z)\, dz, \qquad x > 0, \tag{8.9.2}$$

where $x = (v^2/2)$ (note that x is not a spatial coordinate) and

$$u(t, x) = \text{const} \int_x^\infty \frac{F(t, v)}{\sqrt{v - x}}\, dv.$$

Equation (8.9.2), called the **Tjon–Wu equation** [Barnsley and Cormille, 1981], is nonlinear because of the presence of $u(t, y - z) u(t, z)$ in the integrand on the right-hand side. Thus the considerations of this chapter are of no help in studying the behavior of $u(t, x)$ as $t \to \infty$.

However, note that $\exp(-x)$ is a solution of (8.9.2), a fact that we can use to study a linear problem. Here we will investigate the situation where a small number of particles with an arbitrary velocity distribution f are introduced into a gas, containing many more particles, at equilibrium, so that $u_*(x) = \exp(-x)$. We want to know what the eventual distribution of velocities of the small number of particles tends to.

Thus, on the right-hand side of (8.9.2), we set $u(t, y - z) = u_*(y - z) = \exp[-(y - z)]$, so the resulting **linear Tjon–Wu equation** is of the form

$$\frac{\partial u(t, x)}{\partial t} + u(t, x) = \int_x^\infty \frac{dy}{y} \int_0^y e^{-(y-z)} u(t, z)\, dz, \qquad x > 0. \tag{8.9.3}$$

Equation (8.9.3) is a special case of the linear Boltzmann equation of this chapter with a Markov operator defined by

$$Pf(x) = \int_x^\infty \frac{dy}{y} \int_0^y e^{-(y-z)} f(z) \, dz \qquad (8.9.4)$$

for $f \in L^1((0, \infty))$. Using the definition of the exponential integral,

$$-\mathrm{Ei}(-x) \equiv \int_x^\infty (e^{-y}/y) \, dy,$$

equation (8.9.4) may be rewritten as

$$Pf(x) = \int_0^\infty K(x, y) f(y) \, dy, \qquad (8.9.5)$$

where

$$K(x, y) = \begin{cases} -e^y \mathrm{Ei}(-y) & 0 < x \le y \\ -e^y \mathrm{Ei}(-x) & 0 < y < x. \end{cases} \qquad (8.9.6)$$

To examine the behavior of the solutions $u(t, x)$ of (8.9.3) as $t \to \infty$, we have a number of potential aids available. First, from the preceding section, if $\inf_y K_m(x, y) > 0$ for some m, then we could determine $\lim_{t \to \infty} \hat{P}_t f$. However, $\inf_y K(x, y) = 0$, and further composition of the kernel with itself leads to analytically complex results. Second, if we were able to find a $g(x) \ge K_m(x, y)$ for some m, then the results of the preceding section could be applied. However, the maximum of $K(x, y)$ in y occurs at $y = x$ and $-\exp(x)\mathrm{Ei}(-x)$ is not integrable. As before, compositions of $K(x, y)$ become so complicated that it is difficult to work with them.

A third alternative is the following. Note that $f(x) = \exp(-x)$ is a fixed point of (8.9.4). If we can show that $\exp(-x)$ is the unique fixed point of (8.9.4), then we may apply Corollary 8.7.3 to show that

$$\lim_{t \to \infty} u(t, x) = \lim_{t \to \infty} \hat{P}_t f(x) = e^{-x} \qquad (8.9.7)$$

for all densities $f \in D((0, \infty))$.

From $Pf = f$ and (8.9.4), we have

$$f(x) = \int_x^\infty \frac{dy}{y} \int_0^y e^{-(y-z)} f(z) \, dz, \qquad (8.9.8)$$

which must be solved for f. Since the right-hand side of (8.9.8) is differentiable, f must be differentiable. Its first derivative is

$$\frac{df(x)}{dx} = -\frac{1}{x} \int_0^x e^{-(x-z)} f(z) \, dz.$$

Multiply both sides by $x \exp(x)$ and differentiate again to obtain the nonlinear second-order differential equation

$$x \frac{d^2 f}{dx^2} + (x + 1) \frac{df}{dx} + f = 0. \qquad (8.9.9)$$

We know that one solution of (8.9.9) is $f_1(x) = \exp(-x)$, and a second independent solution may be determined using the d'Alembert reduction method [Kamke, 1959]. This simply consists of substituting $f(x) = g(x)\exp(-x)$ into (8.9.9) and solving the resulting equation for $g(x)$. Once g is determined then the second independent solution of (8.9.8) is $f_2(x) = g(x)\exp(-x)$.

Making this substitution and simplifying gives

$$x\frac{d^2g}{dx^2} + (1 - x)\frac{dg}{dx} = 0,$$

which is a first-order equation in dg/dx, easily solved to give

$$\frac{dg}{dx} = \frac{1}{x}e^x$$

as a particular solution. Thus

$$g(x) = \text{Ei}(x),$$

and the second solution of (8.9.9) is

$$f_2(x) = e^{-x}\text{Ei}(x).$$

Therefore, the general solution of (8.9.9) is

$$f(x) = C_1 e^{-x} + C_2 e^{-x}\text{Ei}(x). \tag{8.9.10}$$

Since we are searching for an $f \in D((0, \infty))$, we must determine C_1 and C_2 such that $f \geq 0$ and

$$\int_0^\infty f(x)\,dx = C_1 + C_2 \int_0^\infty e^{-x}\text{Ei}(x)\,dx = 1.$$

However, $\exp(-x)\text{Ei}(x)$ is not integrable, so we must have $C_1 = 1$, $C_2 = 0$, and thus the unique normalized solution of equation (8.9.9) is

$$f_*(x) = e^{-x}. \tag{8.9.11}$$

Hence f_* is also the unique normalized solution of (8.9.8).

Therefore, since $Pf = f$ has a unique nonnegative normalized solution $f_* \in D$ given by (8.9.11), which is also positive, by Corollary 8.7.3, all solutions of the linear Tjon–Wu equation have the limit

$$\lim_{t \to \infty} u(t, x) = e^{-x} \tag{8.9.12}$$

for all initial conditions $u(0, x) = f(x)$, $f \in D((0, \infty))$.

This illustration of applying the tools developed in this chapter to deal with the Tjon–Wu equation is meant to show their power. Given the integro-differential equation (8.9.3), we have been able to show the global convergence of its solutions by examining only the fixed points of the right-hand side. This led to a second-order ordinary differential equation that was easily solved, in spite of its nonlinearity. Finally, once the solution was available and shown to satisfy the requirements of Corollary 8.7.3, then the asymptotic behavior of $u(t, x)$, for *all* initial conditions, was also known.

Exercises

8.1. Let $(\Omega, \mathcal{F}, \text{prob})$ be a probability space and let $A, B \in \mathcal{F}$. Define $\bar{A} = \Omega \setminus A$ and $\bar{B} = \Omega \setminus B$. Prove that the independence of events A, B implies the independence of events $\bar{A}$, B as well as A, $\bar{B}$ and $\bar{A}$, $\bar{B}$.

8.2. Let $(\Omega, \mathcal{F}, \text{prob})$ be a probability space and $A_1, A_2, \ldots$ be a sequence of events. Define $\xi_n = 1_{A_n}$, $n = 1, 2, \ldots$. Prove that $A_1, A_2, \ldots$ are independent if and only if the random variables ξ_n are independent.

8.3. Let $\{N_t\}_{t \geq 0}$ be a Poisson process and $S: R \to R$ a nonsingular mapping. Consider the following procedure: In a time $t > 0$ a point $x \in R$ is transformed into $S(x) + N_t$. Given an initial density distribution function f of the initial point x find the density $u(t, x)$ of $S(x) + N_t$. (As in Section 8.3 assume that the position x of the initial point and the process N_t are independent.) Prove that $u(t, x)$ satisfies the differential equation

$$\frac{\partial u(t, x)}{\partial t} = -\lambda u(t, x) + \lambda u(t, x - 1) \qquad \text{for } t > 0, \, x \in R,$$

which does not depend explicitly on S. [λ is defined in (8.2.8c).] Explain this paradox.

8.4. Derive formula (8.5.5) for the solution of the linear Boltzmann equation by the use of the Phillips perturbation theorem.

8.5. Consider the linear Boltzmann equation (8.5.1) and corresponding semigroup $\{\hat{P}_t\}_{t \geq 0}$. Assuming that $P: L^1 \to L^1$ is a constrictive operator, prove that

$$\lim_{t \to \infty} \hat{P}_t f$$

exists for every $f \in L^1$.

8.6. Again consider the linear Boltzmann equation (8.5.1) and assume that $P: L^1(X, \mathcal{A}, \mu) \to L^1(X, \mathcal{A}, \mu)$ is sweeping with respect to a family $\mathcal{A}_* \subset \mathcal{A}$. Prove that the semigroup $\{\hat{P}_t\}_{t \geq 0}$ is sweeping with respect to the same family.

8.7. The nonlinear Tjon–Wu equation (8.9.2) may be written in the form

$$\frac{du}{dt} = -u + P(u, u),$$

where $P: L^1(R^+) \times L^1(R^+) \to L^1(R^+)$ is the bilinear operator defined by

$$P(f,g)(x) = \int_x^\infty \frac{dy}{y} \int_0^y f(y-z)g(z)\, dz.$$

Verify that the series

$$u(t) = e^{-t} \sum_{n=0}^\infty (1 - e^{-t})^n u_n$$

with

$$u_n = \frac{1}{n} \sum_{k=0}^{n-1} P(u_k, u_{n-1-k}), \qquad u_0 = f \in \mathcal{D}(R^+)$$

is uniformly convergent on compact subintervals of R^+ and satisfies the nonlinear Tjon–Wu equation with the initial condition $u(0) = f$ (Kielek, 1988).

9
Entropy

The concept of entropy was first introduced by Clausius and later used in a different form by L. Boltzmann in his pioneering work on the kinetic theory of gases published in 1866. Since then, entropy has played a pivotal role in the development of many areas in physics and chemistry and has had important ramifications in ergodic theory. However, the Boltzmann entropy is different from the Kolmogorov–Sinai–Ornstein entropy [Walters, 1975; Parry, 1981] that has been so successfully used in solving the problem of isomorphism of dynamical systems, and which is related to the work of Shannon [see Shannon and Weaver, 1949].

In this short chapter we consider the Boltzmann entropy of sequences of densities $\{P^n f\}$ and give conditions under which the entropy may be constant or increase to a maximum. We then consider the inverse problem of determining the behavior of $\{P^n f\}$ from the behavior of the entropy.

9.1 Basic Definitions

If $(X, \mathcal{A}, \mu)$ is an arbitary measure space and $P: L^1 \to L^1$ a Markov operator, then under certain circumstances valuable information concerning the behavior of $\{P^n f\}$ (or, in the continuous time case, $\{P^t f\}$) can be obtained from the behavior of the sequence

$$H(P^n f) = \int_X \eta(P^n f(x)) \mu(dx), \tag{9.1.1}$$

where $\eta(u)$ is some function appropriately defined for $u \geq 0$.

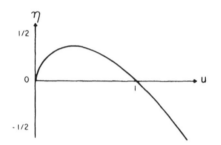

FIGURE 9.1.1. Plot of function $\eta(u) = -u \log u$.

The classical work of Boltzmann on the statistical properties of dilute gases suggested that the function η should be of the form

$$\eta(u) = -u \log u, \quad \eta(0) = 0, \quad (\log u = \log_e u), \qquad (9.1.2)$$

and gives us our definition of entropy.

Definition 9.1.1. If $f \geq 0$ and $\eta(f) \in L^1$ then the **entropy** of f is defined by

$$H(f) = \int_X \eta(f(x))\mu(dx). \qquad (9.1.3)$$

Remark 9.1.1. If $\mu(X) < \infty$, then the integral (9.1.3) is always well defined for every $f \geq 0$. In fact, the integral over the positive parts of $\eta(f(x))$,

$$[\eta(f(x))]^+ = \max[0, \eta(f(x))].$$

is always finite. Thus $H(f)$ is either finite or equal to $-\infty$.   □
Since we take $\eta(0) = 0$, the function $\eta(u)$ is continuous for all $u \geq 0$. The graph of η is shown in Figure 9.1.1. One of the most important properties of η is that it is convex. To see this, note that

$$\eta''(u) = -1/u$$

so $\eta''(u) < 0$ for all $u > 0$. From this it follows immediately that the graph of η always lies below the tangent line, or

$$\eta(u) \leq (u - v)\eta'(v) + \eta(v) \qquad (9.1.4)$$

for every $u, v > 0$. Combining (9.1.4) with the definition of η given in equation (9.1.2) leads to the **Gibbs inequality**

$$u - u \log u \leq v - u \log v \qquad \text{for } u, v > 0, \qquad (9.1.5)$$

which we shall have occasion to use frequently.

If f and g are two densities such that $\eta(f(x))$ and $f(x)\log g(x)$ are integrable, then from (9.1.5) we have the useful integral inequality

$$-\int_X f(x)\log f(x)\mu(dx) \le -\int_X f(x)\log g(x)\mu(dx). \qquad (9.1.6)$$

and the equality holds only for $f = g$. Inequality (9.1.6) is often of help in proving some extremal properties of $H(f)$ as shown in the following.

Proposition 9.1.1. *Let $\mu(X) < \infty$, and consider all the possible densities f defined on X. Then, in the family of all such densities, the maximal entropy occurs for the constant density*

$$f_0(x) = 1/\mu(X), \qquad (9.1.7)$$

and for any other f the entropy is strictly smaller.

Proof. Pick an arbitrary $f \in D$ so that the entropy of f is given by

$$H(f) = -\int_X f(x)\log f(x)\mu(dx)$$

and, by inequality (9.1.6),

$$H(f) \le -\int_X f(x)\log f_0(x)\mu(dx)$$
$$= -\log\left[\frac{1}{\mu(X)}\right]\int_X f(x)\mu(dx)$$

or

$$H(f) \le -\log\left[\frac{1}{\mu(X)}\right],$$

and the equality is satisfied only for $f = f_0$. However, the entropy of f_0 is simply

$$H(f_0) = -\int_X \frac{1}{\mu(X)}\log\left[\frac{1}{\mu(X)}\right]\mu(dx) = -\log\left[\frac{1}{\mu(X)}\right],$$

so $H(f) \le H(f_0)$ for all $f \in D$. $\blacksquare$

If $\mu(X) = \infty$, then there are no constant densities and this proposition fails. However, if additional constraints are placed on the density, then we may obtain other results for maximal entropies as illustrated in the following two examples.

Example 9.1.1. Let $X = [0,\infty)$ and consider all possible densities f such that the first moment of f is given by

$$\int_0^\infty xf(x)\,dx = 1/\lambda. \qquad (9.1.8)$$

Then the density

$$f_0(x) = \lambda e^{-\lambda x} \tag{9.1.9}$$

maximizes the entropy.

The proof proceeds as in Proposition 9.1.1. From inequality (9.1.6) we have, for arbitrary $f \in D$ satisfying (9.1.8),

$$H(f) \leq - \int_0^\infty f(x) \log(\lambda e^{-\lambda x}) \, dx$$

$$= - \log \lambda \int_0^\infty f(x) \, dx + \int_0^\infty \lambda x f(x) \, dx$$

$$= - \log \lambda + 1.$$

Also, however, with f_0 given by (9.1.9),

$$H(f_0) = - \int_0^\infty \lambda e^{-\lambda x} \log(\lambda e^{-\lambda x}) \, dx = - \log \lambda + 1$$

and thus $H(f) \leq H(f_0)$ for all $f \in D$ satisfying (9.1.8).   □

Example 9.1.2. For our next example take $X = (-\infty, \infty)$ and consider all possible densities $f \in D$ such that the second moment of f is finite, that is,

$$\int_{-\infty}^\infty x^2 f(x) \, dx = \sigma^2. \tag{9.1.10}$$

Then the maximal entropy is achieved for the Gaussian density

$$f_0(x) = \frac{1}{\sqrt{2\pi\sigma^2}} \exp\left(-\frac{x^2}{2\sigma^2}\right). \tag{9.1.11}$$

As before, we calculate that, for arbitrary $f \in D$ satisfying (9.1.10),

$$H(f) \leq - \int_{-\infty}^\infty f(x) \log\left[\frac{1}{\sqrt{2\pi\sigma^2}} \exp\left(-\frac{x^2}{2\sigma^2}\right)\right] dx$$

$$= - \log\left[\frac{1}{\sqrt{2\pi\sigma^2}}\right] \int_{-\infty}^\infty f(x) \, dx + \frac{1}{2\sigma^2} \int_{-\infty}^\infty x^2 f(x) \, dx$$

$$= \tfrac{1}{2} - \log\left[\frac{1}{\sqrt{2\pi\sigma^2}}\right].$$

Further

$$H(f_0) = - \int_{-\infty}^\infty f_0(x) \log f_0(x) \, dx = \tfrac{1}{2} - \log\left[\frac{1}{\sqrt{2\pi\sigma^2}}\right]$$

so that the entropy is maximized with the Gaussian density (9.1.11).   □

These two examples are simply special cases covered by the following simple statement.

Proposition 9.1.2. *Let $(X, \mathcal{A}, \mu)$ be a measure space. Assume that a sequence $g_1, \ldots, g_m$ of measurable functions is given as well as two sequences of real constants $\bar{g}_1, \ldots, \bar{g}_m$ and $\nu_1, \ldots, \nu_m$ that satisfy*

$$\bar{g}_i = \frac{\int_X g_i(x) \exp[-\nu_i g_i(x)] \mu(dx)}{\int_X \prod_{i=1}^m \exp[-\nu_i g_i(x)] \mu(dx)},$$

where all of the integrals are finite. Then the maximum of the entropy $H(f)$ for all $f \in D$, subject to the conditions

$$\bar{g}_i = \int_X g_i(x) f(x) \mu(dx), \qquad i = 1, \ldots, m$$

occurs for

$$f_0(x) = \frac{\prod_{i=1}^m \exp[-\nu_i g_i(x)]}{\int_X \prod_{i=1}^m \exp[-\nu_i g_i(x)] \mu(dx)}.$$

Proof. For simplicity, set

$$Z = \int_X \prod_{i=1}^m \exp[-\nu_i g_i(x)] \mu(dx)$$

so

$$f_0(x) = Z^{-1} \prod_{i=1}^m \exp[-\nu_i g_i(x)].$$

From inequality (9.1.6), we have

$$H(f) \leq -\int_X f(x) \log f_0(x) \mu(dx)$$

$$= -\int_X f(x) \left[-\log Z - \sum_{i=1}^m \nu_i g_i(x) \right] \mu(dx)$$

$$= \log Z + \sum_{i=1}^m \nu_i \int_X f(x) g_i(x) \mu(dx)$$

$$= \log Z + \sum_{i=1}^m \nu_i \bar{g}_i.$$

Furthermore, it is easy to show that

$$H(f_0) = \log Z + \sum_{i=1}^m \nu_i \bar{g}_i$$

and thus $H(f) \leq H(f_0)$. $\blacksquare$

Remark 9.1.2. Note that if $m = 1$ and $g(x)$ is identified as the energy of a system, then the maximal entropy occurs for

$$f_0(x) = Z^{-1}e^{-\nu g(x)},$$

which is just the **Gibbs canonical distribution function**, with the **partition function** Z given by

$$Z = \int_X e^{-\nu g(x)} \mu(dx).$$

Further, the maximal entropy

$$H(f_0) = \log Z + \nu \bar{g}$$

is just the thermodynamic entropy. As is well known, all of the results of classical thermodynamics can be derived with the partition function Z and the preceding entropy $H(f_0)$. Indeed, the contents of Proposition 9.1.2 have been extensively used by Jaynes [1957] and Katz [1967] in an alternative formulation and development of classical and quantum statistical mechanics.   □

Thus, the simple Gibbs inequality has far-reaching implications in pure mathematics as well as in more applied fields. Another inequality that we will have occasion to use often is the **Jensen inequality**: If $\eta(u)$, $u \geq 0$ is a function such that $\eta'' \leq 0$ (i.e., the graph of η is convex), $P: L^p \to L^p$ $1 \leq p \leq \infty$, is a linear operator such that $P1 = 1$, and $Pf \geq 0$ for all $f \geq 0$, then for every $f \in L^p$, $f \geq 0$,

$$\eta(Pf) \geq P\eta(f) \qquad \text{whenever } P\eta(f) \text{ exists.} \qquad (9.1.12)$$

The proof of this result is difficult and requires many specialized techniques. However, the following considerations provide some insight into why it is true. Let $\eta(y)$ be a convex function defined for $y \geq 0$. Pick u, v, and z such that $0 \leq u \leq z \leq v$. Since $z \in [u, v]$ there exist nonnegative constants, α and β, with $\alpha + \beta = 1$, such that

$$z = \alpha u + \beta v.$$

Further, from the convexity of η it is clear that $\eta(z) \geq r$, where

$$r = \alpha \eta(u) + \beta \eta(v).$$

Thus $\eta(z) \geq r$ gives

$$\eta(\alpha u + \beta v) \geq \alpha \eta(u) + \beta \eta(v).$$

Further, it is easy to verify by induction that for every sequence $0 \leq u_1 < u_2 \cdots$

$$\eta\left(\sum_i \alpha_i u_i\right) \geq \sum_i \alpha_i \eta(u_i), \tag{9.1.13}$$

where $\alpha_i \geq 0$ and $\sum_i \alpha_i = 1$. Now suppose we have a linear operator $P: R^n \to R^n$ satisfying $P1 = 1$. Since P is linear its coordinates must be of the form

$$(Pf)_i = \sum_{j=1}^n k_{ij} f_j,$$

where $f = (f_1, \ldots, f_n)$ and $\sum_j k_{ij} = 1$, $k_{ij} \geq 0$. By applying inequality (9.1.13) to $(Pf)_i$, we have

$$\eta((Pf)_i) \geq \sum_{j=1}^n k_{ij} \eta(f_j) = P(\eta f)_i,$$

or, suppressing the coordinate index,

$$\eta(Pf) \geq P\eta(f).$$

In an arbitrary (not necessarily finite dimensional) space the proof of the Jensen inequality is much more difficult, but still uses (9.1.13) as a starting point.

The final inequality we will have occasion to use is a direct consequence of integrating inequality (9.1.13) over the entire space X, namely,

$$H\left(\alpha_i \sum_i f_i\right) \geq \sum_i \alpha_i H(f_i), \tag{9.1.14}$$

where again $\alpha_i \geq 0$ and $\sum_i \alpha_i = 1$.

9.2 Entropy of $P^n f$ when P is a Markov Operator

We are now in a position to examine the behavior of the entropy $H(P^n f)$ when P is a Markov operator. We begin with the following theorem.

Theorem 9.2.1. *Let $(X, \mathcal{A}, \mu)$ be a finite measure space $[\mu(X) < \infty]$ and $P: L^1 \to L^1$ a Markov operator. If P has a constant stationary density $[P1 = 1]$, then*

$$H(Pf) \geq H(f) \tag{9.2.1}$$

for all $f \geq 0$, $f \in L^1$.

Proof. Integrating Jensen's inequality (9.1.12) over the entire space X

gives

$$\int_X \eta(Pf(x))\mu(dx) \geq \int_X P\eta(f(x))\mu(dx)$$
$$= \int_X \eta(f(x))\mu(dx)$$

since P preserves the integral. However, the left-most integral is $H(Pf)$, and the last integral is $H(f)$, so that (9.2.1) is proved. ∎

Remark 9.2.1. For a finite measure space, we know that the maximal entropy H_{max} is $-\log[1/\mu(X)]$, so that

$$-\log[1/\mu(X)] \geq H(P^n f) \geq H(f).$$

This, in conjunction with Theorem 9.2.1, tells us that in a finite measure space when P has a constant stationary density, the entropy never decreases and is bounded above by $-\log[1/\mu(X)]$. Thus, in this case the entropy $H(P^n f)$ always converges as $n \to \infty$, although not necessarily to the maximum. Note further that, if we have a normalized measure space, then $\mu(X) = 1$ and $H_{max} = 0$. □

Remark 9.2.2. In the case of a Markov operator without a constant stationary density, it may happen that the sequence $H(P^n f)$ is not increasing as n increases. As a simple example consider the quadratic transformation $S(x) = 4x(1-x)$. The Frobenius–Perron operator for S, derived in Section 1.2, is

$$Pf(x) = \frac{1}{4\sqrt{1-x}}\left\{ f\left(\frac{1}{2} - \frac{1}{2}\sqrt{1-x}\right) + f\left(\frac{1}{2} + \frac{1}{2}\sqrt{1-x}\right)\right\}$$

and it is easy to verify that

$$f_*(x) = \frac{1}{\pi\sqrt{x(1-x)}}$$

is a stationary density for P. Take as an initial density $f = 1$, so $H(f) = 0$ and

$$Pf(x) = \frac{1}{2\sqrt{1-x}}.$$

Then

$$H(Pf) = -\int_0^1 \frac{1}{2\sqrt{1-x}} \log\left(\frac{1}{2\sqrt{1-x}}\right) dx = (\log 2) - 1.$$

Clearly $H(Pf) < H(f) = 0$. □

It is for this reason that it is necessary to introduce the concept of conditional entropy for Markov operators with nonconstant stationary densities.

Definition 9.2.1. Let $f, g \in D$ be such that $\operatorname{supp} f \subset \operatorname{supp} g$. Then the **conditional entropy** of f with respect to g is defined by

$$H(f \mid g) = \int_X g(x) \eta \left[\frac{f(x)}{g(x)}\right] \mu(dx) = - \int_X f(x) \log \left[\frac{f(x)}{g(x)}\right] \mu(dx). \quad (9.2.2)$$

Remark 9.2.3. Since g is a density and $\eta(x) = -x \log x$ is bounded $(\sup \eta < \infty)$ the integral $H(f \mid g)$ is always defined, that is, it is either finite or equal to $-\infty$. In some sense, which is suggested by the equation (9.2.2), the value $H(f \mid g)$ measures the deviation of f from the density g. □

The conditional entropy $H(f \mid g)$ has two properties, which we will use later. They are

1. If $f, g \in D$, then, by inequality (9.1.6), $H(f \mid g) \leq 0$. The equality holds if and only if $f = g$.

2. If g is the constant density, $g = 1$, then $H(f \mid 1) = H(f)$. Thus the conditional entropy $H(f \mid g)$ is a generalization of the entropy $H(f)$.

For $f, g \in D$, the condition $\operatorname{supp} f \subset \operatorname{supp} g$ implies $\operatorname{supp} Pf \subset \operatorname{supp} Pg$ (see Exercise 3.10), and given $H(f \mid g)$ we may evaluate $H(Pf \mid Pg)$ through the following.

Theorem 9.2.2. Let $(X, \mathcal{A}, \mu)$ be an arbitrary measure space and $P: L^1 \to L^1$ a Markov operator. Then

$$H(Pf \mid Pg) \geq H(f \mid g) \qquad \text{for } f, g \in D, \operatorname{supp} f \subset \operatorname{supp} g. \quad (9.2.3)$$

Remark 9.2.4. Note from this theorem that if g is a stationary density of P, then $H(Pf \mid Pg) = H(Pf \mid g)$ and thus

$$H(Pf \mid g) \geq H(f \mid g).$$

Thus the conditional entropy with respect to a stationary density is always increasing and bounded above by zero. It follows that $H(P^n f \mid g)$ always converges, but not necessarily to zero, as $n \to \infty$. □

Proof of Theorem 9.2.2. Here we give the proof of Theorem 9.2.2 only in the case when $Pg > 0$, $g > 0$, and the function f/g is bounded. [Consult Voigt (1981) for the full proof.] Take $g \in L^1$ with $g > 0$. Define an operator $R: L^\infty \to L^\infty$ by

$$Rh = P(hg)/Pg \qquad \text{for } h \in L^\infty,$$

where hg denotes multiplication, not composition. R has the following properties:

1. $Rh \geq 0$ for $h \geq 0$; and

2. $R1 = Pg/Pg = 1$.

Thus R satisfies the assumptions of Jensen's inequality, giving

$$\eta(Rh) \geq R\eta(h). \tag{9.2.4}$$

Setting $h = f/g$ the left-hand side of (9.2.4) may be written in the form

$$\eta(Rh) = -(Pf/Pg)\log(Pf/Pg)$$

and the right-hand side is given by

$$R\eta(h) = (1/Pg)P[(\eta \circ h)g] = -(1/Pg)P[f\log(f/g)].$$

Hence inequality (9.2.4) becomes

$$-Pf\log(Pf/Pg) \geq -P[f\log(f/g)].$$

Integrating this last inequality over the space X, and remembering that P preserves the integral, we have

$$H(Pf \mid Pg) \geq - \int_X P\left\{ f(x) \log\left[\frac{f(x)}{g(x)}\right] \right\} \mu(dx)$$

$$= - \int_X f(x) \log\left[\frac{f(x)}{g(x)}\right] \mu(dx) = H(f \mid g),$$

which finishes the proof. $\square$

9.3   Entropy $H(P^n f)$ when P is a Frobenius–Perron Operator

Inequalities (9.2.1) and (9.2.3) of Theorems 9.2.1 and 9.2.2 are not strong. In fact, the entropy may not increase at all during successive iterations of f. This is always the case when P is the Frobenius–Perron operator corresponding to an invertible transformation, which leads to the following theorem.

Theorem 9.3.1. *Let $(X, \mathcal{A}, \mu)$ be a finite measure space and $S \colon X \to X$ be an invertible measure-preserving transformation. If P is the Frobenius–Perron operator corresponding to S, then $H(P^n f) = H(f)$ for all n.*

Proof. If S is invertible and measure preserving, then by equation (3.2.10) we have $Pf(x) = f(S^{-1}(x))$ since $J^{-1} \equiv 1$. If P_1 is the Frobenius–Perron operator corresponding to S^{-1}, we also have $P_1 f(x) = f(S(x))$. Thus $P_1 P f = P P_1 f = f$, so $P_1 = P^{-1}$. From Theorem 9.2.1 we also have

$$H(P_1 P f) \geq H(P f) \geq H(f),$$

but, since $P_1 P f = P^{-1} P f = f$, we conclude that $H(P f) = H(f)$, so $H(P^n f) = H(f)$ for all n. ∎

Remark 9.3.1. For any discrete or continuous time system that is invertible and measure preserving the entropy is always constant. In particular, for a continuous time system evolving according to the set of differential equations $\dot{x} = F(x)$, the entropy is constant if $\text{div } F = 0$ [see equation (7.8.18)]. Every Hamiltonian system satisfies this condition. □

However, for noninvertible (irreversible) systems this is not the case, and we have the following theorem.

Theorem 9.3.2. Let $(X, \mathcal{A}, \mu)$ be a measure space, $\mu(X) = 1$, $S: X \to X$ a measure-preserving transformation, and P the Frobenius–Perron operator corresponding to S. If S is exact then

$$\lim_{n \to \infty} H(P^n f) = 0$$

for all $f \in D$ such that $H(f) > -\infty$.

Proof. Assume initially that f is bounded, that is, $0 \leq f \leq c$. Then

$$0 \leq P^n f \leq P^n c = c P^n 1 = c.$$

Without any loss of generality, we can assume that $c > 1$. Further, since $\eta(u) \leq 0$ for $u \geq 1$, we have [note $\mu(X) = 1$ and $H_{\max} = 0$]

$$0 \geq H(P^n f) \geq \int_{A_n} \eta(P^n f(x)) \mu(dx), \tag{9.3.1}$$

where

$$A_n = \{x : 1 \leq P^n f(x) \leq c\}.$$

Now, by the mean value theorem [using $\eta(1) = 0$], we obtain

$$\left| \int_{A_n} \eta(P^n f(x)) \mu(dx) \right| = \int_{A_n} |\eta(P^n f(x)) - \eta(1)| \mu(dx)$$
$$\leq k \int_{A_n} |P^n f(x) - 1| \mu(dx)$$
$$\leq k \int_X |P^n f(x) - 1| \mu(dx) = \|P^n f - 1\|,$$

where

$$k = \sup_{1 \leq u \leq c} |\eta'(u)|.$$

Since S is exact, from Theorem 4.4.1, we have $\|P^n f - 1\| \to 0$ as $n \to \infty$ for all $f \in D$ and thus

$$\lim_{n \to \infty} \int_{A_n} \eta(P^n f(x)) \mu(dx) = 0.$$

From inequality (9.3.1), it follows that $H(P^n f)$ converges to zero.

Now relax the assumption that f is bounded and write f in the form

$$f = f_1 + f_2,$$

where

$$f_1(x) = \begin{cases} 0 & \text{if } f(x) > c \\ f(x) & \text{if } 0 \le f(x) \le c \end{cases}$$

and $f_2 = f - f_1$. Fixing $\varepsilon > 0$, we may choose c sufficiently large so that

$$\|f_2\| < \varepsilon \quad \text{and} \quad H(f_2) > -\varepsilon.$$

Write $P^n f$ in the form

$$P^n f = (1 - \delta) P^n \left(\frac{1}{1 - \delta} f_1 \right) + \delta P^n \left(\frac{1}{\delta} f_2 \right),$$

where $\delta = \|f_2\|$. Now $f_1/(1 - \delta)$ is a bounded density, and so from the first part of our proof we know that for n sufficiently large

$$H\left(P^n \left(\frac{1}{1 - \delta} f_1 \right) \right) > -\varepsilon.$$

Furthermore,

$$\delta H \left(P^n \left(\frac{1}{\delta} f_2 \right) \right) = H(P^n f_2) - \log \left(\frac{1}{\delta} \right) \int_X P^n f_2(x) \mu(dx)$$

$$= H(P^n f_2) - \|f_2\| \log \left(\frac{1}{\delta} \right)$$

$$= H(P^n f_2) + \delta \log \delta.$$

Since $H(P^n f_2) \ge H(f_2) > -\varepsilon$, this last expression becomes

$$\delta H \left(P^n \left(\frac{1}{\delta} f_2 \right) \right) \ge -\varepsilon + \delta \log \delta.$$

Combining these results and inequality (9.1.14), we have

$$H(P^n f) \ge (1 - \delta) H \left(P^n \left(\frac{1}{1 - \delta} f_1 \right) \right) + \delta H \left(P^n \left(\frac{1}{\delta} f_2 \right) \right)$$

$$\ge -\varepsilon(1 - \delta) - \varepsilon + \delta \log \delta$$

$$= -2\varepsilon + \delta\varepsilon + \delta \log \delta. \tag{9.3.2}$$

Since $\mu(X) = 1$, we have $H(P^n f) \leq 0$. Further since $\delta < \varepsilon$ and ε is arbitrary, the right-hand side of (9.3.2) is also arbitrarily small, and the theorem is proved. ■

Example 9.3.1. We wish to compare the entropy of the baker transformation

$$S(x,y) = \begin{cases} (2x, \frac{1}{2}y), & 0 \leq x < \frac{1}{2}, 0 \leq y \leq 1 \\ (2x - 1, \frac{1}{2}y + \frac{1}{2}), & \frac{1}{2} \leq x \leq 1, 0 \leq y \leq 1, \end{cases}$$

originally introduced in Example 4.1.3, with that of the dyadic transformation. Observe that the x-coordinate of the baker transformation is transformed by the dyadic transformation

$$S_1(x) = 2x \qquad (\text{mod } 1).$$

From our considerations of Chapter 4, we know that the baker transformation is invertible and measure preserving. Thus by Theorem 9.3.1 it follows that the entropy of the sequence $\{P^n f\}$, where P is the Frobenius–Perron operator corresponding to the baker transformation, is constant for every density f.

Conversely, the dyadic transformation S_1 is exact. Hence, from Theorem 9.3.2, the entropy of $\{P_1^n f\}$, where P_1 is the Frobenius–Perron operator corresponding to S_1, increases to zero for all bounded initial densities f. □

Remark 9.3.2. Observe that in going from the baker to the dyadic transformation, we are going from an invertible (reversible) to a noninvertible (irreversible) system through the loss of information about the y-coordinate. This loss of information is accompanied by an alteration of the behavior of the entropy. An analogous situation occurs in statistical mechanics where, in going from the Liouville equation to the Boltzmann equation, we also lose coordinate information and go from a situation where entropy is constant (Liouville equation) to one in which the entropy increases to its maximal value (Boltzmann H theorem). □

9.4   Behavior of $P^n f$ from $H(P^n f)$

In this section we wish to see what aspects of the eventual behavior of $P^n f$ can be deduced from $H(P^n f)$. This is a somewhat difficult problem, and the major stumbling block arises from the fact that η changes its sign. Thus, because of the integration in the definition of the entropy, it is difficult to determine f or its properties from $H(f)$. However, by use of the spectral representation Theorem 5.3.1 for Markov operators, we are able to circumvent this problem.

In our first theorem we wish to show that, if $H(P^n f)$ is bounded below, then P is constrictive. This is presented more precisely in the following theorem.

Theorem 9.4.1. *Let $(X, \mathcal{A}, \mu)$ be a measure space, $\mu(X) < \infty$, and $P\colon L^1 \to L^1$ a Markov operator such that $P1 = 1$. If there exists a constant $c > 0$ such that for every bounded $f \in D$*

$$H(P^n f) \geq -c \qquad \text{for } n \text{ sufficiently large,}$$

then P is constrictive.

Proof. Observe that $P1 = 1$ implies that Pf is bounded for bounded f. Thus, to prove our theorem, it is sufficient to show that the set $\mathcal{F}$ of all bounded $f \in D$ that satisfy

$$H(f) \geq -c$$

is weakly precompact.

We will use criterion 3 of Section 5.1 to demonstrate the weak precompactness of $\mathcal{F}$. Since $\|f\| = 1$ for all $f \in D$, the first part of the criterion is satisfied. To check the second part take $\varepsilon > 0$. Pick $l = e^{-1}\mu(X)$, $N = \exp[2(c + l)/\varepsilon]$ and $\delta = \varepsilon/2N$, and take a set $A \subset X$ such that $\mu(A) < \delta$. Then

$$\int_A f(x)\mu(dx) = \int_{A_1} f(x)\mu(dx) + \int_{A_2} f(x)\mu(dx), \qquad (9.4.1)$$

where

$$A_1 = \{x \in A\colon f(x) \leq N\}$$

and

$$A_2 = \{x \in A\colon f(x) > N\}.$$

The first integral on the right-hand side of (9.4.1) clearly satisfies

$$\int_{A_1} f(x)\mu(dx) \leq N\delta = \varepsilon/2.$$

In evaluating the second integral, note that from $H(f) \geq -c$, it follows that

$$\int_{A_2} f(x) \log f(x)\mu(dx) \leq c - \int_{X\setminus A_2} f(x) \log f(x)\mu(dx)$$

$$\leq c + \int_{X\setminus A_2} \eta_{\max}\mu(dx)$$

$$\leq c + (1/e)\mu(X) = c + l.$$

Therefore

$$\int_{A_2} f(x) \log N\mu(dx) < c + l$$

or

$$\int_{A_2} f(x)\mu(dx) < \frac{c+l}{\log N} = \frac{\epsilon}{2}.$$

Thus

$$\int_A f(x)\mu(dx) < \epsilon$$

and $\mathcal{F}$ is weakly precompact. Thus, by Definition 5.3.3, the operator P is constrictive. ∎

Before stating our next theorem, consider the following. Let $(X, \mathcal{A}, \mu)$ be a finite measure space, $S: X \to X$ a nonsingular transformation, and P the Frobenius–Perron operator corresponding to S. Assume that for some $c > 0$ the condition

$$H(P^n f) \geq -c$$

holds for every bounded $f \in D$ and n sufficiently large. Since P is a Markov operator and is constrictive, we may write Pf in the form given by the spectral decomposition Theorem 5.3.1, and, for every initial f, the sequence $\{P^n f\}$ will be asymptotically periodic.

Theorem 9.4.2. *Let $(X, \mathcal{A}, \mu)$ be a normalized measure space, $S: X \to X$ a measure-preserving transformation, and P the Frobenius–Perron operator corresponding to S. If*

$$\lim_{n \to \infty} H(P^n f) = 0$$

for all bounded $f \in D$, then S is exact.

Proof. It follows from Theorem 9.4.1 that P is constrictive. Furthermore, since S is measure preserving, we know that P has a constant stationary density. From Proposition 5.4.2 we, therefore, have

$$P^{n+1} f(x) = \sum_{i=1}^{r} \lambda_{\alpha^{-n}(i)}(f)\bar{1}_{A_i}(x) + Q_n f(x) \qquad \text{for } f \in L^1.$$

If we can demonstrate that $r = 1$, then from Theorem 5.5.2 we will have shown S to be exact.

Pick

$$f(x) = [1/\mu(A_1)]1_{A_1}(x)$$

as an initial f. If τ is the asymptotic period of $P^n f$, then we must have

$$P^{n\tau} f(x) = [1/\mu(A_1)]1_{A_1}(x).$$

However, by assumption,

$$\lim_{n \to \infty} H(P^n f) = 0,$$

and, since the sequence $\{H(P^{n r} f)\}$ is a constant sequence, we must have

$$H([1/\mu(A_1)]1_{A_1}) = 0.$$

Note that, by Proposition 9.1.1, $H(f) = 0$ only if

$$f(x) = 1_X(x).$$

So, clearly, we must have

$$[1/\mu(A_1)]1_{A_1}(x) = 1_X(x).$$

This is possible if and only if A_1 is the entire space X, and thus $r = 1$. Hence S is exact. ∎

This theorem in conjunction with Theorem 9.3.2 tells us that the convergence of $H(P^n f)$ to zero as $n \to \infty$ is both necessary and sufficient for the exactness of measure-preserving transformations. If the transformation is not measure preserving then an analogous result using the conditional entropy may be proved.

To see this, suppose we have an arbitrary measure space $(X, \mathcal{A}, \mu)$ and a nonsingular transformation $S: X \to X$. Let P be the Frobenius–Perron operator corresponding to S and $g \in D$ $(g > 0)$ the stationary density of P so $Pg = g$. Since S is not measure preserving, our previous results cannot be used directly in examining the exactness of S.

However, consider the new measure space $(X, \mathcal{A}, \tilde{\mu})$, where

$$\tilde{\mu}(A) = \int_A g(x)\mu(dx) \qquad \text{for } A \in \mathcal{A}.$$

Since $Pg = g$, therefore $\tilde{\mu}$ is an invariant measure. Thus, in this new space the corresponding Frobenius–Perron operator $\tilde{P}$ is defined by

$$\int_A \tilde{P}h(x)\tilde{\mu}(dx) = \int_{S^{-1}(A)} h(x)\tilde{\mu}(dx) \qquad \text{for } A \in \mathcal{A}$$

and satisfies $\tilde{P}1 = 1$. This may be rewritten as

$$\int_A [\tilde{P}h(x)]g(x)\mu(dx) = \int_{S^{-1}(A)} h(x)g(x)\mu(dx).$$

However, we also have

$$\int_{S^{-1}(A)} h(x)g(x)\mu(dx) = \int_A P(h(x)g(x))\mu(dx)$$

so that $(\tilde{P}h)g = P(hg)$ or

$$\tilde{P}h = (1/g)P(hg).$$

Furthermore, by induction,

$$\tilde{P}^n h = (1/g) P^n(hg).$$

In this new space $(X, \mathcal{A}, \tilde{\mu})$, we may also calculate the entropy $\tilde{H}(\tilde{P}^n h)$ as

$$\tilde{H}(\tilde{P}^n h) = -\int_X \tilde{P}^n h(x) \log[\tilde{P}^n h(x)] \tilde{\mu}(dx)$$

$$= -\int_X \frac{1}{g(x)} P^n(h(x)g(x)) \log\left[\frac{P^n(h(x)g(x))}{g(x)}\right] g(x)\mu(dx)$$

$$= H(P^n(hg) \mid g).$$

Observe that $h \in D(X, \mathcal{A}, \tilde{\mu})$ is equivalent to

$$h \geq 0 \quad \text{and} \quad \int_X h(x)g(x)\mu(dx) = 1,$$

which is equivalent to $hg \in D(X, \mathcal{A}, \mu)$. Set $f = hg$, so

$$\tilde{H}(\tilde{P}^n h) = H(P^n f \mid g).$$

We may, therefore, use our previous theorems to examine the exactness of S in the new space $(X, \mathcal{A}, \tilde{\mu})$ or its asymptotic stability in the original space $(X, \mathcal{A}, \mu)$, that is, S is statistically stable in $(X, \mathcal{A}, \mu)$ if and only if

$$\lim_{n \to \infty} H(P^n f \mid g) = 0 \qquad (9.4.2)$$

for all $f \in D$ such that f/g is bounded.

Example 9.4.1. Consider the linear Boltzmann equation [equation (8.3.8)]

$$\frac{\partial u(t, x)}{\partial t} + u(t, x) = Pu(t, x),$$

with the initial condition $u(0, x) = f(x)$, which we examined in Chapter 8. There we showed that the solution of this equation was given by

$$u(t, x) = e^{t(P-I)} f(x) = \hat{P}_t f(x),$$

and $e^{t(P-I)}$ is a semigroup of Markov operators. From Theorem 9.2.2 we know immediately that the conditional entropy $H(\hat{P}_t f \mid f_*)$ is continually increasing for every f_* that is a stationary density of P. Furthermore, by (9.4.2) and Corollary 8.7.3, if $f_*(x) > 0$ and f_* is the unique stationary density of P, then

$$\lim_{t \to \infty} H(\hat{P}_t f \mid f_*) = H(f_* \mid f_*) = 0.$$

Thus, in the case in which f_* is positive and unique, the conditional entropy for the solutions of the linear Boltzmann equation always achieves its maximal value.

Exercises

9.1. Let $X = \{(x_1, \ldots, x_k) \in R^k : x_1 \geq 0, \ldots, x_k \geq 0\}$. Consider the family $\mathcal{F}_{m_1 \cdots m_k}$ of densities $f : X \to R^+$ such that

$$\int_0^\infty \cdots \int_0^\infty x_i f(x_1, \ldots, x_k) dx_1 \cdots dx_k = m_i > 0, \qquad i = 1, \ldots, k.$$

Find the density in $\mathcal{F}_{m_1 \cdots m_k}$ that maximizes the entropy.

9.2. Let $X = \{(x, y) \in R^2 : y \geq \alpha |x|\}$ where α is a constant. Consider the family $\mathcal{F}_{m\alpha}$ of densities $f : X \to R^+$ such that

$$\int \int_X y f(x, y) \, dx dy = m > 0.$$

Show that for $\alpha > 0$ there is a density in $\mathcal{F}_{m\alpha}$ having the maximal entropy and that for $\alpha \leq 0$ the entropy in $\mathcal{F}_{m\alpha}$ is unbounded.

9.3. Consider the space $X = \{1, \ldots, N\}$ with the counting measure. In this space $\mathcal{D}(X)$ consists of all probabilistic vectors $(f_1 = f(1), \ldots, f_N = f(N))$ satisfying

$$f_k \geq 0, \qquad \sum_{k=1}^N f_k = 1.$$

Show that $f_k = 1/N$, $k = 1, \ldots, N$ maximizes the entropy. For which vector is the entropy minimal?

9.4. Consider the heat equation

$$\frac{\partial u}{\partial t} = \frac{\delta^2}{2} \frac{\partial^2 u}{\partial x^2} \qquad \text{for } t > 0, x \in R,$$

and prove that every positive solution $u(t, x)$ corresponding to the bounded initial $u(0, x) = f(x)$, $f \in D$ with compact support, satisfies

$$\frac{d}{dt} H(u) = \int_{-\infty}^{+\infty} u \left(\frac{\partial}{\partial x} \ln u \right)^2 dx \geq 0.$$

9.5. Consider the differential equation

$$\frac{\partial u}{\partial t} = \frac{\delta^2}{2} \frac{\partial^2 u}{\partial x^2} - \frac{\partial}{\partial x} (b(x)u) \qquad \text{for } t > 0, \ 0 \leq x \leq 1$$

with the boundary value conditions

$$u_x(t, 0) = u_x(t, 1) = 0 \qquad \text{for } t > 0.$$

Assume that b is a C^2 function and that $b(0) = b(1) = 0$. Without looking for the explicit formula for the solutions (which, for arbitrary b, is difficult) prove the following properties:

(a) For every solution

$$\int_0^1 u(t,x)\,dx = \text{const.}$$

(b) For every two positive normalized solutions u_1 and u_2

$$\frac{d}{dx}H(u_1 \mid u_2) = \int_0^1 u_1 \left(\frac{\partial}{\partial x}\ln\frac{u_1}{u_2}\right)^2 dx \geq 0,$$

(Risken, 1984; Sec. 6.1.)

9.6. Write a program called CONDENT (conditional entropy) to study the value

$$H(f \mid g) = -\int_0^1 f(x)\log\left[\frac{f(x)}{g(x)}\right]dx \qquad \text{for } f,g \in D([0,1]).$$

Compare for different pairs of sequences $\{f_n\}$, $\{g_n\} \subset D([0,1])$ the asymptotic behavior of

$$\|f_n - g_n\|_{L^1} \quad \text{and} \quad H(f_n \mid g_n).$$

9.7. Let $(X, \mathcal{A}, \mu)$ be a measure space. Prove that for every two sequences $\{f_n\}$, $\{g_n\} \subset D$ the convergence $H(f_n \mid g_n) \to 0$ implies $\|f_n - g_n\|_{L^1} \to 0$. Is the converse implication also true? Exercise 9.6 can be helpful in guessing the proper answer (Loskot and Rudnicki, 1991).

9.8. Consider a density $f_*: R^3 \to R^+$ of the form

$$f_*(x) = \alpha \, \exp(-\beta|x|^2 + kx),$$

where $|x|^2 = x_1^2 + x_2^2 + x_3^2$ and $kx = k_1x_1 + k_2x_2 + k_3x_3$. Assume that a sequence of densities $f_n \subset D(R^3)$ satisfies

$$\int_{R^3} g_i(x)f_n(x)\,dx = \int_{R^3} g_i(x)f_*(x)\,dx, \qquad i = 0,1,2,3,$$

with $g_0(x) = |x|^2$ and $g_i(x) = x_i$, $i = 1,2,3$. Prove that the convergence $H(f_n) \to H(f_*)$ implies $\|f_n - f_*\| \to 0$ (Elmroth, 1984; Loskot and Rudnicki, 1991).

10

Stochastic Perturbation of Discrete Time Systems

We have seen two ways in which uncertainty (and thus probability) may appear in the study of strictly deterministic systems. The first was the consequence of following a random distribution of initial states, which, in turn, led to a development of the notion of the Frobenius–Perron operator and an examination of its properties as a means of studying the asymptotic properties of flows of densities. The second resulted from the random application of a transformation S to a system and led naturally to our study of the linear Boltzmann equations.

In this chapter we consider yet another source of probabilistic distributions in deterministic systems. Specifically, we examine discrete time situations in which at each time the value $x_{n+1} = S(x_n)$ is reached with some error. An extremely interesting situation occurs when this error is small and the system is "primarily" governed by a deterministic transformation S. We consider two possible ways in which this error might be small: Either the error occurs rather rarely and is thus small on the average, or the error occurs constantly but is small in magnitude. In both cases, we consider the situation in which the error is independent of $S(x_n)$ and are, thus, led to first recall the notion of independent random variables in the next section and to explore some of their properties in Sections 10.2 and 10.3.

10.1   Independent Random Variables

Let $(\Omega, \mathcal{F}, \text{prob})$ be a probability space. A finite sequence of random variables $(\xi^1, \ldots, \xi^k)$ is called a **k-dimensional random vector**. Equivalently, we could say that a random vector $\xi = (\xi^1, \ldots, \xi^k)$ is a measurable transformation from Ω into R^k. Measurability means that for every Borel subset $B \subset R^k$ the set

$$\{\xi \in B\} = \xi^{-1}(B)$$

belongs to $\mathcal{F}$.

Thus, having a k-dimensional random vector $(\xi_1, \ldots, \xi_k)$, we may consider two different kinds of densities: the density of each random component ξ_i and the joint density function for the random vector $(\xi_1, \ldots, \xi_k)$. Let the density of ξ_i be denoted by $f_i(x)$, and the joint density of $\xi = (\xi_1, \ldots, \xi_k)$ be $f(x_1, \ldots, x_k)$. Then by definition, we have

$$\int_{B_i} f_i(x)\, dx = \text{prob}\{\xi_i \in B_i\}, \quad \text{for } B_i \subset R, \qquad (10.1.1)$$

and

$$\int \cdots \int_B f(x_1, \ldots, x_k)\, dx_1 \cdots dx_k = \text{prob}\{(\xi_1, \ldots, \xi_k) \in B\}, \qquad \text{for } B \subset R^k,$$

where B_i and B are Borel subsets of R and R^k, respectively. In this last integral take

$$B = B_1 \times \underbrace{R \times \cdots \times R}_{k-1 \text{ times}}$$

so that we have

$$\text{prob}\{(\xi_1, \ldots, \xi_k) \in B\} = \text{prob}\{\xi_1 \in B_1\}$$
$$= \int_{B_1} \left\{ \int \cdots \int_{R^{k-1}} f(x, x_2, \ldots, x_k)\, dx_2 \cdots dx_k \right\} dx. \quad (10.1.2)$$

By comparing (10.1.1) with (10.1.2), we see immediately that

$$f_1(x) = \int \cdots \int_{R^{k-1}} f(x, x_2, \ldots, x_k)\, dx_2 \cdots dx_k. \qquad (10.1.3)$$

Thus, having the joint density function f for $(\xi_1, \ldots, \xi_k)$, we can always find the density for ξ_1 from equation (10.1.3). In an entirely analogous fashion,

f_2 can be obtained by integrating $f(x_1, x, \ldots, x_k)$ over $x_1, x_3, \ldots, x_k$. The same procedure will yield each of the densities f_i.

However, the converse is certainly not true in general since, having the density f_i of each random variable ξ_i $(i = 1, \ldots, k)$, it is not usually possible to find the joint density f of the random vector $(\xi_1, \ldots, \xi_k)$. The one important special case in which this construction is possible occurs when $\xi_1, \ldots, \xi_k$ are independent random variables. Thus, we have the following theorem.

Theorem 10.1.1. *If the random variables $\xi_1, \ldots, \xi_k$ are independent and have densities $f_1, \ldots, f_k$, respectively, then the joint density function for the random vector $(\xi_1, \ldots, \xi_k)$ is given by*

$$f(x_1, \ldots, x_k) = f_1(x_1) \cdots f_k(x_k), \tag{10.1.4}$$

where the right-hand side is a product.

Proof. Consider a Borel set $B \subset R^k$ of the form

$$B = B_1 \times \cdots \times B_k, \tag{10.1.5}$$

where $B_1, \ldots, B_k \subset R$ are Borel sets. Then

$$\text{prob}\{(\xi_1, \ldots, \xi_k) \in B\} = \text{prob}\{\xi_1 \in B_1, \ldots, \xi_k \in B_k\},$$

and, since the random variables $\xi_1, \ldots, \xi_k$ are independent,

$$\text{prob}\{(\xi_1, \ldots, \xi_k) \in B\} = \text{prob}\{\xi_1 \in B_1\} \cdots \text{prob}\{\xi_k \in B_k\}.$$

With this equation and (10.1.1), we obtain

$$\text{prob}\{(\xi_1, \ldots, \xi_k) \in B\} = \int_{B_1} f_1(x_1)\, dx_1 \cdots \int_{B_k} f_k(x_k)\, dx_k$$

$$= \int \cdots \int_B f_1(x_1) \cdots f_k(x_k)\, dx_1 \cdots dx_k. \tag{10.1.6}$$

Since, by definition, sets of the form (10.1.5) are generators of the Borel subsets in R^k, it is clear that (10.1.6) must hold for arbitrary Borel sets $B \subset R^k$. By the definition of the joint density, this implies that $f_1(x_1) \cdots f_k(x_k)$ is the joint density for the random vector $(\xi_1, \ldots, \xi_k)$. ∎

As a simple application of Theorem 10.1.1, we consider two independent random variables ξ_1 and ξ_2 with densities f_1 and f_2, respectively. We wish to obtain the density of $\xi_1 + \xi_2$. Observe that, by Theorem 10.1.1, the random vector (ξ_1, ξ_2) has the joint density $f_1(x_1)f_2(x_2)$. Thus, for an arbitrary Borel set $B \subset R$, we have

$$\text{prob}\{\xi_1 + \xi_2 \in B\} = \iint\limits_{x_1 + x_2 \in B} f_1(x_1) f_2(x_2) \, dx_1 dx_2,$$

or, setting $x = x_1 + x_2$ and $y = x_2$,

$$\text{prob}\{\xi_1 + \xi_2 \in B\} = \iint\limits_{B \times R} f_1(x - y) f_2(y) \, dx \, dy$$

$$= \int_B \left\{ \int_{-\infty}^{\infty} f_1(x - y) f_2(y) \, dy \right\} dx.$$

From the definition of a density, this last equation shows that

$$f(x) = \int_{-\infty}^{\infty} f_1(x - y) f_2(y) \, dy \qquad (10.1.7)$$

is the density of $\xi_1 + \xi_2$.

Remark 10.1.1. From the definition of the density, it follows that, if ξ has a density f, then $c\xi$ has a density $(1/|c|)f(x/c)$. To see this, write

$$\text{prob}\{c\xi \in A\} = \text{prob}\left\{ \xi \in \frac{1}{c} A \right\} = \int_{(1/c)A} f(y) \, dy = \frac{1}{|c|} \int_A f\left(\frac{x}{c}\right) dx.$$

Thus, from (10.1.7), if ξ_1 and ξ_2 are independent and have densities f_1 and f_2, respectively, then $(c_1\xi_1 + c_2\xi_2)$ has the density

$$f(x) = \frac{1}{|c_1 c_2|} \int_{-\infty}^{\infty} f_1\left(\frac{x - y}{c_1}\right) f_2\left(\frac{y}{c_2}\right) dy. \quad \Box \qquad (10.1.8)$$

10.2 Mathematical Expectation and Variance

In previous chapters we have, on numerous occasions, used the concept of mathematical expectation in rather specialized situations without specifically noting that it was, indeed, the mathematical expectation that was involved. We now wish to explicitly introduce this concept in its general sense.

Let $(\Omega, \mathcal{F}, \text{prob})$ be a probability space and let $\xi : \Omega \to R$ be a random variable. Then we have the following definition.

Definition 10.2.1. If ξ is integrable with respect to the measure "prob," then the **mathematical expectation** (or **mean value**) of ξ is given by

$$E(\xi) = \int_{\Omega} \xi(\omega) \, \text{prob}(d\omega).$$

Remark 10.2.1. By definition, $E(\xi)$ is the average value of ξ. A more illuminating interpretation of $E(\xi)$ is given by the law of large numbers [see equation (10.3.4)]. □

In the case when ξ is a constant, $\xi = c$, then it is trivial to derive $E(c)$. Since $\text{prob}\{\Omega\} = 1$ for any constant c, we have

$$E(c) = c \int_\Omega \text{prob}(d\omega) = c. \tag{10.2.1}$$

Now we show how the mathematical expectation may be calculated via the use of a density function. Let $h: R^k \to R$ be a Borel measurable function, that is, $h^{-1}(\Delta)$ is a Borel subset of R^k for each interval Δ. Further, let $\xi = (\xi_1, \ldots, \xi_k)$ be a random vector with the joint density function $f(x_1, \ldots, x_k)$. Then we have the following theorem.

Theorem 10.2.1. *If hf is integrable, that is,*

$$\int \cdots \int_{R^k} h(x_1 \cdots x_k) f(x_1 \cdots x_k) \, dx_1 \cdots dx_k < \infty,$$

then the mathematical expectation of the random variable $h \circ \xi$ exists and is given by

$$E(h \circ \xi) = \int \cdots \int_{R^k} h(x_1, \ldots, x_k) f(x_1, \ldots, x_k) \, dx_1 \cdots dx_k. \tag{10.2.2}$$

Proof. First assume that h is a simple function, that is,

$$h(x) = \sum_{i=1}^n \lambda_i 1_{A_i}(x) \qquad x = (x_1, \ldots, x_k),$$

where the A_i are mutually disjoint Borel subsets of R^k such that $\cup_i A_i = R^k$. Then

$$h(\xi(\omega)) = \sum_{i=1}^n \lambda_i 1_{A_i}(\xi(\omega)) = \sum_{i=1}^n \lambda_i 1_{\xi^{-1}(A_i)}(\omega),$$

and, by the definition of the Lebesgue integral,

$$E(h \circ \xi) = \int_\Omega h(\xi(\omega)) \, \text{prob}(d\omega) = \sum_{i=1}^n \lambda_i \, \text{prob}\{\xi^{-1}(A_i)\}.$$

Further, since f is the density for ξ, we have

$$\text{prob}\{\xi^{-1}(A_i)\} = \text{prob}\{\xi \in A_i\} = \int_{A_i} f(x) \, dx, \qquad dx = dx_1 \cdots dx_k.$$

As a consequence,

$$E(h \circ \xi) = \sum_{i=1}^{n} \lambda_i \int_{A_i} f(x)\,dx = \int_{R^k} \sum_{i=1}^{n} \lambda_i 1_{A_i}(x) f(x)\,dx$$

$$= \int_{R^k} h(x) f(x)\,dx.$$

Thus, for the h that are simple functions, equality (10.2.2) is proved. For an arbitrary h, hf integrable, we can find a sequence $\{h_n\}$ of simple functions converging to h and such that $|h_n| \leq |h|$. From equality (10.2.2), already proved for simple functions, we thus have

$$E(h_n \circ \xi) = \int_{R^k} h_n(x) f(x)\,dx.$$

By the Lebesgue dominated convergence theorem, since $|h_n f| \leq |h| f$, it follows that

$$\int_{\Omega} h(\xi(\omega))\operatorname{prob}(d\omega) = \int_{R^k} h(x) f(x)\,dx,$$

which completes the proof. ∎

In the particular case that $k = 1$ and $h(x) = x$, we have from equation (10.2.2)

$$E(\xi) = \int_{-\infty}^{\infty} x f(x)\,dx. \tag{10.2.3}$$

Thus, if $f(x)$ is taken to be the mass density of a rod of infinite length, then $E(\xi)$ gives the center of mass of the rod.

From Definition 10.2.1, it follows that, for every sequence of random variables $\xi_1, \ldots, \xi_k$ and constants $\lambda_1, \ldots, \lambda_k$, we have

$$E\left(\sum_{i=1}^{k} \lambda_i \xi_i\right) = \sum_{i=1}^{k} \lambda_i E(\xi_i) \tag{10.2.4}$$

since the mathematical expectation is simply a Lebesgue integral on the probability space $(\Omega, \mathcal{F}, \operatorname{prob})$. Moreover, the mathematical expectation of $\sum_i \lambda_i \xi_i$ exists whenever all of the $E(\xi_i)$ exist.

We now turn to a consideration of the variance, starting with a definition.

Definition 10.2.2. Let $\xi: \Omega \to R$ be a random variable such that $m = E(\xi)$ exists. Then the **variance** of ξ is

$$D^2(\xi) = E((\xi - m)^2) \tag{10.2.5}$$

if the corresponding integral is finite.

Thus the variance of a random variable ξ is just the average value of the square of the deviation of ξ away from m. By the additivity of the mathematical expectation, equation (10.2.5) may also be written as

$$D^2(\xi) = E(\xi^2) - 2mE(\xi) + m^2 = E(\xi^2) - m^2. \tag{10.2.6}$$

If ξ has a density $f(x)$, then by the use of equation (10.2.2), we can also write

$$D^2(\xi) = \int_{-\infty}^{\infty} (x - m)^2 f(x)\, dx$$

whenever the integral on the right-hand side exists. Finally, we note that for any constant λ,

$$D^2(\lambda\xi) = E(\lambda^2(\xi - m)^2) = \lambda^2 D^2(\xi).$$

Since in any application there is a certain inconvenience in the fact that $D^2(\xi)$ does not have the same dimension as ξ, it is sometimes more convenient to use the **standard deviation** of ξ, defined by

$$\sigma(\xi) = \sqrt{D^2(\xi)}.$$

For our purposes here, two of the most important properties of the mathematical expectation and variance of a random variable ξ are contained in the next theorem.

Theorem 10.2.2. Let $\xi_1, \ldots, \xi_k$ be independent random variables such that $E(\xi_i)$, $D^2(\xi_i)$, $i = 1, \ldots, k$ exist. Then

$$E(\xi_1 \cdots \xi_k) = E(\xi_1) \cdots E(\xi_k) \tag{10.2.7}$$

and

$$D^2(\xi_1 + \cdots + \xi_k) = D^2(\xi_1) + \cdots + D^2(\xi_k). \tag{10.2.8}$$

Proof. The proof is easy even in the general case. However, to illustrate again the usefulness of (10.2.2), we will prove this theorem in the case when all the ξ_i have densities. Thus, assume that ξ_i has density f_i, $i = 1, \ldots, k$, and pick $h(x_1, \ldots, x_k) = x_1 \cdots x_k$. Since the $\xi_1, \ldots, \xi_k$ are independent random variables, by Theorem 10.1.1, the joint density function for the random vector $(\xi_1, \ldots, \xi_k)$ is

$$f_1(x_1) \cdots f_k(x_k).$$

Hence, by equation (10.2.2),

$$E(\xi_1 \cdots \xi_k) = \int \cdots \int_{R^k} x_1 \cdots x_k f_1(x_1) \cdots f_k(x_k) dx_1 \cdots dx_k$$

$$= \int_{-\infty}^{\infty} x_1 f_1(x_1) dx_1 \cdots \int_{-\infty}^{\infty} x_k f_k(x_k)\, dx_k$$

$$= E(\xi_1) \cdots E(\xi_k),$$

and (10.2.7) is therefore proved.

Now set $E(\xi_i) = m_i$, so that

$$D^2(\xi_1 + \cdots + \xi_k) = E((\xi_1 + \cdots + \xi_k - m_1 - \cdots - m_k)^2)$$

$$= E\left(\sum_{i,j=1}^{k} (\xi_i - m_i)(\xi_j - m_j)\right).$$

Since the $\xi_1, \ldots, \xi_k$ are independent, $(\xi_1 - m_1), \ldots, (\xi_k - m_k)$ are also independent. Therefore, by (10.2.4) and (10.2.7), we have

$$D^2(\xi_1 + \cdots + \xi_k) = \sum_{i=1}^{k} E((\xi_i - m_i)^2) + \sum_{i \neq j} E((\xi_i - m_i)(\xi_j - m_j))$$

$$= \sum_{i=1}^{k} D^2(\xi_i) + \sum_{i \neq j} (E(\xi_i) - m_i)(E(\xi_j) - m_j).$$

Since $E(\xi_i) = m_i$, equation (10.2.8) results immediately. ∎

Remark 10.2.2. In Theorem 10.2.2, it is sufficient to assume that the ξ_i are mutually independent, that is, ξ_i is independent of ξ_j, for $i \neq j$. □

To close this section on mathematical expectation and variance, we give two versions of the Chebyshev inequality, originally introduced in a special context in Section 5.7.

Theorem 10.2.3. *If ξ is nonnegative and $E(\xi)$ exists, then*

$$\text{prob}\{\xi \geq a\} \leq E(\xi)/a \quad \text{for every } a > 0. \quad (10.2.9)$$

If ξ is arbitrary but such that $m = E(\xi)$ and $D^2(\xi)$ exist, then

$$\text{prob}\{|\xi - m| \geq \varepsilon\} \leq D^2(\xi)/\varepsilon^2 \quad \text{for every } \varepsilon > 0. \quad (10.2.10)$$

Proof. By the definition of mathematical expectation,

$$E(\xi) = \int_{\Omega} \xi(\omega)\,\text{prob}(d\omega) \geq \int_{\{\omega : \xi(\omega) \geq a\}} \xi(\omega)\,\text{prob}(d\omega)$$

$$\geq a \int_{\{\omega : \xi(\omega) \geq a\}} \text{prob}(d\omega) = a\,\text{prob}\{\xi \geq a\},$$

which proves (10.2.9). [This is, of course, analogous to equation (5.7.9).]
Now replace ξ by $(\xi - m)^2$ and a by ε^2 in (10.2.9) to give

$$\text{prob}\{(\xi - m)^2 \geq \varepsilon^2\} \leq (1/\varepsilon^2)E((\xi - m)^2) = (1/\varepsilon^2)D^2(\xi),$$

which is equivalent to (10.2.10) and completes the proof. ∎

10.3 Stochastic Convergence

There are several different ways in which the convergence of a sequence $\{\xi_n\}$ of random variables may be defined. For example, if $\xi_n \in L^p(\Omega, \mathcal{F}, \text{prob})$, then we may define both strong and weak convergences of $\{\xi_n\}$ to ξ in $L^p(\Omega)$ space, as treated in Section 2.3.

In probability theory some of these types of convergence have special names. Thus, strong convergence of $\{\xi_n\}$ in $L^2(\Omega)$, defined by the relation

$$\lim_{n\to\infty} \|\xi_n - \xi\|_{L^2(\Omega)} = 0, \qquad (10.3.1)$$

is denoted by

$$\text{l.i.m.}\, \xi_n = \xi$$

and called **convergence in mean**.

A second type of convergence useful in the treatment of probabilistic phenomena is given in the following definition.

Definition 10.3.1. A sequence $\{\xi_n\}$ of random variables is said to be **stochastically convergent** to the random variable ξ if, for every $\varepsilon > 0$,

$$\lim_{n\to\infty} \text{prob}\{|\xi_n - \xi| \geq \varepsilon\} = 0. \qquad (10.3.2)$$

The stochastic convergence of $\{\xi_n\}$ to ξ is denoted by

$$\text{st-lim}\, \xi_n = \xi. \qquad (10.3.3)$$

Note that in terms of L^p norms, the mathematical expectation and variance of a random variable may be written as

$$E(|\xi|^p) = \int_\Omega |\xi|^p \, \text{prob}(d\omega) = \|\xi\|_{L^p(\Omega)}^p$$

and

$$D^2(\xi) = \int_\Omega |\xi - m|^2 \, \text{prob}(d\omega) = \|\xi - m\|_{L^2(\Omega)}^2.$$

This observation allows us to derive a connection between stochastic convergence and strong convergence from the Chebyshev inequality, as contained in the following proposition.

Proposition 10.3.1. *If a sequence $\{\xi_n\}$ of random variables, $\xi_n \in L^p(\Omega)$, is strongly convergent in $L^p(\Omega)$ to ξ, then $\{\xi_n\}$ is stochastically convergent to ξ. Thus, convergence in mean implies stochastic convergence.*

Proof. We only consider $p < \infty$, since for $p = \infty$ the proposition is trivial. Applying the Chebyshev inequality (10.2.9) to $|\xi_n - \xi|^p$, we have

$$\text{prob}\{|\xi_n - \xi|^p \geq \varepsilon^p\} \leq (1/\varepsilon^p) E(|\xi_n - \xi|^p)$$

or, equivalently,

$$\text{prob}\{|\xi_n - \xi| \geq \varepsilon\} \leq (1/\varepsilon^p)\|\xi_n - \xi\|^p_{L^p(\Omega)},$$

which completes the proof. ■

A third type of convergence useful for random variables is defined next.

Definition 10.3.2. A sequence $\{\xi_n\}$ of random variables is said to converge **almost surely** to ξ (or to converge to ξ with probability 1) if

$$\lim_{n \to \infty} \xi_n(\omega) = \xi(\omega)$$

for almost all ω. Equivalently, this condition may be written as

$$\text{prob}\{\lim_{n \to \infty} \xi_n(\omega) = \xi(\omega)\} = 1.$$

Remark 10.3.1. For all of the types of convergence we have defined (strong and weak L^p convergence, convergence in mean, stochastic convergence, and almost sure convergence), the limiting function is determined up to a set of measure zero. That is, if ξ and $\bar{\xi}$ are both limits of the sequence $\{\xi_n\}$, then ξ and $\bar{\xi}$ differ only on a set of measure zero. □

We now show the connection between almost sure and stochastic convergence with the following proposition.

Proposition 10.3.2. *If a sequence of random variables $\{\xi_n\}$ converges almost surely to ξ, then it also converges stochastically to ξ.*

Proof. Set

$$\eta_n(\omega) = \min(1, |\xi_n(\omega) - \xi(\omega)|).$$

Clearly, $|\eta_n| \leq 1$. If $\{\xi_n\}$ converges almost surely to ξ, then $\{\eta_n\}$ converges to zero almost surely, and, by the Lebesgue dominated convergence theorem,

$$\lim_{n \to \infty} \|\eta_n\|_{L^1(\Omega)} = \lim_{n \to \infty} \int_\Omega \eta_n(\omega)\,\text{prob}(d\omega) = 0.$$

By Proposition 10.3.1 this implies that $\{\eta_n\}$ converges stochastically to zero. Since in the definition of stochastic convergence it suffices to consider only $\varepsilon < 1$, it then follows that

$$\text{prob}\{|\xi_n - \xi| \geq \varepsilon\} = \text{prob}\{\eta_n \geq \varepsilon\} \qquad \text{for } 0 < \varepsilon < 1.$$

Thus the stochastic convergence of $\{\eta_n\}$ to zero implies the stochastic convergence of $\{\xi_n\}$ to ξ and the proof is complete. ■

As a simple illustration of the usefulness of the concept of stochastic convergence, we prove the simplest version of the law of large numbers given in the next theorem.

Theorem 10.3.1 (Weak law of large numbers). *Let $\{\xi_n\}$ be a sequence of independent random variables with*

$$E(\xi_n) = m_n$$

and

$$M = \sup_n D^2(\xi_n) < \infty.$$

Then

$$\text{prob}\left\{\left|\frac{1}{n}\sum_{i=1}^{n}(\xi_i - m_i)\right| \geq \varepsilon\right\} < \frac{M}{n\varepsilon^2}$$

for every $\varepsilon > 0$. In particular, if $m_1 = m_2 = \cdots = m_n = m$, then

$$\text{st-lim}\,\frac{1}{n}\sum_{i=1}^{n}\xi_i = m. \tag{10.3.4}$$

Proof. Set

$$\eta_n = \frac{1}{n}\sum_{i=1}^{n}\xi_i.$$

Since the ξ_i are indepenent random variables,

$$D^2(\eta_n) = D^2\left(\frac{1}{n}\sum_{i=1}^{n}\xi_i\right) = \frac{1}{n^2}\sum_{i=1}^{n}D^2(\xi_i) \leq \frac{nM}{n^2} = \frac{M}{n}$$

and, clearly,

$$E(\eta_n) = \frac{1}{n}\sum_{i=1}^{n}m_i.$$

Thus, by the Chebyshev inequality (10.2.10),

$$\text{prob}\{|\eta_n - E(\eta_n)| \geq \varepsilon\} \leq (1/\varepsilon^2)D^2(\eta_n) \leq M/n\varepsilon^2,$$

which completes the proof, as equation (10.3.4) is a trivial consequence. ∎

Equation (10.3.4) is a precise statement of our intuitive notion that the mathematical expectation or mean value of a random variable may be obtained by averaging the results of many independent experiments.

The term "weak law of large numbers" specifically refers to equation (10.3.4) because stochastic convergence is weaker than other types of convergence for which similar results can be proved. One of the most famous versions of these is the so-called **strong law of large numbers**, as contained in the Kolmogorov theorem.

Theorem 10.3.2 (Kolmogorov). *Let $\{\xi_n\}$ be a sequence of independent random variables with*

$$E(\xi_n) = m_n \quad and \quad M = \sup_n D^2(\xi_n) < \infty.$$

Then

$$\lim_{n \to \infty} \frac{1}{n} \sum_{i=1}^{n} (\xi_i - m_i) = 0$$

with probability 1.

We will not give a proof of the Kolmogorov theorem [see Breiman (1968) for the proof] as it is not used in our studies of the flow of densities. We stated it only because of its close correspondence to the Birkhoff individual ergodic Theorem 4.2.4, which also deals with the pointwise convergence of averages. To illustrate this correspondence, consider the sequence

$$f(x), f(S(x)), \dots , \tag{10.3.5}$$

which appears in the Birkhoff theorem, as a sequence of random variables on the probability space $(X, \mathcal{A}, \text{prob})$, where

$$\text{prob}(A) = \mu(A)/\mu(X).$$

These variables (10.3.5) are, in general, highly dependent since $S(x)$ is a function of x, $S^2(x) = S(S(x))$ is a function of x and $S(x)$, and so on.

The reason that a probabilistic treatment of deterministic systems is often more difficult than problems in classical probability theory is directly related to the absence of independent random variables in the former. It is only in some special circumstances that independence may appear in deterministic systems under certain limiting cases, such as mixing and exactness.

However, independence appears in a natural way in the definition of perturbed dynamical systems which we consider in the following Sections 10.4–10.6. There we use the notion of independent random vectors, which is an immediate generalization of the definition of independent random variables. Let $\xi_n : \Omega \to R^{k_n}$, with $n = 1, 2, \dots$ be a sequence (finite or not) of random vectors. We say that the ξ_n are **independent** if, for every sequence of Borel sets $B_n \subset R^{k_n}$, the events

$$\{\xi_1 \in B_1\}, \{\xi_2 \in B_2\}, \dots$$

are independent. Observe that in this definition the ξ_n may have different dimensions for different n.

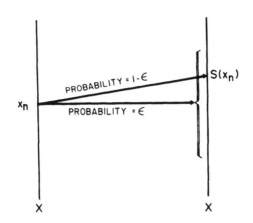

FIGURE 10.4.1. Schematic representation of the operation of a deterministic system with a randomly applied stochastic perturbation.

10.4 Discrete Time Systems with Randomly Applied Stochastic Perturbations

In this section we consider the asymptotic behavior of a nonsingular transformation when a stochastic perturbation is randomly applied.

Let $(X, \mathcal{A}, \mu)$ be a measure space, and $S: X \to X$ a nonsingular transformation with associated Frobenius–Perron operator P. The following rules apply to the evolution of the point $x \in X$: At the nth instant of time we do not know the precise location of x although we do know the density $f_n(x)$. At the next instant of time $(n+1)$, the point moves with probability $(1-\varepsilon)$ to the next location $S(x_n)$. However, there is a probability ε that this new location will not be given by $S(x_n)$ but rather by a random variable, independent of x_n, with density $g(x)$. This process can be visualized as shown in Figure 10.4.1. To make it more precise, we follow the ideas of Chapter 8 in which we derived the linear Boltzmann equation. Thus, consider space $(X, \mathcal{A}, \mu_f)$, where f is the density of the initial position of the point, and probability space $(\Omega, \mathcal{F}, \mathrm{prob})$ related to the perturbations. With these two spaces we define the product space

$$\Omega \times X = \{(\omega, x): \omega \in \Omega, x \in X\}$$

and the product measure

$$\mathrm{Prob}(C \times A) = \mathrm{prob}(C)\mu_f(A), \qquad \text{for } C \in \mathcal{F}, A \in \mathcal{A}$$

(see Theorem 2.2.2).

To describe the perturbations, consider a sequence of independent random vectors

$$\xi_0, \eta_0, \xi_1, \eta_1, \ldots,$$

such that each $\eta_m: \Omega \to R$ takes only two values, 1 and 0, with the following probabilities:

$$\text{prob}(\eta_n = 1) = 1 - \varepsilon, \qquad \text{prob}(\eta_n = 0) = \varepsilon,$$

and each ξ_n has the same density g. Then the equation

$$x_{n+1} = \eta_n S(x_n) + (1 - \eta_n)\xi_n, \qquad n = 0, 1, \ldots \qquad (10.4.1)$$

gives the precise description of our intuitively introduced behavior of the sequence of $\{x_n\}$. Denote the density of x_n by f_n. Our task now is to derive a relation between f_n and f_{n+1}, assuming that the initial density $f_0 = f$ is given.

Note that

$$\begin{aligned}
\text{Prob}\{x_{n+1} \in A\} &= \text{Prob}\{x_{n+1} \in A \text{ and } \eta_n = 0\} \\
&\quad + \text{Prob}\{x_{n+1} \in A \text{ and } \eta_n = 1\}. \qquad (10.4.2)
\end{aligned}$$

Since, from (10.4.1), $x_{n+1}(\omega, x) = \xi_n(\omega)$ if $\eta_n(\omega) = 0$, and $x_{n+1}(\omega, x) = S(x_n)$ if $\eta_n(\omega) = 1$, we can rewrite equation (10.4.2) as

$$\begin{aligned}
\text{Prob}\{x_{n+1} \in A\} &= \text{Prob}\{\xi_n \in A \text{ and } \eta_n = 0\} \\
&\quad + \text{Prob}\{S(x_n) \in A \text{ and } \eta_n = 1\}.
\end{aligned}$$

Since the events $\{\xi_n \in A\}$ and $\{\eta_n = 0\}$ are independent of each other and independent of the initial vector x_0, we have

$$\begin{aligned}
\text{Prob}\{\xi_n \in A \text{ and } \eta_n = 0\} &= \text{prob}\{\xi_n \in A \text{ and } \eta_n = 0\} \\
&= \text{prob}\{\xi_n \in A\} \text{prob}\{\eta_n = 0\} \\
&= \varepsilon \int_A g(x)\mu(dx).
\end{aligned}$$

Further, since x_n is dependent only on $\xi_1, \ldots, \xi_{n-1}$ and $\eta_1, \ldots, \eta_{n-1}$, we have

$$\text{Prob}\{S(x_n) \in A \text{ and } \eta_n = 1\} = \text{Prob}\{S(x_n) \in A\} \text{prob}\{\eta_n = 1\}.$$

Finally, since x_n has the density f_n by assumption, this last formula implies

$$\text{Prob}\{S(x_n) \in A \text{ and } \eta_n = 1\} = (1 - \varepsilon) \int_{S^{-1}(A)} f_n(x)\mu(dx).$$

Thus, combining the foregoing probabilities, we have

$$\text{Prob}\{x_{n+1} \in A\} = (1 - \varepsilon) \int_{S^{-1}(A)} f_n(x)\mu(dx) + \varepsilon \int_A g(x)\mu(dx).$$

Using the definition of the Frobenius–Perron operator P corresponding to S, this may be rewritten as

$$\text{Prob}\{x_{n+1} \in A\} = \int_A [(1-\varepsilon)Pf_n(x) + \varepsilon g(x)]\mu(dx)$$

for all Borel sets $A \subset R$. Hence, if x_n has density f_n, then this demonstrates that x_{n+1} also has a density f_{n+1} given by

$$f_{n+1} = (1-\varepsilon)Pf_n + \varepsilon g. \tag{10.4.3}$$

We want to write the right-hand side of equation (10.4.3) in the form of a linear operator, and so we define $P_\varepsilon : L^1 \to L^1$ by

$$P_\varepsilon f = (1-\varepsilon)Pf + \varepsilon g \int_X f(x)\mu(dx) \tag{10.4.4}$$

for all $f \in L^1$. Using the definition of P_ε, we may rewrite equation (10.4.3) in the form

$$f_{n+1} = P_\varepsilon f_n. \tag{10.4.5}$$

Our goal is to deduce as much as possible concerning the asymptotic behavior of $P_\varepsilon^n f_0$ for $f_0 \in D$.

The first result is contained in the following proposition.

Proposition 10.4.1. *Let the operator $P_\varepsilon : D \to D$ be defined by equation (10.4.4). Then $\{P_\varepsilon^n\}$ is asymptotically stable.*

Proof. The proof is trivial. From the definition of P_ε in (10.4.4), we have

$$P_\varepsilon^n f = P_\varepsilon(P_\varepsilon^{n-1} f) \geq \varepsilon g \int_X f(x)\mu(dx) = \varepsilon g$$

for all $f \in D$. Thus, εg is a nontrivial lower-bound function for $P_\varepsilon^n f$. Further since P_ε is clearly a Markov operator, we have, by Theorem 5.6.2, that $\{P_\varepsilon^n\}$ is asymptotically stable. ∎

Remark 10.4.1. This simple result tells us that given any nonsingular transformation, the addition of even the smallest stochastic perturbation ensures that the system will be asymptotically stable regardless of the character of the deterministic system in the unperturbed case. □

However, much more can be determined about this stochastically perturbed deterministic system. We have the following result that explicitly gives the stationary density for P_ε.

Proposition 10.4.2. *Let the operator $P_\varepsilon : L^1 \to L^1$ be defined by equation (10.4.4). Then for $\varepsilon > 0$, the unique stationary density of P_ε is given by*

$$f_*^\varepsilon = \varepsilon \sum_{k=0}^{\infty} (1-\varepsilon)^k P^k g. \tag{10.4.6}$$

Proof. Since $\|(1-\varepsilon)^k P^k g\| \le (1-\varepsilon)^k \|g\|$ the series in (10.4.6) is absolutely convergent. Substitution of (10.4.6) into

$$P_\varepsilon f = (1 - \varepsilon)Pf + \varepsilon g \qquad (10.4.7)$$

shows that $P_\varepsilon f_*^\varepsilon = f_*^\varepsilon$. ∎

Remark 10.4.2. It may happen that the limit of stationary densities f_*^ε defined by equation (10.4.6) may not exist as $\varepsilon \to 0$. As a simple example consider $S: R^+ \to R^+$ given by $S(x) = \frac{1}{2}x$. In this case, the kth iterate of the Frobenius–Perron operator is given by

$$P^k g(x) = 2^k g(2^k x)$$

and, thus,

$$f_*^\varepsilon(x) = \varepsilon \sum_{k=0}^{\infty} (1 - \varepsilon)^k 2^k g(2^k x).$$

Now pick an arbitrarily small $h > 0$ and integrate f_*^ε over $[0, h]$:

$$\int_0^h f_*^\varepsilon(x)\, dx = \varepsilon \sum_{k=0}^{\infty} 2^k (1 - \varepsilon)^k \int_0^h g(2^k x)\, dx$$

$$= \varepsilon \sum_{k=0}^{\infty} (1 - \varepsilon)^k \int_0^{h2^k} g(y)\, dy.$$

For $\delta > 0$ arbitrarily small, we can always find an m such that, for all $k > m$,

$$\int_0^{h2^k} g(y)\, dy \ge \int_0^\infty g(y)\, dy - \delta = 1 - \delta$$

so

$$\int_0^h f_*^\varepsilon(x)\, dx \ge (1 - \delta)\varepsilon \sum_{k=m}^{\infty} (1 - \varepsilon)^k = (1 - \delta)(1 - \varepsilon)^m.$$

Thus, holding δ and m fixed, assume $f_*^0(x) = \lim_{\varepsilon \to 0} f_*^\varepsilon(x)$ so that

$$\int_0^h f_*^0(x)\, dx \ge 1 - \delta \qquad \text{for every } h > 0.$$

Now it follows directly that $f_*^0 \in D$ cannot exist, for, if it did, then

$$\lim_{h \to 0} \int_0^h f_*^0(x)\, dx = 0,$$

which is a contradiction. □

Theorem 10.4.1. *Let $(X, \mathcal{A}, \mu)$ be a measure space, $S: X \to X$ a nonsingular transformation, P the Frobenius–Perron operator corresponding to*

S, and P_ε the operator defined by equation (10.4.4) with unique stationary density f_*^ε given by (10.4.6). If the limit (strong or weak in L^1)

$$f_*^0(x) = \lim_{\varepsilon \to 0} f_*^\varepsilon(x)$$

exists, then f_*^0 is a stationary density for P.

Proof. Since f_*^ε is a stationary density for P_ε, we have $P_\varepsilon f_*^\varepsilon = f_*^\varepsilon$ or, more explicitly,

$$(1 - \varepsilon)Pf_*^\varepsilon + \varepsilon g = f_*^\varepsilon.$$

Under the assumption that f_*^0 exists, we immediately have $Pf_*^0 = f_*^0$, finishing the proof. ∎

Remark 10.4.3. In this context it is interesting to note that if f_*^0 exists it may depend on g. A simple example comes from $S: R \to R$ given by $S(x) = x$. Then $Pg = g$ for all $g \in D$ and

$$f_*^\varepsilon = g\varepsilon \sum_{k=0}^{\infty} (1 - \varepsilon)^k = g$$

so $f_*^0 = g$. □

Although this example shows that it is quite possible for f_*^0 to depend on g, the following theorem gives sufficient conditions for not only the existence of f_*^0 but also its value.

Theorem 10.4.2. Let $(X, \mathcal{A}, \mu)$ be a finite measure space, $S: X \to X$ a measure-preserving ergodic transformation, and P_ε the operator defined by equation (10.4.4) with the unique stationary density f_*^ε given by (10.4.6). Then $f_*^0 = \lim_{\varepsilon \to 0} f_*^\varepsilon$ exists and is given by

$$f_*^0 = 1/\mu(X).$$

Although the proof of this theorem is straightforward, we will not give it in detail. It suffices to note that the proof is very similar to those of Section 5.2 and 8.7, and the point of similarity resides in the fact that the series representations for $P^n f$ and $\hat{P}_t f$ in each of those sections have coefficients that sum to 1. Exactly the same situation occurs in the explicit representation (10.4.4) for f_*^ε since

$$\varepsilon \sum_{k=0}^{\infty} (1 - \varepsilon)^k = 1.$$

10.5   Discrete Time Systems with Constantly Applied Stochastic Perturbations

In Section 10.4 we examined the asymptotic behavior of deterministic systems with a randomly applied stochastic perturbation. We now turn our attention to deterministic systems with constantly applied stochastic perturbations. Such dynamical systems have been considered by Kifer [1974] and Boyarsky [1984], and in a physical context by Feigenbaum and Hasslacher [1982].

Specifically consider the process defined by

$$x_{n+1} = S(x_n) + \xi_n, \tag{10.5.1}$$

where $S: R^d \to R^d$ is a measurable, though not necessarily nonsingular, transformation and $\xi_0, \xi_1, \ldots$ are independent random vectors each having the same density g. We let the density of x_n be denoted by f_n, and desire a relation connecting f_{n+1} and f_n.

Assume $f_n \in D$. By (10.5.1), x_{n+1} is the sum of two independent random vectors: $S(x_n)$ and ξ_n. Note that $S(x_n)$ and ξ_n are clearly independent since, in calculating $x_1, \ldots, x_n$, we only need $\xi_0, \ldots, \xi_{n-1}$. Let $h: R^d \to R$ be an arbitrary, bounded, measurable function. It is easy to find the mathematical expectation of $h(x_{n+1})$ since, by Theorem 10.2.1,

$$E(h(x_{n+1})) = \int_{R^d} h(x) f_{n+1}(x) \, dx. \tag{10.5.2}$$

Furthermore, because of (10.5.1) and the fact that the joint density of (x_n, ξ_n) is just $f_n(y)g(z)$, we also have

$$E(h(x_{n+1})) = E(h(S(x_n) + \xi_n))$$
$$= \int_{R^d} \int_{R^d} h(S(y) + z) f_n(y) g(z) \, dy \, dz.$$

By a change of variables, this can be rewritten as

$$E(h(x_{n+1})) = \int_{R^d} \int_{R^d} h(x) f_n(y) g(x - S(y)) \, dx \, dy. \tag{10.5.3}$$

Equating (10.5.2) and (10.5.3), and using the fact that h was an arbitrary, bounded, measurable function, we immediately obtain

$$f_{n+1}(x) = \int_{R^d} f_n(y) g(x - S(y)) \, dy. \tag{10.5.4}$$

Remark 10.5.1. Our derivation of (10.5.4), though mathematically precise, is somewhat different from the usual method. Our reasons for this are threefold. First we were able to avoid the introduction of the concept of conditional probabilities. Second, the technique provides a clear proof that if $f_n(x)$ exists then $f_{n+1}(x)$ must also exist. To see this, take $h(x) = 1_A(x)$ in (10.5.3), so (10.5.3) becomes

$$\text{prob}\{x_{n+1} \in A\} = \int_A \int_{R^d} f_n(y)g(x - S(y))\,dx\,dy$$

and, thus, by the definition of density, if f_n exists then f_{n+1} also exists and is given by (10.5.4). Finally, we have introduced this method of obtaining (10.5.4) because we use it later in deriving the Fokker–Planck equation that describes the evolution of densities for continuous time systems in the presence of a stochastic process.   □

From our equation (10.5.4), we define an operator $\bar{P}: L^1 \to L^1$ by

$$\bar{P}f(x) = \int_{R^d} f(y)g(x - S(y))\,dy \qquad (10.5.5)$$

for $f \in L^1$. That $\bar{P}$ is a Markov operator is quite easy to prove. Note first that if we set $K(x,y) = g(x-S(y))$, then, for $g \in D$, K is a stochastic kernel (Section 5.7) and $\bar{P}$ is a Markov operator. Thus, in examining the behavior of the systems forming the subject of this section, we have available all of the tools developed in Section 5.7.

Remark 10.5.2. In the special case in which $d = 1$ and $S = \lambda x$, equation (10.5.5) reduces to that considered in Example 5.7.2 with $a = 1$ and $b = -\lambda$. □

However, because of the characteristics of the function g identified as a kernel, we can prove more than in Section 5.7. We start by stating and proving a result for the asymptotic periodicity of $\{P^n\}$.

Theorem 10.5.1. *Let the operator* $\bar{P}: L^1(R^d) \to L^1(R^d)$ *be defined by* (10.5.5) *and let* $g \in D$. *If there exists a Liapunov function* [*see* (5.7.8)] $V: R^d \to R^d$ *such that, with* $\alpha < 1$,

$$\int_{R^d} g(x - S(y))V(x)\,dx \leq \alpha V(y) + \beta \qquad \text{for all } y \in R^d,$$

then the operator $\bar{P}$ *is constrictive and, as a consequence, for every* $f \in L^1$ *the sequence* $\{\bar{P}^n\}$ *is asymptotically periodic.*

Proof. We will use Theorem 5.7.2 in the proof, noting that now the stochastic kernel is explicitly given by $K(x,y) = g(x - S(y))$.

We first verify that (5.7.19) holds. Since g is integrable, for every $\lambda > 0$ there is a $\delta > 0$ such that

$$\int_A g(x)\,dx < \lambda \qquad \text{for } \mu(A) < \delta.$$

In particular,

$$\int_E K(x,y)\,dx = \int_E g(x - S(y))\,dx = \int_{E-S(y)} g(x)\,dx < \lambda$$

for $\mu(E - S(y)) = \mu(E) < \delta$. Thus, (5.7.19) holds uniformly for all bounded sets B.

Further, from (10.5.5) and the assumptions of the theorem we have

$$\int_{R^d} V(x)\bar{P}f(x)\,dx = \int_{R^d} V(x)\,dx \int_{R^d} f(y)g(x - S(y))\,dy$$

$$\leq \alpha \int_{R^d} V(y)f(y)\,dy + \beta,$$

so inequality (5.7.11) also holds. Thus, by Theorem 5.7.2 we have shown that $\bar{P}$ is constrictive. ■

Theorem 10.5.1 implies that for a very broad class of transformations the addition of a stochastic perturbation will cause the limiting sequence of densities to become asymptotically periodic. For some transformations this would not be at all surprising. For example, the addition of a small stochastic perturbation to a transformation with exponentially stable periodic orbits will induce asymptotic periodicity. However it is surprising that even in a transformation S that has no particularly interesting behavior from a density standpoint, the addition of noise may result in asymptotic periodicity. We may easily illustrate this through an example on $[0, 1]$, since this makes numerical experiments feasible.

Example 10.5.1 (Lasota and Mackey, 1987). Consider the transformation

$$x_{n+1} = S(x_n) = \alpha x_n + \lambda \quad \text{mod } 1, 0 < \alpha < 1, 0 < \lambda < 1, \quad (10.5.6)$$

which is an example of a general class of transformations considered by Keener [1980]. From Keener's general results, for $0 < \alpha < 1$ there exists an uncountable set Λ such that for all $\lambda \in \Lambda$ the rotation number corresponding to (10.5.6) is irrational. As a consequence, for these λ the sequence $\{x_n\}$ is not periodic and the invariant limiting set

$$\bigcap_{k=0}^{\infty} S^k([0, 1]) \quad (10.5.7)$$

is a Cantor set. The proof of Keener's general result offers a constructive tool for numerically determining values of λ that approximate elements of Λ.

The perturbed dynamical system

$$x_{n+1} = \alpha x_n + \lambda + \xi_n \quad \text{mod } 1 \quad (10.5.8)$$

leads to an integral operator for which it is easy to verify the conditions of Theorem 5.7.2 (see Exercises 10.3 and 10.4).

For illustration we pick $\alpha = \frac{1}{2}$. Keener's results show that $\lambda = \frac{17}{30}$ is close to an element of Λ such that the invariant limiting set (10.5.7) is a Cantor set and the sequence $\{x_n\}$ is not periodic. Using the explicit transformation (10.5.8), where the ξ_n are random numbers uniformly distributed on $[0, \theta]$, in Figure 10.5.1 we show the eventual limiting behavior of the sequence $\{P^n f\}$ of densities for $\theta = \frac{1}{15}$ and an initially uniform density on $[0, 1]$. It is clear that $P^{13} f(x)$ is the same as $P^{10} f(x)$, and $P^{14} f(x)$ is identical to $P^{11} f(x)$. Thus, in this example we have a noise induced period three asymptotic periodicity.

Theorem 10.5.1 also implies that $\bar{P}$ has a stationary density f_* since this is a consequence of the spectral decomposition Theorem 5.3.3. This does not, of course, guarantee the uniqueness of f_*, but a simple assumption concerning the positivity of g will not only ensure uniqueness of f_* but also asymptotic stability. More specifically, we have the following result.

Corollary 10.5.1. *If $\bar{P}$ given by (10.5.5) satisfies the conditions of Theorem 10.5.1, and $g(x) > 0$, then $\{P^n\}$ is asymptotically stable.*

Proof. We start with the observation that for every fixed x the product $g(x - S(y))P^{n-1} f(y)$, considered as a function of y, does not vanish everywhere. As a consequence,

$$\bar{P}^n f(x) = \int_{R^d} g(x - S(y))\bar{P}^{n-1} f(y)\, dy > 0 \qquad \text{for all } x \in R^d, n \geq 1, f \in D.$$

The asymptotic stability of $\{\bar{P}^n\}$ is thus proved by applying Theorem 5.6.1. ∎

It is interesting that we may also prove the uniqueness of a stationary density f_* of $\bar{P}$ defined by (10.5.5) without the rather restrictive conditions required by Corollary 10.5.1.

Theorem 10.5.2. *Let the operator $\bar{P}: L^1 \to L^1$ be defined by equation (10.5.5) and let $g \in D$. If $g(y) > 0$ for all $y \in R^d$ and if a stationary density f_* for $\bar{P}$ exists, then f_* is unique.*

Proof. Assume there are two stationary densities for $\bar{P}$, namely, f_1 and f_2. Set $f = f_1 - f_2$, so we clearly have

$$\bar{P}f = f. \tag{10.5.9}$$

We may write $f = f^+ - f^-$ by definition, so that, if $f_1 \neq f_2$, then neither f^+ nor f^- are zero. Since $\bar{P}f^+ = f^+$ (by Proposition 3.1.3), from (10.5.5) we have

$$\bar{P}f^+(x) = \int_{R^d} g(x - S(y))f^+(y)\, dy \tag{10.5.10}$$

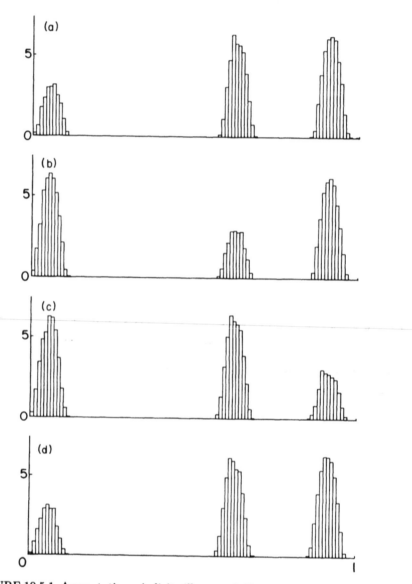

FIGURE 10.5.1. Asymptotic periodicity illustrated. Here we show the histograms obtained after iterating 10^4 initial points uniformly distributed on $[0,1]$ with $\alpha = \frac{1}{2}$, $\lambda = \frac{17}{30}$, and $\theta = \frac{1}{15}$ in equation (10.5.8). In (a) $n = 10$; (b) $n = 11$; (c) $n = 12$; and (d) $n = 13$. The correspondence of the histograms for $n = 10$ and $n = 13$ indicates that, with these parameter values, numerically the sequence of densities has period 3.

and similarly for $\bar{P}f^-$. Since f^+ is not identically zero and g is strictly positive, the integral in (10.5.10) is a nonzero function for every x, and, thus, $\bar{P}f^+(x) > 0$ for all x. Clearly, too, $\bar{P}f^-(x) > 0$ for all x and, thus, the supports of $\bar{P}f^+$ and $\bar{P}f^-$ are not disjoint. By Proposition 3.1.2, then, we must have $\|\bar{P}f\| < \|f\|$, which contradicts equality (10.5.9). Thus, f_1 and f_2 must be identical almost everywhere if they exist. ∎

Remark 10.5.3. It certainly may happen that there is no solution to $\bar{P}f = f$ in D. As a simple example, consider $S(x) = x$ for all $x \in R$. Take g to be the Gaussian density

$$g(x) = \frac{1}{\sqrt{2\pi}} \exp(-x^2/2),$$

so the operator $\bar{P}$ defined in (10.5.5) becomes

$$\bar{P}f(x) = \int_{-\infty}^{\infty} \frac{1}{\sqrt{2\pi}} \exp[-(x-y)^2/2] f(y)\, dy.$$

Note that $\bar{P}f(x)$ is simply the solution $u(t, x)$ of the heat equation (7.4.13) with $\sigma^2 = 1$ at time $t = 1$, assuming an initial condition $u(0, y) = f(y)$. Since this solution is given by a semigroup of operators [cf. equations (7.4.11) and (7.9.9)], it can be shown that

$$\bar{P}^n f(x) = \frac{1}{\sqrt{2\pi n}} \int_{-\infty}^{\infty} \exp[-(x-y)^2/2n] f(y)\, dy$$

$$\leq \frac{1}{\sqrt{2\pi n}} \int_{-\infty}^{\infty} f(y)\, dy = \frac{1}{\sqrt{2\pi n}}.$$

Thus $\bar{P}^n f$ converges uniformly to zero as $n \to \infty$ for all $f \in D$, and there is no solution to $\bar{P}f = f$. □

If these conditions for the existence and uniqueness of stationary densities of $\bar{P}$ are strengthened somewhat, we can prove that $\{\bar{P}^n\}$ is asymptotically stable. In fact from our results of Theorem 5.7.1, we have the following corollary.

Corollary 10.5.2. Let the operator $\bar{P}: L^1 \to L^1$ be defined by equation (10.5.5) and let $g \in D$. If there is a Liapunov function $V: R^d \to R$ such that

$$\int_{R^d} g(x - S(y)) V(x)\, dx \leq \alpha V(y) + \beta \qquad \text{for all } y \in R^d, \qquad (10.5.11)$$

for some nonnegative constants α, β, $\alpha < 1$, and

$$\int_{R^d} \inf_{|y| \leq r} g(x - S(y))\, dx > 0 \qquad (10.5.12)$$

for every $r > 0$, then $\{\bar{P}^n\}$ is asymptotically stable.

Remark 10.5.4. Note that condition (10.5.12) is automatically satisfied if $g: R^d \rightarrow R$ is positive and continuous and $S: R^d \rightarrow R^d$ is continuous because

$$\inf_{|y| \leq r} g(x - S(y)) = \min_{|y| \leq r} g(x - S(y)) > 0$$

for every $x \in R^d$. $\square$

Example 10.5.2. Consider a point moving through R^d whose trajectory is determined by

$$x_{n+1} = S(x_n) + \xi_n,$$

where $S: R^d \rightarrow R^d$ is continuous and satisfies

$$|S(x)| \leq \lambda |x|, \qquad \text{for } |x| \geq M, \qquad (10.5.13)$$

where $\lambda < 1$ and $M > 0$ are given constants. Assume that $\xi_0, \xi_1, \ldots$ are independent random variables with the same density g, which is continuous and positive, and such that $E(\xi_n)$ exists. Then $\{\bar{P}^n\}$ defined by (10.5.5) is asymptotically stable.

To show this, it is enough to confirm that condition (10.5.11) is satisfied. Set $V(x) = |x|$, so

$$\int_{R^d} g(x - S(y))V(x)\, dx = \int_{R^d} g(x - S(y))|x|\, dx$$

$$= \int_{R^d} g(x)|x + S(y)|\, dx$$

$$\leq \int_{R^d} g(x)(|x| + |S(y)|)\, dx$$

$$= |S(y)| + \int_{R^d} g(x)|x|\, dx.$$

From (10.5.13) we also have

$$|S(y)| \leq \lambda |y| + \max_{|x| \leq M} |S(x)|$$

so that

$$\int_{R^d} g(x - S(y))V(x)\, dx \leq \lambda |y| + \max_{|x| \leq M} |S(x)| + \int_{R^d} g(x)|x|\, dx.$$

Thus, since $E(\xi_n)$ exists, equation (10.5.11) is satisfied with $\alpha = \lambda$ and

$$\beta = \int_{R^d} g(x)|x|\, dx + \max_{|x| \leq M} |S(x)|. \quad \square$$

It is important to note that throughout it has not been necessary to require that S be a nonsingular transformation. Indeed, one of the goals of this section was to demonstrate that the addition of random perturbations to a singular transformation may lead to interesting results.

However, if S is nonsingular, then the Frobenius–Perron operator P corresponding to S exists and allows us to rewrite (10.5.5) in an alternate form that will be of use in the following section. By definition,

$$\bar{P}f(x) = \int_{R^d} g(x - S(y))f(y)\, dy.$$

Assume S is nonsingular, therefore the Frobenius–Perron and Koopman operators corresponding to S exist. Let $h_x(y) = g(x - y)$, so we can write $\bar{P}f$ as

$$\bar{P}f(x) = \int_{R^d} h_x(S(y))f(y)\, dy = \langle f, Uh_x \rangle = \langle Pf, h_x \rangle,$$

or, more explicitly,

$$\bar{P}f(x) = \int_{R^d} g(x - y)Pf(y)\, dy. \tag{10.5.14}$$

By a change of variables, (10.5.14) may also be written as

$$\bar{P}f(x) = \int_{R^d} g(y)Pf(x - y)\, dy. \tag{10.5.15}$$

Remark 10.5.5. Observe that for $d = 1$, equations (10.5.14) and (10.5.15) could also be obtained as an immediate consequence of equation (10.1.7) applied to equation (10.5.1) since ξ_n and $S(x_n)$ are independent. $\square$

10.6 Small Continuous Stochastic Perturbations of Discrete Time Systems

This section examines the behavior of the system

$$x_{n+1} = S(x_n) + \varepsilon\xi_n, \qquad \varepsilon > 0, \tag{10.6.1}$$

where $S: R^d \to R^d$ is measurable and nonsingular. As in the preceding section, we assume the ξ_n to be independent random variables each having the same density g.

Since the variables $\varepsilon\xi_n$ have the density $(1/\varepsilon)g(x/\varepsilon)$, see Remark 10.1.1, equation (10.5.15) takes the form

$$P_\varepsilon f(x) = \frac{1}{\varepsilon}\int_{R^d} g\left(\frac{y}{\varepsilon}\right)Pf(x - y)\, dy \tag{10.6.2}$$

and gives the recursive relation

$$f_{n+1} = P_\epsilon f_n \tag{10.6.3}$$

that connects successive densities f_n of x_n.

The operator P_ϵ can also be written, via a change of variables, as

$$P_\epsilon f(x) = \int_{R^d} g(y) P f(x - \epsilon y) \, dy. \tag{10.6.4}$$

Since

$$\int_{R^d} g(y) P f(x) \, dy = P f(x),$$

we should expect that in some sense $\lim_{\epsilon \to 0} P_\epsilon f(x) = P f(x)$. To make this more precise, we state the following theorem.

Theorem 10.6.1. *For the system defined by equation (10.6.1)*

$$\lim_{\epsilon \to 0} \| P_\epsilon f - P f \| = 0, \qquad \text{for all } f \in L^1,$$

where P is the Frobenius–Perron operator corresponding to S and P_ϵ is given by (10.6.4).

Proof. Since P and P_ϵ are linear we may restrict ourselves to $f \in D$. Write

$$P f(x) = \int_{R^d} g(y) P f(x) \, dy,$$

then

$$P_\epsilon f(x) - P f(x) = \int_{R^d} g(y) [P f(x - \epsilon y) - P f(x)] \, dy.$$

Pick an arbitrarily small $\delta > 0$. Since g and Pf are both integrable functions on R^d, there must exist an $r > 0$ such that

$$\int_{|y| \geq r} g(y) \, dy \leq \frac{\delta}{4} \quad \text{and} \quad \int_{|x| \geq r/2} P f(x) \, dx \leq \frac{\delta}{4}.$$

To calculate the norm of $P_\epsilon f - P f$,

$$\| P_\epsilon f - P f \| \leq \int_{R^d} \int_{R^d} g(y) |P f(x - \epsilon y) - P f(x)| \, dx \, dy,$$

we split the integral into two parts,

$$\| P_\epsilon f - P f \| \leq I_1 + I_2,$$

where

$$I_1 = \int_{R^d} \int_{|y| \leq r} g(y) |P f(x - \epsilon y) - P f(x)| \, dx \, dy,$$

and

$$I_2 = \int_{R^d} \int_{|y| \geq r} g(y) |Pf(x - \varepsilon y) - Pf(x)| \, dx \, dy.$$

We consider each in turn.

With respect to I_1, note that, since the function Pf is integrable, by Corollary 5.1.1, we may assume

$$\int_{R^d} |Pf(x - \varepsilon y) - Pf(x)| \, dx \leq \frac{\delta}{2}$$

for $\varepsilon \leq \varepsilon_0$ with ε_0 sufficiently small. Hence

$$I_1 \leq \frac{\delta}{2} \int_{|y| \leq r} g(y) \, dy \leq \frac{\delta}{2} \int_{R^d} g(y) \, dy = \frac{\delta}{2}.$$

In examining I_2, we use the triangle inequality to write

$$I_2 \leq \int_{R^d} \int_{|y| \geq r} g(y) Pf(x - \varepsilon y) \, dx \, dy + \int_{R^d} \int_{|y| \geq r} g(y) Pf(x) \, dx \, dy.$$

Change the variables in the first integral to $v = y$ and $z = x - \varepsilon y$, then

$$\int_{R^d} \int_{|y| \geq r} g(y) Pf(x - \varepsilon y) \, dx \, dy = \int_{R^d} \int_{|v| \geq r} g(v) Pf(z) \, dz \, dv$$

$$= \int_{|v| \geq r} g(v) \, dv \leq \frac{\delta}{4}.$$

Further, we also have

$$\int_{R^d} \int_{|y| \geq r} g(y) Pf(x) \, dx \, dy \leq \frac{\delta}{4}$$

so that $I_2 \leq \delta/2$.

Thus

$$\|P_\varepsilon f - Pf\| \leq \delta \qquad \text{for any } \varepsilon \leq \min\left(\tfrac{1}{2}, \varepsilon_0\right),$$

that is,

$$\lim_{\varepsilon \to 0} \|P_\varepsilon f - Pf\| = 0 \quad \blacksquare$$

As an immediate consequence of Theorem 10.6.1 we have the following corollary.

Corollary 10.6.1. *Suppose that S and g are given and that for every small ε, $0 < \varepsilon < \varepsilon_0$, the operator P_ε, defined by (10.6.4), has a stationary density f_ε. If the limit*

$$f_* = \lim_{\varepsilon \to 0} f_\varepsilon$$

exists then f_* is a stationary density for the Frobenius–Perron operator corresponding to S.

Proof. Write

$$P_\varepsilon f_* = f_\varepsilon + P_\varepsilon (f_* - f_\varepsilon).$$

Since P_ε is contractive,

$$\|P_\varepsilon (f_* - f_\varepsilon)\| \leq \|f_* - f_\varepsilon\|.$$

Thus $f_\varepsilon + P_\varepsilon (f_* - f_\varepsilon) \to f_*$ as $\varepsilon \to 0$ and, as a consequence, $P_\varepsilon f_* \to f_*$. However, Theorem 10.6.1 also tells us that $P_\varepsilon f_* \to P f_*$, so $P f_* = f_*$. ∎

10.7 Discrete Time Systems with Multiplicative Perturbations

Up to now in this chapter we have confined our attention to situations in which a discrete time system is perturbed in an additive fashion, for example, (10.4.1), (10.5.1), and (10.6.1). We now turn to a consideration of the influence of perturbations that appear in a multiplicative way. Since in many applied problems this arises because of noise in parameters, it is also known as parametric noise.

Specifically, we examine a process

$$x_{n+1} = \xi_n S(x_n) \tag{10.7.1}$$

where $S: R^+ \to R^+$ is continuous and positive a.e. and, as before, the ξ_n are independent random variables, each distributed with the same density g. We denote the density of x_n by f_n, and our first task in the study of (10.7.1) will be to derive a relation connecting f_{n+1} and f_n.

Using exactly the same approach employed in Section 10.5, let $h: R^+ \to R^+$ be an arbitrary bounded and Borel measurable function. The expectation of $h(x_{n+1})$ is given by

$$E(h(x_{n+1})) = \int_0^\infty h(x) f_{n+1}(x)\, dx. \tag{10.7.2}$$

However, using (10.7.1) we also have

$$
\begin{aligned}
E(h(x_{n+1})) &= E(\xi_n S(x_n)) \\
&= \int_0^\infty \int_0^\infty h(zS(y)) f_n(y) g(z)\, dy\, dz \\
&= \int_0^\infty \int_0^\infty h(x) f_n(y) g\left(\frac{x}{S(y)}\right) \frac{1}{S(y)}\, dx\, dz, \tag{10.7.3}
\end{aligned}
$$

where we used a change of variables $z = x/S(y)$ in passing from the second to third lines of (10.7.3). Equating (10.7.2) and (10.7.3), and using the fact that h was arbitrary by assumption, we arrive at

$$f_{n+1}(x) = \int_0^\infty f_n(y) g\left(\frac{x}{S(y)}\right) \frac{1}{S(y)} \, dy, \qquad (10.7.4)$$

which is the desired relation.

From (10.7.4) we may also write $f_{n+1} = \bar{P} f_n$ where the operator $\bar{P}$, given by

$$\bar{P} f(x) = \int_0^\infty f(y) g\left(\frac{x}{S(y)}\right) \frac{1}{S(y)} \, dy, \qquad (10.7.5)$$

is a Markov operator with a stochastic kernel

$$K(x, y) = g\left(\frac{x}{S(y)}\right) \frac{1}{S(y)}. \qquad (10.7.6)$$

Our first result is related to the generation of asymptotic periodicity by multiplicative noise. Though originally formulated by Horbacz [1989a], the proof we give is different from the original.

Theorem 10.7.1. *Let the Markov operator* $\bar{P}: L^1(R^+) \to L^1(R^+)$ *be defined by* (10.7.5). *Assume that* $g \in D$,

$$0 < S(x) \le \alpha x + \beta \qquad \text{for } x \ge 0, \qquad (10.7.7)$$

and

$$\alpha m < 1 \qquad \text{with } m = \int_0^\infty x g(x) \, dx, \qquad (10.7.8)$$

where α *and* β *are nonnegative constants. Then* $\bar{P}$ *is constrictive. As a consequence the sequence* $\{P^n\}$ *is asymptotically periodic.*

Proof. Once again we will employ Theorem 5.7.2 in the proof. We first show that (5.7.11) holds for the kernel (10.7.6) with $V(x) = x$. We have

$$\int_0^\infty x P f(x) \, dx = \int_0^\infty x \, dx \int_0^\infty g\left(\frac{x}{S(y)}\right) \frac{1}{S(y)} f(y) \, dy$$

$$= \int_0^\infty f(y) \, dy \int_0^\infty g\left(\frac{x}{S(y)}\right) \frac{x}{S(y)} \, dx.$$

Using the change of variables $z = x/S(y)$ and then (10.7.7) we obtain

$$\int_0^\infty x P f(x) \, dx = \int_0^\infty f(y) S(y) \, dy \int_0^\infty z g(z) \, dz$$

$$= m \int_0^\infty f(y) S(y) \, dy \le \alpha m \int_0^\infty y f(y) \, dy + \beta m.$$

Thus, we have verified inequality (5.7.11).

We next show that the kernel $K(x,y)$ given by (10.7.6) satisfies inequality (5.7.19) of Theorem 5.7.2. Fix an arbitrary positive $\lambda < 1$ and choose a bounded set $B \subset R^+$. Since g is uniformly integrable there must be a $\delta_1 > 0$ such that

$$\int_F g(x)\, dx \leq \lambda \qquad \text{for } \mu(F) < \delta_1.$$

Define

$$\delta = \delta_1 \inf_{y \in B} S(y).$$

Then for $\mu(E) < \delta$ we have $\mu(E/S(y)) < \delta_1$ and

$$\int_E K(x,y)\, dx = \int_E g\left(\frac{x}{S(y)}\right) \frac{1}{S(y)}\, dx$$
$$= \int_{E/S(y)} g(x)\, dx \leq \lambda \qquad \text{for } y \in B \text{ and } \mu(E) < \delta,$$

and all of the conditions of Theorem 5.7.2 are satisfied. Thus $\bar{P}$ is constrictive and a simple application of the spectral decomposition Theorem 5.3.1 finishes the proof. ∎

We close with a second theorem concerning asymptotic stability induced by multiplicative noise.

Theorem 10.7.2. *If the Markov operator $\bar{P}: L^1(R^+) \to L^1(R^+)$ defined by (10.7.5) satisfies the conditions of Theorem 10.7.1 and, in addition, $g(x) > 0$, then $\{P^n\}$ is asymptotically stable.*

Proof. Note that, for fixed x, the quantity

$$g\left(\frac{x}{S(y)}\right) \frac{1}{S(y)} \bar{P}^{n-1} f(y),$$

as a function of y, does not vanish everywhere. Consequently,

$$\bar{P}^n f(x) = \int_0^\infty g\left(\frac{x}{S(y)}\right) \frac{1}{S(y)} \bar{P}^{n-1} f(y)\, dy > 0$$
$$\text{for all } x \in R^+, n \geq 1, f \in D,$$

and Theorem 5.6.1 finishes the proof of the asymptotic stability of $\{P^n\}$. ∎

Theorems 10.7.1 and 10.7.2 illustrate the behaviors that may be induced by multiplicative noise in discrete time systems. A number of other results concerning asymptotic periodicity and asymptotic stability induced by multiplicative noise may be proved, but rather than giving these we refer the reader to Horbacz [1989a,b].

Exercises

10.1. Let $\xi_n: \Omega \to R^{d_n}$, $n = 1, 2, \ldots$, be a sequence of independent random vectors, and let $\varphi_n: R^{d_n} \to R^{d_n}$ be a sequence of Borel measurable functions. Prove that the random vectors $\eta_n(\omega) = \varphi_n(\xi_n(\omega))$ are independent.

10.2. Replace inequality (10.7.7) in Theorem 10.7.1 by

$$0 \le S(x) \le \alpha x, \qquad \alpha < 1,$$

and show that in this case the sequence $\{P^n\}$ is sweeping to zero. Formulate an analogous sufficient condition for sweeping to $+\infty$.

10.3. Let $S: [0, 1] \to [0, 1]$ be a measurable transformation and let $\{\xi_n\}$ be a sequence of independent random variables each having the same density g. Consider the process defined by

$$x_{n+1} = S(x_n) + \xi_n \qquad (\text{mod } 1),$$

and denote by f_n the density of the distribution of x_n. Find an explicit expression for the Markov operator $P: L^1([0, 1]) \to L^1([0, 1])$ such that $f_{n+1} = Pf_n$.

10.4. Under the assumptions of the previous exercise, show that $\{P^n\}$ is asymptotically periodic. Find sufficient conditions for the asymptotic stability of $\{P^n\}$.

10.5. Consider the dynamical system (10.7.1) on the unit interval. Assume that $S: [0, 1] \to [0, 1]$ is continuous and that $\xi_n: \Omega \to [0, 1]$ are independent random variables with the same density $g \in D([0, 1])$. Introduce the corresponding Markov operator and reformulate Theorems 10.7.1 and 10.7.2 in this case.

10.6. As a specific example of the dynamical system (10.7.1) on the unit interval (see the previous exercise), consider the quadratic map $S(x) = \alpha x(1 - x)$ and ξ_n having a density $g \in D([0, 1])$ such that

$$0 \le g(x) \le Kx^r, \qquad 0 \le x \le 1.$$

Show that for every $\alpha \in (1, 4]$ there is a $K > 0$ and $r > 0$ such that $\{P^n\}$ is asymptotically stable (Horbacz, 1989a).

10.7. Consider the system

$$x_{n+1} = S(x_n) + \xi_n$$

with additive noise. Note that with the definitions $y = e^x$, $T = e^S$, and $\eta = e^\xi$ this can be rewritten in the alternative form

$$y_{n+1} = \eta_n T(\ln y_n)$$

as if there were multiplicative noise. Using this transformation, discuss the results for multiplicative noise that can be obtained from the theorems and corollaries of Section 10.5.

10.8. As a counterpoint to the previous examples, note that if

$$x_{n+1} = \xi_n S(x_n)$$

and we set $y = \ln x$, $\eta = \ln \xi$, and $T = \ln S$, then

$$y_{n+1} = T(e^{y_n}) + \eta_n$$

results. Examine the results for additive noise that can be obtained using this technique on the theorems of Section 10.7 pertaining to multiplicative noise.

11

Stochastic Perturbation of Continuous Time Systems

In this chapter continuous time systems in the presence of noise are considered. This leads us to examine systems of stochastic differential equations and to a derivation of the forward Fokker–Planck equation, describing the evolution of densities for these systems. We close with some results concerning the asymptotic stability of solutions to the Fokker–Planck equation.

11.1 One-Dimensional Wiener Processes (Brownian Motion)

In this and succeeding sections of this chapter, we turn to a consideration of continuous time systems with stochastic perturbations. We are specifically interested in the behavior of the system

$$\frac{dx}{dt} = b(x) + \sigma(x)\xi, \tag{11.1.1}$$

where $\sigma(x)$ is the amplitude of the perturbation and $\xi = dw/dt$ is known as a "white noise" term that may be considered to be the time derivative of a Wiener process. The system (11.1.1) is the continuous time analog of the discrete time problem with a constantly applied stochastic perturbation considered in Section 10.5.

The consideration of continuous time problems such as (11.1.1) will offer new insight into the possible behavior of systems, but at the expense of introducing new concepts and techniques. Even though the remainder of this chapter is written to be self-contained, it does not constitute an exhaustive treatment of stochastic differential equations such as (11.1.1). A definitive treatment of this subject may be found in Gikhman and Skorokhod [1969].

In this section and the material following, we will denote stochastic processes by $\{\xi(t)\}$, $\{\eta(t)\},\dots$ as well as $\{\xi_t\}$, $\{\eta_t\},\dots$, depending on the situation. Remember that in this notation $\xi(t)$ or ξ_t denote, for fixed t, a random variable, namely, a measurable function $\xi_t\colon \Omega \to R$. Thus $\xi(t)$ and ξ_t, are really abbreviations for $\xi(t,w)$ and $\xi_t(w)$, respectively. The symbol ξ will be reserved for white noise stochastic processes (to be described later), whereas η will be used for other stochastic processes.

Let a probability space $(\Omega, \mathcal{F}, \text{prob})$ be given. We start with a definition.

Definition 11.1.1. A stochastic process $\{\eta(t)\}$ is called **continuous** if, for almost all w (except for a set of probability zero), the sample path $t \to \eta(t,w)$ is a continuous function.

A Wiener process can now be defined as follows.

Definition 11.1.2. A one-dimensional normalized Wiener process (or **Brownian motion**) $\{w(t)\}_{t\geq 0}$ is a continuous stochastic process with independent increments such that

(a) $w(0) = 0$; and

(b) for every s,t, $0 \leq s < t$, the random variable $w(t) - w(s)$ has the Gaussian density

$$g(t-s,x) = \frac{1}{\sqrt{2\pi(t-s)}} \exp[-x^2/2(t-s)]. \qquad (11.1.2)$$

Figure 11.1.1a shows a sample path for a process approximating a Wiener process.

It is clear that a Wiener process has stationary increments since $w(t) - w(s)$ and $w(t+t') - w(s+t')$ have the same density function (11.1.2). Further, since $w(t) = w(t) - w(0)$, the random variable $w(t)$ has the density

$$g(t,x) = \frac{1}{\sqrt{2\pi t}} \exp(-x^2/2t). \qquad (11.1.3)$$

An easy calculation shows

$$E((w(t)-w(s))^n) = \frac{1}{\sqrt{2\pi(t-s)}} \int_{-\infty}^{\infty} x^n \exp[-x^2/2(t-s)]\, dx$$

$$= \begin{cases} 1.3\cdots(n-1)(t-s)^{n/2} & \text{for } n \text{ even} \\ 0 & \text{for } n \text{ odd} \end{cases} \qquad (11.1.4)$$

and thus, in particular,

$$E(w(t) - w(s)) = 0 \qquad (11.1.5)$$

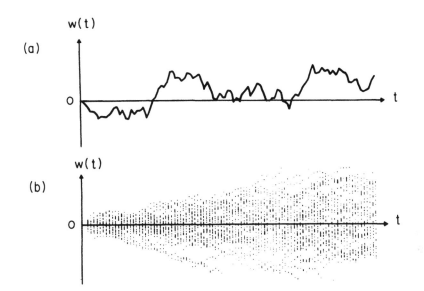

FIGURE 11.1.1. A process approximating a Wiener process. In (a) we show a single sample path for this process. In (b) we superimpose the points of many sample paths to show the progressive increase in the variance.

and

$$D^2(w(t) - w(s)) = (t - s). \tag{11.1.6}$$

This last equation demonstrates that the variance of a Wiener process increases linearly with t.

Remark 11.1.1. The adjective **normalized** in our definition of the Wiener process is used because $D^2(w(t)) = t$. It is clear that multiplication of a normalized Wiener process by a constant $\sigma > 0$ again yields a process with properties similar to those of Definition 11.1.2, but now with the density

$$\frac{1}{\sqrt{2\pi\sigma^2 t}} \exp(-x^2/2\sigma^2 t)$$

and with the variance $\sigma^2 t$. These processes are also called Wiener processes. From this point on we will always refer to a normalized Wiener process as a Wiener process. □

In Figure 11.1.1b we have drawn a number of sample paths for a process approximating a Wiener process. Note that as time increases they all seem to be bounded by a convex envelope. This is due to the fact that the standard deviation of a Wiener process, from (11.1.6), increases as $\sqrt{t}$, that is,

$$[D^2(w(t))]^{1/2} = \sqrt{t}.$$

The highly irregular behavior of these individual trajectories is such that magnification of any part of the trajectory by a factor α^2 in the time direction and α in the x direction yields a picture indistinguishable from the original trajectory. This procedure can be repeated as often as one wishes, and, indeed, the sample paths of a Wiener process are fractal curves [Mandelbrot, 1977]. To obtain some insight into the origin of this behavior consider the absolute value of the differential quotient

$$\left|\frac{\Delta w}{\Delta t}\right| = \frac{1}{|\Delta t|}|w(t_0 + \Delta t) - w(t_0)|.$$

We have

$$E\left(\left|\frac{\Delta w}{\Delta t}\right|\right) = \frac{1}{|\Delta t|}E(|w(t_0 + \Delta t) - w(t_0)|)$$

and, since the density of $w(t_0 + \Delta t) - w(t_0)$ is given by (11.1.3),

$$E(|w(t_0 + \Delta t) - w(t_0)|) = \frac{1}{\sqrt{2\pi\Delta t}}\int_{-\infty}^{\infty}|x|\exp(-x^2/2\Delta t)dx$$

$$= \sqrt{2\Delta t/\pi}$$

or

$$E\left(\left|\frac{\Delta w}{\Delta t}\right|\right) = \sqrt{\frac{2}{\pi}}\frac{1}{\sqrt{\Delta t}}.$$

Thus the mathematical expectation of $|\Delta w/\Delta t|$ goes to infinity, with a speed proportional to $(\Delta t)^{-1/2}$, when $|\Delta t| \to 0$. This is the origin of the irregular behavior shown in Figure 11.1.1.

Extending the foregoing argument, it can be proved that the sample paths of a Wiener process are not differentiable at any point almost surely. Thus, the white noise term $\xi = dw/dt$ in (11.1.1) does not exist as a stochastic process. However, since we do wish ultimately to consider (11.1.1) with such a perturbation, we must inquire how this can be accomplished. As shown in following sections, this is simply done by formally integrating (11.1.1) and treating the resulting system,

$$x(t) = \int_0^t b(x(s))\, ds + \int_0^t \sigma(x(s))\, dw(s) + x^0.$$

However, this approach leads to the new problem of defining what the integrals on the right-hand side mean, which will be dealt with in Section 11.3.

To obtain further insight into the nature of the process $w(t)$, examine the alternative sequence $\{z_n\}$ of processes, defined by

$$z_n(t) = w(t_{i-1}^n) + \frac{t - t_{i-1}^n}{t_i^n - t_{i-1}^n}[w(t_i^n) - w(t_{i-1}^n)] \qquad \text{for } t \in [t_{i-1}^n, t_i^n],$$

where $t_i^n = i/n$, $n = 1, 2, \ldots$, $i = 0, 1, 2, \ldots$. In other words, z_n is obtained by sampling the Wiener process $w(t)$ at times t_i^n and then applying a linear interpolation between t_i^n and t_{i+1}^n. Any sample path of the process $\{z_n(t)\}$ is differentiable, except at the points t_i^n, and the derivative $\eta_n = z_n'$ is given by

$$\eta_n(t) = n[w(t_i^n) - w(t_{i-1}^n)], \qquad \text{for } t \in (t_{i-1}^n, t_i^n).$$

The process $\eta_n(t)$ is piecewise constant. The heights of the individual segments are independent, have a mean value zero, and variance $D^2 \eta_n(t) = n$. Thus, the variance grows linearly with n. If we look at this process approximating white noise, we see that it consists of a sequence of independent impulses of width $(1/n)$ and variance n. For very large n we will see peaks of almost all possible sizes uniformly spread along the t-axis.

Note that the random variable $z_n(t)$ for fixed t and large n is the sum of many independent increments. Thus the density of $z_n(t)$ must be close to a Gaussian by the central limit theorem. The limiting process $w(t)$ will, therefore, also have a Gaussian density, which is why we assumed that $w(t)$ had a Gaussian density in Definition 11.1.2.

Historically, Wiener processes (or Brownian motion) first became of interest because of the findings of the English biologist Brown, who observed the microscopic movement of pollen particles in water due to the random collisions of water molecules with the particles. The impulses coming from these collisions are almost ideal realizations of the process of white noise, somewhat similar to our process $\eta_n(t)$ for large n.

In other applications, however, much slower processes are admitted as "white noise" perturbations, for example, waves striking the side of a large ship or the influence of atmospheric turbulence on an airplane. In the example of the ship, the reason that this assumption is a valid approximation stems from the fact that waves of quite varied energies strike both sides of the ship almost independently with a frequency much larger than the free oscillation frequency of the ship.

Example 11.1.1. Having defined a one-dimensional Wiener process $\{w(t)\}_{t \geq 0}$, it is rather easy to construct an exact, continuous time, semi-dynamical system that corresponds to the partial differential equation

$$\frac{\partial u}{\partial t} + s \frac{\partial u}{\partial s} = \frac{1}{2} u. \qquad (11.1.7)$$

Our arguments follow those of Rudnicki [1985], which generalize results of Lasota [1981], Brunovsky [1983], and Brunovsky and Komornik [1984].

The first step in this process is to construct the Wiener measure. Let X be the space of all continuous functions $x: [0, 1] \to R$ such that $x(0) = 0$. We are going to define some special subsets of X that are called cylinders. Thus, given a sequence of real numbers,

$$0 < s_1 < \cdots < s_n \leq 1,$$

and a sequence of Borel subsets of R,

$$A_1, \ldots, A_n,$$

we define the corresponding **cylinder** by

$$C(s_1, \ldots, s_n; A_1, \ldots, A_n) = \{x \in X : x(s_i) \in A_i, i = 1, \ldots, n\}. \quad (11.1.8)$$

Thus the cylinder defined by (11.1.8) is the set of all functions $x \in X$ pasing through the set A_i at s_i (see Figure 11.1.2). The **Wiener measure** μ_w of the cylinders (11.1.8) is defined by

$$\mu_w(C(s_1, \ldots, s_n; A_1, \ldots, A_n))$$
$$= \text{prob}\{w(s_1) \in A_1, \ldots, w(s_n) \in A_n\}. \quad (11.1.9)$$

To derive an explicit formula for μ_w, consider a transformation $y = F(x)$ of R^n into itself given by

$$y_1 = x_1, y_2 = x_2 - x_1, \ldots, y_n = x_n - x_{n-1} \quad (11.1.10)$$

and set $A = A_1 \times \cdots \times A_n$. Then the condition

$$(w(s_1), \ldots, w(s_n)) \in A$$

is equivalent to the requirement that the random vector

$$(w(s_1), w(s_2) - w(s_1), \ldots, w(s_n) - w(s_{n-1})) \quad (11.1.11)$$

belong to $F(A)$. Since $\{w(t)\}_{t \geq 0}$ is a random process with independent increments, the density function of the random vector (11.1.11) is given by

$$g(s_1, y_1)g(s_2 - s_1, y_2), \ldots, g(s_n - s_{n-1}, y_n),$$

where, by the definition of the Wiener process [see equation (11.1.3)],

$$g(s, y) = \frac{1}{\sqrt{2\pi s}} \exp(-y^2/2s). \quad (11.1.12)$$

Thus we have

$$\text{prob}\{w(s_1) \in A_1, \ldots, w(s_n) \in A_n\}$$

$$= \int \cdots \int_{F(A)} g(s_1, y_1)g(s_2 - s_1, y_2) \cdots g(s_n - s_{n-1}, y_n) dy_1 \cdots dy_n.$$

Using the variables defined in (11.1.10), this becomes

$$\text{prob}\{w(s_1) \in A_1, \ldots, w(s_n) \in A_n\}$$

$$= \int_{A_1} \cdots \int_{A_n} g(s_1, x_1)g(s_2 - s_1, x_2 - x_1)$$
$$\cdots g(s_n - s_{n-1}, x_n - x_{n-1}) dx_1 \cdots dx_n.$$

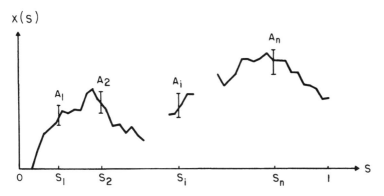

FIGURE 11.1.2. Schematic representation of implications of the cylinder definition [equation (11.1.8).]

By combining this expression with equations (11.1.9) and (11.1.12), we obtain the famous formula for the Wiener measure:

$$\mu_w(C(s_1, \ldots, s_n; A_1, \ldots, A_n))$$
$$= \frac{1}{\sqrt{(2\pi)^n (s_1 - s_0) \cdots (s_n - s_{n-1})}}$$
$$\cdot \int_{A_1} \cdots \int_{A_n} \exp\left[-\frac{1}{2} \sum_{k=1}^{n} \frac{(x_k - x_{k-1})^2}{s_k - s_{k-1}}\right] dx_1 \cdots dx_n. \quad (11.1.13)$$

(We assume, for simplicity, that $s_0 = x_0 = 0$.)

To extend the definition of μ_w, we can define the σ-algebra $\mathcal{A}$ to be the smallest σ-algebra of the subsets of X that contains all the cylinders defined by (11.1.8) for arbitrary n. By definition, the Wiener measure μ_w is the (unique) extension of μ_w, given by (11.1.13) on cylinders, to the entire σ-algebra $\mathcal{A}$. The proof that μ_w given by (11.1.13) on cylinders can be extended to the entire σ-algebra is technically difficult, and we omit it. However, note that if a Wiener process $\{w(t)\}_{t \geq 0}$ is given, then it is a direct consequence of our construction of the Wiener measure for cylinders that

$$\mu_w(E) = \text{prob}(\tilde{w} \in E) \qquad \text{for } E \in \mathcal{A}, \qquad (11.1.14)$$

where $\tilde{w}$ is the restriction of w to the interval $[0, 1]$. (Incidentally, from this equation, it also follows that the assumption that a Wiener process $\{w(t)\}_{t \geq 0}$ exists is not trivial, but, in fact, is equivalent to the existence of the Wiener measure.)

With the element of the measure space $(X, \mathcal{A}, \mu_w)$ defined, we now turn to a definition of the semidynamical system $\{S_t\}_{t \geq 0}$ corresponding to (11.1.7). With the initial condition

$$u(0, s) = x(s), \qquad (11.1.15)$$

equation (11.1.7) has the solution

$$u(t, s) = e^{t/2}x(se^{-t}).$$

Thus, if we set

$$S_t x(s) = e^{t/2}x(se^{-t}), \tag{11.1.16}$$

this equation defines $\{S_t\}_{t \geq 0}$.

We first show that $\{S_t\}_{t \geq 0}$ preserves the Wiener measure μ_w. Since the measures μ_w on cylinders generate the Wiener measure on the entire σ-algebra $\mathcal{A}$, we will only verify the measure-preservation condition

$$\mu_w(S_t^{-1}(C)) = \mu_w(C) \tag{11.1.17}$$

for cylinders. First observe that for every $\alpha \in (0, 1]$,

$$\mu_w(C(\alpha^2 s_1, \ldots, \alpha^2 s_n; \alpha A_1, \ldots, \alpha A_n))$$
$$= \mu_w(C(s_1, \ldots, s_n; A_1, \ldots, A_n)). \tag{11.1.18}$$

This follows directly from equation (11.1.13) if we set $y_i = \alpha x_i$ in the integral on the right-hand side. Further, from (11.1.16), it is clear that $(S_t x)(s_i) \in A_i$ if and only if $x(s_i e^{-t}) \in e^{-t/2} A_i$. Thus,

$$S_t^{-1}(C(s_1, \ldots, s_n; A_1, \ldots, A_n))$$
$$= \{x \in X : (S_t x)(s_i) \in A_i, i = 1, \ldots, n\}$$
$$= C(e^{-t}s_1, \ldots, e^{-t}s_n; e^{-t/2}A_1, \ldots, e^{-t/2}A_n).$$

From this relation and (11.1.18) with $\alpha = e^{-t/2}$, we immediately obtain (11.1.17), thereby verifying that $\{S_t\}_{t \geq 0}$ preserves the Wiener measure μ_w.

To demonstrate the exactness of $\{S_t\}_{t \geq 0}$, we will be content to show that

$$\lim_{t \to \infty} \mu_w(S_t(C)) = 1 \qquad \text{if } \mu_w(C) > 0 \tag{11.1.19}$$

for cylinders. In this case we have

$$S_t(C) = S_t(C(s_1, \ldots, s_n; A_1, \ldots, A_n))$$
$$= \{S_t x : x \in C\} = \{e^{t/2}x(se^{-t}) : x \in C\}.$$

Set $y(s) = e^{t/2}x(se^{-t})$ so this becomes

$$S_t(C) = S_t(C(s_1, \ldots, s_n; A_1, \ldots, A_n))$$
$$= \{y \in X : y(s) = e^{t/2}x(se^{-t}), x(s_i) \in A_i, i = 1, \ldots, n\} \tag{11.1.20}$$

Since $s \in [0, 1]$, and, thus, $se^{-t} \in [0, e^{-t}]$, the conditions $x(s_i) \in A_i$ are irrelevant for $s_i > e^{-t}$. Thus

$$S_t(C(s_1, \ldots, s_n; A_1, \ldots, A_n)) = C(s_1 e^t, \ldots, s_k e^t; e^{t/2}A_1, \ldots, e^{t/2}A_k)$$

where $k = k(t)$ is the largest integer $k \leq n$ such that $s_k \leq e^{-t}$. Once t becomes sufficiently large, that is, $t > -\log s_1$, then from (11.1.20) we see that the last condition $x_1 \in A_1$ disappears and we are left with

$$S_t(C(s_1, \ldots, s_n; A_1, \ldots, A_n)) = \{y \in X : y(s) = e^{t/2} x(se^{-t})\}.$$

However, since X is the space of all possible continuous functions $x : [0, 1] \to R$, the set on the right-hand side is just X and, as a consequence,

$$\mu_w(S_t(C(s_1, \ldots, s_n; A_1, \ldots, A_n))) = 1 \qquad \text{for } t > -\log s_1,$$

which proves equation (11.1.19) for cylinders.

In the general case, for an arbitrary $C \in \mathcal{A}$ the demonstration that (11.1.19) holds is more difficult, but the outline of the argument is as follows. Starting with the equality

$$\mu_w(S_t(C)) = \mu_w(S_t^{-1} S_t(C)),$$

and using the fact that the family $\{S_t^{-1} S_t(C)\}_{t \geq 0}$ is increasing with t, we obtain

$$\lim_{t \to \infty} \mu_w(S_t(C)) = \mu_w(B), \qquad (11.1.21)$$

where

$$B = \bigcup_{t \geq t_0} S_t^{-1} S_t(C) \qquad (11.1.22)$$

and t_0 is an arbitrary nonnegative number. From (11.1.22), it follows that

$$B \in \mathcal{A}_\infty = \bigcap_{t \geq 0} S_t^{-1}(\mathcal{A}).$$

From the Blumenthal zero-one law [see Remark 11.2.1] it may be shown that the σ-algebra $\mathcal{A}_\infty$ contains only trivial sets. Thus, since $\mu_w(B) \geq \mu_w(C)$, we must have $\mu_w(B) = 1$ whenever $\mu_w(C) > 0$. Thus (11.1.19) follows immediately from (11.1.21).

A proof of exactness may also be carried out for equations more general than the linear version (11.1.7). The nonlinear equation

$$\frac{\partial u}{\partial t} + c(s) \frac{\partial u}{\partial s} = f(s, u), \qquad (11.1.23)$$

has been used to model the dynamics of a population of cells undergoing simulaneous proliferation and maturation [Lasota, Mackey, and Ważewska-Czyżewska, 1981; Mackey and Dörmer, 1982], where s is the maturation variable. When the coefficients c and f satisfy some additional conditions, it can be shown that all the solutions of (11.1.23) with the initial condition (11.1.15) converge to the same limit if $x(0) > 0$. However, if $x(0) = 0$, then the solutions of (11.1.23) will exhibit extremely irregular behavior that can be identified with the exactness of the semidynamical system $\{S_t\}_{t\geq 0}$ corresponding to $u(t,s)$. This latter situation $[x(0) = 0]$ corresponds to the destruction of the most primitive cell type (maturity $= 0$), and in such situations the erratic behavior corresponding to exactness of $\{S_t\}_{t\geq 0}$ is noted clinically. $\square$

11.2 d-Dimensional Wiener Processes (Brownian Motion)

In considering d-dimensional Wiener processes we will require an extension of our definition of independent sets. Suppose we have a finite sequence

$$\mathcal{F}_1, \mathcal{F}_2, \ldots, \mathcal{F}_n, \qquad \mathcal{F}_i \subset \mathcal{F} \tag{11.2.1}$$

of σ-algebras. We define the independence of (11.2.1) as follows.

Definition 11.2.1. A sequence (11.2.1) consists of **independent** σ-**algebras** if all possible sequences of sets $A_1, \ldots, A_n$ such that

$$A_1 \in \mathcal{F}_1, \ldots, A_n \in \mathcal{F}_n$$

are independent.

Further, for every random variable ξ we denote by $\mathcal{F}(\xi)$ the σ-algebra of all events of the form $\{\omega: \xi(\omega) \in B\}$, where the B are Borel sets, or, more explicitly,

$$\mathcal{F}(\xi) = \{\xi^{-1}(B): B \text{ is a Borel set}\}.$$

Having a stochastic process $\{\eta(t)\}_{t\in\Delta}$ on an interval Δ, we denote the smallest σ-algebra that contains all sets of the form

$$\{\omega: \eta(t,\omega) \in B\}, t \in \Delta, \ B \text{ is a Borel set,}$$

by $\mathcal{F}(\eta(t): t \in \Delta)$.

With this notation we can restate our definition of independent random variables as follows. The random variables $\xi_1, \ldots, \xi_n$ are independent if $\mathcal{F}(\xi_1), \ldots, \mathcal{F}(\xi_n)$ are independent. In an analogous fashion, stochastic processes $\{\eta_1(t)\}_{t\in\Delta_1}, \ldots, \{\eta_n(t)\}_{t\in\Delta_n}$ are independent, if

$$\mathcal{F}(\eta_1(t): t \in \Delta_1), \ldots, \mathcal{F}(\eta_n(t): t \in \Delta_n)$$

are independent.

Finally, having m random variables $\xi_1, \ldots, \xi_m$ and n stochastic processes $\{\eta_1(t)\}_{t\in\Delta_1}, \ldots, \{\eta_n(t)\}_{t\in\Delta_n}$, we say that they are independent if the σ-algebras

$$\mathcal{F}(\xi_1), \ldots, \mathcal{F}(\xi_m), \mathcal{F}(\eta_1(t): t \in \Delta_1), \ldots, \mathcal{F}(\eta_n(t): t \in \Delta_n)$$

are independent. We will also say that a stochastic process $\{\eta(t)\}_{t\in\Delta}$ and a σ-algebra $\mathcal{F}_0$ are independent if $\mathcal{F}(\eta(t):t\in\Delta)$ and $\mathcal{F}_0$ are independent. Now it is straightforward to define a d-dimensional Wiener process.

Definition 11.2.2. A d-dimensional vector valued process

$$w(t) = \{w_1(t),\ldots,w_d(t)\}, \qquad t \geq 0$$

is a d-**dimensional Wiener process (Brownian motion)** if its components $\{w_1(t)\}_{t\geq 0},\ldots,\{w_d(t)\}_{t\geq 0}$ are one-dimensional independent Wiener processes (Brownian motion).

From this definition it follows that for every fixed t the random variables $w_1(t),\ldots,w_d(t)$ are independent. Thus, it is an immediate consequence of Theorem 10.1.1 that the joint density of the random vector $(w_1(t),\ldots,w_d(t))$ is given by

$$g(t,x_1,\ldots,x_d) = g(t,x_1)\cdots g(t,x_d)$$
$$= \frac{1}{(2\pi t)^{d/2}}\exp\left[-\frac{1}{2t}\sum_{i=1}^{d}x_i^2\right]. \qquad (11.2.2)$$

The joint density g has the following properties:

$$\int_{R^d}\cdots\int g(t,x_1,\ldots,x_d)\,dx_1\cdots dx_d = 1, \qquad (11.2.3)$$

$$\int_{R^d}\cdots\int x_i g(t,x_1,\ldots,x_d)\,dx_1\cdots dx_d = 0, \qquad i = 1,\ldots,d, \qquad (11.2.4)$$

and

$$\int_{R^d}\cdots\int x_i x_j g(t,x_1,\ldots,x_d)\,dx_1\cdots dx_d = \delta_{ij}t, \qquad i,j = 1,\ldots,d,$$
$$(11.2.5)$$

where δ_{ij} is the Kronecker delta ($\delta_{ij} = 0$, $i \neq j$, $\delta_{ii} = 1$).

Remark 11.2.1. The family $\mathcal{F}(w(u):0 \leq u \leq t)$ of σ-algebras generated by the Wiener process (or d-dimensional Wiener process) has the interesting property that it is right-hand continuous. We have (modulo zero)

$$\mathcal{F}(w(u):0 \leq u \leq t) = \bigcap_{h>0}\mathcal{F}(w(u):0 \leq u \leq t+h). \qquad (11.2.6)$$

In particular at $t = 0$, since $w(0) = 0$ and the σ-algebra generated by $w(0)$ is trivial, we can see from equality (11.2.6) that the product

$$\bigcap_{h>0} \mathcal{F}(w(u): 0 \leq u \leq h)$$

contains only sets of measure zero or one. The last statement is referred to as the **Blumenthal zero-one law** (Friedman [1975]). □

11.3   The Stochastic Itô Integral: Development

To understand what is meant by a solution to the stochastic differential equation (11.1.1), it is necessary to introduce the concept of the stochastic Itô integral. In this section we offer a simple but precise definition of this integral and calculate some specific cases so that a comparison with the usual Lebesgue integral may be made.

Let a probability space $(\Omega, \mathcal{F}, \text{prob})$ be given, and let $\{w(t)\}_{t \geq 0}$ be a one-dimensional Wiener process. If $\{\eta(t)\}_{t \in [\alpha, \beta]}$ is another stochastic process defined for $t \in [\alpha, \beta]$, $\alpha \geq 0$, we wish to know how to interpret the integral

$$\int_\alpha^\beta \eta(t) \, dw(t). \tag{11.3.1}$$

Proceeding naively from the classical rules of calculus would suggest that (11.3.1) should be replaced by

$$\int_\alpha^\beta \eta(t) w'(t) \, dt.$$

However, this integral is only defined if $w(t)$ is a differentiable function, which we have already observed is not the case for a Wiener process.

Another possibility suggested by classical analysis is to consider (11.3.1) as the limit of approximating sums $\bar{s}$ of the form

$$\bar{s} = \sum_{i=1}^k \eta(\bar{t}_i)[w(t_i) - w(t_{i-1})], \tag{11.3.2}$$

where

$$\alpha = t_0 < t_1 < \cdots < t_k = \beta$$

is a partition of the interval $[\alpha, \beta]$ and the intermediate points $\bar{t}_i \in [t_i, t_{i+1}]$. This turns out to be a more fruitful idea but has the surprising consequence that the limit of the approximating sums $\bar{s}$ of the form (11.3.2) depends on the choice of the intermediate points $\bar{t}_i$, in sharp contrast to the situation for the Riemann and Stieltjes integrals. This occurs because $w(t)$, at fixed ω, is not a function of bounded variation.

With these preliminary remarks in mind, we now proceed to develop some concepts of use in the definition of the Itô integral.

Definition 11.3.1. A family $\{\mathcal{F}_t\}$, $\alpha \leq t \leq \beta$, of σ-algebras contained in $\mathcal{F}$ is called **nonanticipative** if the following three conditions are satisfied:

(1) $\mathcal{F}_u \subset \mathcal{F}_t$ for $u \leq t$, so $\mathcal{F}_t$ increases as t increases;

(2) $\mathcal{F}_t \supset \mathcal{F}(w(u){:}\,\alpha \leq u \leq t)$, so $w(u)$, $\alpha \leq u \leq t$, is measurable with respect to $\mathcal{F}_t$;

(3) $w(t+h) - w(t)$ is independent of $\mathcal{F}_t$ for $h \geq 0$, so all pairs of sets A_1, A_2 such that $A_1 \in \mathcal{F}_t$ and $A_2 \in \mathcal{F}(w(t+h) - w(t))$ are independent.

From this point on we will assume that a Wiener process $w(t)$ and a family of nonanticipative σ-algebras $\{\mathcal{F}_t\}$, $\alpha \leq t \leq \beta$, are given.
We next define a fourth condition.

Definition 11.3.2. A stochastic **process** $\{\eta(t)\}$, $\alpha \leq t \leq \beta$, is called **non-anticipative** with respect to $\{\mathcal{F}_t\}$ if

(4) $\mathcal{F}_t \supset \mathcal{F}\{\eta(u){:}\,\alpha \leq u \leq t\}$, so $\eta(u)$ is measurable with respect to $\mathcal{F}_t$.

For every random process $\{\eta(t)\}$, $\alpha \leq t \leq \beta$, we define the **Itô sum s** by

$$s = \sum_{i=1}^{k} \eta(t_{i-1})[w(t_i) - w(t_{i-1})]. \tag{11.3.3}$$

Note that in the definition of the Itô sum (11.3.3), we have specified the intermediate points $\bar{t}_i$ of (11.3.2) to be the left end of each interval, $\bar{t}_i = t_{i-1}$. For a given Itô sum s, we define

$$\delta(s) = \max_i (t_i - t_{i-1})$$

and call a sequence of Itô sums $\{s_n\}$ **regular** if $\delta(s_n) \to 0$ as $n \to \infty$.
We now define the Itô integral as follows.

Definition 11.3.3. Let $\{\eta(t)\}$, $\alpha \leq t \leq \beta$, be a nonanticipative stochastic process. If there exists a random variable ζ such that

$$\zeta = \text{st-lim } s_n \tag{11.3.4}$$

for every regular sequence of the Itô sums $\{s_n\}$, then we say that ζ is the **Itô integral** of $\{\eta(t)\}$ on the interval $[\alpha, \beta]$ and denote it by

$$\zeta = \int_{\alpha}^{\beta} \eta(t)\, dw(t). \tag{11.3.5}$$

Remark 11.3.1. It can be proved that for every continuous nonanticipative process the limit (11.3.4) always exists. □

Remark 11.3.2. Definition 11.3.1 of a nonanticipative σ-algebra is complicated, and the reason for introducing each element of the definition, as well as the implication of each, may appear somewhat obscure. Condition (1) is easy, for it merely means that the σ-algebra $\mathcal{F}_t$ of events grows as time proceeds. The second condition ensures that $\mathcal{F}_t$ contains all of the events that can be described by the Wiener process $w(s)$ for times $s \in [\alpha, t]$. Finally, condition (3) says that no information concerning the behavior of the process $w(u) - w(t)$ for $u > t$ can influence calculations involving the probability of the events in $\mathcal{F}_t$. Definition 11.3.2 gives to a stochastic process $\eta(u)$ the same property that condition (2) of Definition 11.3.1 gives to $w(u)$. Thus, all of the information that can be obtained from $\eta(u)$ for $u \in [\alpha, t]$ is contained in $\mathcal{F}_t$.

Taken together, these four conditions ensure that the integrand $\eta(t)$ of the Itô integral (11.3.5) does not depend on the behavior of $w(t)$ for times greater than β and aid in the proof of the convergence of the Itô approximating sums. Further, the nonanticipatory assumption plays an important role in the proof of the existence and uniqueness of solutions to stochastic differential equations since it guarantees that the behavior of a solution in a time interval $[0, t]$ is not influenced by the Wiener process for times larger than t. □

Example 11.3.1. For our first example of the calculation of a specific Itô integral, we take

$$\int_0^T dw(t).$$

In this case the integrand of (11.3.5) is $\eta(t) \equiv 1$. Thus $\mathcal{F}(\eta(t): 0 \le t \le T)$ is a trivial σ-algebra that contains the whole space Ω and the empty set $\emptyset$. To see this, note that, if $1 \in B$, then $\{\omega: \eta(t) \in B\} = \Omega$ and, if $1 \notin B$ then $\{\omega: \eta(t) \in B\} = \emptyset$. This trivial σ-algebra $\{\emptyset, \Omega\}$ is contained in any other σ-algebra, and thus condition (4) of Definition 11.3.2 is satisfied.

By definition

$$s = \sum_{i=1}^k [w(t_i) - w(t_{i-1})] = w(t_k) - w(t_0) = w(T)$$

and, thus,

$$\int_0^T dw(t) = w(T). \quad □$$

Example 11.3.2. In this example we will evaluate

$$\int_0^T w(t)\, dw(t),$$

which is not as trivial as our previous example.

In this case, $\eta(t) = w(t)$, so that condition (4) of Definition 11.3.2 follows from condition (2) of Definition 11.3.1. The Itô sum,

$$s = \sum_{i=1}^{k} w(t_{i-1})[w(t_i) - w(t_{i-1})],$$

may be rewritten as

$$s = \tfrac{1}{2}\sum_{i=1}^{k}[w^2(t_i) - w^2(t_{i-1})] - \tfrac{1}{2}\sum_{i=1}^{k}[w(t_i) - w(t_{i-1})]^2$$

$$= \tfrac{1}{2}w^2(T) - \tfrac{1}{2}\sum_{i=1}^{k}\gamma_i, \tag{11.3.6}$$

where

$$\gamma_i = [w(t_i) - w(t_{i-1})]^2.$$

To evaluate the last summation in (11.3.6), observe that, from the Chebyshev inequality (10.2.10),

$$\text{prob}\left\{\left|\tfrac{1}{2}\sum_{i=1}^{k}\gamma_i - \tfrac{1}{2}\sum_{i=1}^{k}m_i\right| \geq \varepsilon\right\} \leq \frac{1}{\varepsilon^2}D^2\left(\tfrac{1}{2}\sum_{i=1}^{k}\gamma_i\right)$$

$$= \frac{1}{4\varepsilon^2}\sum_{i=1}^{k}D^2(\gamma_i), \tag{11.3.7}$$

where $m_i = E(\gamma_i)$. Further, by (11.1.4),

$$E(\gamma_i) = E([w(t_i) - w(t_{i-1})]^2) = t_i - t_{i-1}$$

and, by equations (10.2.6) and (11.1.4),

$$D^2(\gamma_1) \leq E(\gamma_i^2) = E\big([w(t_i) - w(t_{i-1})]^4\big) = 3(t_i - t_{i-1})^2.$$

Thus,

$$\tfrac{1}{2}\sum_{i=1}^{k}m_i = \tfrac{1}{2}\sum_{i=1}^{k}(t_i - t_{i-1}) = \frac{T}{2}$$

and

$$\sum_{i=1}^{k}D^2(\gamma_i) \leq 3\sum_{i=1}^{k}(t_i - t_{i-1})^2 \leq 3T\max_{i}(t_i - t_{i-1}).$$

Setting $\delta(s) = \max_i(t_i - t_{i-1})$ as before and using (11.3.7), we finally obtain

$$\text{prob}\left\{\left|\tfrac{1}{2}\sum_{i=1}^{k}\gamma_i - \frac{T}{2}\right| \geq \varepsilon\right\} \leq \frac{3T}{4\varepsilon^2}\delta(s)$$

or, from (11.3.6),

$$\text{prob}\left\{\left|s - \left(\frac{w^2(T)}{2} - \frac{T}{2}\right)\right| \geq \varepsilon\right\} \leq \frac{3T}{4\varepsilon^2}\delta(s).$$

If $\{s_n\}$ is a regular sequence, then $\delta(s_n)$ converges to zero as $n \to \infty$ and

$$\text{st-lim } s_n = \tfrac{1}{2}w^2(T) - \tfrac{1}{2}T.$$

Thus we have shown that

$$\int_0^T w(t)\, dw(t) = \tfrac{1}{2}w^2(T) - \tfrac{1}{2}T,$$

clearly demonstrating that the stochastic Itô integral does not obey the usual rules of integration.   ☐

This last example illustrates the fact that the calculation of stochastic integrals is, in general, not an easy matter and requires many analytical tools that may vary from situation to situation. What is even more interesting is that the sufficient conditions for the existence of stochastic integrals related to the construction of nonanticipative σ-algebras are quite complicated in comparison with the Lebesgue integration of deterministic functions.

Remark 11.3.3. From Example 11.3.2, it is rather easy to demonstrate how the choice of the intermediate point $\bar{t}_i$ influences the value of the integral. For example, picking $\bar{t}_i = \tfrac{1}{2}(t_{i-1} + t_i)$, we obtain, in place of the Itô sum, the **Stratonovich sum**,

$$s = \sum_{i=1}^{k} w\left(\tfrac{1}{2}(t_{i-1} + t_i)\right)[w(t_i) - w(t_{i-1})]$$

$$= \tfrac{1}{2}w^2(T) - \tfrac{1}{2}\sum_{i=1}^{k}\gamma_i + \tfrac{1}{2}\sum_{i=1}^{k}\rho_i,$$

where

$$\gamma_i = \left[w(t_i) - w\left(\tfrac{1}{2}(t_{i-1} + t_i)\right)\right]^2$$

and

$$\rho_i = \left[w\left(\tfrac{1}{2}(t_{i-1} + t_i)\right) - w(t_{i-1})\right]^2.$$

Since the variables $\gamma_1, \ldots, \gamma_k$ are independent as are $\rho_1, \ldots, \rho_k$, we may use the Chebyshev inequality as in the previous example to show that

$$\text{st-lim }\sum_{i=1}^{k}\gamma_i = \tfrac{1}{2}T = \text{st-lim }\sum_{i=1}^{k}\rho_i.$$

Thus the Stratonovich sums $\{s_n\}$ converge to $\tfrac{1}{2}w^2(T)$, and the Stratonovich integral gives a result more in accord with our experience from calculus.

However, the use of the Stratonovich integral in solving stochastic differential equations leads to other more serious problems.   □

To close this section, we extend our definition of the Itô integral to the multidimensional case. If $G(t) = (\eta_{ij}(t))$, $i, j = 1, \ldots, d$ is a $d \times d$ matrix of continuous stochastic processes, defined for $\alpha \le t \le \beta$, and $w(t) = (w_i(t))$, $i = 1, \ldots, d$, is a d-dimensional Wiener process, then

$$\int_\alpha^\beta G(t)\, dw(t) = \begin{pmatrix} \zeta_1 \\ \vdots \\ \zeta_d \end{pmatrix}, \qquad (11.3.8)$$

where

$$\zeta_i = \sum_{j=1}^d \int_\alpha^\beta \eta_{ij}(t)\, dw_j(t)$$

defines the Itô integral. Thus, equation (11.3.8) is integrated term by term. In this case the family $\{\mathcal{F}_t\}$ of nonanticipative σ-algebras must satisfy conditions (2) and (3) of Definition 11.3.1 with respect to all $\{w_i(t)\}$, $i = 1, \ldots, d$, and condition (4) of Definition 11.3.2 must be satisfied by all $\{\eta_{ij}(t)\}$, $i, j = 1, \ldots, d$.

11.4 The Stochastic Itô Integral: Special Cases

In the special case when the integrand of the Itô integral does not depend on ω, that is to say, it is not a stochastic process, the convergence of the approximating sums is quite strong. This section is devoted to an examination of this situation and one in which we are simply integrating a stochastic process with respect to t.

Before stating our first proposition, we note that, if $f \colon [\alpha, \beta] \to R$ is a continuous function, then every regular sequence $\{s_n\}$ of approximating sums

$$s_n = \sum_{i=1}^{k_n} f(\tilde{t}_i^n)[w(t_i^n) - w(t_{i-1}^n)], \qquad \tilde{t}_i^n \in [t_{i-1}^n, t_i^n]$$

converges in the mean [i.e., strongly in $L^2(\Omega)$] to the integral

$$\zeta = \int_\alpha^\beta f(t)\, dw(t). \qquad (11.4.1)$$

Although we will not prove this assertion, it suffices to say that the proof proceeds in a fashion similar to the proof of the following proposition.

Proposition 11.4.1. *If $f \colon [\alpha, \beta] \to R$ is a continuous function, then*

$$E\left(\int_\alpha^\beta f(t)\, dw(t) \right) = 0 \qquad (11.4.2)$$

and

$$D^2\left(\int_\alpha^\beta f(t)\,dw(t)\right) = \int_\alpha^\beta [f(t)]^2\,dt. \qquad (11.4.3)$$

Proof. Set

$$s = \sum_{i=1}^k f(t_{i-1})[w(t_i) - w(t_{i-1})] = \sum_{i=1}^k f(t_{i-1})\Delta w_i,$$

where $\Delta w_i = w(t_i) - w(t_{i-1})$. Then

$$s^2 = \sum_{i,j=1}^k f(t_{i-1})f(t_{j-1})\Delta w_i \Delta w_j.$$

We have immediately that

$$E(s) = \sum_{i=1}^k f(t_{i-1})E(\Delta w_i) = 0$$

and, since $w(t)$ is a Wiener process with independent increments,

$$E(\Delta w_i \Delta w_j) = \begin{cases} E(\Delta w_i)E(\Delta w_j) = 0 & \text{if } i \neq j \\ E(\Delta w_i^2) = t_i - t_{i-1} & \text{if } i = j. \end{cases}$$

We also have

$$D^2(s) = E(s^2) = \sum_{i,j=1}^k f(t_{i-1})f(t_{j-1})E(\Delta w_i \Delta w_j)$$

$$= \sum_{i=1}^k [f(t_{i-1})]^2 (t_i - t_{i-1}).$$

Thus for any regular sequence $\{s_n\}$,

$$\lim_{n\to\infty} E(s_n) = 0 \qquad (11.4.4)$$

and

$$\lim_{n\to\infty} D^2(s_n) = \int_\alpha^\beta [f(t)]^2\,dt. \qquad (11.4.5)$$

Since, from the remarks preceding the proposition, $\{s_n\}$ converges in mean to the integral ζ given in equation (11.4.1), we have $\lim_{n\to\infty} E(s_n) = E(\zeta)$ and $\lim_{n\to\infty} D^2(s_n) = D^2(\zeta)$, which, by (11.4.4) and (11.4.5), completes the proof. ∎

A second special case of the stochastic integral occurs when the integrand is a stochastic process but it is desired to have the integral only with respect to time. Hence we wish to consider

$$\zeta = \int_\alpha^\beta \eta(t)\, dt \qquad (11.4.6)$$

when $\{\eta(t)\}$, $\alpha \le t \le \beta$, is a given stochastic process. To define (11.4.6) we consider approximating sums of the form

$$\bar{s} = \sum_{i=1}^k \eta(\bar{t}_i)(t_i - t_{i-1}),$$

corresponding to the partition

$$\alpha = t_0 < t_1 < \cdots < t_k = \beta$$

with arbitrary intermediate points $\bar{t}_i \in [t_{i-1}, t_i]$. We now have the following definition.

Definition 11.4.1. If every regular $[\delta(\bar{s}_n) \to 0]$ sequence $\{\bar{s}_n\}$ of approximating sums is stochastically convergent and

$$\zeta = \text{st-lim } \bar{s}_n, \qquad (11.4.7)$$

then this common limit is called the **integral** of $\eta(t)$ on $[\alpha, \beta]$ and is denoted by (11.4.6).

Observe that, when $\eta(t, \omega)$ possesses continuous sample paths, that is, it is a continuous function of t, the limit

$$\lim_{n \to \infty} \bar{s}_n(\omega)$$

exists as the classical Riemann integral. Thus when $\{\eta(t)\}$, $\alpha \le t \le \beta$, is a continuous stochastic process, this limit exists for almost all ω. Further, since, by Proposition 10.3.2, almost sure convergence implies stochastic convergence, the limit (11.4.7) must exist.

There is an interesting connection between the Itô integral (11.3.5) and the integral of (11.4.6) reminiscent of the classical "integration by parts" formula. It can be stated formally as follows.

Proposition 11.4.2. If $f: [\alpha, \beta] \to R$ is differentiable with a continuous derivative f', then

$$\int_\alpha^\beta f(t)\, dw(t) = -\int_\alpha^\beta f'(t)w(t)\, dt + f(\beta)w(\beta) - f(\alpha)w(\alpha). \qquad (11.4.8)$$

Proof. Since the integrals in (11.4.8) both exist we may pick special approximating sums of the form

$$\bar{s}_n = \sum_{i=1}^{k_n} f'(\bar{t}_i^n) w(\bar{t}_i^n)(t_i^n - t_{i-1}^n), \tag{11.4.9}$$

where the intermediate points $\bar{t}_i$ are chosen in such a way that

$$f(t_i^n) - f(t_{i-1}^n) = f'(\bar{t}_i^n)(t_i^n - t_{i-1}^n).$$

Substituting this expression into (11.4.9), we may rewrite $\bar{s}_n$ as

$$\begin{aligned}
\bar{s}_n &= \sum_{i=1}^{k_n} [f(t_i^n) - f(t_{i-1}^n)] w(\bar{t}_i^n) \\
&= - \sum_{i=1}^{k_n-1} [w(\bar{t}_{i+1}^n) - w(\bar{t}_i^n)] f(t_i^n) + f(t_{k_n}^n) w(\bar{t}_{k_n}^n) \\
&\quad - f(t_1^n) w(\bar{t}_1^n). \tag{11.4.10}
\end{aligned}$$

The sum on the right-hand side of (11.4.10) corresponds to the partition

$$\bar{t}_1^n < \cdots < \bar{t}_{k_n}^n$$

that does not contain intervals $(\alpha, \bar{t}_1^n)$ and $(\bar{t}_{k_n}^n, \beta)$. Setting $\bar{t}_0^n = \alpha$ and $\bar{t}_{k_n}^n = \beta$, we may rewrite (11.4.10) in the form

$$\bar{s}_n = -s_n + w(\beta)f(\beta) - w(\alpha)f(\alpha), \tag{11.4.11}$$

where

$$s_n = \sum_{i=1}^{k_n-1} [w(\bar{t}_{i+1}^n) - w(\bar{t}_i^n)] f(t_i^n).$$

The sequence $\{\bar{s}_n\}$ converges to

$$\int_\alpha^\beta f'(t) w(t)\, dt,$$

whereas $\{s_n\}$ converges, by our remarks preceding Proposition 11.4.1, to the integral

$$\int_\alpha^\beta f(t)\, dw(t).$$

Thus, passing to the limit in (11.4.11) finishes the proof. ∎

Remark 11.4.1. In our short development of the Itô integral and presentation of its main properties, we have restricted ourselves to the special situation where the integrand is a continuous stochastic process. This allowed us to define the Itô integral in a relatively simple and direct way as the limit of the Itô sums (11.3.3). Generally, such an approach is inconvenient because of the restrictive nature of the assumption concerning the continuity of the integrand. Usually, the definition of the Itô integral is given in a more sophisticated manner. It is first defined for stochastic processes that are piecewise constant in time, and then, by using a limiting procedure in $L^2(\Omega)$ the definition is extended to a quite general class of integrands. An exhaustive treatment of this procedure may be found in Gikhman and Skorokhod [1969, 1975]. ☐

11.5   Stochastic Differential Equations

All the material developed in the previous sections was a necessary prelude to be able to study the stochastic differential equation

$$\frac{dx}{dt} = b(x) + \sigma(x)\xi \tag{11.5.1}$$

with initial condition

$$x(0) = x^0, \tag{11.5.2}$$

where

$$b(x) = \begin{pmatrix} b_1(x) \\ \vdots \\ b_d(x) \end{pmatrix} \quad \text{and} \quad \sigma(x) = \begin{pmatrix} \sigma_{11}(x) & \cdots & \sigma_{1d}(x) \\ \vdots & & \vdots \\ \sigma_{d1}(x) & \cdots & \sigma_{dd}(x) \end{pmatrix}$$

are given functions of x and

$$x(t) = \begin{pmatrix} x_1(t) \\ \vdots \\ x_d(t) \end{pmatrix}$$

is the unknown. In (11.5.1), the "white noise" vector

$$\xi = \begin{pmatrix} \frac{dw_1}{dt} \\ \vdots \\ \frac{dw_d}{dt} \end{pmatrix}$$

should be considered, from a mathematical point of view, as a pure symbol much like the letters "dt" in the notation for the derivative. However, from an application standpoint, ξ denotes a very specific process consisting of "infinitely" many independent, or random, impulses as discussed in Section 11.1. We assume that the initial vector x^0 and the Wiener process $\{w(t)\}$

are independent. To examine the solution of equations (11.5.1) and (11.5.2), we formally integrate (11.5.1) over the interval $[0, t]$ to give

$$x(t) = \int_0^t b(x(s)) \, ds + \int_0^t \sigma(x(s)) \, dw(s) + x^0. \tag{11.5.3}$$

Since the integrals that appear on the right-hand side of (11.5.3) are defined from our considerations of the previous sections, we are close to a formal definition of the solution.

First, however, it is necessary to choose a specific family of nonanticipative σ-algebras $\{\mathcal{F}_t\}_{t \geq 0}$. We may, for example, assume that $\mathcal{F}$ is the smallest σ-algebra containing all events of the form $\{w: w(u, \omega) \in B\}$ and $(x^0)^{-1}(B)$ for $0 \leq u \leq t$ and Borel sets B, that is, $\mathcal{F}_t$ is the smallest σ-algebra with respect to which $w(u)$, $0 \leq u \leq t$, and x^0 are measurable. This family is nonanticipative since conditions (1) and (2) of Definition 11.3.1 are evidently satisfied, and condition (3) follows from the fact that $\{w(t)\}$ is a process with independent increments and that x^0 and $\{w(t)\}$ are independent.

With this family of nonanticipative σ-algebras, we define the solution to equations (11.5.1) and (11.5.2).

Definition 11.5.1. A continuous stochastic process $\{x(t)\}_{t \geq 0}$ is called a **solution** of equations (11.5.1) and (11.5.2) if:

(a) $\{x(t)\}$ is nonanticipative, that is, it satisfies condition (4) of Definition 11.3.2; and

(b) For every $t \geq 0$, equation (11.5.3) is satisfied with probability 1.

It is well known from the theory of ordinary differential equations that it is necessary to assume some special conditions on the right-hand side in order to guarantee the existence and uniqueness of a solution. It is interesting that analogous conditions are also sufficient for stochastic differential equations. Thus we have the following theorem.

Theorem 11.5.1. *If $b(x)$ and $\sigma(x)$ satisfy the Lipschitz conditions*

$$|b(x) - b(y)| \leq L|x - y|, \qquad x, y \in R^d \tag{11.5.4}$$

and

$$|\sigma(x) - \sigma(y)| \leq L|x - y|, \qquad x, y \in R^d \tag{11.5.5}$$

with some constant L, then the initial value problem, equations (11.5.1) and (11.5.2), has a unique solution $\{x(t)\}_{t \geq 0}$.

Theorem 11.5.1 can be proved by the method of successive approximations as can the corresponding result for ordinary differential equations.

Thus a sequence $\{x^i(t)\}_{t \geq 0}$ of stochastic processes would be defined with $x(t = 0) = x^0$ and

$$x^i(t) = \int_0^t b(x^{i-1}(s))\, ds + \int_0^t \sigma(x^{i-1}(s))\, dw(s) + x^0.$$

Then, using the Lipschitz conditions (11.5.4) and (11.5.5), it is possible to evaluate the series

$$x(t) = \sum_{i=1}^{\infty} [x^i(t) - x^{i-1}(t)] + x^0$$

in $L^2(\Omega)$ norm by a convergent series of the form,

$$\sum_{n=0}^{\infty} k^n t^n / n!,$$

and to prove that $x(t)$ is, indeed, the desired solution. We omit the details as this proof is quite complicated, but a full proof may be found in Gikhman and Skorokhod [1969].

An alternative way to generate an approximating solution is to use the Euler linear extrapolation formula. Suppose that the solution $x(t)$ is given on the interval $[0, t_0]$. Then for values $t_0 + \Delta t$ larger than, but close to, t_0, we write

$$x(t_0 + \Delta t) = x(t_0) + b(x(t_0))\Delta t + \sigma(x(t_0))\Delta w, \qquad (11.5.6)$$

where $\Delta w = w(t_0 + \Delta t) - w(t_0)$. (Observe that for an ordinary differential equation, this equation defines a ray tangent to the solution on $[0, t_0]$ at t_0.) In particular, when an interval $[0, T]$ is given, we may take a partition

$$0 = t_0 < \cdots < t_n = T$$

and define

$$\Delta x(t_i) = b(x(t_{i-1}))\Delta t_i + \sigma(x(t_{i-1}))\Delta w_i, \qquad (11.5.7)$$

where $\Delta x(t_i) = x(t_i) - x(t_{i-1})$, $\Delta t_i = t_i - t_{i-1}$, $\Delta w_i = w(t_i) - w(t_{i-1})$, and $x(t_0) = x^0$.

It is evident that in some respects this approach is much simpler than the method of successive approximations, since no knowledge concerning the Itô integral is even necessary. Indeed, S. Bernstein employed this technique in his original investigations into stochastic differential equations, so we will call equations (11.5.6) and (11.5.7) the **Euler–Bernstein equations**.

Example 11.5.1. The oldest and best-known example of a stochastic differential equation is probably the **Langevin equation**.

$$\frac{dx}{dt} = -bx + \sigma\xi, \qquad x(0) = x^0, \qquad (11.5.8)$$

where x is a scalar and the coefficients b and σ are constant.

By definition, the solution of (11.5.8) satisfies

$$x(t) = -b \int_0^t x(s) \, ds + \sigma \int_0^t dw(s) + x^0$$

or, using our calculations of Example 11.3.1,

$$x(t) = -b \int_0^t x(s) \, ds + \sigma w(t) + x^0. \qquad (11.5.9)$$

Equation (11.5.9) is rather easy to deal with since it does not contain an Itô integral, and, since the one integral that does appear exists for almost w taken separately, we may use the usual rules of calculus.

Setting

$$z(t) = \int_0^t x(s) \, ds, \qquad (11.5.10)$$

equation (11.5.9) becomes, for almost all w,

$$\frac{dz}{dt} = -bz(t) + \sigma w(t) + x^0.$$

For fixed w, this is an ordinary differential equation and, thus,

$$z(t) = \int_0^t e^{-b(t-s)} (\sigma w(s) + x^0) \, ds. \qquad (11.5.11)$$

Combining equations (11.5.9) through (11.5.11) after some manipulation, yields

$$x(t) = x^0 e^{-bt} - b\sigma \int_0^t e^{-b(t-s)} w(s) \, ds + \sigma w(t).$$

Using the integration by parts formula (11.4.8), this becomes

$$x(t) = x^0 e^{-bt} + \sigma \int_0^t e^{-b(t-s)} dw(s). \qquad (11.5.12)$$

From (11.5.12) and (11.4.2), it follows that

$$E(x(t)) = e^{-bt} E(x^0)$$

and, taking note of the independence of x^0 and $w(t)$,

$$D^2(x(t)) = e^{-2bt} D^2(x^0) + \sigma^2 D^2 \left(\int_0^t e^{-b(t-s)} dw(s) \right).$$

With (11.4.3), this finally reduces to

$$D^2(x(t)) = e^{-2bt} D^2(x^0) + \sigma^2 \int_0^t e^{-2b(t-s)} ds$$

$$= e^{-2bt} D^2(x^0) + (\sigma^2/2b)[1 - e^{-2bt}].$$

Thus, for the Langevin equation,

$$\lim_{t \to \infty} D^2(x(t)) = \sigma^2/2b.$$

This asymptotic property of the variance is a special case of a more general result that we will establish in the next section where we examine uses of the Fokker–Planck equation.   □

11.6   The Fokker–Planck (Kolmogorov Forward) Equation

The preceding sections were aimed at obtaining an understanding of the dynamical system

$$\frac{dx}{dt} = b(x) + \sigma(x)\xi \tag{11.6.1}$$

with

$$x(0) = x^0 \tag{11.6.2}$$

under a stochastic perturbation ξ. This required us to first introduce the abstract concept of nonanticipative σ-algebras. Then we had to define the Itô integral, which is generally quite difficult to calculate. Finally we gave the solution to equations (11.6.1)–(11.6.2) in terms of a general formula, generated by the method of successive approximations, which contains infinitely many Itô integrals.

In this section we extend this to a discussion of the density function of the process $x(t)$, which is a solution of (11.6.1) and (11.6.2). This density is defined as the function $u(t, x)$ that satisfies

$$\text{prob}\{x(t) \in B\} = \int_B u(t, z)\, dz. \tag{11.6.3}$$

The uniqueness of $u(t, x)$ follows immediately from Proposition 2.2.1, but the existence requires some regularity conditions on the coefficients $b(x)$ and $\sigma(x)$, which are given in the following. We will also show how $u(t, x)$ can be found without any knowledge concerning the solution $x(t)$ of the stochastic differential equations (11.6.1) with (11.6.2). It will turn out that $u(t, x)$ is given by the solution of a partial differential equation, known as the Fokker–Planck (or Kolmogorov forward) equation and that it is completely specified by the coefficients $b(x)$ and $\sigma(x)$ of equation (11.6.1). Now set

$$a_{ij}(x) = \sum_{k=1}^{d} \sigma_{ik}(x)\sigma_{jk}(x). \tag{11.6.4}$$

From (11.6.4) it is clear that $a_{ij} = a_{ji}$ and, thus, the quadratic form,

$$\sum_{i,j=1}^{d} a_{ij}(x)\lambda_i\lambda_j, \tag{11.6.5}$$

is symmetric. Further, since

$$\sum_{i,j=1}^{d} a_{ij}(x)\lambda_i\lambda_j = \sum_{k=1}^{d} \left(\sum_{i=1}^{d} \sigma_{ik}(x)\lambda_i\right)^2,$$

(11.6.5) is nonnegative.

We are now ready to state the main theorem of this section, which gives the Fokker–Planck equation.

Theorem 11.6.1. *If the functions σ_{ij}, $\partial\sigma_{ij}/\partial x_k$, $\partial^2\sigma_{ij}/\partial x_k\partial x_l$, b_i, $\partial b_i/\partial x_j$, $\partial u/\partial t$, $\partial u/\partial x_i$, and $\partial^2 u/\partial x_i\partial x_j$ are continuous for $t > 0$ and $x \in R^d$, and if b_i, σ_{ij} and their first derivatives are bounded, then $u(t,x)$ satisfies the equation*

$$\frac{\partial u}{\partial t} = \frac{1}{2}\sum_{i,j=1}^{d} \frac{\partial^2}{\partial x_i\partial x_j}(a_{ij}u) - \sum_{i=1}^{d} \frac{\partial}{\partial x_i}(b_i u), \qquad t > 0, x \in R^d. \tag{11.6.6}$$

Equation (11.6.6) is called the **Fokker–Planck equation** or **Kolmorgorov forward equation**.

Remark 11.6.1. In Theorem 11.6.1 we assumed $\partial b_i/\partial x_j$ and $\partial\sigma_{ij}/\partial x_k$ were bounded since this implies the Lipschitz conditions (11.5.4) and (11.5.5) which, in turn, guarantee the existence and uniqueness of the solution to the stochastic equations (11.6.1) with (11.6.2). In order to assure the existence and differentiability of u, it is sufficient, for example, that a_{ij} and b_i, together with their derivatives up to the third order, are continuous, bounded, and satisfy the uniform parabolicity condition (11.7.5). □

Proof of Theorem 11.6.1. We will use the Euler–Bernstein approximation formula (11.5.6) in the proof of this theorem as it allows us to derive (11.6.6) in an extremely simple and transparent fashion.

Thus let $t_0 > 0$ be arbitrary, and let $x(t)$ be the solution to equations (11.6.1) and (11.6.2) on the interval $[0, t_0]$. Define $x(t)$ on $[t_0, t_0 + \varepsilon]$ by

$$x(t_0 + \Delta t) = x(t_0) + b(x(t_0))\Delta t + \sigma(x(t_0))[w(t_0 + \Delta t) - w(t_0)], \tag{11.6.7}$$

where $0 \le \Delta t \le \varepsilon$ and ε is a positive number. We assume (and this is the only additional assumption needed for simplifying the proof) that $x(t)$, extended according to (11.6.7), has a density $u(t, x)$ for $0 \le t \le t_0 + \varepsilon$ and that for $t = t_0$, $u_t(t, x)$ exists. Observe that at the point $t = t_0$, $u(t, x)$ (and $u_t(t, x)$) is simultaneously the density (and its derivative) for the exact and for the extended solution.

Now let $h: R^d \to R$ be a C^3 function with compact support. We wish to calculate the mathematical expectation of $h(x(t_0 + \Delta t))$. First note that since $u(t_0 + \Delta t, x)$ is the density of $x(t_0 + \Delta t)$, we have, by (10.2.2),

$$E(h(x(t_0 + \Delta t))) = \int_{R^d} h(x) u(t_0 + \Delta t, x)\, dx. \qquad (11.6.8)$$

However, using equation (11.6.7), we may write the random variable $h(x(t_0 + \Delta t))$ in the form

$$h(x(t_0 + \Delta t)) = h(Q(x(t_0), w(t_0 + \Delta t) - w(t_0))), \qquad (11.6.9)$$

where

$$Q(x, y) = x + b(x)\Delta t + \sigma(x)y.$$

The variables $x(t_0)$ and $\Delta w(t_0) = w(t_0 + \Delta t) - w(t_0)$ are independent for each $0 \le \Delta t \le \varepsilon$ since $x(t_0)$ is $\mathcal{F}_{t_0}$-measurable and $\Delta w(t_0)$ is independent with respect to $\mathcal{F}_{t_0}$. Thus the random vector $(x(t_0), \Delta w(t_0))$ has the joint density

$$u(t_0, x) g(\Delta t, y),$$

where g is given by (11.1.3). As a consequence, the mathematical expectation of (11.6.9) is given by

$$\int_{R^d} \int_{R^d} h(Q(x, y)) u(t_0, x) g(\Delta t, y)\, dx\, dy$$

$$= \int_{R^d} \int_{R^d} h(x + b(x)\Delta t + \sigma(x)y) u(t_0, x) g(\Delta t, y)\, dx\, dy.$$

From this and (11.6.8), we obtain

$$\int_{R^d} h(x) u(t_0 + \Delta t, x)\, dx$$

$$= \int_{R^d} \int_{R^d} h(x + b(x)\Delta t + \sigma(x)y) u(t_0, x) g(\Delta t, y)\, dx\, dy.$$

By developing h in a Taylor expansion, we have

$$\int_{R^d} h(x) u(t_0 + \Delta t, x) dx = \int_{R^d} \int_{R^d} \left\{ h(x) + \sum_{i=1}^{d} \frac{\partial h}{\partial x_i} [b_i(x)\Delta t + (\sigma(x)y)_i] \right.$$

$$+ \frac{1}{2} \sum_{i,j=1}^{d} \frac{\partial^2 h}{\partial x_i \partial x_j} [b_i(x)\Delta t + (\sigma(x)y)_i]$$

$$\left. \cdot [b_j(x)\Delta t + (\sigma(x)y)_j] + r(\Delta t) \right\}$$

$$\cdot u(t_0, x) g(\Delta t, y)\, dx\, dy, \qquad (11.6.10)$$

where $r(\Delta t)$ denotes the remainder and $(\sigma(x)y)_i$ is the ith coordinate of the vector $\sigma(x)y$.

On the right-hand side of (11.6.10) we have a finite collection of integrals that we will first integrate with respect to y. Observe that

$$(\sigma(x)y)_i(\sigma(x)y)_j = \sum_{k,l=1}^{d} \sigma_{ik}(x)\sigma_{jl}(x)y_k y_l.$$

By equation (11.2.3)

$$\int_{R^d} g(\Delta t, y)\,dy = 1,$$

whereas from (11.2.4)

$$\int_{R^d} (\sigma(x)y)_i g(\Delta t, y)\,dy = 0.$$

Finally, from (11.2.5), we have

$$\int_{R^d} (\sigma(x)y)_i(\sigma(x)y)_j g(\Delta t, y)\,dy = a_{ij}(x)\Delta t,$$

where a_{ij} is as defined in (11.6.4). By combining all of these results, we can write equation (11.6.10) as

$$\int_{R^d} h(x)[u(t_0 + \Delta t, x) - u(t_0, x)]\,dx$$

$$= \Delta t \int_{R^d} \left\{ \sum_{i=1}^{d} \frac{\partial h}{\partial x_i} b_i(x) \right.$$

$$\left. + \frac{1}{2} \sum_{i,j=1}^{d} \frac{\partial^2 h}{\partial x_i \partial x_j} a_{ij}(x) \right\} u(t_0, x)\,dx + R(\Delta t), \quad (11.6.11)$$

where the new remainder $R(\Delta t)$ is

$$R(\Delta t) = \frac{1}{2} \int_{R^d} \sum_{i,j=1}^{d} \frac{\partial^2 h}{\partial x_i \partial x_j} b_i(x)b_j(x)(\Delta t)^2 u(t_0, x)\,dx$$

$$+ \int_{R^d} \int_{R^d} r(\Delta t) u(t_0, x)g(\Delta t, y)\,dx\,dy. \quad (11.6.12)$$

It is straightforward to show that $R(\Delta t)/\Delta t$ goes to zero as $\Delta t \to 0$. The first integral on the right-hand side of (11.6.12) contains $(\Delta t)^2$, so this is easy. The second integral may be evaluated by using the classical formula for the remainder $r(\Delta t)$:

$$r(\Delta t) = \sum_{i,j,k=1}^{d} \frac{\partial^3 h}{\partial x_i \partial x_j \partial x_k}\bigg|_z [b_i \Delta t + (\sigma y)_i]$$

$$\cdot [b_j \Delta t + (\sigma y)_j][b_k \Delta t + (\sigma y)_k].$$

The third derivatives of h are evaluated at some intermediate point z, which is irrelevant because we only use the fact that these derivatives are bounded since h is of compact support.

All of the components appearing in $r(\Delta t)$ can be evaluated by terms of the form

$$M(\Delta t)^3, M(\Delta t)^2 |y_i|, M(\Delta t)|y_i y_j|, M|y_i y_j y_k|,$$

where M is a constant. To evaluate $R(\Delta t)$ we must integrate these terms with respect to x and y. Using

$$\int_{-\infty}^{\infty} |z|^n g(\Delta t, z)\, dz = \alpha_n (\Delta t)^{n/2},$$

where the constants α_n depend only on n, integration of $M(\Delta t)^3$ again gives $M(\Delta t)^3$ since $u(t_0, x)$ and $g(\Delta t, y)$ are both densities. Integration of $M(\Delta t)^2 |y_i|$ gives $M(\Delta t)^2 C_i (\Delta t)^{1/2}$, where $C_i = \alpha_1$. Analogously, integration of the third term gives $M(\Delta t) C_{ij}(\Delta t)$, whereas the fourth yields $M C_{ijk}(\Delta t)^{3/2}, where C_{ij}$ depends on α_1 and α_2, and C_{ijk} depends on α_1, α_2, and α_3. All these terms divided by Δt approach zero as $\Delta t \to 0$.

Returning to (11.6.11), dividing by Δt and passing to the limit as $\Delta t \to 0$, we obtain

$$\int_{R^d} h(x) \frac{\partial u}{\partial t}\, dx = \int_{R^d} \left\{ \sum_{i=1}^{d} \frac{\partial h}{\partial x_i} b_i(x) + \frac{1}{2} \sum_{i,j=1}^{d} \frac{\partial^2 h}{\partial x_i \partial x_j} a_{ij}(x) \right\} u(t_0, x)\, dx.$$

$$(11.6.13)$$

Since h has compact support we may easily integrate the right-hand side of (11.6.13) by parts. Doing this and shifting all terms to the left-hand side, we finally have

$$\int_{R^d} h(x) \left\{ \frac{\partial u(t_0, x)}{\partial t} + \sum_{i=1}^{d} \frac{\partial}{\partial x_i} [b_i(x) u(t_0, x)] \right.$$

$$\left. - \frac{1}{2} \sum_{i,j=1}^{d} \frac{\partial^2}{\partial x_i \partial x_j} [a_{ij}(x) u(t_0, x)] \right\} dx = 0. \qquad (11.6.14)$$

Since $h(x)$ is a C^3 function with compact support, but otherwise arbitrary, the integral condition (11.6.14), which is satisfied for every such h implies that the term in braces vanishes. This completes the proof that $u(t_0, x)$ satisfies equation (11.6.6.). ∎

Remark 11.6.2. To deal with the stochastic differential equations (11.6.1) with (11.6.2), we were forced to introduce many abstract and difficult concepts. It is ironic that, once we pass to a consideration of the density function $u(t, x)$ of the random process $x(t)$, all this material becomes unnecessary, as we must only insert the appropriate coefficients a_{ij} and b_i into the Fokker–Planck equation (11.6.6)! □

11.7 Properties of the Solutions of the Fokker–Planck Equation

As we have shown in the previous section, the density function $u(t,x)$ of the solution $x(t)$ of the stochastic differential equation (11.6.1) with (11.6.2) satisfies the partial differential equation (11.6.6). Moreover, if the initial condition $x(0) = x^0$, which is a random variable, has a density f then $u(0,x) = f(x)$. Thus, to understand the behavior of the densities $u(t,x)$, we must study the initial-value (Cauchy) problem:

$$\frac{\partial u}{\partial t} = \frac{1}{2} \sum_{i,j=1}^{d} \frac{\partial^2}{\partial x_i \partial x_j}[a_{ij}(x)u] - \sum_{i=1}^{d} \frac{\partial}{\partial x_i}[b_i(x)u], \quad t > 0, x \in R^d, \quad (11.7.1)$$

$$u(0,x) = f(x), \qquad x \in R^d. \tag{11.7.2}$$

Observe that equation (11.7.1) is of second order and may be rewritten in the form

$$\frac{\partial u}{\partial t} = \frac{1}{2} \sum_{i,j=1}^{d} a_{ij}(x) \frac{\partial^2 u}{\partial x_i \partial x_j} + \sum_{i=1}^{d} \tilde{b}_i(x) \frac{\partial u}{\partial x_i}$$
$$+ \tilde{c}(x)u, \qquad t > 0, x \in R^d, \tag{11.7.3}$$

where

$$\tilde{b}_i(x) = -b_i(x) + \sum_{j=1}^{d} \frac{\partial a_{ij}(x)}{\partial x_j}$$

and

$$\tilde{c}(x) = \frac{1}{2} \sum_{i,j=1}^{d} \frac{\partial^2 a_{ij}(x)}{\partial x_i \partial x_j} - \sum_{i=1}^{d} \frac{\partial b_i}{\partial x_i}. \tag{11.7.4}$$

As was shown in Section 11.6, the quadratic form

$$\sum_{i,j=1}^{d} a_{ij}(x)\lambda_i\lambda_j,$$

corresponding to the term of (11.7.3) with second-order derivatives, is always nonnegative. We will assume the somewhat stronger inequality,

$$\sum_{i,j=1}^{d} a_{ij}(x)\lambda_i\lambda_j \geq \rho \sum_{i=1}^{d} \lambda_i^2, \tag{11.7.5}$$

where ρ is a positive constant, holds. This is called the **uniform parabolicity condition**.

It is known that, if the coefficients a_{ij}, $\tilde{b}_i$, and $\tilde{c}_i$ are smooth and satisfy the growth conditions

$$|a_{ij}(x)| \leq M, \quad |\tilde{b}_i(x)| \leq M(1+|x|), \quad |\tilde{c}(x)| \leq M(1+|x|^2), \qquad (11.7.6)$$

then the classical solution of the Cauchy problem, equations (11.7.2) and (11.7.3), is unique and given by the integral formula

$$u(t,x) = \int_{R^d} \Gamma(t,x,y) f(y)\, dy, \qquad (11.7.7)$$

where the kernel Γ, called the **fundamental solution**, is independent of the initial density function f.

However, we are more interested in studying equation (11.7.1) than (11.7.3) which plays an ancillary role in our considerations. To this end, we start with the following.

Definition 11.7.1. Let $f: R^d \to R$ be a continuous function. A function $u(t,x)$, $t > 0$, $x \in R^d$ is called a **classical solution** of equation (11.7.1) with the initial condition (11.7.2) if it satisfies the following conditions:

(a) For every $T > 0$ there is a $c > 0$ and $\alpha > 0$ such that

$$|u(t,x)| \leq ce^{\alpha|x|^2} \qquad \text{for } 0 < t \leq T, x \in R^d.$$

(b) $u(t,x)$ has continuous derivatives u_t, u_x, $u_{x_i x_j}$ and satisfies equation (11.7.3) for every $t > 0$, $x \in R^d$; and

(c) $\lim\limits_{t \to 0} u(t,x) = f(x)$. $\qquad\qquad (11.7.8)$

Condition (a) is necessary because for functions which grow faster than $e^{\alpha|x|^2}$, the Cauchy problem, even for the heat equation $u_t = \frac{1}{2}\sigma^2 u_{xx}$, is not uniquely determined. Condition (b) is obvious, and (c) is necessary since (11.7.3) is satisfied only for $t > 0$ and, thus, the values of $u(t,x)$ for $t > 0$ must be related to the initial condition $u(0,x) = f(x)$.

The existence and uniqueness or solutions for the initial value (Cauchy) problems (11.7.1)–(11.7.2) or (11.7.3)–(11.7.2) are given in every standard textbook on parabolic equations. General results may be found in Friedman [1964], Eidelman [1969], Chabrowski [1970], and Bessala [1975].

To state a relatively simple existence and uniqueness theorem, we require the next definition.

Definition 11.7.2. We say that the coefficients a_{ij} and b_i of equation (11.7.1) are **regular for the Cauchy problem** if they are C^4 functions such that the corresponding coefficients a_{ij}, $\tilde{b}_i$, and $\tilde{c}$ of equation (11.7.3) satisfy the uniform parabolicity condition (11.7.5) and the growth conditions (11.7.6).

The theorem that ensures the existence and uniqueness of classical solutions may be stated as follows.

Theorem 11.7.1. *Assume that the coefficients a_{ij} and b_i are regular for the Cauchy problem and that f is a continuous function satisfying the inequality $|f(x)| \leq ce^{\alpha|x|^2}$ with constants $c > 0$ and $\alpha > 0$. Then there is a unique classical solution of (11.7.1)–(11.7.2) which is given by (11.7.7). The kernel $\Gamma(t, x, y)$, defined for $t > 0$, $x, y \in R^d$, is continuous and differentiable with respect to t, is twice differentiable with respect to x_i, and satisfies (11.7.3) as a function of (t, x) for every fixed y. Further, in every strip $0 < t \leq T$, $x \in R$, $|y| \leq r$, Γ satisfies the inequalities*

$$0 < \Gamma(t, x, y) \leq \Phi(t, x - y), \qquad \left|\frac{\partial \Gamma}{\partial t}\right| \leq \Phi(t, x - y),$$

$$\left|\frac{\partial \Gamma}{\partial x_i}\right| \leq \Phi(t, x - y), \qquad \left|\frac{\partial^2 \Gamma}{\partial x_i \partial x_j}\right| \leq \Phi(t, x - y), \qquad (11.7.9)$$

where

$$\Phi(t, x - y) = kt^{-(n+2)/2} \exp[-\delta(x - y)^2/t] \qquad (11.7.10)$$

and the constants δ and k depend on T and r.

The explicit construction of the fundamental solution Γ for general coefficients a_{ij}, $\bar{b}_i$, and $\tilde{c}$ is usually impossible. It is easy only for some special cases, such as the heat equation,

$$u_t = (\sigma^2/2)u_{xx}.$$

In this case, Γ is the familiar kernel

$$\Gamma(t, x, y) = \frac{1}{\sqrt{2\pi\sigma^2 t}} \exp[-(x - y)^2/2\sigma^2 t].$$

Nevertheless, the properties (11.7.9) of Γ given in Theorem 11.7.1 allow us to deduce some very interesting properties of the solution $u(t, x)$.

Let f be a continuous function with bounded support, say on the ball $B_r = \{x : |x| \leq r\}$, and let u be the corresponding solution of equations (11.7.2) and (11.7.3). Then, from the first inequality (11.7.9), we have

$$|u(t, x)| \leq \int_{B_r} \Gamma(t, x, y)|f(y)|\, dy \leq M \int_{B_r} \Phi(t, x - y)\, dy$$

where $M = \max_y |f|$. Further, since $|x - y|^2 \geq \frac{1}{2}x^2 - r^2$ for $|y| \leq r$, we have

$$\int_{B_r} \Phi(t, x - y)\, dy \leq kt^{-(n+2)/2} \exp\left(-\delta \left[\tfrac{1}{2}|x|^2 - r^2\right]/t\right) |B_r|$$

and, consequently,

$$|u(t, x)| \leq Kt^{-(n+2)/2} \exp\left(-\tfrac{1}{2}\delta|x|^2/t\right),$$

where $K = kMe^{\delta r^2}|B_r|$. By using the remaining inequalities (11.7.9), we may derive analogous inequalities for the derivatives of u as summarized in

$$|u|, |u_t|, |u_{x_i}|, |u_{x_i x_j}| \leq Kt^{-(n+2)/2} \exp\left(-\tfrac{1}{2}\delta|x|^2/t\right). \qquad (11.7.11)$$

These inequalities are quite important for they allow us to multiply equation (11.7.3) by any function that increases more slowly than $\exp(-\tfrac{1}{2}|x|^2)$ decreases (e.g., $x, x^2, \ldots, e^{rx}$), and then to integrate term by term to, for example, calculate the moments of $u(t, x)$.

Example 11.7.1. Again consider the Langevin equation

$$\frac{dx}{dt} = -bx + \sigma\xi$$

first introduced in Example 11.5.1. The corresponding Fokker–Planck equation is

$$\frac{\partial u}{\partial t} = \tfrac{1}{2}\sigma^2 \frac{\partial^2 u}{\partial x^2} + b\frac{\partial}{\partial x}(xu). \qquad (11.7.12)$$

Multiply (11.7.12) by x^n and integrate to obtain

$$\frac{d}{dt} \int_{-\infty}^{\infty} x^n u \, dx = \tfrac{1}{2}\sigma^2 \int_{-\infty}^{\infty} x^n \frac{\partial^2 u}{\partial x^2} \, dx + b \int_{-\infty}^{\infty} x^n \frac{\partial}{\partial x}(xu) \, dx.$$

Since by our foregoing discussion, u and its derivatives decay exponentially as $|x| \to \infty$, we can integrate by parts to give

$$\frac{d}{dt} \int_{-\infty}^{\infty} x^n u \, dx = \tfrac{1}{2}\sigma^2 n(n-1) \int_{-\infty}^{\infty} x^{n-2} u \, dx - nb \int_{-\infty}^{\infty} x^n u \, dx. \qquad (11.7.13)$$

Let

$$m_n(t) = \int_{-\infty}^{\infty} x^n u(t, x) \, dx$$

be the nth moment of the function $u(t, x)$. From (11.7.13) we thus have an infinite system of ordinary differential equations in the moments,

$$\frac{dm_0}{dt} = 0, \qquad \frac{dm_1}{dt} = -bm_1,$$

$$\frac{dm_n}{dt} = \tfrac{1}{2}\sigma^2 n(n-1)m_{n-2} - nbm_n, \qquad n \geq 2,$$

which can be solved sequentially. Assuming that the initial function f is a density, we have

$$m_0(t) = m_0(0) = \int_{-\infty}^{\infty} f \, dx = 1,$$

$$m_1(t) = C_1 e^{-bt}, \qquad C_1 = m_1(0),$$

$$m_2(t) = \frac{\sigma^2}{2b} + C_2 e^{-2bt}, \qquad C_2 = m_2(0) - \frac{\sigma^2}{2b},$$

$$m_3(t) = C_3 e^{-3bt} + \frac{3C_1\sigma^2}{2b}(e^{-bt} - e^{-3bt}), \qquad C_3 = m_3(0).$$

Successive formulas for higher moments become progressively more complicated. However, it is straightforward to demonstrate inductively that

$$\lim_{t\to\infty} m_n(t) = \begin{cases} 1 \cdot 3 \cdot 5 \cdots (n-1)\left(\frac{\sigma^2}{2b}\right)^{n/2}, & \text{for } n \text{ even} \\ 0, & \text{for } n \text{ odd.} \end{cases}$$

Thus the limiting moments are the same as the moments of the Gaussian density

$$g_{\sigma b}(x) = \sqrt{b/\pi\sigma^2}\, \exp(-bx^2/\sigma^2).$$

At the end of the next section it will become clear that not only do the moments of the solution of equation (11.7.12) converge to the moments of the Gaussian density $g_{\sigma b}$, but also that $u(t,x) \to g_{\sigma b}(x)$ as $t \to \infty$. $\square$

Remark 11.7.1. A comparison of the discrete and continuous time systems with stochastic perturbations considered here reveals a close analogy between the dynamical laws, equations (10.4.1) and (10.5.1), and the stochastic differential equation (11.6.1) as well as between equations (10.4.3) and (10.5.4), for the evolution of densities, and the Cauchy problem, equations (11.7.1) and (11.7.2). $\square$

11.8   Semigroups of Markov Operators Generated by Parabolic Equations

In this section we examine the solutions of the Fokker–Planck equation as a flow of densities governed by a semigroup of Markov operators. We start with the following definition.

Definition 11.8.1. Assume that the coefficients a_{ij} and b_i, of (11.7.1) are regular for the Cauchy problem. Then, for every $f \in L^1$, not necessarily continuous, the function

$$u(t,x) = \int_{R^d} \Gamma(t,x,y)f(y)\,dy \tag{11.8.1}$$

will be called a **generalized solution** of the Cauchy problem (11.7.1) and (11.7.2).

Since $\Gamma(t,x,y)$, as a function of (t,x), satisfies (11.7.1) for $t > 0$, $u(t,x)$ has the same property. However, if f is discontinuous, then condition (11.7.8) might not hold at a point of discontinuity.

Having a generalized solution, we define a family of operators $\{P_t\}_{t\geq 0}$ by

$$P_0 f(x) = f(x), \qquad P_t f(x) = \int_{R^d} \Gamma(t, x, y) f(y)\, dy. \qquad (11.8.2)$$

We will now show that, from the properties of Γ stated in Theorem 11.7.1, we obtain the following corollary.

Corollary 11.8.1. *The family of operators $\{P_t\}_{t\geq 0}$ is a stochastic semigroup, that is,*

(1) $P_t(\lambda_1 f_1 + \lambda_2 f_2) = \lambda_1 P_t f_1 + \lambda_2 P_t f_2,\ f_1 f_2 \in L^1$;

(2) $P_t f \geq 0$ *for* $f \geq 0$;

(3) $\|P_t f\| = \|f\|$ *for* $f \geq 0$;

(4) $P_{t_1+t_2} f = P_{t_1}(P_{t_2} f),\ f \in L^1$.

Proof. Properties (1) and (2) follow immediately from equation (11.8.1) since the right-hand side is an integral operator with a positive kernel.

To verify (3), first assume that f is continuous with compact support. By multiplying the Fokker–Planck equation by a C^2 bounded function $h(x)$ and integrating, we obtain

$$\int_{R^d} h(x) u_t dx = \int_{R^d} h(x) \left\{ \frac{1}{2} \sum_{i,j=1}^{d} \frac{\partial^2}{\partial x_i \partial x_j}(a_{ij} u) - \sum_{i=1}^{d} \frac{\partial}{\partial x_i}(b_i u) \right\} dx$$

and integration by parts gives

$$\int_{R^d} h(x) u_t\, dx = \int_{R^d} \left\{ \frac{1}{2} \sum_{i,j=1}^{d} a_{ij} \frac{\partial^2 h}{\partial x_i \partial x_j} + \sum_{i=1}^{d} b_i \frac{\partial h}{\partial x_i} \right\} u\, dx.$$

Setting $h = 1$, we have

$$\frac{d}{dt} \int_{R^d} u\, dx = \int_{R^d} u_t\, dx = 0.$$

Since $u \geq 0$ for $f \geq 0$, we have

$$\frac{d}{dt}\|u\| = 0 \qquad \text{for } t > 0.$$

Further, the initial condition (11.7.8), inequality (11.7.11) and the boundedness of u imply, by the Lebesgue dominated convergence theorem, that $\|P_t f\|$ is continuous at $t = 0$. This proves that $\|P_t f\|$ is constant for all $t \geq 0$. If $f \in L^1$ is an arbitrary function, we can choose a sequence $\{f_k\}$ of continuous functions with compact support that converges strongly to f. Now,

$$|(\|P_t f\| - \|f\|)| \leq |(\|P_t f\| - \|P_t f_k\|)| + |(\|P_t f_k\| - \|f_k\|)| + \|f_k - f\|. \quad (11.8.3)$$

Since, as we just showed, P_t preserves the norm, the term $\|P_t f\| - \|f_k\|$ is zero. To evaluate the first term, note that

$$|(\|P_t f\| - \|P_t f_k\|)| \leq \|P_t f - P_t f_k\|$$
$$\leq \int_{R^d} \Gamma(t, x, y) \|f - f_k\| \, dy \leq M_t \|f - f_k\|,$$

where $M_t = \sup_{x,y} \Gamma$. Thus the right-hand side of (11.8.3) converges strongly to zero as $k \to \infty$. Since the left-hand side is independent of k, we have $\|P_t f\| = \|f\|$, which completes the proof of (3). As we know, conditions (1)–(3) imply that $\|P_t f\| \leq \|f\|$ for all f and, thus, the operators P_t are continuous.

Finally to prove (4), again assume f is a continuous function with compact support and set $\bar{u}(t, x) = u(t + t_1, x)$. An elementary calculation shows that $\bar{u}(t, x)$ satisfies the Fokker–Planck equation with the initial condition $\bar{u}(0, x) = u(t_1, x)$. Thus, by the uniqueness of solutions to the Fokker–Planck equation,

$$u(t + t_1, x) = P_t u(t_1, x)$$

and, at the same time,

$$u(t + t_1, x) = P_{t+t_1} f(x) \quad \text{and} \quad u(t_1, x) = P_{t_1} f(x).$$

From these it is immediate that

$$P_{t+t_1} f = P_t(P_{t_1} f),$$

which proves (4) for all continuous f with compact support. If $f \in L^1$ is arbitrary, we again pick a sequence $\{f_k\}$ of continuous functions with compact supports that converges strongly to f and for which

$$P_{t_2+t_1} f_k = P_{t_2}(P_{t_1} f_k)$$

holds. Since the P_t have been shown to be continuous, we may pass to the limit of $k \to \infty$ and obtain (4) for arbitrary f. ∎

Remark 11.8.1. In developing the material of Theorems 11.6.1, 11.7.1, and Corollary 11.8.1, we have passed from the description of $u(t, x)$ as the density of the random variable $x(t)$, through a derivation of the Fokker–Planck equation for $u(t, x)$ and then shown that the solutions of the Fokker–Planck equation define a stochastic semigroup $\{P_t\}_{t \geq 0}$. This semigroup describes the behavior of the semi-dynamical system, equations (11.6.1) and

(11.6.2). In actuality, our proof of Theorem 11.6.1 shows that the right-hand side of the Fokker–Planck equation is the infinitesimal operator for $P_t f$, although our results were not stated in this fashion. Further, Theorem 11.7.1 and Corollary 11.8.1 give the construction of the semigroup generated by this infinitesimal operator. □

Remark 11.8.2. Observe that, when the stochastic perturbation disappears ($\sigma_{ij} = 0$), then the Fokker–Planck equation reduces to the Liouville equation and $\{P_t\}$ is simply the semigroup of Frobenius–Perron operators corresponding to the dynamical system

$$\frac{dx_i}{dt} = b_i(x), \qquad i = 1, \ldots, d. \quad \square$$

11.9 Asymptotic Stability of Solutions of the Fokker–Planck Equation

As we have seen, the fundamental solution Γ may be extremely useful. However, since a formula for Γ is not available in the general case, it is not of much use in the determination of asymptotic stability properties of $u(t,x)$. Thus, we would like to have other techniques available, and in this section we develop the use of Liapunov functions for this purpose, following Dłotko and Lasota [1983].

Here, by a **Liapunov function** we mean any function $V\colon R^d \to R$ that satisfies the following four properties:

(1) $V(x) \geq 0$ for all x;

(2) $\lim_{|x|\to\infty} V(x) = \infty$;

(3) V has continuous derivatives $(\partial V/\partial x_i)$, $(\partial^2 V/\partial x_i \partial x_j)$, $i, j = 1, \ldots, d$; and

(4) $V(x) \leq \rho e^{\delta|x|}$, $\left|\dfrac{\partial V(x)}{\partial x_i}\right| \leq \rho e^{\delta|x|}$, and $\left|\dfrac{\partial^2 V(x)}{\partial x_i \partial x_j}\right| \leq \rho e^{\delta|x|}$ (11.9.1)

for some constants ρ, δ.

Conditions (1)–(4) are not very restrictive, for example, any positive definite quadratic form (of even order m)

$$V(x) = \sum_{i_1,\ldots,i_m=1}^{d} a_{i_1\ldots i_m} x_{i_1} \cdots x_{i_m}$$

is a Liapunov function. Our main purpose will be to use a Liapunov function

V that satisfies the differential inequality

$$\sum_{i,j=1}^{d} a_{ij}(x)\frac{\partial^2 V}{\partial x_i \partial x_j} + \sum_{i=1}^{d} b_i(x)\frac{\partial V}{\partial x_i} \leq -\alpha V(x) + \beta \qquad (11.9.2)$$

with positive constants α and β. Specifically, we can state the following theorem.

Theorem 11.9.1. *Assume that the coefficients a_{ij} and b_i of equation (11.7.1) are regular for the Cauchy problem and that there is a Liapunov function V satisfying (11.9.2). Then the stochastic semigroup $\{P_t\}_{t\geq 0}$ defined by the generalized solution of the Fokker–Planck equation and given in (11.8.2) is asymptotically stable.*

Proof. The proof is similar to that of Theorem 5.7.1. First pick a continuous density f with compact support and then consider the mathematical expectation of V calculated with respect to the solution u of equations (11.7.1) and (11.7.2). We have

$$E(V \mid u) = \int_{R^d} V(x)u(t,x)\,dx. \qquad (11.9.3)$$

By inequalities (11.7.11) and (11.9.1), $u(t,x)V(x)$ and $u_t(t,x)V(x)$ are integrable. Thus, differntiation of (11.9.3) with respect to t gives

$$\frac{dE(V \mid u)}{dt} = \int_{R^d} V(x)u_t(t,x)\,dx$$

$$= \int_{R^d} V(x)\left\{\frac{1}{2}\sum_{i,j=1}^{d}\frac{\partial^2}{\partial x_i \partial x_j}[a_{ij}(x)u] - \sum_{i=1}^{d}\frac{\partial}{\partial x_i}[b_i(x)u]\right\} dx.$$

Integrating by parts and using the fact that the products uV, $u_{x_i}V$, and uV_{x_i} vanish exponentially as $|x| \to \infty$, we obtain

$$\frac{dE(V \mid u)}{dt} = \int_{R^d}\left\{\frac{1}{2}\sum_{i,j=1}^{d} a_{ij}(x)\frac{\partial^2 V}{\partial x_i \partial x_j} + \sum_{i=1}^{d} b_i(x)\frac{\partial V}{\partial x_i}\right\} u(t,x)\,dx.$$

From this and inequality (11.9.2), we have

$$\frac{dE(V \mid u)}{dt} \leq -\alpha E(V \mid u) + \beta.$$

To solve this differential inequality, multiply through by $e^{\alpha t}$, which gives

$$\frac{d}{dt}[E(V \mid u)e^{\alpha t}] \leq \beta e^{\alpha t}.$$

Since $E(V \mid u)$ at $t = 0$ equals $E(V \mid f)$, integration on the interval $[0, t]$ yields

$$E(V \mid u)e^{\alpha t} - E(V \mid f) \le (\beta/\alpha)(e^{\alpha t} - 1)$$

or

$$E(V \mid u) \le e^{-\alpha t}E(V \mid f) + (\beta/\alpha)(1 - e^{-\alpha t}).$$

Since $E(V \mid f)$ is finite, we can find a $t_0 = t_0(f)$ such that

$$E(V \mid u) \le (\beta/\alpha) + 1 \qquad \text{for } t \ge t_0.$$

Now let $G_q = \{x: V(x) < q\}$. From the Chebyshev inequality (5.7.9), we have

$$\int_{G_q} u(t, x)\, dx \ge 1 - \frac{E(V \mid u)}{q},$$

and taking $q > 1 + (\beta/\alpha)$ gives

$$\int_{G_q} u(t, x)\, dx \ge 1 - \frac{1}{q}\left[1 + \frac{\beta}{\alpha}\right] = \varepsilon > 0$$

for $t \ge t_0$. Since $V(x) \to \infty$ as $|x| \to \infty$, there is an $r > 0$ such that $V(x) \ge q$ for $|x| \ge r$. Thus the set G_q is contained in the ball B_r and, as a consequence,

$$u(t, x) = \int_{R^d} \Gamma(1, x, y)u(t - 1, y)dy \ge \int_{B_r} \Gamma(1, x, y)u(t - 1, y)\, dy$$

$$\ge \inf_{|y| \le r} \Gamma(1, x, y) \int_{B_r} u(t - 1, y)\, dy \ge \varepsilon \inf_{|y| \le r} \Gamma(1, x, y),$$

$$\text{for } t \ge t_0 + 1, x \in R^d. \qquad (11.9.4)$$

Since $\Gamma(1, x, y)$ is strictly positive and continuous, the function

$$h(x) = \varepsilon \inf_{|y| \le r} \Gamma(1, x, y)$$

is also positive. From (11.9.4), we have

$$P_t f(x) = u(t, x) \ge h \qquad \text{for } t \ge t_0 + 1,$$

which shows that $\{P_t\}$ has a nontrivial lower-bound function. Hence, by Theorem 7.4.1, the proof is complete. ∎

When $\{P_t\}$ is asymptotically stable, the next problem is to determine the limiting function

$$\lim_{t \to \infty} P_t f(x) = u_*(x), \qquad f \in D. \qquad (11.9.5)$$

This may be accomplished by using the following proposition.

Proposition 11.9.1. *If the assumptions of Theorem 11.9.1 are satisfied, then the limiting function u of (11.9.5) is the unique density satisfying the elliptic equation*

$$\frac{1}{2} \sum_{i,j=1}^{d} \frac{\partial^2}{\partial x_i \partial x_j} [a_{ij}(x)u] - \sum_{i=1}^{d} \frac{\partial}{\partial x_i} [b_i(x)u] = 0. \tag{11.9.6}$$

Proof. Assume that $\bar{u}(x)$ is a solution of (11.9.6). To prove the uniqueness of $\bar{u}(x)$, note that, because $\bar{u}$ is a solution of (11.9.6), it follows that $u(t, x) = \bar{u}(x)$ is a time-independent solution of the Fokker-Planck equation (11.7.1). Thus, by Theorem 11.9.1,

$$\bar{u}(x) = \lim_{t \to \infty} u(t, x) = u_*(x)$$

and $u_*(x) = \bar{u}(x)$ is unique.

Next we show that u_* satisfies (11.9.6). Let $f \in D(R^d)$ be a continuous function with compact support. We have $u(t + s, x) = P_t u(s, x)$, or

$$u(t + s, x) = \int_{R^d} \Gamma(t, x, y) u(s, y) \, dy.$$

Passing to the limit as $s \to \infty$, we obtain

$$u_*(x) = \int_{R^d} \Gamma(t, x, y) u_*(y) \, dy.$$

Since Γ is a fundamental solution of the Fokker–Planck equation, $u_*(x)$ is also a solution, and, since $u_*(x)$ is independent of t, it must satisfy equation (11.9.6). Thus the proof is complete. ∎

Example 11.9.1. Again consider the Langevin equation

$$\frac{dx}{dt} = -bx + \sigma \xi$$

and the corresponding Fokker–Planck equation

$$\frac{\partial u}{\partial t} = \frac{1}{2} \sigma^2 \frac{\partial^2 u}{\partial x^2} + b \frac{\partial}{\partial x}(xu).$$

Inequality (11.9.2) becomes

$$\frac{1}{2} \sigma^2 \frac{\partial^2 V}{\partial x^2} - bx \frac{\partial V}{\partial x} \leq -\alpha V + \beta,$$

which is satisfied with $V(x) = x^2$, $\alpha = 2b$, and $\beta = \sigma^2$. Thus all solutions $u(t, x)$, such that $u(0, x) = f(x)$ is a density, converge to the unique (nonnegative and normalized) solution u_* of

$$\frac{1}{2}\sigma^2 \frac{d^2 u}{dx^2} + b\frac{d}{dx}(xu) = 0. \tag{11.9.7}$$

The function

$$u_*(x) = g_{\sigma b}(x) = (1/\sigma)\sqrt{b/\pi}\,\exp(-bx^2/\sigma^2),$$

which is the Gaussian density with mean zero and variance $\sigma^2/2b$, satisfies (11.9.7), and, by Proposition 11.9.1, it is the unique solution. □

Example 11.9.2. Next consider the system of stochastic differential equations

$$\frac{dx}{dt} = Bx + \sigma\xi, \tag{11.9.8}$$

where $B = (b_{ij})$ and $\sigma = (\sigma_{ij})$ are constant matrices. Assume that the matrix (a_{ij}) with

$$a_{ij} = \sum_{k=1}^{d} \sigma_{ik}\sigma_{jk}$$

is nonsingular and that the unperturbed system

$$\frac{dx}{dt} = Bx \tag{11.9.9}$$

is stable asymptotically, that is, all solutions converge to zero as $t \to \infty$.

The Fokker–Planck equation corresponding to (11.9.8) has the form

$$\frac{\partial u}{\partial t} = \frac{1}{2}\sum_{i,j=1}^{d} a_{ij}\frac{\partial^2 u}{\partial x_i \partial x_j} - \sum_{i=1}^{d}\frac{\partial}{\partial x_i}[b_i(x)u]$$

where

$$b_i(x) = \sum_{j=1}^{d} b_{ij}x_j.$$

Since the coefficients a_{ij} are constant and the matrix (a_{ij}) is nonsingular, this guarantees that the uniform parabolicity condition (11.7.5) is satisfied. All of the remaining conditions appearing in Theorem 11.7.1 in this case are obvious. Since (11.9.9) is asymptotically stable, the real parts of all eigenvalues of B are negative and from the classical results of Liapunov stability theory there is a Liapunov function V such that

$$\sum_{i=1}^{d} b_i(x)\frac{\partial V}{\partial x_i} \leq -\alpha V(x), \tag{11.9.10}$$

where V is a positive definite quadratic form

$$V(x) = \sum_{i',j'=1}^{d} k_{i'j'} x_{i'} x_{j'}. \tag{11.9.11}$$

Differentiating (11.9.11) with respect to x_i and then x_j, multiplying by $\frac{1}{2} a_{ij}$, summing over i and j, and adding the result to (11.9.10) gives

$$\frac{1}{2} \sum_{i,j=1}^{d} a_{ij} \frac{\partial^2 V}{\partial x_i \partial x_j} + \sum_{i=1}^{d} b_i(x) \frac{\partial V}{\partial x_i} \leq -\alpha V(x) + \sum_{i,j=1}^{d} a_{ij} k_{ij}.$$

Thus inequality (11.9.2) is satisfied. Hence the semigroup $\{P_t\}$ generated by the perturbed system (11.9.8) is asymptotically stable.

To summarize, if the unperturbed system (11.9.9) is asymptotically stable, then any stochastic perturbation with a nonsingular matrix (a_{ij}) leads to a stochastic semigroup that is also asymptotically stable. In this case the limiting density is also Gaussian and can be found by the method of undetermined coefficients by substituting

$$u(x) = c \exp \left(\sum_{i,j=1}^{d} \rho_{ij} x_j x_j \right)$$

into the equation

$$\frac{1}{2} \sum_{i,j=1}^{d} a_{ij} \frac{\partial^2 u}{\partial x_i \partial x_j} - \sum_{i=1}^{d} \frac{\partial}{\partial x_i} \left(u \sum_{j=1}^{d} b_{ij} x_j \right) = 0. \quad \square$$

Example 11.9.3. Consider the second-order system

$$m \frac{d^2 x}{dt^2} + \beta \frac{dx}{dt} + F(x) = \sigma \xi \tag{11.9.12}$$

with constant coefficients m, β, and σ. Equation (11.9.12) describes the dynamics of many mechanical and electrical systems in the presence of "white noise." In the mechanical interpretation, m would be the mass of a body whose position is x, β is a friction coefficient, and $F = \partial \phi / \partial x$ is a conservative force (with a corresponding potential function ϕ) acting on the body. Introducing the velocity $v = dx/dt$ as a new variable, equation (11.9.12) is equivalent to the system

$$\frac{dx}{dt} = v \quad \text{and} \quad m \frac{dv}{dt} = -\beta v - F(x) + \sigma \xi. \tag{11.9.13}$$

The Fokker–Planck equation corresponding to (11.9.13) is

$$\frac{\partial u}{\partial t} = \frac{\sigma^2}{2m^2} \frac{\partial^2 u}{\partial v^2} - \frac{\partial}{\partial x}(vu) + \frac{1}{m} \frac{\partial}{\partial v}\{[\beta v + F(x)]u\}. \tag{11.9.14}$$

Unfortunately, the asymptotic stability of the solutions of (11.9.14) cannot be studied by Theorem 11.9.1 as the quadratic form associated with the second-order term is

$$0 \cdot \lambda_1^2 + (\sigma^2/m^2) \cdot \lambda_2^2,$$

which is clearly not positive definite. Using some sophisticated techniques, it is possible to prove that the solutions to some parabolic equations with associated semidefinite quadratic forms are asymptotically stable. However, in this example we wish only to derive the steady-state solution to (11.9.14) and to bypass the question of asymptotic stability.

In a steady state, $(\partial u/\partial t) = 0$, so (11.9.14) becomes

$$\frac{\sigma^2}{2m^2} \frac{\partial^2 u}{\partial v^2} - \frac{\partial}{\partial x}(vu) + \frac{1}{m} \frac{\partial}{\partial v}\{[\beta v + F(x)]u\} = 0,$$

which may be written in the alternate form

$$\left(\frac{\beta}{m} \frac{\partial}{\partial v} - \frac{\partial}{\partial x}\right)\left[vu + \frac{\sigma^2}{2m\beta} \frac{\partial u}{\partial v}\right] + \frac{\partial}{\partial v}\left[\frac{1}{m}F(x)u + \frac{\sigma^2}{2m\beta} \frac{\partial u}{\partial x}\right] = 0.$$

Set $u(x,v) = X(x)V(v)$, so that the last equation becomes

$$\left(\frac{\beta}{m} \frac{\partial}{\partial v} - \frac{\partial}{\partial x}\right)\left[X\left(vV + \frac{\sigma^2}{2m\beta} \frac{dV}{dv}\right)\right] + \left[\frac{1}{m}F(x)X + \frac{\sigma^2}{2m\beta} \frac{dX}{dx}\right]\frac{dV}{dv} = 0,$$

which will certainly be satisfied if X and V satisfy

$$\frac{dX}{dx} + \frac{2\beta}{\sigma^2}F(x)X = 0 \tag{11.9.15}$$

and

$$\frac{dV}{dv} + \frac{2m\beta}{\sigma^2}vV = 0, \tag{11.9.16}$$

respectively.

Integrating equations (11.9.15) and (11.9.16) and combining the results gives

$$u(x,v) = c\exp\left\{-(2\beta/\sigma^2)\left[\tfrac{1}{2}mv^2 + \phi(x)\right]\right\}. \tag{11.9.17}$$

The constant c in (11.9.17) is determined from the normalization condition

$$\int_{-\infty}^{\infty}\int_{-\infty}^{\infty} u(x,v)\,dx\,dv = 1.$$

The velocity integration is easily carried out and we have

$$c = c_1\sqrt{\beta m/\pi\sigma^2},$$

where

$$\frac{1}{c_1} = \int_{-\infty}^{\infty} \exp[(-2\beta/\sigma^2)\phi(x)] \, dx. \tag{11.9.18}$$

Thus (11.9.17) becomes

$$u(x,v) = c_1\sqrt{\beta m/\pi\sigma^2} \exp\left\{-(2\beta/\sigma^2)\left[\tfrac{1}{2}mv^2 + \phi(x)\right]\right\}. \tag{11.9.19}$$

The interesting feature of (11.9.19) is that the right-hand side may be written as the product of two functions, one dependent on v and the other on x. This can be interpreted to mean that in the steady state the positions and velocities are independent. Furthermore, observe that for every ϕ for which the integral (11.9.18) is convergent, $u(x,v)$, as given by (11.9.19), is a well-defined solution of the steady-state equation and that the distribution of velocities is Maxwellian, independent of the nature of the potential function ϕ. The Maxwellian nature of the velocity distribution is a natural consequence of the characteristics of the noise perturbation term in the force balance equation (11.9.13). $\square$

11.10   An Extension of the Liapunov Function Method

A casual inspection of the proofs of Theorems 5.7.1 and 11.9.1 shows that they are based on the same idea: We first prove that the mathematical expectation $E(V \mid P_t f)$ is bounded for large t and then show, by the Chebyshev inequality, that the density $P_t f$ is concentrated on some bounded region. With these facts we are then able to construct a lower-bound function. This technique may be formalized as follows.

Let a stochastic semigroup $\{P_t\}_{t\geq 0}$, $P_t: L^1(G) \to L^1(G)$, be given, where G is an unbounded measurable subset of R^d. Further, let $V: G \to R$ be a continuous nonnegative function such that

$$\lim_{|x|\to\infty} V(x) = \infty. \tag{11.10.1}$$

Also set, as before,

$$E(V \mid P_t f) = \int_G V(x) P_t f(x) \, dx. \tag{11.10.2}$$

With these definitions it is easy to prove the following proposition.

Proposition 11.10.1. *Assume there exists a linearly dense subset $D_0 \subset D(G)$ and a constant $M < \infty$ such that*

$$E(V \mid P_t f) \leq M \tag{11.10.3}$$

for every $f \in D_0$ and sufficiently large t, say $t \geq t_1(f)$. Let r be such that $V(x) \geq M + 1$ for $|x| \geq r$ and $x \in G$. If, for some $t_0 > 0$, there is a nontrivial function h_r with $h_r \geq 0$ and $\|h_r\| > 0$ such that

$$P_{t_0} f \geq h_r \qquad \text{for } f \in D \tag{11.10.4}$$

whose support is contained in the ball $B_r = \{x \in R^d : |x| \leq r\}$, then the stochastic semigroup $\{P_t\}_{t \geq 0}$ is asymptotically stable.

Proof. Pick $f \in D_0$. From the Chebyshev inequality and (11.10.3), it follows that

$$\int_{G_a} P_t f(x)\, dx \geq 1 - \frac{M}{a}, \qquad \text{for } t \geq t_1, \tag{11.10.5}$$

where $G_a = \{x \in G : V(x) < a\}$. Pick $a = M + 1$ so $V(x) \geq a$ for $|x| \geq r$. Then $G_a \subset B_r$ and

$$P_t f = P_{t_0} P_{t-t_0} f \geq P_{t_0} f_t = \|f_t\| P_{t_0} \tilde{f}, \tag{11.10.6}$$

where $f_t = (P_{t-t_0} f) 1_{G_a}$ and $\tilde{f} = f_t / \|f_t\|$. From (11.10.5), we have

$$\|f_t\| = \int_{G_a} P_{t-t_0} f(x)\, dx \geq 1 - \frac{M}{a} \qquad \text{for } t \geq t_0 + t_1,$$

and, by (11.10.4), $P_{t_0} \tilde{f} \geq h_r$. Thus, using (11.10.6)., we have shown that

$$[1 - (M/a)] h_r$$

is a lower-bound function for the semigroup $\{P_t\}_{t \geq 0}$. Since, by assumption, h_r is a nontrivial function and we took $a > M$, then it follows that the lower-bound function for the semigroup $\{P_t\}_{t \geq 0}$ is also nontrivial. Application of Theorem 7.4.1 completes the proof. $\square$

Example 11.10.1. As an example of the application of Proposition 11.10.1, we will first prove the asymptotic stability of the semigroup generated by the integro-differential equation

$$\frac{\partial u(t, x)}{\partial t} + u(t, x) = \frac{\sigma^2}{2} \frac{\partial^2 u}{\partial x^2} + \int_{-\infty}^{\infty} K(x, y) u(t, y)\, dy,$$

$$t > 0, x \in R \tag{11.10.7}$$

with the initial condition

$$u(0, x) = \phi(x), \qquad x \in R, \tag{11.10.8}$$

which we first considered in Example 7.9.1. As in that example, we assume that K is a stochastic kernel, but we also assume that K satisfies

$$\int_{-\infty}^{\infty} |x| K(x, y)\, dx \leq \alpha |y| + \beta \qquad \text{for } y \in R \tag{11.10.9}$$

where α and β are nonnegative constants and $\alpha < 1$.

To slightly simplify an intricate series of calculations we assume, without any loss of generality, that $\sigma = 1$. (This is equivalent to defining a new $\bar{x} = x/\sigma$.) Our proof of the asymptotic stability of the stochastic semigroup, corresponding to equations (11.10.7) and (11.10.8), follows arguments given by Jama [1986] in verifying (11.10.3) and (11.10.4) of Proposition 11.10.1.

From Example 7.9.1, we know that the stochastic semigroup $\{P_t\}_{t\geq 0}$ generated by equations (11.10.7) and (11.10.8) is defined by (with $\sigma^2 = 1$)

$$P_t \phi = e^{-t} \sum_{n=0}^{\infty} T_n(t)\phi, \tag{11.10.10}$$

where

$$T_n(t)f = \int_0^t T_0(t-\tau)PT_{n-1}(\tau)f \, d\tau,$$

$$T_0(t)f(x) = \int_{-\infty}^{\infty} g(t, x-y)f(y) \, dy \tag{11.10.11}$$

and

$$Pf(x) = \int_{-\infty}^{\infty} K(x,y)f(y) \, dy, \quad g(t,x) = \frac{1}{\sqrt{2\pi t}} \exp(-x^2/2t). \tag{11.10.12}$$

Let $f \in D(R)$ be a continuous function with compact support. Define

$$E(t) = E(|x| \, P_t f) = \int_{-\infty}^{\infty} |x| P_t f(x) \, dx,$$

which may be rewritten using (11.10.10) as

$$E(t) = e^{-t} \sum_{n=0}^{\infty} e_n(t),$$

where

$$e_n(t) = \int_{-\infty}^{\infty} |x| T_n(t)f(x) \, dx.$$

We are going to show that $E(t)$, as given here, satisfies condition (11.10.3). If we set

$$f_{nt} = PT_{n-1}(t)f \quad \text{and} \quad q_{n\tau}(t) = \int_{-\infty}^{\infty} |x| T_0(t-\tau)f_{nt}(x) \, dx$$

then, using (11.10.11), we may write $e_n(t)$ as

$$e_n(t) = \int_0^t q_{n\tau}(t) \, d\tau. \tag{11.10.13}$$

Using the second relation in equations (11.10.11), $q_{n\tau}(t)$ can be written as

$$q_{n\tau}(t) = \int_{-\infty}^{\infty} f_{n\tau}(y) \left[\int_{-\infty}^{\infty} |x| g(t-\tau, x-y) \, dx \right] dy.$$

Since $|x| \leq |x-y| + |y|$, it is evident that

$$\int_{-\infty}^{\infty} |x| g(t-\tau, x-y) \, dx \leq \sqrt{\frac{2(t-\tau)}{\pi}} + |y| \tag{11.10.14}$$

and, as a consequence,

$$q_{n\tau}(t) \leq \int_{-\infty}^{\infty} |y| f_{n\tau}(y) \, dy + \sqrt{\frac{2(t-\tau)}{\pi}} \int_{-\infty}^{\infty} f_{n\tau}(y) \, dy. \tag{11.10.15}$$

By using equation (7.9.18) from the proof of the Phillips perturbation theorem and noting that P is a Markov operator (since K is a stochastic kernel) and $\|f\| = 1$, we have

$$\int_{-\infty}^{\infty} f_{n\tau}(y) \, dy = \|PT_{n-1}(\tau)f\| = \|T_{n-1}(\tau)f\| \leq \frac{\tau^{n-1}}{(n-1)!}. \tag{11.10.16}$$

Furthermore, from equations (11.10.9) and (7.9.18),

$$\int_{-\infty}^{\infty} |y| f_{n\tau}(y) \, dy = \int_{-\infty}^{\infty} \int_{-\infty}^{\infty} |y| K(y,z) T_{n-1}(\tau) f(z) \, dy \, dz$$

$$\leq \alpha \int_{-\infty}^{\infty} |z| T_{n-1}(\tau) f(z) \, dz + \beta \int_{-\infty}^{\infty} T_{n-1}(\tau) f(z) \, dz$$

$$\leq \alpha e_{n-1}(\tau) + \beta \frac{\tau^{n-1}}{(n-1)!}.$$

Substituting this and (11.10.16) into (11.10.15) gives

$$q_{n\tau}(t) \leq \alpha e_{n-1}(\tau) + \left[\beta + \sqrt{\frac{2(t-\tau)}{\pi}} \right] \frac{\tau^{n-1}}{(n-1)!}$$

so that (11.10.13) becomes

$$e_n(t) \leq \alpha \int_0^t e_{n-1}(\tau) \, d\tau + \beta \frac{t^n}{n!} + \sqrt{\frac{2}{\pi}} \int_0^t \sqrt{t-\tau} \frac{\tau^{n-1}}{(n-1)!} \, d\tau,$$

$$n = 1, 2, \ldots . \tag{11.10.17}$$

To obtain $e_0(t)$ we again use (11.10.14) to give

$$e_0(t) = \int_{-\infty}^{\infty} |x| T_0(t) f(x) \, dx = \int_{-\infty}^{\infty} \int_{-\infty}^{\infty} |x| g(t, x-y) f(y) \, dx \, dy$$

$$\leq \sqrt{\frac{2t}{\pi}} + m_1, \quad m_1 = \int_{-\infty}^{\infty} |y| f(y) \, dy. \tag{11.10.18}$$

With equations (11.10.17) and (11.10.18) we may now proceed to examine $E(t)$. Sum (11.10.17) from $n = 1$ to m and add (11.10.18). This gives

$$\sum_{n=0}^{m} e_n(t) \le m_1 + \sqrt{\frac{2t}{\pi}} + \beta e^t + \sqrt{\frac{2}{\pi}} \int_0^t \sqrt{t-\tau}\, e^\tau\, d\tau + \int_0^t \sum_{n=0}^{m} e_n(\tau)\, d\tau,$$

where we used the fact that

$$\sum_{n=1}^{m} \frac{t^n}{n!} \le \sum_{n=0}^{\infty} \frac{t^n}{n!} = e^t.$$

Define $E_m(t) = e^{-t} \sum_{n=0}^{m} e_n(t)$; hence we can write

$$E_m(t) \le m_1 e^{-t} + \rho + \alpha \int_0^t e^{-(t-\tau)} E_m(\tau)\, d\tau, \qquad (11.10.19)$$

where

$$\rho = \beta + \max_t \left[\sqrt{\frac{2t}{\pi}} e^{-t} \right] + \int_0^\infty \sqrt{u}\, e^{-u}\, du.$$

To solve the integral inequality (11.10.19), it is enough to solve the corresponding equality and note that $E_m(t)$ is below this solution [Walter, 1970]. This process leads to

$$E_m(t) \le [\rho/(1-\alpha)] + m_1 e^{-(1-\alpha)t},$$

or passing to the limit as $m \to \infty$,

$$E(t) \le [\rho/(1-\alpha)] + m_1 e^{-(1-\alpha)t}. \qquad (11.10.20)$$

Since the constant ρ does not depend on f, (11.10.20) proves that the semigroup $\{P_t\}_{t\ge0}$, generated by (11.10.7) and (11.10.8), satisfies equation (11.10.3) with $V(x) = |x|$.

Next we verify equation (11.10.4). Assume that $f \in D(R)$ is supported on $[-r, r]$. Then we have

$$
\begin{aligned}
P_1 f &\ge e^{-1} T_0(1) f = e^{-1} \frac{1}{\sqrt{2\pi}} \int_{-r}^{r} f(y) \exp\left[-\tfrac{1}{2}(x-y)^2\right] dy \\
&\ge \frac{1}{\sqrt{2\pi}} \exp[-(x^2 + r^2 + 1)] \int_{-r}^{r} f(y)\, dy \\
&= \frac{1}{\sqrt{2\pi}} \exp[-(x^2 + r^2 + 1)],
\end{aligned}
$$

and the function on the right-hand side is clearly nontrivial.

Thus we have shown that the semigroup $\{P_t\}_{t\geq 0}$ generated by equations (11.10.7) and (11.10.8) is asymptotically stable, and therefore the solution with every initial condition $\phi \in D$ converges to the same limit. □

Example 11.10.2. Using a quite analogous approach, we now prove the asymptotic stability generated by the equation

$$\frac{\partial u(t,x)}{\partial t} + c\frac{\partial u(t,x)}{\partial x} + u(t,x) = \int_x^\infty K(x,y)u(t,y)\,dy \qquad (11.10.21)$$

with the conditions

$$u(t,0) = 0 \quad \text{and} \quad u(0,x) = \phi(x) \qquad (11.10.22)$$

(see Example 7.9.2). However, in this case some additional constraints on kernel K will be introduced at the end of our calculations. The necessity of these constraints is related to the fact that the smoothing properties of the semigroup generated by the infinitesimal operator (d^2/dx^2) of the previous example are not present now (see Example 7.4.1). Rather, in the present example the operator (d/dx) generates a semigroup that merely translates functions (see Example 7.4.2). Thus, in general the properties of equations (11.10.7) and (11.10.21) are quite different in spite of the fact that we are able to write the explicit equations for the semigroups generated by both equations using the formulas of the Phillips perturbation theorem. Our treatment follows that of Dłotko and Lasota [1986].

To start, we assume K is a stochastic kernel and satisfies

$$\int_0^t xK(x,y)\,dx \leq \alpha y + \beta \qquad \text{for } y > 0, \qquad (11.10.23)$$

where α and β are nonnegative constants and $\alpha < 1$. In the Chandrasekhar–Münch equation, $K(x,y) = \psi(x/y)/y$, and (11.10.23) is automatically satisfied since

$$\int_0^y xK(x,y)\,dx = \int_0^y (x/y)\psi(x/y)\,dx = y\int_0^1 z\psi(z)\,dz$$

and

$$\int_0^1 z\psi(z)\,dz < \int_0^1 \psi(z)\,dz = 1.$$

As in the preceding example, the semigroup $\{P_t\}_{t\geq 0}$ generated by equations (11.10.21) and (11.10.22) is given by equations (11.10.10) and (11.10.11), but now (assuming $c = 1$ for ease of calculations)

$$T_0(t)f(x) = 1_{[0,\infty)}(x - t)f(x - t) \qquad (11.10.24)$$

and

$$Pf(x) = \int_x^\infty K(x,y)f(y)\,dy. \qquad (11.10.25)$$

To verify condition (11.10.3), assume that $f \in D([0, \infty))$ is a continuous function with compact support contained in $(0, \infty)$ and consider

$$E(t) = \int_0^\infty x P_t f(x) \, dx.$$

By using notation similar to that introduced in Example 11.10.1, we have

$$E(t) = e^{-t} \sum_{n=0}^\infty e_n(t), \qquad e_n(t) = \int_0^\infty x T_n(t) f(x) \, dx,$$

and

$$e_n(t) = \int_0^t q_{n\tau}(\tau) \, d\tau, \qquad q_{n\tau}(t) = \int_0^\infty x T_0(t - \tau) P f_{n\tau}(x) \, dx,$$

where $f_{n\tau} = T_{n-1}(\tau) f$. From equations (11.10.24) and (11.10.25), we have

$$q_{n\tau}(t) = \int_{t-\tau}^\infty x \left[\int_{x-t+\tau}^\infty K(x - t + \tau, y) f_{n\tau}(y) \, dy \right] dx,$$

or, setting $x - t + \tau = z$ and using (11.10.23),

$$q_{n\tau}(t) = \int_0^\infty \left[\int_z^\infty z K(z, y) f_{n\tau}(y) \, dy \right] dz$$

$$+ (t - \tau) \int_0^\infty \left[\int_z^\infty K(z, y) f_{n\tau}(y) \, dy \right] dz$$

$$\leq \alpha \int_0^\infty y f_{n\tau}(y) \, dy + \beta \int_0^\infty f_{n\tau}(y) \, dy$$

$$+ (t - \tau) \int_0^\infty \left[\int_z^\infty K(z, y) f_{n\tau}(y) \, dy \right] dz.$$

Since K is stochastic and

$$\|f_{n\tau}\| = \|T_{n-1}(\tau) f\| \leq \tau^{n-1}/(n - 1)!,$$

this inequality reduces to

$$q_{n\tau}(t) \leq \alpha e_{n-1}(\tau) + [\beta + t - \tau][\tau^{n-1}/(n - 1)!], \qquad n = 1, 2, \ldots.$$

Thus

$$e_n(t) \leq \alpha \int_0^t e_{n-1}(\tau) d\tau + \beta \frac{t^n}{n!} + \int_0^t (t - \tau) \frac{\tau^{n-1}}{(n - 1)!} \, d\tau. \qquad (11.10.26)$$

Further,

$$e_0(t) = \int_0^\infty x T_0(t) f(x) \, dx = \int_t^\infty x f(x - t) \, dx$$

$$= \int_0^\infty z f(z) \, dz + t \int_0^\infty f(z) \, dz$$

or

$$e_0(t) = m_1 + t, \qquad m_1 = \int_0^\infty z f(z)\, dz. \qquad (11.10.27)$$

Observe the similarity between equations (11.10.26)–(11.10.27) and equations (11.10.17)–(11.10.18). Thus, proceeding as in Example 11.10.1, we again obtain (11.10.20) with

$$\rho = \beta + \int_0^\infty u e^{-u}\, du + \max_t(te^{-t}).$$

Thus we have shown that the semigroup generated by equations (11.10.21)–(11.10.22) satisfies condition (11.10.3).

However, the proof that (11.10.4) holds is more difficult for the reasons set out at the beginning of this example. To start, pick $r > 0$ as in Proposition 11.10.1, that is,

$$r = M + 1 = [\rho/(1 - \alpha)] + 1.$$

For an arbitrary $f \in D([0, r])$ and $t_0 > 0$, we have

$$P_{t_0} f(x) \geq e^{-t_0} T_1(t_0) f(x) = e^{-t_0} \int_0^{t_0} T_0(t_0 - \tau) P T_0(\tau) f(x)\, d\tau$$

$$= e^{-t_0} \int_0^{t_0} \Big[1_{[0,\infty)}(x - t_0 + \tau)$$

$$\cdot \int_{x-t_0+\tau}^\infty K(x - t_0 + \tau, y)$$

$$\cdot 1_{[0,\infty)}(y - \tau) f(y - \tau) dy \Big] d\tau.$$

In particular, for $0 \leq x \leq t_0$,

$$P_{t_0} f(x) \geq e^{-t_0} \int_{t_0-x}^{t_0} \Big[\int_\tau^\infty K(x - t_0 + \tau, y) f(y - \tau)\, dy \Big] d\tau.$$

Now set $z = y - \tau$ and $s = x - t_0 + \tau$ and remember that $f \in D([0, r])$ to obtain

$$P_{t_0} f(x) \geq e^{-t_0} \int_0^x \Big[\int_0^r K(s, z + s + t_0 - x) f(z)\, dz \Big] ds$$

$$\geq h_r(x) \int_0^r f(z)\, dz = h_r(x) \qquad \text{for } 0 \leq x \leq t_0,$$

where

$$h_r(x) = e^{-t_0} \inf_{0 \leq z \leq r} \int_0^x K(s, z + s + t_0 - x)\, ds.$$

It is therefore clear that $h_r \geq 0$, and it is easy to find a sufficient condition for h_r to be nontrivial. For example, if $K(s, u) = \psi(s/u)/u$, as in the Chandrasekhar–Münch equation, then

$$h_r(x) = e^{-t_0} \inf_z \int_0^x \psi\left(\frac{s}{z + s + t_0 - x}\right) \frac{ds}{z + s + t_0 - x}.$$

If we set $q = s/(z + s + t_0 - x)$ in this expression, then

$$h_r(x) = e^{-t_0} \inf_z \int_0^{x/(z+t_0)} \frac{\psi(q)}{1 - q} \, dq$$

$$\geq e^{-t_0} \int_0^{x/(r+t_0)} \psi(q) \, dq.$$

Since $\psi(q)$ is a density, we have

$$\lim_{t_0 \to \infty} \int_0^{x/(r+t_0)} \psi(q) \, dq = 1$$

uniformly for $x \in [t_0 - 1, t_0]$. Thus, for some sufficiently large t_0, we obtain

$$h_r(x) \geq e^{-t_0} \int_0^{x/(r+t_0)} \psi(q) \, dq > 0 \qquad \text{for } x \in [t_0 - 1, t_0],$$

showing that h_r is a nontrivial function. Therefore all the assumptions of Proposition 11.10.1 are satisfied and the semigroup $\{P_t\}_{t \geq 0}$ generated by the Chandrasekhar–Münch equation is asymptotically stable. $\square$

11.11   Sweeping for Solutions of the Fokker–Planck Equation

As we have seen in Section 11.9, semigroups generated by the Fokker–Planck equation may, for some value of the coefficients, be asymptotically stable. The example provided was the Langevin equation. On the other hand, the heat equation, perhaps the simplest Fokker–Planck equation, generates a sweeping semigroup. In this and the next section we develop a technique to distinguish between these two possibilities.

We return to equation (11.7.1) with the initial condition (11.7.2) and consider the stochastic semigroup $\{P_t\}_{t \geq 0}$ given by equations (11.8.2) generated by these conditions. We say that $\{P_t\}_{t \geq 0}$ is **sweeping** if it is sweeping with respect to the family $\mathcal{A}_c$ of all compact subsets of R^d. Thus, $\{P_t\}_{t \geq 0}$ is sweeping if

$$\lim_{t \to \infty} \int_A P_t f(x) \, dx = \lim_{t \to \infty} \int_A u(t, x) \, dx = 0, \qquad \text{for } f \in D, A \in \mathcal{A}_c.$$

$$(11.11.1)$$

In this section, we understand a **Bielecki function** to be any function $V: R^d \to R$ that satisfies the following three conditions:

(1) $V(x) > 0$ for all x;

(2) V has continuous derivatives

$$\frac{\partial V}{\partial x_i}, \quad \frac{\partial^2 V}{\partial x_i \partial x_j}, \qquad i, j = 1, \ldots, d;$$

and

(3)

$$V(x) \le \rho e^{\delta|x|}, \quad \left| \frac{\partial V(x)}{\partial x_i} \right| \le \rho e^{\delta|x|}, \quad \text{and} \quad \left| \frac{\partial^2 V(x)}{\partial x_i x_j} \right| \le \rho e^{\delta|x|}, \tag{11.11.2}$$

for some constants ρ and δ.

From condition (1) and the continuity of V it follows that

$$\inf_{x \in A} V(x) > 0 \qquad \text{for } A \in \mathcal{A}_c,$$

and consequently our new definition of a Bielecki function is completely consistent with the general definition given in Section 5.9.

With these preliminaries we are in a position to state an analog of Theorem 11.9.1, which gives a sufficient sweeping condition for semigroups generated by the Fokker–Planck equation.

Theorem 11.11.1. *Assume that the coefficients a_{ij} and b_i of equation (11.7.1) are regular for the Cauchy problem, and that there is a Bielecki function $V: R^d \to R$ satisfying the inequality*

$$\frac{1}{2} \sum_{i,j=1}^{d} a_{ij}(x) \frac{\partial^2 V}{\partial x_i \partial x_j} + \sum_{i=1}^{d} b_i(x) \frac{\partial V}{\partial x_i} \le -\alpha V(x), \tag{11.11.3}$$

with a constant $\alpha > 0$. Then the semigroup $\{P_t\}_{t \ge 0}$ generated by (11.7.1)–(11.7.2) is sweeping.

Proof. The proof proceeds exactly as the proof of Theorem 11.9.1, but is much shorter. First we pick a continuous density f with compact support and consider the mathematical expectation (11.9.3). Using inequality (11.11.3), we obtain

$$\frac{dE(V \mid u)}{dt} \le -\alpha E(V \mid u),$$

and, consequently,

$$E(V \mid P_t f) = E(V \mid u) \leq e^{-\alpha t} E(V \mid f).$$

Since $e^{-\alpha t} < 1$ for $t > 0$, Proposition 7.11.1 completes the proof. ∎

Example 11.11.1. Consider the stochastic equation

$$\frac{dx}{dt} = bx + \sigma\xi \tag{11.11.4}$$

where b and σ are positive constants and ξ is a white noise perturbation. Equation (11.11.4) differs from the Langevin equation because the coefficient of x is positive. The Fokker–Planck equation corresponding to (11.11.4) is

$$\frac{\partial u}{\partial t} = \frac{\sigma^2}{2}\frac{\partial^2 u}{\partial x^2} - b\frac{\partial(xu)}{\partial x}. \tag{11.11.5}$$

Now the inequality (11.11.3) becomes

$$\frac{\sigma^2}{2}\frac{\partial^2 V}{\partial x^2} + bx\frac{\partial V}{\partial x} \leq -\alpha V. \tag{11.11.6}$$

Pick a Bielecki function of the form $V(x) = e^{-\varepsilon x^2}$ and substitute it into (11.11.6) to obtain

$$2\varepsilon(\varepsilon\sigma^2 - b)x^2 - \varepsilon\sigma^2 \leq -\alpha.$$

This inequality is satisfied for arbitrary positive $\varepsilon \leq b/\sigma^2$ and $\alpha \leq \varepsilon\sigma^2$. This demonstrates that for $b > 0$ the semigroup $\{P_t\}_{t\geq0}$ generated by equation (11.11.5) is sweeping.

11.12 Foguel Alternative for the Fokker–Planck Equation

Stochastic semigroups generated by the Fokker–Planck equation are especially easy to study using the Foguel alternative introduced in Section 7.12. This is due to the fact that these semigroups are given by the integral formula (11.8.2).

We have the following.

Theorem 11.12.1. *Assume that the coefficients a_{ij} and b_i of equation (11.7.1) are regular for the Cauchy problem. Further assume that all stationary nonnegative solutions of equation (11.7.1) are of the form $cu_*(x)$ where $u_*(x) > 0$ a.e. and c is a nonnegative constant. Then the semigroup $\{P_t\}_{t\geq0}$ generated by equations (11.7.1)–(11.7.2) is either asymptotically stable or sweeping. Asymptotic stability occurs when*

$$I \equiv \int_{R^d} u_*(x)\,dx < \infty \tag{11.12.1}$$

and sweeping when $I = \infty$.

Proof. We are going to use Theorem 7.12.1 in the proof, sequentially verifying conditions (a), (b), and (c).

First we are going to show that the kernel $\Gamma(t, x, y)$ in equation (11.7.7) is stochastic for each $t > 0$. We already know that Γ is positive and that $\{P_t\}_{t \geq 0}$ is stochastic.

Furthermore, for each $f \in L^1(R^d)$ we have

$$\int_{R^d} f(y)\, dy = \int_{R^d} P_t f(x)\, dx = \int_{R^d} \int_{R^d} \Gamma(t, x, y) f(y)\, dx\, dy,$$

and consequently

$$\int_{R^d} \left[\int_{R^d} \Gamma(t, x, y)\, dx - 1 \right] f(y)\, dy = 0.$$

Since $f \in L^1(R^d)$ is arbitrary, this implies

$$\int_{R^d} \Gamma(t, x, y)\, dx = 1 \qquad \text{for } t > 0, y \in R^d.$$

Thus, Γ is a stochastic kernel and condition (a) of Theorem 7.12.1 is satisfied.

In verifying condition (b), note that according to the definition of the semigroup $\{P_t\}_{t \geq 0}$ the function

$$u(t, x) = P_t u_*(x)$$

is a solution of equations (11.7.1) and (11.7.2) with $f = u_*$. Since u_* is a stationary solution and the Cauchy problem is uniquely solvable, we have

$$u_*(x) = P_t u_*(x) \qquad \text{for } t \geq 0.$$

Thus, condition (b) of Theorem 7.12.1 is satisfied for $f_* = u_*$.

To verify (c) simply observe that the positivity of Γ implies that $P_t f(x) > 0$ for every $t > 0$ and $f \in D$. Thus, supp $P_t f = R^d$ and P_t is expanding for every $t > 0$. This completes the proof. ■

It is rather easy to illustrate the general theory developed above with a simple example in one dimension. Consider the stochastic differential equation

$$\frac{dx}{dt} = b(x) + \sigma(x)\xi \tag{11.12.2}$$

where σ, b, and x are scalar functions, and ξ is a one-dimensional white noise. The corresponding Fokker–Planck equation is of the form

$$\frac{\partial u}{\partial t} = \frac{1}{2} \frac{\partial^2 [\sigma^2(x) u]}{\partial x^2} - \frac{\partial [b(x) u]}{\partial x}. \tag{11.12.3}$$

Assume that $a(x) \equiv \sigma^2(x)$ and $b(x)$ are regular for the Cauchy problem, and that

$$xb(x) \le 0 \qquad \text{for } |x| \ge r, \tag{11.12.4}$$

where r is a positive constant. This last condition simply means that the interval $[-r, r]$ is attracting (or at least not repelling) for trajectories of the unperturbed equation $\dot{x} = b(x)$.

To find a stationary solution of (11.12.3) we must solve the differential equation

$$\frac{1}{2} \frac{d^2[\sigma^2(x)u]}{dx^2} - \frac{d[b(x)u]}{dx} = 0,$$

or

$$\frac{dz}{dx} = \frac{2b(x)}{\sigma^2(x)} z + c_1$$

where $z = \sigma^2 u$ and c_1 is a constant. A straightforward calculation gives

$$z(x) = e^{B(x)} \left\{ c_2 + c_1 \int_0^x e^{-B(v)} \, dy \right\},$$

where c_2 is a second constant and

$$B(x) = \int_0^x \frac{2b(y)}{\sigma^2(y)} \, dy.$$

The solution $z(x)$ will be positive if and only if

$$c_2 + c_1 \int_0^x e^{-B(v)} \, dy > 0 \qquad \text{for } -\infty < x < \infty. \tag{11.12.5}$$

From condition (11.12.4) it follows that the integral

$$\int_0^x e^{-B(v)} \, dy$$

converges to $+\infty$ if $x \to +\infty$ and to $-\infty$ if $x \to -\infty$. This shows that for $c_1 \ne 0$ inequality (11.12.5) cannot be satisfied. Thus, the unique (up to a multiplicative constant) positive stationary solution of equation (11.12.3) is given by

$$u_*(x) = \frac{c}{\sigma^2(x)} e^{B(x)}$$

with $c > 0$. Applying Theorem 11.12.1 to equation (11.12.3) we obtain the following.

Corollary 11.12.1. *Assume that the coefficients $a = \sigma^2$ and b of equation (11.12.3) are regular for the Cauchy problem and that inequality (11.12.4) is satisfied. If*

$$I = \int_{-\infty}^{\infty} \frac{1}{\sigma^2(x)} e^{B(x)} \, dx < \infty,$$

then the semigroup $\{Pt\}_{t\geq 0}$ generated by equation (11.12.3) is asymptotically stable. If $I = \infty$, then $\{Pt\}_{t\geq 0}$ is sweeping.

Example 11.12.1. Consider the differential equation (11.12.2) with $\sigma \equiv 1$ and

$$b(x) = -\frac{\lambda x}{1 + x^2},$$

where $\lambda \geq 0$ is a constant. Then

$$B(x) = -\int_0^x \frac{2\lambda y}{1 + y^2}\, dy = -\lambda \ln(1 + x^2),$$

and

$$u_*(x) = ce^{-\lambda \ln(1+x^2)} = \frac{c}{(1 + x^2)^\lambda}.$$

The function u_* is integrable on R only for $\lambda > \frac{1}{2}$, and thus the semigroup $\{P_t\}_{t\geq 0}$ is asymptotically stable for $\lambda > \frac{1}{2}$ and sweeping for $0 \leq \lambda \leq \frac{1}{2}$. This example shows that even though the origin $x = 0$ is attracting in the unperturbed system, asymptotic stability may vanish in a perturbed system whenever the coefficient of the attracting term is not sufficiently strong.

Remark 11.12.1. In Corollary 11.12.1, the conditions (11.12.4) may be replaced by the less restrictive assumption

$$\int_0^\infty e^{-B(x)}\, dx = \int_{-\infty}^0 e^{-B(x)}\, dx = \infty. \tag{11.12.6}$$

Exercises

11.1. Let $\{w(t)\}_{t\geq 0}$ be a one-dimensional Wiener process defined on a complete probabilistic measure space. Show that for every $t_0 \geq 0$, $r > 0$, and $M > 0$ the probability of the event

$$\left\{ \left| \frac{w(t_0 + h) - w(t_0)}{h} \right| \leq M \text{ for } 0 < h \leq r \right\}$$

is equal to zero. Using this, show that for every fixed $t_0 \geq 0$ the probability of the event

$$\{w'(t_0) \text{ exists}\}$$

is equal to zero.

11.2. Generalize the previous result and show that the probability of the event

$$\{w'(t) \text{ exists at least for one } t \geq 0\}$$

is equal to zero.

11.3. Show that every regular sequence $\{s_n\}$ of Itô approximation sums for the integral

$$\int_0^T w(t)\,dw(t)$$

converges to $\frac{1}{2}w^2(T) - \frac{1}{2}T$ not only stochastically but also in the mean [i.e., strongly in $L^2(\Omega)$].

11.4. Consider the stochastic differential equation

$$\frac{dx}{dt} = x(c + 2x^2 - x^4) + \sigma\xi, \qquad t > 0, \xi \in R,$$

where c and $\sigma > 0$ are constant and ξ is normalized white noise. Show that the corresponding stochastic semigroup $\{P_t\}_{t \geq 0}$ is asymptotically stable (Mackey, Longtin, and Lasota, 1990).

11.5. Show that the stochastic semigroup $\{P_t\}_{t \geq 0}$ defined in Exercise 7.8 is asymptotically stable for an arbitrary stochastic kernel K (Jama, 1986).

11.6. A stochastic semigroup $\{P_t\}_{t \geq 0}$ is called **weakly (strongly) mixing** if, for every $f_1, f_2 \in D$ the difference $P_t f_1 - P_t f_2$ converges weakly (strongly) to zero in L^1. Show that the stochastic semigroup $\{T_t\}_{t \geq 0}$ given by equation (7.9.9), corresponding to the heat equation, is strongly mixing.

11.7. Consider equation (11.12.3) with $b(x) = x/(1 + x^2)$ and $\sigma \equiv 1$. Prove that the stochastic semigroup $\{P_t\}_{t \geq 0}$ corresponding to this equation satisfies

$$\int_{-\infty}^{+\infty} (\arctan x)P_t f(x)\,dx = \text{constant} \qquad \text{for } f \in L^1,$$

and is not weakly mixing (Brzeźniak and Szafirski, 1991).

11.8. Consider the semigroup $\{P_t\}_{t \geq 0}$ defined in the previous exercise. Show that the limit

$$\lim_{t \to \infty} H(P_t f_1 \mid P_t f_2), \qquad f_1, f_2 \in D$$

depends on the choice of f_1 and f_2, where H denotes the conditional entropy, cf. Chapter 9.

12
Markov and Foias Operators

Throughout this book we have studied the asymptotic behavior of densities. However, in some cases the statistical properties of dynamical systems are better described if we use a more general notion than a density, namely, a measure. In fact, the sequences (or flows) of measures generated by dynamical systems simultaneously generalize the notion of trajectories and the sequences (or flows) of densities. They are of particular value in studying fractals.

The study of the evolution of measures related to dynamical systems is difficult. It is more convenient to study them by use of functionals on the space $C_0(X)$ of continuous functions with bounded support. Thus, we start in Section 12.1 by examining the relationship between measures and linear functionals given by the Riesz representation theorem, and then look at weak and strong convergence notions for measures in Section 12.2. After defining the notions of Markov and Foias operators on measures (Section 12.3 and 12.4, respectively), we study the behavior of dynamical systems with stochastic perturbations. Finally, we apply these results to the theory of fractals in Section 12.8.

12.1 The Riesz Representation Theorem

Let $X \subset R^d$ be a nonempty closed set which, in general, is unbounded. We denote by $\mathcal{B} = \mathcal{B}(X)$ the σ-algebra of Borel subsets of X. A measure $\mu: \mathcal{B} \to R^+$ will be called **locally finite** if it is finite on every bounded measurable subset of X, that is,

$$\mu(A) < \infty \qquad \text{for } A \in \mathcal{B}, A \text{ bounded.}$$

Of course, every locally finite measure μ is σ-finite, since X may be written as a countable sum of bounded sets:

$$X = \bigcup_{n=1}^{\infty} X_n, \tag{12.1.1}$$

where

$$X_n = \{x \in X \colon |x| \leq n\}.$$

The space of all locally finite measures on X will be denoted by $\mathcal{M} = \mathcal{M}(X)$. The subspace of $\mathcal{M}$ which contains only finite or probabilistic measures will be denoted by $\mathcal{M}_{\text{fin}}$ and $\mathcal{M}_1$, respectively. We say that a measure μ is supported on a set A if $\mu(X \setminus A) = 0$. Observe that the set A on which μ is supported is in general not unique, since if B is measurable and contains A, then $X \setminus A \supset X \setminus B$ and consequently $\mu(X \setminus B) = 0$. The elements of $\mathcal{M}_1$ are often called **distributions**.

In general, the smallest measurable set on which a measure μ is supported does not exist. However, this difficulty may be partially avoided. Denote by $B_r(x)$ a ball in X with center located at $x \in X$ and radius r, that is,

$$B_r(x) = \{y \in X \colon |y - x| < r\}.$$

Let $\mu \in \mathcal{M}$. We define the **support of the measure** μ by setting

$$\operatorname{supp} \mu = \{x \in X \colon \mu(B_\varepsilon(x)) > 0 \text{ for every } \varepsilon > 0\}.$$

It is easy to verify that $\operatorname{supp} \mu$ is a closed set. Observe that it also has the property that if A is a closed set and μ is supported on A, then $A \supset \operatorname{supp} \mu$. To see this, assume that $x \notin A$. Since $X \setminus A$ is an open set, there exists a ball $B_\varepsilon(x)$ contained in $X \setminus A$. Thus,

$$\mu(B_\varepsilon(x)) \leq \mu(X \setminus A) = 0,$$

and $x \notin \operatorname{supp} \mu$. This shows that $x \notin A$ implies $x \notin \operatorname{supp} \mu$, and consequently $A \supset \operatorname{supp} \mu$.

From the above arguments it follows that the support of a measure μ can be equivalently defined as the smallest closed set on which μ is supported. (The adjective closed is important here.)

It should also be noted that the definition of the support of a measure μ does not coincide exactly with the definition of the support of an element $f \in L^1$. The main difference is that $\operatorname{supp} \mu$ is defined precisely for every single point, but $\operatorname{supp} f$ is not (see Remarks 3.12 and 3.13).

We will often discuss measures that are supported on finite or countable sets. Perhaps the simplest of these is the δ-Dirac measure defined by

$$\delta_{x_0}(A) = \begin{cases} 1 & \text{if } x_0 \in A, \\ 0 & \text{if } x_0 \notin A. \end{cases} \qquad (12.1.2)$$

Another important class of measures are those absolutely continuous with respect to the standard Borel measure on X. According to Definition 3.1.4 every measure that is absolutely continuous with respect to the Borel measure is given by

$$\mu(A) = \int_A f(x)\, dx \qquad \text{for } A \in \mathcal{B}, \qquad (12.1.3)$$

where $f \in L^1(X)$ and $f \geq 0$.

Let $C_0 = C_0(X)$ be the space of all continuous functions $h: X \to R$ with compact support. Our goal is to study the relationship between locally finite measures on X and linear functionals on C_0. We start with the following.

Definition 12.1.1. A mapping $\varphi: C_0 \to R$ is called a **linear functional** if

$$\varphi(\lambda_1 h_1 + \lambda_2 h_2) = \lambda_1 \varphi(h_1) + \lambda_2 \varphi(h_2) \text{ for } \lambda_1, \lambda_2 \in R; h_1, h_2 \in C_0. \quad (12.1.4)$$

A linear functional is **positive** if $\varphi(h) \geq 0$ for every $h \in C_0$ with $h \geq 0$.

It is easy to define a linear functional corresponding to a locally finite measure μ. Namely, we may write

$$\varphi(h) = \int_X h(x)\mu(dx) \qquad \text{for } h \in C_0. \qquad (12.1.5)$$

Since the support of h is bounded and μ is finite on bounded sets, this integral is always well defined. Further, from the known properties of integrals (see Section 2.2) it follows that condition (12.1.4) is satisfied and that $\varphi(h) \geq 0$ for $h \geq 0$. Thus by (12.1.5) every measure $\mu \in \mathcal{M}$ defines a positive linear functional on C_0 in a natural way.

It is surprising that formula (12.1.5) gives all positive functionals on C_0. Namely, the following celebrated **Riesz representation theorem** holds.

Theorem 12.1.1. *For every positive linear functional $\varphi: C_0 \to R$ there is a unique measure $\mu \in \mathcal{M}$ such that condition (12.1.5) is satisfied.*

The proof can be found in Halmös [1974].

Observe that Theorem 12.1.1 is somewhat similar to the Radon–Nikodym theorem. In the Radon–Nikodym theorem, a measure is represented by integrals with a given density. In the Riesz theorem a functional is represented by integrals with a given measure. However, it should be noted that in the Riesz theorem even the uniqueness of the measure μ is not obvious. Namely, we cannot substitute the characteristic function $h = 1_A$ of a measurable set $A \subset X$ into formula (12.1.5) to find an explicit value of $\mu(A)$. In general, except for some trivial cases like $A = \emptyset$ or $A = X$, the characteristic function 1_A is not continuous and $\varphi(1_A)$ is not defined.

The Riesz theorem allows us to also characterize finite probabilistic measures by the use of corresponding functionals. Consider first the simplest case when X is bounded. Then $h = 1_X$ has bounded support and is continuous on X. From (12.1.5) it follows immediately that

$$\varphi(1_X) = \mu(X).$$

Thus, probabilistic measures correspond to those functionals for which

$$\varphi(1_X) = 1. \tag{12.1.6}$$

In the case when X is bounded it is not necessary to characterize the finite measures, since every locally finite measure is automatically finite.

However, if X is unbounded we cannot substitute $h = 1_X$ into (12.1.5) and we must use a more sophisticated method. Namely, let $\{h_n\}$ with $h_n \in C_0$ be a sequence of functions such that

$$0 \le h_1 \le h_2 \le \cdots, \quad \lim_{n \to \infty} h_n(x) = 1 \quad \text{for } x \in X. \tag{12.1.7}$$

Substituting h_n into (12.1.5) we obtain

$$\varphi(h_n) = \int_X h_n(x)\mu(dx)$$

which by the Lebesgue monotone convergence theorem (see Remark 2.2.4) gives

$$\lim_{n \to \infty} \varphi(h_n) = \int_X 1\mu(dx) = \mu(X). \tag{12.1.8}$$

Thus, probabilistic measures correspond to functionals φ such that

$$\lim_{n \to \infty} \varphi(h_n) = 1 \tag{12.1.9}$$

for any sequence $\{h_n\}$ with $h_n \in C_0$ satisfying conditions (12.1.7). Further, from equality (12.1.8) it follows that the validity of condition (12.1.9) does not depend on the particular choice of the sequence $\{h_n\}$. In other words, if (12.1.9) holds for one sequence $\{h_n\}$ with $h_n \in C_0$ and satisfying (12.1.7), then it is also valid for any other sequence of the same type.

Using (12.1.8) we may also characterize finite measures. Namely, all of the finite measures correspond to functionals such that

$$\lim_{n \to \infty} \varphi(h_n) < \infty \tag{12.1.10}$$

for any sequence $\{h_n\}$ with $h_n \in C_0$ and satisfying conditions (12.1.7). Again the validity of (12.1.10) does not depend on the choice of $\{h_n\}$.

Example 12.1.1. Consider a δ-Dirac measure $\mu = \delta_{x_0}$ supported on the point set $\{x_0\}$ and given by conditions (12.1.2). Then formula (12.1.5) implies

$$\varphi(h) = \int_X h(x)\mu(dx) = \int_{\{x_0\}} h(x)\mu(dx) = h(x_0).$$

Thus, the functional that corresponds to the δ-Dirac measure supported on $\{x_0\}$ is simply the map that adjoins to each function $h \in C_0$ its value at x_0. This observation is, incidentally, the starting point for the Schwartz [1966] approach to the theory of generalized functions. $\square$

Example 12.1.2. Consider an absolutely continuous measure μ with a density f. In this case, formula (12.1.5) gives

$$\varphi(h) = \int_X h(x)\mu(dx) = \int_X h(x)f(x)\,dx = \langle h, f \rangle.$$

Thus, the functional corresponding to an absolutely continuous measure is given by a scalar product. $\square$

12.2   Weak and Strong Convergence of Measures

In Section 2.3 we introduced the notions of the weak and strong convergence of sequences of L^p functions. In a somewhat similar (but not identical!) way we may introduce the concepts of weak and strong convergence of sequences of measures. We start from the definition of weak convergence, since it is quite simple and natural.

Definition 12.2.1. Let $\{\mu_n\}$ with $\mu_n \in \mathcal{M}$ be a sequence of measures and let $\mu \in \mathcal{M}$. We say that $\{\mu_n\}$ is **weakly convergent** to μ if

$$\lim_{n \to \infty} \int_X h(x)\mu_n(dx) = \int_X h(x)\mu(dx) \qquad \text{for every } h \in C_0. \qquad (12.2.1)$$

Before giving examples of weak convergence, observe that in the case when μ_n and μ are absolutely continuous, and have densities f_n and f, respectively, condition (12.2.1) reduces to

$$\langle h, f_n \rangle = \int_X h(x)f_n(x)\,dx \to \int_X h(x)f(x)\,dx = \langle h, f \rangle \qquad \text{for } h \in C_0. \tag{12.2.2}$$

This looks quite similar to condition (2.3.2) in Definition 2.3.1 for the weak convergence of a sequence $\{f_n\}$ of functions in L^p space. However, there is an important difference between conditions (2.3.2) and (12.2.2). Namely, in (2.3.2) the space of "test functions" g is larger and we must verify (2.3.2) for all g which belong to the space $L^{p'}$ adjoint to L^p. In (12.2.2) all the "test functions" h belong to C_0 and thus are continuous with compact supports.

To simplify the notation we will quite often use the notion of scalar product for measures. Thus, we write

$$\langle h, \mu \rangle = \int_X h(x)\mu(dx).$$

In this notation the weak convergence of measures has an especially simple form. Namely, $\{\mu_n\}$ converges to μ weakly if

$$\lim_{n \to \infty} \langle h, \mu_n \rangle = \langle h, \mu \rangle \qquad \text{for } h \in C_0. \tag{12.2.3}$$

Example 12.2.1. Let $X = R$ and let $\mu_n = \delta_{x_n}$ be a sequence of δ-Dirac measures supported at points $x_n \in R$. Assume that $\{x_n\}$ converges to x_* and denote by $\mu_* = \delta_{x_*}$ the δ-Dirac measure supported at x_*. We have

$$\langle h, \mu_n \rangle = h(x_n) \quad \text{and} \quad \langle h, \mu_* \rangle = h(x_*) \qquad \text{for } h \in C_0.$$

For each fixed $h \in C_0$, from the continuity of h the sequence $\{h(x_n)\}$ converges to $h(x_*)$. Consequently, the sequence of measures $\{\mu_n\}$ converges weakly to μ_*. $\square$

Example 12.2.2. Let $X = R$ and let $\{\mu_n\}$ be a sequence of measures with Gaussian densities

$$f_n(x) = \frac{1}{\sqrt{2\pi\sigma_n^2}} \exp\left\{-\frac{x^2}{2\sigma_n^2}\right\}, \qquad n = 1, 2, \ldots . \tag{12.2.4}$$

Assume that $\sigma_n \to 0$ as $n \to \infty$ and denote by $\mu_* = \delta_0$ the δ-Dirac measure supported at $x = 0$. We have

$$|\langle h, \mu_n \rangle - \langle h, \mu_* \rangle| = \left| \int_R h(x) f_n(x) dx - h(0) \right|$$

$$= \left| \int_R h(x) f_n(x) dx - \int_R f_n(x) h(0) dx \right|$$

$$\le \int_R |h(x) - h(0)| f_n(x) dx.$$

Choose an $\varepsilon > 0$. Let $r > 0$ be such that $|h(x) - h(0)| \le \varepsilon$ for $|x| \le r$. Then

$$|\langle h, \mu_n - \mu_* \rangle| \le \int_{|x| \le r} |h(x) - h(0)| f_n(x)\, dx$$

$$+ \int_{|x| \ge r} |h(x) - h(0)| f_n(x)\, dx$$

$$\le \varepsilon + 2M \int_{|x| \ge r} f_n(x)\, dx,$$

where $M = \max |h(x)|$. Using (12.2.4) and setting $x/\sigma_n = y$ we finally have

$$|\langle h, \mu_n - \mu_* \rangle| \le \varepsilon + 4M \frac{1}{\sqrt{2\pi}} \int_{r/\sigma_n}^{\infty} \exp\left(-\frac{x^2}{2}\right) dx.$$

Since the sequence $\{\sigma_n\}$ converges to zero the last integral also converges to zero which implies that

$$\lim_{n \to \infty} \langle h, \mu_n - \mu_* \rangle = 0.$$

Thus, the Gaussian measures converge weakly to a δ-Dirac measure when the standard deviations go to zero. $\square$

Example 12.2.3. As in the previous example, let the μ_n be Gaussian measures with densities given by (12.2.4). This time, however, assume that $\sigma_n \to \infty$. Denote by μ_* the measure identically equal to zero. We have

$$|\langle h, \mu_n - \mu_* \rangle| = |\langle h, \mu_n \rangle| \le \int_{-\infty}^{+\infty} |h(x)||f_n(x)| \, dx$$

$$= \int_K |h(x)||f_n(x)| \, dx, \qquad (12.2.5)$$

where K is the support of h. Let $[a, b]$ denote a bounded interval which contains K. From (12.2.4) and (12.2.5) it follows immediately that

$$|\langle h, \mu_n - \mu_* \rangle| \le \max |h| \frac{1}{\sqrt{2\pi\sigma_n^2}} \int_a^b \exp\left\{-\frac{x^2}{2\sigma_n^2}\right\} dx$$

$$\le \frac{\max |h|}{\sqrt{2\pi\sigma_n^2}} (b - a).$$

Since the sequence $\{\sigma_n\}$ converges to infinity, the integrals on the right-hand side converge to zero. This shows that the Gaussian measures converge weakly to zero when the standard deviations go to infinity. Observe, however, that in this case the sequence of densities $\{f_n\}$ does not converge weakly in L^1 to $f_* \equiv 0$. In fact, setting $g \equiv 1$ in (2.2.3) we have

$$\langle g, f_n \rangle = \int_R f_n(x) \, dx = 1, \qquad \langle g, f_* \rangle = 0,$$

and the sequence $\{\langle g, f_n \rangle\}$ does not converge to $\langle g, f_* \rangle$. $\square$

The weak convergence of $\{\mu_n\}$ to μ does not imply the convergence of $\{\mu_n(A)\}$ to $\{\mu(A)\}$ for all measurable sets A. However, it is easy to obtain some inequalities between $\mu(A)$ and $\mu_n(A)$ for large n and some special sets A.

We say that $G \subset X$ is open in X if $X \setminus G$ is a closed set. For example the ball

$$B_r(x) = \{y \in X : |x - y| < r\}$$

is open in X since

$$X \setminus B_r(x) = \{y \in X : |x - y| \geq r\}$$

is a closed set.

Theorem 12.2.1. *Assume that a sequence* $\mu_n \in \mathcal{M}_{fin}$ *converges weakly to* $\mu \in \mathcal{M}_{fin}$. *Then*

$$\liminf_{n \to \infty} \mu_n(G) \geq \mu(G) \qquad \text{for } G \subset X, \ G \text{ open in } X. \qquad (12.2.6)$$

Proof. Since G is open in X there exists a sequence of compact sets $F_1 \subset F_2 \subset \cdots$ such that

$$G = \bigcup_{k=1}^{\infty} F_k.$$

Thus, $\lim \mu(F_k) = \mu(G)$ and for any given $\varepsilon > 0$ there is a set F_k such that $\mu(F_k) \geq \mu(G) - \varepsilon$. Let $h \in C_0(X)$ be such that $0 \leq h \leq 1$ and

$$h(x) = \begin{cases} 1 & \text{for } x \in F_k, \\ 0 & \text{for } x \in X \setminus G. \end{cases}$$

Since F_k and $X \setminus G$ are closed and disjoint, the function h always exists. Evidently $h \leq 1_G$ and

$$\langle h, \mu_n \rangle \leq \mu_n(G)$$

which gives, in the limit,

$$\langle h, \mu \rangle \leq \liminf_{n \to \infty} \mu_n(G).$$

On the other hand, $h \geq 1_{F_k}$ and

$$\langle h, \mu \rangle \geq \mu(F_k) \geq \mu(G) - \varepsilon.$$

Consequently,

$$\liminf_{n \to \infty} \mu_n(G) \geq \mu(G) - \varepsilon.$$

Since $\varepsilon > 0$ was arbitrary this completes the proof. ∎

Remark 12.2.1. It is easy to observe that in general inequality in (12.2.6) cannot be replaced by the equality. In fact, let $X = R$, $\mu_n = \delta_{1/n}$, $\mu = \delta_0$ and $A = (0, 1)$. In this case the sequence $\{\mu_n\}$ converges weakly to μ, but $\mu_n(A) = 1$, $\mu(A) = 0$, and the inequality (12.2.6) is strong. □

Now we are going to show how the Riesz representation theorem may be used to show that a given sequence of measures is convergent.

Theorem 12.2.2. *Let a sequence of measures $\{\mu_n\}$, $\mu_n \in \mathcal{M}$, be given. If for each $h \in C_0$ the sequence $\{\langle h, \mu_n \rangle\}$ is convergent, then there is a unique measure μ such that $\{\mu_n\}$ converges weakly to μ.*

Proof. Define

$$\varphi(h) = \lim_{n \to \infty} \langle h, \mu_n \rangle \qquad \text{for } h \in C_0.$$

Evidently φ is a linear positive functional. Thus, according to Theorem 12.1.1 there is a unique measure μ such that

$$\varphi(h) = \langle h, \mu \rangle \qquad \text{for } h \in C_0.$$

From this and the definition of φ, it follows that the sequence $\{\mu_n\}$ converges to μ weakly. ∎

Remark 12.2.2. In the special case when the μ_n are probabilistic measures the use of Theorem 12.2.2 can be greatly simplified. Namely, it is not necessary to verify the convergence of sequences $\{\langle h, \mu_n \rangle\}$ for all $h \in C_0$. Let $C_* \subset C_0$ be a dense subset of C_0 which means that for every $h \in C_0$ and $\varepsilon > 0$ there is $g \in C_*$ such that

$$\sup_{x \in X} |g(x) - h(x)| \leq \varepsilon.$$

Then the convergence of sequences $\{\langle g, \mu_n \rangle\}$ for $g \in C_*$ implies the convergence of $\{\langle h, \mu_n \rangle\}$ for $h \in C_0$. In fact, the inequality

$$|\langle h, \mu_n - \mu_m \rangle| \leq |\langle g, \mu_n - \mu_m \rangle| + 2 \sup |g - h|$$

and the Cauchy condition for all sequences $\{\langle g, \mu_n \rangle\}$ imply the Cauchy condition for $\{\langle h, \mu_n \rangle\}$. □

We close this section by introducing the concept of the strong convergence of measures. First we need to define the **distance between two measures** $\mu_1, \mu_2 \in \mathcal{M}_{\text{fin}}$. Let $(X_1, \ldots, X_n)$ be a measurable partition of X, that is

$$X = \bigcup_{i=1}^{n} X_i, \quad X_i \cap X_j = \emptyset \qquad \text{for } i \neq j, \ X_i \in \mathcal{B}.$$

We set

$$\|\mu_1 - \mu_2\| = \sup \sum_{i=1}^{n} |\mu_1(X_i) - \mu_2(X_i)|, \qquad (12.2.7)$$

where the supremum is taken over all possible measurable partitions of X (with arbitrary n). The value $\|\mu_1 - \mu_2\|$ is the desired distance. In the special case where $\mu = \mu_1$ is arbitrary and $\mu_2 \equiv 0$ we have

$$\|\mu\| = \sup \sum_{i=1}^{n} \mu(X_i) = \mu(X). \tag{12.2.8}$$

This value will be called the **norm of the measure** μ. It is the distance from μ to zero. The norm of a probabilistic measure is equal to 1.

Definition 12.2.2. We say that a sequence $\{\mu_n\}$, $\mu_n \in \mathcal{M}_{\text{fin}}$ is **strongly convergent** to a measure $\mu \in \mathcal{M}_{\text{fin}}$ if

$$\lim_{n \to \infty} \|\mu_n - \mu\| = 0. \tag{12.2.9}$$

Before passing to examples of strong convergence, we will calculate the norm $\|\mu_1 - \mu_2\|$ in the case when the measures μ_1 and μ_2 are absolutely continuous with Radon–Nikodym derivatives f_1 and f_2, respectively. We have

$$\mu_1(X_i) - \mu_2(X_i) = \int_{X_i} (f_1(x) - f_2(x))\,dx.$$

Substituting this into (12.2.7) we obtain immediately

$$\|\mu_1 - \mu_2\| = \sup \sum_{i=1}^{n} \left| \int_{X_i} (f_1(x) - f_2(x))\,dx \right|$$

$$\leq \sup \sum_{i=1}^{n} \int_{X_i} |f_1(x) - f_2(x)|\,dx$$

$$= \int_X |f_1(x) - f_2(x)|\,dx. \tag{12.2.10}$$

Now let

$$X_1 = \{x\colon f_1(x) \geq f_2(x)\}, \qquad X_2 = \{x\colon f_1(x) < f_2(x)\}.$$

Then (X_1, X_2) is a partition of X and, consequently,

$$\|\mu_1 - \mu_2\| \geq |\mu_1(X_1) - \mu_2(X_1)| + |\mu_1(X_2) - \mu_2(X_2)|$$

$$= \int_{X_1} (f_1(x) - f_2(x))\,dx + \int_{X_2} (f_2(x) - f_1(x))\,dx$$

$$= \int_X |f_1(x) - f_2(x)|\,dx.$$

This and (12.2.10) implies

$$\|\mu_1 - \mu_2\| = \int_X |f_1(x) - f_2(x)|\,dx. \tag{12.2.11}$$

From this equality a necessary and sufficient condition for the strong convergence of absolutely continuous measures follows immediately. Namely, if the μ_n are absolutely continuous with densities f_n, and μ is absolutely continuous with density f, then $\{\mu_n\}$ converges strongly to μ if and only if $\|f_n - f\| \to 0$.

Example 12.2.4. Assume $X = R$. Let $x_0 \in X$. Denote by $\mu_0 = \delta_{x_0}$ the δ-Dirac measure supported at x_0. Further, let $\{\mu_n\}$ be a sequence of absolutely continuous measures with densities f_n. Write

$$X_1 = \{x_0\}, \qquad X_2 = X \setminus \{x_0\},$$

where, as usual, $\{x_0\}$ denotes the set that contains only the one point $x = x_0$. We have

$$\mu_0(X_1) = 1, \qquad \mu_0(X_2) = 0$$
$$\mu_n(X_1) = \int_{\{x_0\}} f_n(x)\, dx = 0$$
$$\mu_n(X_2) = \int_{X \setminus \{x_0\}} f_n(x)\, dx = \int_X f_n(x)\, dx = 1.$$

Thus, since (X_1, X_2) is a partition of X

$$\|\mu_n - \mu_0\| \geq |\mu_n(X_1) - \mu_0(X_1)| + |\mu_n(X_2) - \mu_0(X_2)|$$
$$= |0 - 1| + |1 - 0| = 2.$$

This shows that a sequence of absolutely continuous measures cannot converge strongly to a δ-Dirac measure. □

Example 12.2.5. Assume $X = R$ and consider a probabilistic measure μ supported on the set of nonnegative integers $\{0, 1, \ldots\}$. The measure μ may be written in the form

$$\mu = \sum_{k=0}^{\infty} c_k \delta_k, \qquad \sum_{k=0}^{\infty} c_k = 1, \qquad c_k \geq 0, \tag{12.2.12}$$

where δ_k denotes the δ-Dirac measure supported at $x = k$. Further, let $\{\mu_n\}$ be a sequence of similar measures, so

$$\mu_n = \sum_{k=0}^{\infty} c_{k_n} \delta_k, \qquad \sum_{k=0}^{\infty} c_{k_n} = 1, \qquad c_{k_n} \geq 0. \tag{12.2.13}$$

Assume that for each fixed k $(k = 0, 1, \ldots)$ the sequence $\{c_{k_n}\}$ converges to c_k as $n \to \infty$. We are going to show that under this condition the sequence of measures $\{\mu_n\}$ converges strongly to μ. Thus we must evaluate the distance $\|\mu_n - \mu\|$.

From (12.2.12) and (12.2.13) it follows that

$$|\mu_n(X_i) - \mu(X_i)| \le \sum_{k=0}^{\infty} |c_{k_n} - c_k| \delta_k(X_i)$$

for each measurable subset X_i of X. Consequently,

$$\|\mu_n - \mu\| = \sup \sum_{i=1}^{m} |\mu_n(X_i) - \mu(X_i)|$$

$$\le \sup \sum_{i=1}^{m} \sum_{k=0}^{\infty} |c_{k_n} - c_k| \delta_k(X_i)$$

$$\le \sup \sum_{k=0}^{\infty} |c_{k_n} - c_k| \sum_{i=1}^{m} \delta_k(X_i),$$

where the supremum is taken over all partitions $\{X_i\}$ of X. Since for every partition

$$\sum_{i=1}^{m} \delta_k(X_i) = 1 \qquad k = 0, 1, \dots,$$

this gives

$$\|\mu_n - \mu\| \le \sum_{k=0}^{\infty} |c_{k_n} - c_k|. \qquad (12.2.14)$$

Now fix an $\varepsilon > 0$ and choose an integer N such that

$$\sum_{k=N+1}^{\infty} c_k < \frac{\varepsilon}{4}.$$

When N is fixed we can find an integer n_0 such that

$$\sum_{k=0}^{N} |c_{k_n} - c_k| \le \frac{\varepsilon}{4} \qquad \text{for } n \ge n_0.$$

We have, therefore,

$$\sum_{k=N+1}^{\infty} c_{k_n} = 1 - \sum_{k=0}^{N} c_{k_n} \le 1 - \sum_{k=0}^{N} c_k + \sum_{k=0}^{N} |c_{k_n} - c_k|$$

$$= \sum_{k=N+1}^{\infty} c_k + \sum_{k=0}^{N} |c_{k_n} - c_k| \le \frac{\varepsilon}{4} + \frac{\varepsilon}{4} = \frac{\varepsilon}{2}$$

and, finally,

$$\sum_{k=0}^{\infty} |c_{k_n} - c_k| \le \sum_{k=0}^{N} |c_{k_n} - c_k| + \sum_{k=N+1}^{\infty} c_{k_n} + \sum_{k=n+1}^{\infty} c_k$$

$$\le \frac{\varepsilon}{4} + \frac{\varepsilon}{2} + \frac{\varepsilon}{4} = \varepsilon \qquad \text{for } n \ge n_0.$$

From the last inequality and (12.2.14) it follows that $\{\mu_n\}$ is strongly convergent to μ.

As a typical situation described in this example consider a sequence of measures $\{\mu_n\}$ corresponding to the binomial distribution.

$$
c_{k_n} = \begin{cases} \binom{n}{k} p_n^k q_n^{n-k} & \text{if } k = 0, \ldots, n \\ 0 & \text{if } k > n, \end{cases}
$$

where $0 < p_n < 1$ and $q_n = 1 - p_n$. Further, let μ be a measure corresponding to the Poisson distribution

$$
c_k = e^{-\lambda} \frac{\lambda^k}{k!}.
$$

If $p_n = \lambda/n$, then

$$
c_{k_n} = \frac{(n-k+1)\cdots(n-1)n}{k!} \frac{\lambda^k}{n^k} \left(1 - \frac{\lambda}{n}\right)^{n-k}
$$

$$
= \left(\frac{n-k+1}{n}\right) \cdots \left(\frac{n}{n}\right) \left(1 - \frac{\lambda}{k}\right)^{n-k} \frac{\lambda^k}{k!}.
$$

Evidently the first k factors converges to 1 and the $(k+1)$th to $e^{-\lambda}$. Thus, $c_{k_n} \to c_k$ as $n \to \infty$ for every fixed k, and the sequence of measures corresponding to the binomial distribution converges strongly to the measure corresponding to the Poisson distribution. This is a classical result of probability theory known as **Poisson's theorem**, but it is seldom stated in terms of strong convergence. $\square$

12.3   Markov Operators

In Chapter 3 we introduced Markov operators in Definition 3.1.1, taking a Markov operator to be a linear, positive, and norm-preserving mapping on the space L^1. Now we will extend this notion to the space of all finite measures $\mathcal{M}_{\text{fin}}$ and, in particular, to all probabilistic measures $\mathcal{M}_1$. We start from a formal definition of this extension.

Definition 12.3.1. A mapping $P\colon \mathcal{M}_{\text{fin}}(X) \to \mathcal{M}_{\text{fin}}(X)$ will be called a **Markov operator on measures** if it satisfies the following two conditions:

(a) $P(\lambda_1 \mu_1 + \lambda_2 \mu_2) = \lambda_1 P \mu_1 + \lambda_2 P \mu_2$ for $\lambda_1, \lambda_2 \geq 0$, $\mu_1, \mu_2 \in \mathcal{M}_{\text{fin}}$, and

(b) $P\mu(X) = \mu(X)$ for $\mu \in \mathcal{M}_{\text{fin}}$.

Assumption (a) will often be called the **linearity condition**; however, it is restricted to nonnegative λ_i only. Assumption (b) may be written in the form $\|P\mu\| = \|\mu\|$ (see 12.2.8) and will be called the **preservation of the norm**.

In the following we will quite often omit the qualifying phrase "on measures" if this does not lead to a misunderstanding. On the other hand, if it is necessary we will add the words "on densities" for Markov operators described by Definition 3.1.1.

Our first goal is to show how these two definitions of Markov operators are related. Thus, suppose that the Borel measure of the set X is positive (finite or not) and consider one operator $P: \mathcal{M}_{\text{fin}} \to \mathcal{M}_{\text{fin}}$. Assume that it satisfies conditions (a) and (b) of Definition 12.3.1 and that, moreover, for every absolutely continuous μ the measure $P\mu$ is also absolutely continuous.

Take an arbitrary $f \in L^1$, $f \geq 0$, and define

$$\mu_f(A) = \int_A f(x)\, dx \qquad \text{for } A \in \mathcal{B}. \tag{12.3.1}$$

Since $P\mu_f$ is absolutely continuous it can be written in the form

$$P\mu_f(A) = \int_A g(x)\, dx \qquad \text{for } A \in \mathcal{B}, \tag{12.3.2}$$

where g is a Radon–Nikodym derivative with respect to the Borel measure $P\mu_f$ on X. In this way to every $f \in L^1$, $f \geq 0$, we adjoin a unique $g \in L^1$, $g \geq 0$, for which conditions (12.3.1) and (12.3.2) are satisfied. The uniqueness follows immediately from Proposition 2.2.1 or from the Radon–Nikodym theorem. Thus, f is mapped to g. Denote this mapping by $\hat{P}$, so $g = \hat{P}f$. We may illustrate this situation by the diagram

$$
\begin{array}{ccc}
\mathcal{M}_a & \xrightarrow{\;P\;} & \mathcal{M}_a \\[4pt]
\text{IF} \uparrow & & \downarrow \text{RN} \\[4pt]
L^1_+ & \xrightarrow[\hat{P}]{} & L^1_+
\end{array}
\tag{12.3.3}
$$

where $\mathcal{M}_a$ denotes the family of absolutely continuous measures, L^1_+ is the subspace of L^1 which contains nonnegative functions, IF denotes the integral formula (12.3.1), and RN stands for the Radon–Nikodym derivative. The operator $\hat{P}$ is defined as a "shortcut" between L^1_+ and L^1_+ or, more precisely, in such a way that the diagram (12.3.3) commutes. Thus, $\hat{P}$ is the unique operator on densities that corresponds to the operator P on measures.

Substituting (12.3.1) and (12.3.2) with $g = \hat{P}f$ we obtain

$$P\left\{\int_A f(x)\, dx\right\} = \int_A \hat{P}f(x)\, dx \qquad \text{for } A \in \mathcal{B}, f \in L^1_+. \tag{12.3.4}$$

This is the shortest analytical description of P. To understand this formula correctly we must remember that on the left-hand side of the operator P is applied to the measure given by the integral in braces and then the new measure is applied to the set A.

From condition (a) and formula (12.3.4) it follows immediately that $\hat{P}$ satisfies the linearity condition for nonnegative functions, that is,

$$\hat{P}(\lambda_1 f_1 + \lambda_2 f_2) = \lambda_1 \hat{P} f_1 + \lambda_2 \hat{P} f_2 \qquad \text{for } \lambda_1 \lambda_2 \geq 0, f_1, f_2 \in L^1_+. \quad (12.3.5)$$

Further, using (12.3.4) we obtain

$$\|\hat{P}f\| = \int_X \hat{P}f(x)\,dx = P\left\{\int_X f(x)\,dx\right\} = P\mu_f(X)$$

and analogously

$$\|f\| = \int_X f(x)\,dx = \mu_f(X).$$

From condition (b) this implies

$$\|\hat{P}f\| = \|f\| \qquad \text{for } f \in L^1_+. \qquad\qquad (12.3.6)$$

Now we may extend the definition of $\hat{P}$ to the whole space L^1 that contains all integrable (not necessarily nonnegative) functions by setting

$$\hat{P}f = \hat{P}f^+ - \hat{P}f^- \qquad \text{for } f \in L^1. \qquad (12.3.7)$$

Using this extension and condition (12.3.5) one can verify that $\hat{P}$ is a linear operator. Further, from our construction, and in particular from (12.3.4), it follows that $\hat{P}f \geq 0$ for $f \geq 0$. Finally, (12.3.6) shows that $\hat{P}$ preserves the norm of nonnegative functions. We may summarize this discussion with the following.

Proposition 12.3.1. *Let $P: \mathcal{M}_{\text{fin}} \to \mathcal{M}_{\text{fin}}$ be a Markov operator on measures such that for every absolutely continuous measure μ the measure $P\mu$ is also absolutely continuous. Then the corresponding operator $\hat{P}$ defined by formulas (12.3.4) and (12.3.7) is a Markov operator on densities and the diagram (12.3.3) commutes.*

The commutative property of diagram (12.3.3) has an important consequence. Namely, if $\hat{P}$ is the operator on densities corresponding to an operator P on measures, then $(\hat{P})^n$ corresponds to P^n. To prove this consider the following row of n blocked diagrams (12.3.8).

$$\mathcal{M}_a \xrightarrow{P} \mathcal{M}_a \xrightarrow{P} \mathcal{M}_a \cdots \mathcal{M}_a \xrightarrow{P} \mathcal{M}_a$$

$$\uparrow_{\text{IF}} \qquad \text{RN}\downarrow \uparrow_{\text{IF}} \qquad \text{RN}\downarrow \quad \uparrow_{\text{IF}} \qquad \text{RN}\downarrow \qquad (12.3.8)$$

$$L_+^1 \xrightarrow[\hat{P}]{} L_+^1 \xrightarrow[\hat{P}]{} L_+^1 \cdots L_+^1 \xrightarrow[\hat{P}]{} L_+^1$$

Since each of the blocks commutes, the total diagram (12.3.8) also commutes. This shows that $(\hat{P})^n$ corresponds to P^n.

Remark 12.3.1. There is an evident asymmetry in our approach to the definition of Markov operators. In Section 3.1 we defined a Markov operator on the whole space L^1 which contains positive and negative functions $f: X \to R$. Now we have defined a Markov operator on $\mathcal{M}_{\text{fin}}$ which contains only nonnegative functions $\mu: \mathcal{B} \to R$. This asymmetry can be avoided. Namely, we extend the definition of P on the set of signed measures, that is, all possible differences $\mu_1 - \mu_2$, where $\mu_1, \mu_2 \in \mathcal{M}_{\text{fin}}$, by setting

$$P(\mu_1 - \mu_2) = P\mu_1 - P\mu_2.$$

Such an extension is unnecessary for our purposes and leads to some difficulties in calculating integrals, and in the use of the Riesz representation theorem which is more complicated for signed measures on unbounded regions. $\square$

Example 12.3.1. Let $X = R^+$. For a given $\mu \in \mathcal{M}_{\text{fin}}$ define

$$P\mu(A) = \mu([0,1))\delta_0(A) + \mu([1,\infty) \cap A) \qquad (12.3.9)$$

where, as usual, δ_0 denotes the δ-Dirac measure supported at $x = 0$. Evidently, P satisfies the linearity condition (a) of Definition 12.3.1.
    Moreover,

$$P\mu(R^+) = \mu([0,1))\delta_0(R^+) + \mu([1,\infty) \cap R^+)$$
$$= \mu([0,1)) + \mu([1,\infty)) = \mu(R^+),$$

which shows that condition (b) is also satisfied. Thus, (12.3.9) defines a Markov operator on measures.
    The operator P is relatively simple, but it has an interesting property. Namely, if a measure $\mu \in \mathcal{M}_1$ is supported on $[0,1)$, then $P\mu$ is a δ-Dirac measure. If μ is supported on $[1,\infty)$, then $P\mu = \mu$. In other words, P shrinks all of the measures on $[0,1)$ down to the point $x = 0$ and leaves the remaining portion of the measure untouched. In particular, P does not map absolutely continuous measures into absolutely continuous ones, and the corresponding Markov operator $\hat{P}$ on densities cannot be defined. $\square$

Example 12.3.2. Let $X = R$ and let $t > 0$ be a fixed number. For every

$\mu \in \mathcal{M}_{\text{fin}}$ define the measure $P_t\mu$ by

$$P_t\mu(A) = \int_A \left\{ \int_R \frac{1}{\sqrt{2\pi t}} \exp\left(-\frac{(x-y)^2}{2t} \right) \mu(dy) \right\} dx. \qquad (12.3.10)$$

Again the linearity of P_t is obvious, and to verify that P_t is a Markov operator it is sufficient to check the preservation of the norm.

To do this, substitute $A = R$ into (12.3.10) and change the order of integration to obtain

$$P_t\mu(R) = \int_R \left\{ \int_R \frac{1}{\sqrt{2\pi t}} \exp\left(-\frac{(x-y)^2}{2t} \right) dx \right\} \mu(dy).$$

Inside the braces we have the integral of the Gaussian density, and consequently

$$P_t\mu(R) = \int_R 1\mu(dy) = \mu(R),$$

so P_t is a Markov operator.

To understand the meaning of the family of operators $\{P_t\}$, first observe that for every $\mu \in \mathcal{M}_{\text{fin}}$ the measure $P_t\mu$ is given by the integral (12.3.10) and has the Radon–Nikodym derivative

$$g_t(x) = \frac{1}{\sqrt{2\pi t}} \int_R \exp\left\{ -\frac{(x-y)^2}{2t} \right\} \mu(dy). \qquad (12.3.11)$$

If μ is absolutely continuous with density f, we may replace $\mu(dy)$ by $f(y)\,dy$ and in this way obtain an explicit formula for the operator $\hat{P}_t$ on densities corresponding to P_t. Namely,

$$\hat{P}_t f(x) = g_t(x) = \frac{1}{\sqrt{2\pi t}} \int_R \exp\left\{ -\frac{(x-y)^2}{2t} \right\} f(y)\,dy.$$

The function $u(t,x) = g_t(x)$ is the familiar solution (7.4.11), (7.4.12) of the heat equation (7.4.13)

$$\frac{\partial u}{\partial t} = \frac{1}{2} \frac{\partial^2 u}{\partial x^2} \qquad \text{for } t > 0, x \in R,$$

with the initial condition

$$u(0,x) = f(x).$$

It is interesting that $u(t,x) = g_t(x)$ satisfies the heat equation even in the case when μ has no density. This can be verified simply by differentiation of the integral formula (12.3.11). (Such a procedure is always possible since μ is a finite measure and the integrand

$$\frac{1}{\sqrt{2\pi t}} e^{-(x-y)^2/2t}$$

and its derivatives are bounded C^∞ functions for $t \geq \varepsilon > 0$.)

Further, in the case of arbitrary μ the initial condition is also satisfied. Namely, the measures $P_t\mu$ converge weakly to μ as $t \to 0$. To prove this choose an arbitrary $h \in C_0(R)$. Since g_t is the Radon–Nikodym derivative of $P_t\mu$ we have

$$\langle h, P_t\mu \rangle = \int_R h(x)P_t\mu(dx) = \int_R h(x)g_t(x)\,dx$$
$$= \int_R h(x)\left\{ \int_R \frac{1}{\sqrt{2\pi t}} \exp\left(-\frac{(x-y)^2}{2t} \right) \mu(dy) \right\} dx$$

or by changing the order of integration

$$\langle h, P_t\mu \rangle = \int_R v(t,y)\mu(dy) \tag{12.3.12}$$

where

$$v(t,y) = \frac{1}{\sqrt{2\pi t}} \int_R \exp\left(-\frac{(x-y)^2}{2t} \right) h(x)\,dx.$$

Observe that $v(t,y)$ is the solution of the heat equation corresponding to the initial function $h(y)$. Since h is continuous and bounded, this is a classical solution and we have

$$\lim_{t \to 0} v(t,y) = h(y) \qquad \text{for } y \in R.$$

Evidently

$$|v(t,y)| \le \frac{\max |h|}{\sqrt{2\pi t}} \int_R \exp\left(-\frac{(x-y)^2}{2t} \right) dx = \max |h|.$$

Thus by the Lebesgue dominated convergence theorem (see Remark 2.2.4)

$$\lim_{t \to 0} \int_R v(t,y)\mu(dy) = \int_R h(y)\mu(dy).$$

From this and (12.3.12) it follows that $P_t\mu$ converges weakly to μ.

Thus, we can say that the family of measures $\{P_t\mu\}$ describes the transport of the initial measure by the heat equation. From a physical point of view, if $u(t,x) = g_t(x)$ is the temperature at time t at the point x, then

$$P^t\mu(A) = \int_A g_t(x)\,dx$$

is equal (up to multiplicative constraint) to the amount of heat carried by a segment A at time t. In particular, substituting $\mu = \delta_{x_0}$ (the δ-Dirac measure supported at $x = x_0$) in (12.3.11) we obtain

$$u(t,x) = g_t(x) = \frac{1}{\sqrt{2\pi t}} e^{-(x-x_0)^2/2t}.$$

This equation is identical to the fundamental solution $\Gamma(t, x, x_0)$ of the heat equation (see Section 11.7) and it gives a simple physical interpretation of this solution. Namely, $\Gamma(t, x, x_0)$ is the temperature at time t and point x corresponding to the situation in which the initial amount of heat was concentrated at a single point x_0. $\square$

12.4 Foias Operators

At the end of the previous section we have given two examples of Markov operators constructed in two different methods. The goal of the present section is to develop these methods in detail.

Let $X \subset R^d$ be a nonempty closed set. We start from the following

Definition 12.4.1. Let $S: X \to X$ be a Borel measurable transformation. Then the operator $P: \mathcal{M}_{\text{fin}} \to \mathcal{M}_{\text{fin}}$ defined by

$$P\mu(A) = \mu(S^{-1}(A)) \qquad \text{for } A \in \mathcal{B}(X) \tag{12.4.1}$$

is called the **Frobenius–Perron operator on measures** corresponding to S.

Evidently P defined by (12.4.1) is a Markov operator. Now observe how P works on measures supported on a single point. Let $x_0 \in X$ be fixed. Then

$$P\delta_{x_0}(A) = \delta_{x_0}(S^{-1}(A)) = \begin{cases} 0 & \text{if } x_0 \notin S^{-1}(A) \\ 1 & \text{if } x_0 \in S^{-1}(A) \end{cases}$$

or

$$P\delta_{x_0}(A) = \begin{cases} 0 & \text{if } S(x_0) \notin A \\ 1 & \text{if } S(x_0) \in A. \end{cases}$$

Thus, $P\delta_{x_0} = \delta_{S(x_0)}$. By induction we obtain

$$P^n \delta_{x_0} = \delta_{S^n(x_0)}.$$

This shows that the iterates of the Markov operator (12.4.1) can produce a trajectory of a transformation S. To obtain this trajectory it is sufficient to start from a δ-Dirac measure.

We next show that P can also transform densities. Consider the special case when μ is absolutely continuous with density f and S is a nonsingular transformation. Then

$$\mu(A) = \int_A f(x)\, dx,$$

and the right-hand side of (12.4.1) may be written in the form

$$\mu(S^{-1}(A)) = \int_{S^{-1}(A)} f(x)\, dx = \int_A \hat{P}f(x)\, dx,$$

where $\hat{P}$ is the Frobenius–Perron operator on densities corresponding to S. Now equality (12.4.1) may be explicitly written in the form

$$P\left\{\int_A f(x)\,dx\right\} = \int_A \hat{P}f(x)\,dx.$$

This is a special case of formula (12.3.4) and it shows that the Frobenius–Perron operator $\hat{P}$ on densities corresponds, in the sense of diagram (12.3.3), to the Frobenius–Perron operator P on measures.

This correspondence was obtained under the additional assumption that S is nonsingular. For an arbitrary Borel measurable transformation S, the operator P given by (12.4.1) may transform absolutely continuous measures into measures without density.

Example 12.4.1. Let $X = R^+$ and

$$S(x) = \begin{cases} 0 & 0 \le x < 1 \\ x & x \ge 1. \end{cases}$$

Then

$$S^{-1}(A) = S^{-1}(A \cap [0,1)) \cup S^{-1}(A \cup [1,\infty)),$$

where $S^{-1}(A \cup [1,\infty)) = A \cap [1,\infty)$ and

$$S^{-1}(A \cap [0,1)) = \begin{cases} [0,1) & \text{if } 0 \in A \\ \emptyset & \text{if } 0 \notin A. \end{cases}$$

From the last formula it follows that

$$\mu(S^{-1}(A \cap [0,1))) = 1_A(0)\mu([0,1)).$$

Consequently, the Frobenius–Perron operator for S is given by

$$P\mu(A) = \mu(S^{-1}(A)) = 1_A(0)\mu([0,1)) + \mu(A \cap [1,\infty))$$

which is identical with (12.3.9).   □

Now we are going to study a more general, and complicated, situation when the dynamical system includes random perturbations. Thus, we consider the system

$$x_{n+1} = T(x_n, \xi_n) \qquad \text{for } n = 0, 1, \ldots, \tag{12.4.2}$$

where T is a given transformation and the ξ_n are independent random vectors. We make the following assumptions:

(i) T is defined on the subset $X \times W$ of $R^d \times R^k$ with values in X. The set $X \subset R^d$ is closed and $W \subset R^k$ is Borel measurable. For every fixed $y \in W$ the function $T(x,y)$ is continuous in x and for every fixed $x \in X$ it is measurable in y.

(ii) The random vectors $\xi_0, \xi_1, \ldots$, have values in W and have the same distribution, that is, the measure

$$\nu(B) = \text{prob}(\xi_n \in B) \qquad \text{for } B \in \mathcal{B}(W)$$

is the same for all n.

(iii) The initial random vector x_0 has values in X and the vectors $x_0, \xi_0, \xi_1, \ldots$, are independent.

A dynamical system of the form (12.4.2) satisfying conditions (i)–(iii) will be called a **regular stochastic dynamical system**. We emphasize that in studying (12.4.2) it is assumed that the transformation T and the random vectors ξ_n are given. The initial vector x_0 can be arbitrary, but must be such that condition (iii) is satisfied. Observe that in particular if ξ_0, ξ_1 are independent and $x_0 \in X$ is constant (not random) then the vectors $x_0, \xi_1, \xi_1, \ldots$, are also independent. This can be easily verified using the definition of the independence of random vectors and the fact that the value of $\text{prob}(x_0 \in A)$ is either 0 or 1 for x_0 constant.

According to (12.4.2) the random vector x_n is a function of x_0 and ξ_0, $\xi_1, \ldots, \xi_{n-1}$. From this and condition (iii) it follows that x_n and ξ_n are independent. Using this fact we will derive a recurrence formula for the measures

$$\mu_n(A) = \text{prob}(x_n \in A), \qquad A \in \mathcal{B}(X), \qquad (12.4.3)$$

which statistically describe the behavior of the dynamical system (12.4.2).

Thus, choose a bounded Borel measurable function $h: X \to R$ and for some integer $n \geq 0$ consider the random vector $z_{n+1} = h(x_{n+1})$. Observe that

$$\mu_{n+1}(A) = \text{prob}(x_{n+1}^{-1}(B)).$$

Using this equality and the change of variables Theorem 3.2.1, the mathematical expectation $E(z_{n+1})$ can be calculated as follows:

$$E(z_{n+1}) = \int_\Omega h(x_{n+1}(\omega))\text{prob}(d\omega) = \int_X h(x)\text{prob}(x_{n+1}^{-1}(dx))$$

$$= \int_X h(x)\mu_{n+1}(dx) = \langle h, \mu_{n+1} \rangle. \qquad (12.4.4)$$

However, since $z_{n+1} = h(T(x_n, \xi_n))$ we have

$$E(z_{n+1}) = \int_\Omega h(T(x_n(\omega), \xi_n(\omega)))\text{prob}(d\omega)$$

$$= \int\int_{X \times W} h(T(x, y))\text{prob}((x_n, \xi_n)^{-1}(dx\, dy)). \qquad (12.4.5)$$

The independence of the random vectors x_n and ξ_n implies that

$$\text{prob}((x_n, \xi_n) \in A \times B) = \text{prob}(x_n \in A, \xi_n \in B) = \text{prob}(x_n \in A)\text{prob}(\xi_n \in B),$$

or
$$\text{prob}((x_n, \xi_n)^{-1}(A \times B)) = \text{prob}(x_n^{-1}(A))\text{prob}(\xi_n^{-1}(B)),$$

which shows that the measure $\text{prob}((x_n, \xi_n)^{-1}(C))$ is the product of measures

$$\mu_n(A) = \text{prob}(x_n^{-1}(A)) \quad \text{and} \quad \nu(B) = \text{prob}(\xi_n^{-1}(B)).$$

Thus, by the Fubini Theorem 2.2.3, equality (12.4.5) may be rewritten in the form

$$E(z_{n+1}) = \int_X \left\{ \int_W h(T(x,y))\nu(dy) \right\} \mu_n(dx).$$

Equating this expression with (12.4.4) we immediately obtain

$$\langle h, \mu_{n+1} \rangle = \int_X \left\{ \int_W h(T(x,y))\nu(dy) \right\} \mu_n(dx). \tag{12.4.6}$$

This is the desired recurrence formula, derived under the assumption that h is Borel measurable and bounded. The boundness of h asserts that all the integrals appearing in the derivation are well defined and finite, since the measures μ_n, μ_{n+1}, prob,..., were probabilistic. The same derivation can be repeated for unbounded h as long as all the integrals are well defined. In particular the derivation can be made for an arbitrary measurable nonnegative h. However, in this case the integrals on both sides of (12.4.5) could be infinite.

Using (12.4.5) we may calculate the values of $\mu_{n+1}(A)$ for an arbitrary measurable set $A \subset X$. Namely, setting $h = 1_A$ we obtain

$$\mu_{n+1}(A) = \int_X \left\{ \int_W 1_A(T(x,y))\nu(dy) \right\} \mu_n(dx).$$

Now we are in a position to define the Foias operator corresponding to the dynamical system (12.4.2).

Definition 12.4.2. Let a function $T: X \times W \to X$ satisfying condition (i) and a probabilistic measure (supported on W) be given. Then the operator $P: \mathcal{M}_{\text{fin}} \to \mathcal{M}_{\text{fin}}$ given by

$$P\mu(A) = \int_X \left\{ \int_W 1_A(T(x,y))\nu(dy) \right\} \mu(dx) \quad \text{for } \mu \in \mathcal{M}_{\text{fin}}, A \in \mathcal{B}(X)$$
$$\tag{12.4.7}$$

will be called the **Foias operator** corresponding to the dynamical system (12.4.2).

Since ν is a probabilistic measure, it is obvious that P is a Markov operator. Moreover, from the definition of P it follows that $\mu_n = P^n \mu_0$, where $\{\mu_n\}$ denotes the sequence of distributions (12.4.3) described by the dynamical system (12.4.2).

Setting

$$Uh(x) = \int_W h(T(x,y))\nu(dy) \qquad \text{for } x \in X, \qquad (12.4.8)$$

we may rewrite (12.4.7) in the form

$$\langle 1_A, P\mu \rangle = \langle U1_A, \mu \rangle.$$

Due to the linearity of the scalar product this implies

$$\langle g_n, P\mu \rangle = \langle Ug_n, \mu \rangle,$$

where

$$g_n = \sum_{i=1}^{n} \lambda_i 1_{A_i}$$

is a simple function. Further, since every measurable function h can be approximated by a sequence $\{g_n\}$ of simple functions, we obtain in the limit

$$\langle h, P\mu \rangle = \langle Uh, \mu \rangle \qquad (12.4.9)$$

if $\{g_n\}$ and $\{Ug_n\}$ satisfy the conditions of the Lebesgue dominated or Lebesgue monotone convergence theorem. In particular, (12.4.9) is valid if h is Borel measurable and bounded or nonnegative.

From (12.4.9) it follows by an induction argument that

$$\langle h, P^n\mu \rangle = \langle U^n h, \mu \rangle \qquad \text{for } n = 1, 2, \ldots . \qquad (12.4.10)$$

Now define a sequence of functions $T_n(x, y_1, \ldots, y_n)$ by setting

$$T_1(x, y_1) = T(x, y_1), \qquad T_n(x, y_1, \ldots, y_n) = T(T_{n-1}(x, y_1, \ldots, y_{n-1}), y_n).$$

Using this notation we obtain

$$U^n h(x) = \int_W \cdots \int_W h(T_n(x, y_1, \ldots, y_n)\nu(dy_1) \cdots \nu(dy_n) \qquad (12.4.11)$$

from (12.4.8), or, more briefly,

$$U^n h(x) = \int_{W^n} h(T_n(x, y^n))\nu^n(dy^n), \qquad (12.4.12)$$

where $y^n = (y_1, \ldots, y_n)$, $W^n = W \times \cdots \times W$ is the Cartesian product of n sets W and $\nu^n(dy^n) = \nu(dy_1) \cdots \nu(dy_n)$ is the corresponding product measure on W^n.

Equations (12.4.10) and (12.4.12) give convenient tools for studying the asymptotic behavior of the sequence $\{P^n\mu\}$. Moreover, U^n and T_n have a simple dynamical interpretation. Namely, from (12.4.2) and the definition of T_n it follows that

$$x_n = T_n(x_0, \xi_0, \ldots, \xi_{n-1}) \qquad \text{for } n = 1, 2, \ldots,$$

which shows that T_n describes the position of x_n as a function of the initial position x_0 and perturbations. Further, repeating the calculation of the mathematical expectation $E(h(x_n))$ we obtain

$$E(h(x_n)) = \int_\Omega h(x_n(\omega))\text{prob}(d\omega)$$

$$= \int_X h(x)\text{prob}(x_n^{-1}(dx)) = \int_X h(x)\mu_n(dx)$$

or

$$Eh(x_n) = \langle h, P^n\mu_0\rangle = \langle U^nh, \mu_0\rangle. \qquad (12.4.13)$$

In particular, if the starting point x_0 is fixed, corresponding to $\mu_0 = \delta_{x_0}$, we have

$$E(h(x_n)) = U^nh(x_0). \qquad (12.4.14)$$

Thus, U^nh gives the mathematical expectation of $h(x_n)$ as a function of the initial position x_0.

We close this section by discussing the relationship between the Frobenius–Perron and Foias operators. Having a continuous transformation $S\colon X \to X$ we may formally write

$$T(x,y) = S(x) + 0y.$$

In this case (12.4.7) takes the form

$$P\mu(A) = \int_X \left\{ \int_W 1_A(S(x))\nu(dy) \right\} \mu(dx)$$

$$= \int_X 1_A(S(x))\mu(dx) = \mu(S^{-1}(A)),$$

and is identical with (12.4.1). Thus, in the case when $T(x,y)$ does not depend on y the notions of the Foias operator and the Frobenius–Perron operator coincide. Moreover, in this case

$$Uh(x) = \int_W h(S(x))\nu(dy) = h(S(x)), \qquad (12.4.15)$$

and U is the Koopman operator.

It is evident that the operator U given by equation (12.4.8) (with $\nu \in \mathcal{M}_1$), or (12.4.15), maps a bounded function h into a bounded function Uh. Moreover, if S is continuous [or more generally T satisfies condition (i)], then Uh is continuous for continuous bounded h. However, in general, the support of Uh is not bounded for $h \in C_0(X)$.

12.5 Stationary Measures: Krylov–Bogolubov Theorem for Stochastic Dynamical Systems

We begin our study of the asymptotic properties of $\{P^n\mu\}$ by looking for stationary measures.

Definition 12.5.1. A measure $\mu_* \in \mathcal{M}_{\text{fin}}$ is called **invariant** or **stationary** with respect to a Markov operator P if $P\mu_* = \mu_*$. In particular, when P is a Foias operator corresponding to the dynamical system (12.4.2) and $P\mu_* = \mu_*$, we say that μ_* is **stationary** with respect to (12.4.2). A stationary probabilistic measure is called a **stationary distribution**.

If μ_* is a stationary distribution for (12.4.2) and if the initial vector x_0 is distributed according to μ_*, that is,

$$\text{prob}(x_0 \in A) = \mu_*(A) \qquad \text{for } A \in \mathcal{B}(X),$$

then all the vectors x_n have the same property, that is,

$$\text{prob}(x_n \in A) = \mu_*(A) \qquad \text{for } A \in \mathcal{B}(X),\ n = 0, 1, \ldots.$$

Our main result concerning the existence of a stationary distribution is contained in the following.

Theorem 12.5.1. *Let P be the Foias operator corresponding to a regular stochastic dynamical system* (12.4.2). *Assume that there is a $\mu_0 \in \mathcal{M}_1$ having the following property. For every $\varepsilon > 0$ there is a bounded set $B \in \mathcal{B}(X)$ such that*

$$\mu_n(B) = P^n\mu_0(B) \geq 1 - \varepsilon \qquad \text{for } n = 0, 1, 2, \ldots. \tag{12.5.1}$$

Then P has an invariant distribution.

Proof. Define

$$\zeta_n = \frac{1}{n}\sum_{i=0}^{n-1} P^i\mu_0 = \frac{1}{n}\sum_{i=0}^{n-1}\mu_i \qquad \text{for } n = 1, 2, \ldots. \tag{12.5.2}$$

Choose a countable subset $\{h_1, h_2, \ldots\}$ of $C_0(X)$ dense in $C_0(X)$ (see Exercises 12.1 and 12.2). The sequence $\{\langle h_1, \zeta_n\rangle\}$ is bounded since the ζ_n are probabilistic and $|\langle h_1, \zeta_n\rangle| \leq \max |h_1|$. Thus, there is a subsequence $\{\zeta_{1n}\}$ of $\{\zeta_n\}$ such that $\{\langle h_1, \zeta_{1n}\rangle\}$ is convergent. Again, since $\{\langle h_2, \zeta_{1n}\rangle\}$ is bounded we can choose a subsequence $\{\zeta_{2n}\}$ of $\{\zeta_{1n}\}$ such that $\{\langle h_2, \zeta_{2n}\rangle\}$ is convergent. By induction for every integer $k > 1$ we may construct a sequence $\{\zeta_{kn}\}$ such that all sequences $\{\langle h_j, \zeta_{kn}\rangle\}$ for $j = 1, \ldots, k$ are convergent and $\{\zeta_{kn}\}$ is a subsequence of $\{\zeta_{k-1,n}\}$. Evidently the diagonal sequence $\{\zeta_{nn}\}$ has the property that $\{\langle h_j, \zeta_{nn}\rangle\}$ is convergent for every $j = 1, 2, \ldots.$

This procedure of choosing subsequences is known as the Cantor diagonal process [Dunford and Schwartz, 1957, Chapter I.6]. Since the set $\{h_j\}$ is dense in C_0, then according to Remark 12.2.1 the sequence $\{\zeta_{nn}\}$ is weakly convergent to a measure μ_*. It remains to prove that μ_* is probabilistic and invariant.

Without any loss of generality we may assume that the set B in (12.5.1) is compact. Then $X \setminus B$ is open and according to Theorem 12.2.1

$$\mu_*(X \setminus B) \le \liminf_{n \to \infty} \zeta_{nn}(X \setminus B) \le 1 - \inf_n \mu_n(B) \le 1 - (1 - \varepsilon) = \varepsilon.$$

Now we may prove that $\{\langle h, \zeta_{nn} \rangle\}$ converges to $\langle h, \mu_* \rangle$ for every bounded continuous h. Let h be given. Define $h_\varepsilon = h g_\varepsilon$ where $g_\varepsilon \in C_0$ is such that

$$0 \le g_\varepsilon \le 1 \quad \text{and} \quad g_\varepsilon(x) = 1 \quad \text{for } x \in B.$$

Then

$$|\langle h, \mu_* - \zeta_{nn} \rangle| \le |\langle h_\varepsilon, \mu_* - \zeta_{nn} \rangle| + |\langle h(1 - g_\varepsilon), \mu_* - \zeta_{nn} \rangle|$$
$$\le |\langle h_\varepsilon, \mu_* - \zeta_{nn} \rangle| + \sup |h|(\mu_*(X \setminus B) + \zeta_{nn}(X \setminus B))$$

or

$$|\langle h, \mu_* - \zeta_{nn} \rangle| \le |\langle h_\varepsilon, \mu_* - \zeta_{nn} \rangle| + 2\varepsilon \sup |h|.$$

Since $h_\varepsilon \in C_0$ and $\{\zeta_{nn}\}$ converges weakly to μ_* this implies

$$\lim_{n \to \infty} \langle h, \zeta_{nn} \rangle = \langle h, \mu_* \rangle$$

for every bounded continuous h. In particular, setting $h = 1_X$ we obtain

$$\mu_*(X) = \lim_{n \to \infty} \zeta_{nn}(X) = 1,$$

so μ_* is probabilistic.

Now we are ready to prove that μ_* is invariant. The sequence $\{\zeta_{nn}\}$, as a subsequence of $\{\zeta_n\}$, may be written in the form

$$\zeta_{nn} = \frac{1}{k_n} \sum_{i=0}^{k_n-1} P^i \mu_0,$$

where $\{k_n\}$ is a strictly increasing sequence of integers. Thus,

$$P\zeta_{nn} - \zeta_{nn} = \frac{1}{k_n}(P^{k_n} \mu_0 - \mu_0),$$

and, consequently,

$$|\langle Uh, \zeta_{nn} \rangle - \langle h, \zeta_{nn} \rangle| = |\langle h, P\zeta_{nn} \rangle - \langle h, \zeta_{nn} \rangle| \le \frac{1}{k_n} \sup |h|.$$

Passing to the limit we obtain

$$\langle Uh, \mu_* \rangle - \langle h, \mu_* \rangle = 0,$$

or

$$\langle h, P\mu_* \rangle = \langle h, \mu_* \rangle.$$

The last equality holds for every bounded continuous h and in particular for $h \in C_0$. Thus, by the Riesz representation theorem 12.1.1, $P\mu_* = \mu_*$. The proof is completed. ■

Condition (12.5.1) is not only sufficient for the existence of an invariant distribution μ_* but also necessary. To see this, assume that μ_* exists. Let $\{B_k\}$ be an increasing sequence of bounded measurable sets such that $\bigcup_k B_k = X$. Then

$$\lim_{k \to \infty} \mu_*(B_k) = \mu_*(X) = 1.$$

Thus, for every $\varepsilon > 0$ there is a bounded set B_k such that $\mu_*(B_k) \geq 1 - \varepsilon$. Setting $\mu_0 = \mu_*$ we have $\mu_n = \mu_*$ and, consequently,

$$\mu_n(B_k) \geq 1 - \varepsilon \qquad \text{for } n = 0, 1, \ldots .$$

Remark 12.5.1. In the case when X is bounded (and hence compact, because we always assume that X is closed), condition (12.5.1) is automatically satisfied with $B = X$. Thus for a regular stochastic dynamical system there always exists a stationary distribution. In particular, for a continuous transformation $S: X \to X$ of a compact set X there always exists an invariant probabilistic measure. This last assertion is known as the **Krylov–Bogolubov theorem**. It is valid not only when X is a compact subset of R^d, but also for arbitrary compact topological Hausdorff spaces. □

Now we will concentrate on the case when $X \subset R^d$ is unbounded (but closed!), and formulate some sufficient conditions for (12.5.1) based on the technique of Liapunov functions. Recall from (5.7.8) that a Borel measurable function $V: X \to R$ is called a Liapunov function if $V(x) \to \infty$ for $|x| \to \infty$.

Proposition 12.5.1. *Let P be the Foias operator corresponding to a regular stochastic dynamical system (12.4.2). Assume that there is an initial random vector x_0 and a Liapunov function V such that*

$$\sup_n E(V(x_n)) < \infty. \qquad (12.5.3)$$

Then P has an invariant distribution.

Proof. Consider the family of bounded sets

$$B_a = \{x \in X : V(x) \leq a\} \qquad \text{for } a \geq 0.$$

By Chebyshev's inequality (10.2.9) we have

$$\mu_n(X \setminus B_a) = \text{prob}(V(x_n) > a) \leq \frac{E(V(x_n))}{a}$$

or

$$\mu_n(B_a) \geq 1 - \frac{K}{a} \qquad \text{for } n = 0, 1, \dots ,$$

where $K = \sup_n E(V(x_n))$. Thus, for every $\varepsilon > 0$ inequality (12.5.1) is satisfied with $B = B_a$ and $a = K - \varepsilon$. It follows from Theorem 12.5.1 that P has an invariant distribution and the proof is complete. ■

It is easy to formulate a sufficient condition for (12.5.3) related explicitly to properties of the function T of (12.4.2) and the distribution ν. Thus we have the following

Proposition 12.5.2. *Let P be the Foias operator corresponding to a regular stochastic dynamical system (12.4.2). Assume that there exists a Liapunov function V and nonnegative constants α, β, $\alpha < 1$, such that*

$$\int_W V(T(x,y))\nu(dy) \leq \alpha V(x) + \beta \qquad \text{for } x \in X. \qquad (12.5.4)$$

Then P has an invariant distribution.

Proof. By an induction argument from inequality (12.5.4), it follows that

$$\int_{W^n} V(T_n(x,y^n))\nu^n(dy^n) \leq \alpha^n V(x) + \alpha^{n-1}\beta + \cdots + \alpha\beta + \beta$$

$$\leq V(x) + \frac{\beta}{1-\alpha}.$$

Fix an $x_0 \in X$ and define $\mu_0 = \delta_{x_0}$. Then according to (12.4.14) and (12.4.12) we have

$$E(V(x_n)) = U^n V(x_0) = \int_{W^n} V(T_n(x_0, y^n))\nu^n(dy^n) \leq V(x_0) + \frac{\beta}{1-\alpha},$$

which implies (12.5.3), and Proposition 12.5.1 completes the proof. ■

12.6 Weak Asymptotic Stability

In the previous section we developed sufficient conditions for the existence of a stationary measure μ_*. Now we are going to prove conditions that ensure that this measure is asymptotically stable. Since in the space of measures there are two natural notions of convergence (weak and strong), we will introduce two types of asymptotic stability. We will start from the following.

Definition 12.6.1. Let $P: \mathcal{M}_{\text{fin}} \to \mathcal{M}_{\text{fin}}$ be a Markov operator. We say that the sequence $\{P^n\}$ is **weakly asymptotically stable** if P has a unique invariant distribution μ_* and

$$\{P^n\mu\} \text{ converges weakly to } \mu_* \text{ for } \mu \in \mathcal{M}_1. \tag{12.6.1}$$

In the special case that P is a Foias operator corresponding to a stochastic dynamical system (12.4.2) and $\{P^n\}$ is weakly asymptotically stable, we say that the system is **weakly asymptotically stable**.

It may be shown that the uniqueness of the stationary distribution μ_* is a consequence of the condition (12.6.1). To show this, let $\tilde{\mu} \in \mathcal{M}_1$ be another stationary distribution. Then $P^n\tilde{\mu} = \tilde{\mu}$ and from (12.6.1) applied to $\mu = \tilde{\mu}$ we obtain

$$\langle h, \tilde{\mu} \rangle = \langle h, \mu_* \rangle \quad \text{for } h \in C_0(X).$$

By the Riesz representation theorem 12.1.1, this gives $\tilde{\mu} = \mu_*$. On the other hand, condition (12.6.1) does not imply that μ_* is stationary for an arbitrary Markov operator.

Example 12.6.1. Let $X = [0,1]$. Consider the Frobenius–Perron operator P on measures and the Koopman operator U corresponding to the transformation

$$S(x) = \begin{cases} \frac{1}{2}x & x > 0 \\ c & x = 0, \end{cases}$$

where $c \in [0,1]$ is a constant. Now

$$U^n h(x) = h(S^n(x)) + \begin{cases} h\left(\frac{1}{2^n}x\right) & x > 0 \\ h\left(\frac{c}{2^{n-1}}\right) & x = 0. \end{cases}$$

Thus, for every $\mu \in \mathcal{M}_1$ and $h \in C_0(X)$ we have

$$\langle h, P^n\mu \rangle = \langle U^n h, \mu \rangle = \int_{[0,1]} h(S^n(x))\mu(dx)$$

$$= \mu(0)h\left(\frac{c}{2^{n-1}}\right) + \int_{(0,1]} h\left(\frac{x}{2^n}\right)\mu(dx).$$

Since h is continuous this implies

$$\lim_{n \to \infty} \langle h, P^n\mu \rangle = \mu(0)h(0) + \mu((0,1])h(0) = h(0)$$

and consequently $\{P^n\mu\}$ converges to δ_0. On the other hand, $P\delta_0 = \delta_c$ and the system is weakly asymptotically stable only for $c = 0$ when S

is continuous. If $c > 0$ the operator P has no invariant distribution but condition (12.6.1) holds with $\mu_* = \delta_0$. $\square$

Next we give two easily proved criteria for the weak asymptotic stability of a sequence $\{P^n\}$.

Proposition 12.6.1. *Let* $P: \mathcal{M}_{\text{fin}} \to \mathcal{M}_{\text{fin}}$ *be a Markov operator. The sequence* $\{P^n\}$ *is weakly asymptotically stable if P has an invariant distribution and if*

$$\lim_{n \to \infty} \langle h, P^n \mu - P^n \bar{\mu} \rangle = 0 \qquad \text{for } h \in C_0; \ \mu, \bar{\mu} \in \mathcal{M}. \tag{12.6.2}$$

Proof. First assume that $\{P^n\}$ is weakly asymptotically stable. Then by the triangle inequality

$$|\langle h, P^n \mu - P^n \bar{\mu} \rangle| \leq |\langle h, P^n \mu - \mu_* \rangle| + |\langle h, P^n \bar{\mu} - \mu_* \rangle|$$

and (12.6.1) implies (12.6.2). Alternately, if (12.6.2) holds and μ_* is stationary, then substituting $\bar{\mu} = \mu_*$ in (12.6.2) we obtain (12.6.1). ∎
The main advantage of condition (12.6.2) in comparison with (12.6.1) is that in proving the convergence we may restrict the verification to subsets of C_0 and $\mathcal{M}_1$.

Proposition 12.6.2. *Let* $C_* \subset C_0$ *be a dense subset. If condition (12.6.2) holds for every $h \in C_*$ and $\mu, \bar{\mu} \in \mathcal{M}_1$ with bounded supports, then it is satisfied for arbitrary $h \in C_0$ and $\mu, \bar{\mu} \in \mathcal{M}_1$.*

Proof. Choose $\mu, \bar{\mu} \in \mathcal{M}_1$ and fix an $\varepsilon > 0$. Without any loss of generality we may assume that $\varepsilon \leq 1/2$. Since μ and $\bar{\mu}$ are probabilistic, there is a bounded set $B \subset X$ such that

$$\mu(X \setminus B) \leq \varepsilon \quad \text{and} \quad \bar{\mu}(X \setminus B) \leq \varepsilon.$$

Define

$$\rho(A) = \frac{\mu(A \cap B)}{\mu(B)} \quad \text{and} \quad \bar{\rho}(A) = \frac{\bar{\mu}(A \cap B)}{\bar{\mu}(B)} \qquad \text{for } A \in \mathcal{B}(X).$$

Evidently ρ and $\bar{\rho}$ are probabilistic measures with bounded supports. We have $\mu(B) \geq 1 - \varepsilon \geq \frac{1}{2}$, and consequently

$$\begin{aligned}
|\mu(A) - \rho(A)| &\leq 2|\mu(A)\mu(B) - \mu(A \cap B)| \\
&= 2|\mu(A)(1 - \mu(X \setminus B)) - \mu(A \cap B)| \\
&\leq 2|\mu(A) - \mu(A \cap B)| + 2\mu(A)\mu(X \setminus B) \\
&\leq 2\mu(A \setminus B) + 2\mu(X \setminus B) \leq 4\varepsilon \quad \text{for } A \in \mathcal{B}(X).
\end{aligned}$$

In an analogous fashion we may verify that

$$|\bar{\mu}(A) - \bar{\rho}(A)| \leq 4\varepsilon \qquad \text{for } A \in \mathcal{B}(X).$$

Now let a function $g \in C_*$ be given. Then

$$|\langle g, P^n \mu - P^n \bar{\mu} \rangle| = |\langle U^n g, \mu - \bar{\mu} \rangle|$$
$$\leq |\langle U^n g, \rho - \bar{\rho} \rangle| + 8\varepsilon \sup |U^n g|,$$

and, finally,

$$|\langle h, P^n \mu - P^n \bar{\mu} \rangle| \leq |\langle g, P^n \rho - P^n \bar{\rho} \rangle| + 8\varepsilon \max |g|.$$

Since ρ and $\bar{\rho}$ have bounded supports the sequence $\{\langle g, P^n \rho - P^n \bar{\rho}\rangle\}$ converges to zero. Consequently, $\{\langle g, P^n \mu - P^n \bar{\mu}\rangle\}$ converges to zero for every $g \in C_*$ and $\mu, \bar{\mu} \in \mathcal{M}_1$. Now from the inequality

$$|\langle h, P^n \mu - P^n \bar{\mu} \rangle| \leq |\langle g, P^n \mu - P^n \bar{\mu} \rangle| + 2\sup |g - h| \qquad \text{for } g \in C_*, h \in C_0$$

and the density of C_* in C_0, (12.6.3) follows for all $h \in C_0$. Thus the proof is complete. ∎

Now we may establish the main result of this section, which is an effective criterion for the weak asymptotic stability of the stochastic system (12.4.2).

Theorem 12.6.1. *Let P be the Foias operator corresponding to the regular stochastic dynamical system* (12.4.2). *Assume that*

$$E(|T(x, \xi_n) - T(z, \xi_n)|) \leq \alpha |x - z| \qquad \text{for } x, z \in X \qquad (12.6.3)$$

and

$$E(|T(0, \xi_n)|) \leq \beta, \qquad (12.6.4)$$

where E is the mathematical expectation and α, β are nonnegative constants with $\alpha < 1$. Then the system (12.4.2) *is weakly asymptotically stable.*

Before passing to the proof observe that conditions (12.6.3) and (12.6.4) can be rewritten, using the distribution ν appearing in the definition of the stochastic system (12.4.2), in the form

$$\int_W |T(x, y) - T(z, y)| \nu(dy) \leq \alpha |x - z| \qquad (12.6.5)$$

and

$$\int_W |T(0, y)| \nu(dy) \leq \beta. \qquad (12.6.6)$$

Proof of Theorem 12.6.1. From the inequality

$$\int_W |T(x, y)| \nu(dy) \leq \int_W |T(x, y) - T(x, 0)| \nu(dy) + \int_W |T(0, y)| \nu(dy)$$

and conditions (12.6.5) and (12.6.6), inequality (12.5.4) follows immediately if we take $V(x) = |x|$. Thus, according to Proposition 12.5.2 there exists a stationary probabilistic measure μ_*. Using the definition of T_n from Section 12.4 and inequality (12.6.5) we obtain

$$\int_{W^n} |T_n(x, y^n) - T_n(z, y^n)| \nu^n(dy^n)$$

$$= \int_{W^{n-1}} \left\{ \int_W |T(T_{n-1}(x, y^{n-1}), y_n) - T(T_{n-1}(z, y^{n-1}), y_n)| \nu(dy_n) \right\} \nu^{n-1}(dy^{n-1})$$

$$\leq \alpha \int_{W^{n-1}} |T_{n-1}(x, y^{n-1}) - T_{n-1}(z, y^{n-1})| \nu^{n-1}(dy^{n-1})$$

$$\leq \cdots \leq \alpha^n |x - z|. \tag{12.6.7}$$

Now consider the subset C_* of C_0 which consists of functions h satisfying the Lipschitz condition

$$|h(x) - h(z)| \leq k|x - z| \quad \text{for } x, z \in X,$$

where the constant k depends, in general, on h. Further let μ and $\bar{\mu}$ be two distributions with bounded support. Then

$$|\langle h, P^n \mu - P^n \bar{\mu} \rangle| = |\langle U^n h, \mu - \bar{\mu} \rangle|$$

$$= \left| \int_B U^n h(x) \mu(dx) - \int_B U^n h(x) \bar{\mu}(dx) \right|, \tag{12.6.8}$$

where B is a bounded set such that $\mu(B) = \mu(\bar{B}) = 1$. Since the measures μ and $\bar{\mu}$ are probabilistic there exist points $q_n, r_n \in B$ such that

$$\left| \int_B U^n h(x) \mu(dx) - \int_B U^n h(x) \bar{\mu}(dx) \right| \leq |U^n h(q_n) - U^n h(r_n)|.$$

From this and (12.6.8) we have

$$|\langle h, P^n \mu - P^n \bar{\mu} \rangle| \leq |U^n h(g_n) - U^n h(r_n)|$$

$$\leq \int_{W^n} |h(T_n(g_n, y^n)) - h(T^n(r_n, y^n))| \nu^n(dy^n).$$

Using the Lipschitz condition for h and (12.6.7) we finally obtain

$$|\langle h, P^n \mu - P^n \bar{\mu} \rangle| \leq k \int_{W^n} |T_n(q_n, y^n) - T_n(r_n, y^n)| \nu^n(dy^n)$$

$$\leq k\alpha^n |q_n - r_n| \leq k d\alpha^n,$$

where $d = \sup\{|x - z|: x, z \in B\}$. Since $k\,da^n \to 0$ as $n \to \infty$, this implies (12.6.3) for arbitrary $h \in C_*$ and $\mu, \bar{\mu} \in \mathcal{M}_1$ with bounded supports. According to Propositions 12.6.1 and 12.6.2 the proof of the weak asymptotic stability is complete. ■

Remark 12.6.1. When $T(x, y) = S(x)$ does not depend on y, condition (12.6.4) is automatically satisfied with $\beta = |S(0)|$ and inequality (12.6.3) reduces to

$$|S(x) - S(z)| \leq \alpha|x - z| \qquad \text{for } x, z \in X.$$

In this case the statement of Theorem 12.6.1 is close to the Banach contraction principle. However, it still gives something new. Namely, the classical Banach theorem shows that all the trajectories $\{S^n(x_0)\}$ converge to the unique fixed point $x_* = S(x_*)$. From Theorem 12.6.1 it follows also that the measures $\mu(S^{-n}(A))$ (with $\mu \in \mathcal{M}_1$) converge to δ_{x_*} which is the unique stationary distribution. □

12.7 Strong Asymptotic Stability

In Example 12.2.1 we have shown that if a sequence of points $\{x_n\}$ converges to x_*, then the corresponding sequence of measures $\{\delta_{x_n}\}$ converges weakly to δ_{x_*}. In general, this convergence is not strong since $\|\delta_{x_n} - \delta_{x_*}\| = 2$ for $x_n \neq x_*$. Thus, in the space of measures, weak convergence seems to be a more convenient and natural notion than strong convergence. However this is not necessarily true for stochastic dynamical systems in which the perturbations ξ_n are nonsingular. To make this notion precise we introduce the following.

Definition 12.7.1. A measure $\mu \in \mathcal{M}_{\text{fin}}(X)$ is called **nonsingular** if there is an absolutely continuous measure μ_a such that

$$\mu_a(B) \leq \mu(B) \qquad \text{for } B \in \mathcal{B}(X) \tag{12.7.1}$$

and $\mu_a(X) > 0$.

It can be proved that for every measure $\mu \in \mathcal{M}_{\text{fin}}$ there exists a **maximal** absolutely continuous measure μ_a satisfying (12.7.1). The word maximal means that for any other continuous measure $\mu_{a'}$ satisfying $\mu_{a'}(B) \leq \mu(B)$ for all measurable sets B, we also have $\mu_{a'}(B) \leq \mu_a(B)$ for all measurable B. This maximal measure μ_a is called the **absolutely continuous part** of μ. The remaining component, $\mu_s = \mu - \mu_a$, is called the **singular part**. Thus, Definition 12.7.1 may be restated as follows: The measure $\mu \in \mathcal{M}_{\text{fin}}$ is nonsingular if its absolutely continuous part μ_a is not identically equal to zero. We always denote the absolutely continuous and singular parts of any measure by subscripts a and s, respectively. The equation

$$\mu = \mu_a + \mu_s \tag{12.7.2}$$

is called the **Lebesgue decomposition** of the measure μ.

In this section we will exclusively consider regular stochastic dynamical systems of the form

$$x_{n+1} = S(x_n) + \xi_n \qquad \text{for } n = 0, 1, \ldots, \tag{12.7.3}$$

where $S: X \to X$ is a continuous mapping of a closed set $X \subset R^d$ into itself, and $x_0, \xi_0, \xi_1, \ldots,$ are independent random vectors. The values of ξ_n belong to a Borel measurable set $W \subset R^d$ such that

$$x \in X, \ y \in W \quad \text{implies} \quad x + y \in X.$$

This condition is satisfied, for example, when $X = W = R^d$ or $X = W = R^+$. The dynamical system (12.7.3) with additive perturbations reduces to the general form (12.4.2) for $T(x, y) = S(x) + y$. Then equations (12.4.7) and (12.4.8) for the Foias operator P and its adjoint U take the form

$$P\mu(A) = \int_X \left\{ \int_W 1_A(S(x) + y)\nu(dy) \right\} \mu(dx) \qquad \text{for } A \in \mathcal{B}(X) \tag{12.7.4}$$

and

$$Uh(x) = \int_W h(S(x) + y)\nu(dy) \qquad \text{for } x \in X. \tag{12.7.5}$$

Consequently, for the scalar product we obtain

$$\langle h, P^n \mu \rangle = \langle U^n h, \mu \rangle = \int_X \left\{ \int_W h(S(x) + y)\nu(dy) \right\} \mu(dx). \tag{12.7.6}$$

From Proposition 12.5.2 and Theorem 12.6.1 we immediately obtain the following result.

Proposition 12.7.1. *If in the regular stochastic dynamical system (12.7.3) the transformation S and perturbations $\{\xi_n\}$ satisfy the conditions*

$$|S(x)| \le \alpha|x| + \gamma \qquad \text{for } x \in X \tag{12.7.7}$$

and

$$E(|\xi_n|) \le k, \tag{12.7.8}$$

where α, γ, k are nonnegative constants with $\alpha < 1$, then (12.7.3) has a stationary distribution. Moreover, if (12.7.7) is replaced by the stronger condition

$$|S(x) - S(z)| \le \alpha|x - z| \qquad \text{for } x, z \in X, \tag{12.7.9}$$

then (12.7.3) is weakly asymptotically stable.

Proof. The proof is immediate. It is sufficient to verify conditions (12.5.4) and (12.6.5). First observe that (12.7.8) is equivalent to

$$\int_W |y|\nu(dy) \le k.$$

Consequently, setting $T(x, y) = S(x) + y$ and using (12.7.7) and (12.7.8) we obtain

$$\int_W |T(0, y)| \nu(dy) = \int_W |S(0) + y| \nu(dy)$$
$$\leq \int_W |S(0)| \nu(dy) + \int_W |y| \nu(dy)$$
$$\leq |S(0)| + k \leq \gamma + k.$$

This is a special case of (12.6.6) with $\beta = \gamma + k$. Further, (12.7.9) yields

$$\int_W |T(x, y) - T(z, y)| \nu(dy) = \int_W |S(x) - S(z)| \nu(dy)$$
$$= |S(x) - S(z)| \leq \alpha |x - z|,$$

which gives (12.6.5). ∎

We will now show that under rather mild additional assumptions the asymptotic stability guaranteed by Proposition 12.7.1 is, in fact, strong. This is related to an interesting property of the absolutely continuous part μ_{na} of the distribution μ_n. Namely, $\|\mu_{na}\| = \mu_{na}(X)$ increases to 1 as $n \to \infty$. Our first result in this direction is the following.

Proposition 12.7.2. *Let P be the Foias operator corresponding to a regular stochastic dynamical system* (12.7.3) *in which S is a nonsingular transformation. If $\mu \in \mathcal{M}_{fin}$ is absolutely continuous, then $P\mu$ is also.*

Proof. Let f be the Radon–Nikodym derivative of μ. Then equation (12.7.4) gives

$$P\mu(A) = \int_X \left\{ \int_W 1_A(S(x) + y) \nu(dy) \right\} f(x)\, dx$$
$$= \int_W \left\{ \int_X 1_A(S(x) + y) f(x)\, dx \right\} \nu(dy).$$

For fixed $y \in W$ the function $1_A(S(x)+y)$ is the result of the application of the Koopman operator to $1_A(x+y)$. Denoting by P_S the Frobenius–Perron operator (acting on densities) corresponding to S, we may rewrite the last integral to obtain

$$P\mu(A) = \int_W \left\{ \int_X 1_A(x + y) P_S f(x)\, dx \right\} \nu(dy)$$
$$= \int_W \left\{ \int_{X+y} 1_A(x) P_S f(x - y)\, dx \right\} \nu(dy).$$

Inside the braces the integration runs over all x such that $x \in A$ and $x \in X + y$, or, equivalently, $x \in A$ and $x - y \in X$. Thus,

$$P\mu(A) = \int_W \left\{ \int_A 1_X(x-y)P_S f(x-y)\, dx \right\} \nu(dy)$$

$$= \int_A \left\{ \int_W 1_X(x-y)P_S f(x-y)\nu(dy) \right\} dx. \qquad (12.7.10)$$

The function

$$q(x) = \int_W 1_X(x-y)P_S f(x-y)\nu(dy) \qquad (12.7.11)$$

inside the braces of (12.7.10) is the convolution of the element $P_S f \in L^1$ with the measure ν. Thus we have verified that $P\mu$ is an absolutely continuous measure with density q. ∎

From Proposition 12.7.2, an important consequence concerning the behavior of the absolutely continuous part of $P^n \mu$ follows directly. Namely, we have

Corollary 12.7.1. *Let P be the Foias operator corresponding to the regular stochastic system* (12.7.3) *with nonsingular S. Then*

$$(P\mu)_a(X) \geq \mu_a(X) \qquad for\ \mu \in M_{fin}, \qquad (12.7.12)$$

and the sequence $\mu_{na}(X)$ is increasing.

Proof. By the linearity of P we have

$$P\mu = P\mu_a + P\mu_s \geq P\mu_a.$$

Since $(P\mu)_a$ is the maximal absolutely continuous measure which does not exceed $P\mu$, we have $(P\mu)_a \geq P\mu_a$. In particular,

$$(P\mu)_a(X) \geq P\mu_a(X) = \mu_a(X),$$

and the proof is complete. ∎

Proposition 12.7.2 also implies that when S is nonsingular the operator $\hat{P}$ on densities corresponding to P exists. In fact the right hand side of (12.7.11) gives an explicit equation for this operator, that is,

$$\hat{P}f(x) = \int_W 1_X(x-y)P_S f(x-y)\nu(dy). \qquad (12.7.13)$$

If S and ν are both nonsingular, we can say much more about the asymptotic behavior of $(P^n \mu)_a$. This behavior is described as follows.

Theorem 12.7.1. *Let P be the Foias operator corresponding to the regular stochastic system* (12.7.3). *If the transformation S and the distribution ν of random vectors $\{\xi_n\}$ are nonsingular, then*

$$\lim_{n\to\infty} (P^n \mu)_a(X) = 1 \qquad for\ \mu \in M_1. \qquad (12.7.14)$$

Proof. Let g_a be the Radon–Nikodym derivative of the measure ν_a. Using the inequality $\nu \geq \nu_a$ in equation (12.7.4) applied to μ_s, we obtain

$$P\mu_s(A) \geq \int_X \left\{ \int_W 1_A(S(x) + y)\nu_a(dy) \right\} \mu_s(dx)$$

$$= \int_X \left\{ \int_W 1_A(S(x) + y)g_a(y)\,dy \right\} \mu_s(dx)$$

$$= \int_X \left\{ \int_{W+S(x)} 1_A(y)g_a(y - S(x))\,dy \right\} \mu_s(dx).$$

The integration in the braces of the last integral runs over all y such that $y \in A$ and $y \in W + S(x)$, or equivalently all $y \in A$ and $y - S(x) \in W$. Thus, the last inequality may be rewritten in the form

$$P\mu_s(A) \geq \int_X \left\{ \int_A 1_W(y - S(x))g_a(y - S(x))\,dy \right\} \mu_s(dx)$$

$$= \int_A \left\{ \int_X 1_W(y - S(x))g_a(y - S(x))\mu_s(dx) \right\} dy.$$

Setting

$$r(y) = \int_X 1_W(y - S(x))g_a(y - S(x))\mu_s(dx) \quad \text{and} \quad \sigma(A) = \int_A r(y)\,dy$$

we may easily evaluate the measure $P\mu$ from below:

$$P\mu = P\mu_a + P\mu_s \geq P\mu_a + \sigma.$$

The measure $P\mu_a + \sigma$ is absolutely continuous and consequently the absolutely continuous part of $P\mu$ satisfies

$$(P\mu)_a \geq P\mu_a + \sigma.$$

In particular,

$$(P\mu)_a(X) \geq P\mu_a(X) + \sigma(X) = \mu_a(X) + \sigma(X). \qquad (12.7.15)$$

We may easily evaluate $\sigma(X)$ since

$$\sigma(X) = \int_X \left\{ \int_X 1_W(y - S(x))g_a(y - S(x))\,dy \right\} \mu_s(dx)$$

$$= \int_X \left\{ \int_{X-S(x)} 1_W(y)g_a(y)\,dy \right\} \mu_s(dx).$$

In the braces we integrate over all y such that $y \in W$ and $y \in X - S(x)$ or equivalently $y \in W$ and $y + S(x) \in X$. Since $W + X \subset X$ the condition

$y + S(x) \in X$ is always satisfied with $y \in W$ and $x \in X$. Thus,

$$\sigma(X) = \int_X \left\{ \int_W g_a(y)\, dy \right\} \mu_s(dx) = \nu_a(W)\mu_s(X)$$
$$= \nu_a(W)(1 - \mu_a(X)).$$

Set $\nu_a(W) = \varepsilon$ and use (12.7.15) to obtain

$$(P\mu)_a(X) \geq \mu_a(X) + \varepsilon(1 - \mu_a(X)) = \varepsilon + (1 - \varepsilon)\mu_a(X).$$

From this, we obtain by an induction argument

$$(P^n\mu)_a(X) \geq \varepsilon + \varepsilon(1-\varepsilon) + \cdots + \varepsilon(1-\varepsilon)^{n-1} + \varepsilon(1-\varepsilon)^n \mu_a(X) \geq 1 - (1-\varepsilon)^n.$$

Since $\varepsilon = \nu_a(X) > 0$, this completes the proof. ■

Now we are in a position to state our main result concerning the strong asymptotic stability of (12.7.3).

Theorem 12.7.2. *Assume that* (12.7.3) *is a regular stochastic system and that the transformation S and the distribution ν are nonsingular. If* (12.7.3) *is weakly asymptotically stable, then it is also strongly asymptotically stable and the limiting measure μ_* is absolutely continuous.*

Proof. Let P be the Foias operator given by equation (12.7.4) and $\hat{P}$ the corresponding operator (12.7.13) for densities. The proof will be constructed in three steps. First we are going to show that $\hat{P}$ is constrictive. Then we will prove that $r = 1$ in equation (5.3.10) and $\{\hat{P}^n\}$ is asymptotically stable in the sense of Definition 5.6.1. Finally, using Theorem 12.7.1 we will show that $\{P^n\}$ is strongly asymptotically stable.

Step I. Since (12.7.3) is weakly asymptotically stable there exists a stationary measure μ_*. Choose $\varepsilon = \nu_a(X)/3$ and an open bounded set B in X such that

$$\mu_*(B) > 1 - \varepsilon.$$

Now consider an absolutely continuous $\mu_0 \in M_1$ with a density f_0. According to the diagram (12.3.8), for each integer $n \geq 1$ the function $\hat{P}^n f_0$ is the density of $\mu_n = P^n \mu_0$. The sequence $\{\mu_n\}$ converges weakly to μ_* and according to Theorem 12.2.1 there is an integer n_0 such that

$$\int_B \hat{P}^n f_0(x)\, dx = \mu_n(B) \geq 1 - \varepsilon \qquad \text{for } n \geq n_0,$$

or

$$\int_{X \setminus B} \hat{P}^n f_0(x)\, dx \leq \varepsilon \qquad \text{for } n \geq n_0. \tag{12.7.16}$$

Now let $F \subset X$ be a measurable set. We have

$$\int_F \hat{P}^n f_0(x)\, dx = \mu_n(F) = P\mu_{n-1}(F),$$

and from (12.7.4) with $\nu = \nu_a + \nu_s$

$$P\mu_{n-1}(F) = \int_X \left\{ \int_W 1_F(S(x) + y)\nu_a(dy) \right\} \mu_{n-1}(dx)$$
$$+ \int_X \left\{ \int_W 1_F(S(x) + y)\nu_s(dy) \right\} \mu_{n-1}(dx).$$

Since μ_{n-1} is a probabilistic measure and

$$\nu_s(X) = 1 - \nu_a(X) = 1 - 3\varepsilon,$$

this implies

$$P\mu_{n-1}(F) \le \sup_{z \in X} \left\{ \int_W 1_F(y + z)\nu_a(dy) \right\} + 1 - 3\varepsilon.$$

Let g_a be the Radon–Nikodym derivative of ν_a so we may rewrite the last inequality in the form

$$P\mu_{n-1}(F) \le \sup_{z \in X} \left\{ \int_W 1_F(y + z)g_a(y)\, dy \right\} + 1 - 3\varepsilon$$
$$= \sup_{z \in X} \left\{ \int_{W \cap (F-z)} g_a(y)\, dy \right\} + 1 - 3\varepsilon.$$

The standard Borel measure of $W \cap (F - z)$ is smaller than the measure of F. Thus there exists a $\delta > 0$ such that

$$\int_{W \cap (F-z)} g_a(y)\, dy \le \varepsilon \qquad \text{for } F \in \mathcal{B}(X),\ m(F) \le \delta$$

and consequently

$$\int_F \hat{P}^n f_0(x)\, dx = P\mu_{n-1}(F) \le \varepsilon + (1 - 3\varepsilon) = 1 - 2\varepsilon.$$

From this and (12.7.16) we obtain

$$\int_{(X \setminus B) \cup F} \hat{P}^n f_0(x)\, dx \le \varepsilon + (1 - 2\varepsilon) = 1 - \varepsilon \qquad \text{for } n \ge n_0(f)$$

which proves that $\hat{P}$ is a constrictive operator. According to the spectral decomposition theorem [see equation (5.3.10)] the iterates of $\hat{P}$ may be written in the form

$$\hat{P}^n f = \sum_{i=1}^r \lambda_i(f)g_{\alpha^n(i)} + Q_n f \qquad n = 0, 1, \ldots, \qquad (12.7.17)$$

where the densities g_i have disjoint supports and $\hat{P}g_i = g_{\alpha(i)}$.

Step II. Now we are going to prove that $r = 1$ in equation (12.7.17). Let $k = r!$ and g_i be an arbitrary density. Then $\alpha^k(i) = i$ and consequently $\hat{P}^{kn} g_i = g_i$ for all n. Since (12.7.3) is weakly asymptotically stable the sequence $\{\langle h, \hat{P}^{nk} g_i \rangle\}$ converges to $\langle h, \mu_* \rangle$. However, this sequence is constant so

$$\langle h, g_i \rangle = \langle h, \mu_* \rangle \qquad \text{for } h \in C_0. \tag{12.7.18}$$

The last equality implies that g_i is the density of μ_*. Thus, there is only one term in the summation portion of (12.7.17) and g_1 is the invariant density.

Step III. Consider the sequence $\{P^n \mu_0\}$ with an arbitrary $\mu_0 \in \mathcal{M}_1$. Choose an $\varepsilon > 0$. According to Theorem 12.7.1 there exists an integer k such that

$$(P^k \mu_0)_a(X) = \mu_{ka}(X) \geq 1 - \varepsilon.$$

Define $\theta = \mu_{ka}(X)$. Since $\mu_k = \mu_{ka} + \mu_{ks}$ we have

$$\mu_{n+k} - \mu_* = P^n \mu_k - \mu_* = P^n \mu_{ka} - \theta \mu_* + P^n \mu_{ks} - (1 - \theta)\mu_*$$

or

$$\|\mu_{n+k} - \mu_*\| \leq \theta\|P^n(\theta^{-1}\mu_{ka}) - \mu_*\| + \|P^n \mu_{ks}\| + (1 - \theta)\|\mu_*\| \tag{12.7.19}$$

where $\|\cdot\|$ denotes the distance defined by equation (12.2.7). The last two terms are easy to evaluate since

$$\|P^n \mu_{ks}\| = P^n \mu_{ks}(X) = \mu_{ks}(X) = 1 - \theta \leq \varepsilon \tag{12.7.20}$$

and

$$(1 - \theta)\|\mu_*\| = (1 - \theta)\mu_*(X) = 1 - \theta \leq \varepsilon. \tag{12.7.21}$$

The measure $\theta^{-1}\mu_{ka}$ is absolutely continuous and normalized. Denote its density by f_a. $P^n(\theta^{-1}\mu_{kn})$ clearly has density $\hat{P}^n f_a$ and from equation (12.2.11)

$$\|P^n(\theta^{-1}\mu_{ka}) - \mu_*\| = \|\hat{P}^n f_a - g_1\|_{L^1}.$$

Since $\{\hat{P}^n\}$ is asymptotically stable the right-hand side of this equality converges to zero as $n \to \infty$. From this convergence and inequalities (12.7.20) and (12.7.21) applied to (12.7.19), it follows that

$$\lim_{n \to \infty} \|\mu_{n+k} - \mu_n\| = 0.$$

This completes the proof. ∎

12.8   Iterated Function Systems and Fractals

In the previous section we considered a special case of a regular stochastic dynamical system with additive nonsingular perturbations. As we have seen, these systems produce absolutely continuous limiting distributions. In this section we consider another special class in which the set W is finite. We will see that such systems produce limiting measures supported on very special sets—fractals.

Intuitively a system with finite W can be described as follows. Consider N continuous transformations

$$S_i: X \to X \qquad i = 1, \dots, N$$

of a closed nonempty subset $X \subset R^d$. If the initial point $x_0 \in X$ is chosen we toss an N-sided die, and if the number i_0 is drawn we define $x_1 = S_{i_0}(x_0)$. Then we toss up the die again and if the number i_1 is drawn we define $x_2 = S_{i_1}(x_1)$, and so on.

This procedure can be easily formalized. Consider a probabilistic vector

$$(p_1, \dots, p_N), \quad p_i \geq 0, \quad \sum_{i=1}^{N} p_i = 1,$$

and the sequence of independent random variables $\xi_0, \xi_1, \dots$ such that

$$\text{prob}(\xi_n = i) = p_i \qquad \text{for } i = 1, \dots, N.$$

The dynamical system is defined by the formula

$$x_{n+1} = S_{\xi_n}(x_n) \qquad \text{for } n = 0, 1, \dots. \tag{12.8.1}$$

It is clear that in this case $T(x, y) = S_y(x)$ and $W = \{1, \dots, N\}$. The system (12.8.1) is called [Barnsley, 1988] an **iterated function system** **(IFS)**.

Using the general equations (12.4.7) and (12.4.8) it is easy to find explicit formulas for the operators U and P corresponding to an iterated function system. Namely,

$$Uh(x) = \int_W h(T(x, y)) \nu(dy) = \int_W h(S_y(x)) \nu(dy)$$

or

$$Uh(x) = \sum_{i=1}^{N} p_i h(S_i(x)) \qquad \text{for } x \in X. \tag{12.8.2}$$

Further,

$$P\mu(A) = \langle U 1_A, \mu \rangle = \sum_{i=1}^{N} p_i \int_X 1_A(S_i(x)) \mu(dx)$$

or

$$P\mu(A) = \sum_{i=1}^{N} p_i \mu(S_i^{-1}(A)) \qquad \text{for } A \in \mathcal{B}(X). \tag{12.8.3}$$

Now assume that the S_i satisfy the Lipschitz condition

$$|S_i(x) - S_i(z)| \le L_i|x - z| \qquad \text{for } x, z \in X; \ i = 1, \ldots, N, \qquad (12.8.4)$$

where L_i are nonnegative constants. In this case Theorem 12.6.1 implies the following result.

Proposition 12.8.1. *If*

$$\sum_{i=1}^{N} p_i L_i < 1, \qquad (12.8.5)$$

then the iterated function system (12.8.1) is weakly asymptotically stable.

Proof. It is sufficient to verify conditions (12.6.3) and (12.6.4). We have

$$E(|S_{\xi_n}(x) - S_{\xi_n}(z)|) = \sum_{i=1}^{N} p_i |S_i(x) - S_i(z)|$$

$$\le |x - z| \sum_{i=1}^{N} p_i L_i$$

and

$$E(|S_{\xi_n}(0)|) = \sum_{i=1}^{N} p_i |S_i(0)|.$$

Consequently (12.6.3) and (12.6.4) are satisfied with $\alpha = \sum p_i L_i$ and $\beta = \sum p_i |S_i(0)|$, and by Theorem 12.6.1 the proof is complete. ∎

Condition (12.8.5) is automatically satisfied when $L_i < 1$ for $i = 1, \ldots, N$. An iterated function system for which

$$L = \max_i L_i < 1 \quad \text{and} \quad p_i > 0, \qquad i = 1, \ldots, N \qquad (12.8.6)$$

is called **hyperbolic**. Our goal now is to study the structure of the set

$$A_* = \operatorname{supp} \mu_*, \qquad (12.8.7)$$

where μ_* is a stationary distribution for hyperbolic systems. We will show that A_* does not depend on the probabilistic vector $(p_1, \ldots, p_N)$ as long as all the p_i are strictly positive. To show an alternative, nonprobabilistic method of constructing A_*, we introduce a transformation F on the subset of X such that the iterates F^n approximate A_*.

Definition 12.8.1. Let an iterated function system (12.8.1) be given. Then the transformation

$$F(A) = \bigcup_{i=1}^{N} S_i(A) \qquad \text{for } A \subset X \qquad (12.8.8)$$

mapping subsets of X into subsets of X is called the **Barnsley operator** corresponding to (12.8.1).

It is easy to observe that for every compact set $A \subset X$ its image $F(A)$ is also a compact set. In fact, the $S_i(A)$ are compact since the images of compact sets by continuous transformations are compact and the finite union of compact sets is compact. To show the connection between F and the dynamical system (12.8.1) we prove the following.

Proposition 12.8.2. *Let F be the Barnsley operator corresponding to (12.8.1). Moreover, let $\{\mu_n\}$ be the sequence of distributions corresponding to (12.8.1), that is, $\mu_n = P^n \mu_0$. If supp μ_0 is a compact set, then*

$$\text{supp } \mu_n = F^n(\text{supp } \mu_0) \tag{12.8.9}$$

Proof. It is clearly sufficient to verify that supp $\mu_1 = F(\text{supp } \mu_0)$ since the situation repeats. Let $x \in F(\text{supp } \mu_0)$ and $\varepsilon > 0$ be fixed. Then $x = S_j(z)$ for some integer j and $z \in \text{supp } \mu_0$. Consequently, for the ball $B_r(z)$ we have $\mu_0(B_r(z)) > 0$ for every $r > 0$. Further, due to the continuity of S_j there is an $r > 0$ such that

$$B_r(z) \subset S_j^{-1}(B_\varepsilon(x)).$$

This gives

$$\mu_1(B_\varepsilon(x)) = \sum_{i=1}^{n} p_i \mu_0(S_i^{-1}(B_\varepsilon(x)))$$
$$\geq p_j \mu_0(S_j^{-1}(B_\varepsilon(x))) \geq p_j \mu_0(B_r(z)) > 0.$$

Since $\varepsilon > 0$ was arbitrary this shows that $x \in \text{supp } \mu_1$. We have proved the inclusion $F(\text{supp } \mu_0) \subset \text{supp } \mu_1$.

Now, suppose that this inclusion is proper and there is a point $x \in \text{supp } \mu_1$ such that $x \notin F(\text{supp } \mu_0)$. Due to the compactness of $F(\text{supp } \mu_0)$ there must exist an $\varepsilon > 0$ such that the ball $B_\varepsilon(x)$ is disjoint with $F(\text{supp } \mu_0)$. This implies

$$B_\varepsilon(x) \cap S_i(\text{supp } \mu_0) = \emptyset \qquad \text{for } i = 1, \dots, N$$

or

$$S_i^{-1}(B_\varepsilon(x)) \cap \text{supp } \mu_0 = \emptyset \qquad \text{for } i = 1, \dots, N.$$

The last condition implies that

$$\mu_1(B_\varepsilon(x)) = \sum_{i=1}^{N} p_i \mu_0(S_i^{-1}(B_\varepsilon(x)) = 0$$

which contradicts to the assumption that $x \in \text{supp } \mu_1$. This contradiction shows that $F(\text{supp } \mu_0) = \text{supp } \mu_1$. An induction argument completes the proof. ∎

Formula (12.8.9) allows us to construct the supports of μ_n from the support of μ_0 by purely geometrical methods without any use of probabilistic arguments. Now we will show that the set

$$A_* = \text{supp } \mu_*, \qquad\qquad (12.8.10)$$

which is called the **attractor** of the iterated function system, can be obtained as the limit of the sequence of sets

$$A_n = \text{supp } \mu_n = F^n(A_0). \qquad\qquad (12.8.11)$$

To state this fact precisely we introduce the notion of the Hausdorff distance between two sets.

Definition 12.8.2. Let $A_1, A_2 \subset R^d$ be nonempty compact sets and let $r > 0$ be a real number. We say that A_1 approximates A_2 with accuracy r if, for every point $x_1 \in A_1$, there is a point $x_2 \in A_2$ such that $|x_1 - x_2| \le r$ and for every $x_2 \in A$ there is an $x_1 \in A_1$ such that the same inequality holds. The infimum of all r such that A_1 approximates A_2 with accuracy r is called the **Hausdorff distance** between A_1 and A_2 and is denoted by $\text{dist}(A_1, A_2)$.

We say that a sequence $\{A_n\}$ of compact sets converges to a compact set A if

$$\lim_{n \to \infty} \text{dist}(A_n, A) = 0.$$

From the compactness of A it easily follows that the limit of the sequence $\{A_n\}$, if it exists, must be unique. This limit will be denoted by $\lim_{n \to \infty} A_n$.

Example 12.8.1. Let $X = R$, $A = [0, 1]$ and

$$A_n = \left\{ \frac{1}{2^n}, \frac{2}{2^n}, \ldots, \frac{2^n - 1}{2^n} \right\} \qquad \text{for } n = 1, 2, \ldots .$$

Clearly, $A_n \subset [0, 1]$. Moreover for every $x \in [0, 1]$ there is an integer k, $1 \le k \le 2^{n-1}$, such that

$$\left| x - \frac{k}{2^n} \right| \le \frac{1}{2^n}.$$

Thus, A_n approximates A with accuracy $1/2^n$. Moreover, for $x = 0 \in A$ the nearest point in A_n is $1/2^n$. Consequently,

$$\text{dist}(A_n, A) = \frac{1}{2^n}.$$

This example shows that sets which are close in the sense of Hausdorff distance can be quite different from a topological point of view. In fact, each

A_n consists of a finite number of points, whereas $A = [0,1]$ is a continuum. This is a typical situation in the technical reproduction of pictures; on a television screen a picture is composed of a finite number of pixels. $\square$

We have introduced the notion of the distance between compact sets only. We already know that for compact $A_0 = \text{supp } \mu_0$ all the sets $A_n = \text{supp } \mu_n$ are compact. Now we are going to show the compactness of the limiting set $A_* = \text{supp } \mu_*$.

Proposition 12.8.3. *If the iterated function system* (12.8.1) *is hyperbolic and μ_* is the stationary distribution, then the set $A_* = \text{supp } \mu_*$ is compact.*

Proof. Since the support of every measure is a closed set, it is sufficient to verify that A_* is bounded. Further, since μ_* does not depend on μ_0 we may assume that $\mu_0 = \delta_{x_0}$ for an $x_0 \in X$. Define

$$r = \max\{|S_i(x_0) - x_0|: i = 1, \ldots, N\}.$$

Then

$$|S_{i_1} \circ S_{i_2}(x_0) - x_0| \leq |S_{i_1} \circ S_{i_2}(x_0) - S_{i_1}(x_0)| + |S_{i_1}(x_0) - x_0| \leq Lr + r$$

or by induction,

$$|S_{i_1} \circ \cdots \circ S_{i_n}(x_0) - x_0| \leq L^{n-1}r + \cdots + Lr + r \leq \frac{r}{1-L} \qquad (12.8.12)$$

for every sequence of integers $i_1, \ldots, i_n$ with $0 \leq i_k \leq N$. Choose an arbitrary point $z \in X$ such that

$$|z - x_0| \geq \frac{r}{1-L} + 1. \qquad (12.8.13)$$

We are going to prove that $z \notin \text{supp } \mu_*$. Fix an $\varepsilon \in (0,1)$. From inequality (12.2.6) and equation (12.8.3) we obtain

$$\mu_*(B_\varepsilon(z)) \leq \liminf_{n\to\infty} \mu_n(B_\varepsilon(z)) \qquad (12.8.14)$$

$$= \liminf_{n\to\infty} \sum_{i_1,\ldots,i_n} p_{i_1} \cdots p_{i_n} \delta_{x_0}(S_{i_1}^{-1} \circ \cdots \circ S_{i_n}^{-1}(B_\varepsilon(z))).$$

According to (12.8.12) and (12.8.13) we have

$$|z - S_{i_n} \circ \cdots \circ S_{i_1}(x_0)| \geq 1$$

which implies that

$$x_0 \notin S_{i_1}^{-1} \circ \cdots \circ S_{i_n}^{-1}(B_\varepsilon(z)).$$

Thus the right-hand side of (12.8.14) is equal to zero and as a consequence $\mu_*(B_\varepsilon(z)) = 0$. We have proved that $z \notin \text{supp } \mu_*$ and that the support of μ_* is contained in a ball centered at x_0 with radius $1 + r/(1-L)$. $\blacksquare$

Now we formulate a convergence theorem which allows us to construct the set A_* without any use of probabilistic tools.

Theorem 12.8.1. *Let* (12.8.1) *be a hyperbolic system and let F be the corresponding Barnsley operator. Further, let A_* be the support of the invariant distribution. Then*

$$A_* = \lim_{n \to \infty} F^n(A_0) \qquad (12.8.15)$$

whenever $A_0 \subset X$ is a nonempty compact set.

Prooof. We divide the proof into two steps. First we show that the limit of $\{F^n(A_0)\}$ does not depend on the particular choice of A_0, and then we will prove that this limit is equal to supp μ_*.

Step I. Consider two initial compact sets $A_0, Z_0 \subset X$ and the corresponding sequences

$$An = F^n(A_0), \quad Z_n = F^n(Z_0) \qquad n = 0, 1, \dots .$$

We are going to show that dist(A_n, Z_n) converges to zero. Let $r > 0$ be sufficiently large so A_0 and Z_0 are contained in a ball of radius r. Now fix an integer n and a point $x \in A_n$. According to the definition of F^n there exists a sequence of integers $k_1, \dots, k_n$ and a point $u \in A_0$ such that

$$x = S_{k_1} \circ \cdots \circ S_{k_n}(u).$$

Now choose an arbitrary point $v \in Z_0$ and define $z \in Z_n$ by

$$z = S_{k_1} \circ \cdots \circ S_{k_n}(v).$$

Since the S_i are Lipschitzean we have

$$|x - z| \le L^n|u - v| \le 2rL^n.$$

We have proved that for every $x \in A_n$ there is a $z \in Z_n$ such that $|x - z| \le 2rL^n$. Since the assumptions concerning the sets A_0 and Z_0 are symmetric this shows that the distance between A_n and Z_n is smaller than $2rL^n$. Consequently,

$$\lim_{n \to \infty} \text{dist}(A_n, Z_n) = 0. \qquad (12.8.16)$$

Step II. Choose an arbitrary nonempty compact sets $A_0 \subset X$ and define

$$Z_0 = A_* = \text{supp } \mu_*.$$

Since μ_* is invariant we also have

$$Z_n = F^n(Z_0) = A_*.$$

Substituting this into (12.8.16) we obtain (12.8.15) and the proof is complete. ∎

It is worth noting that for systems which are not hyperbolic, equality (12.8.15) may be violated even if condition (12.8.5) is satisfied. In general the set $\lim F^n(A_0)$ is larger than $A_* = \operatorname{supp} \mu_*$.

Example 12.8.2. Let $X = R$, $S_1(x) = x$ and $S_2(x) = 0$ for $x \in R$. Evidently for every probabilistic vector (p_1, p_2) with $p_1 < 1$ the condition (12.8.5) is satisfied. Thus the system is weakly asymptotically stable and there exists unique stationary distribution μ_*. It is easy to guess that $\mu_* = \delta_0$. In fact, according to (12.8.3),

$$P\delta_0(A) = p_1\delta_0(S_1^{-1}(A)) + p_2\delta_0(S_2^{-1}(A)),$$

where $S_1^{-1}(A) = A$ and

$$S_2^{-1}(A) = \begin{cases} R & \text{if } 0 \in A \\ \emptyset & \text{if } 0 \notin A. \end{cases}$$

Therefore

$$P\delta_0(A) = p_1\delta_0(A) + p_2\delta_0(A) = \delta_0(A).$$

On the other hand, for $A_0 = [0, 1]$ we have

$$F(A_0) = S_1(A_0) \cup S_2(A_0) = [0, 1] \cup \{0\} = [0, 1]$$

and by induction

$$F^n(A_0) = [0, 1] \qquad n = 0, 1, \ldots .$$

This sequence does not converge to $A_* = \operatorname{supp} \mu_* = \{0\}$. □

Now we are going to use equation (12.8.15) for the construction of attractors of hyperbolic systems. This procedure can often be simplified using the following result concerning the Barnsley operator (12.8.8).

Proposition 12.8.3. *Assume that the $S_i: X \to X$, $i = 1, \ldots, N$ appearing in equation (12.8.8) are continuous and that $A_0 \subset X$ is a compact set. Denote $A_n = F^n(A_0)$ and assume that $A_* = \lim_{n\to\infty} A_n$ exists. If $A_0 \supset F(A_0)$, then*

$$A_0 \supset A_1 \supset A_2 \supset \cdots \supset A_*. \tag{12.8.17}$$

Proof. The Barnsley operator F is monotonic, that is, $A \subset B$ implies $F(A) \subset F(B)$. Thus from $A_1 \supset A_0$ it follows $F^n(A_1) \supset F^n(A_0)$ or $A_{n+1} \supset A_n$. It remains to prove that $A_n \supset A_*$. Fix an integer n and a point $x \in A_*$. Consider a sequence $\varepsilon_j = 1/j$. Since $\{A_{n+k}\}$ converges to A_* as $k \to \infty$ we can find a set $A_{n+k(j)}$ which approximates A_* with accuracy ε_j. There exists, therefore, $x_j \in A_{n+k(j)}$ such that $|x_j - x| \leq \varepsilon_j$. Evidently $x_j \in A_n$

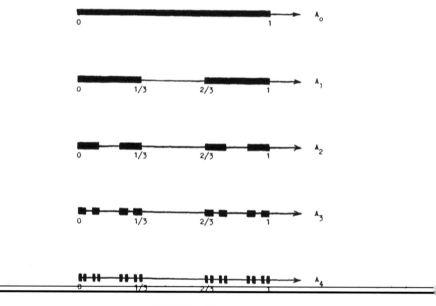

FIGURE 12.8.1.

since, by the first part of the proof, $A_n \supset A_{n+k(j)}$. The set A_n is closed and the conditions $x_j \in A_n$, $x_j \to x$ imply $x \in A_n$. This verifies the inclusion $A_n \supset A_*$ and completes the proof. ∎

Our first example of the construction of an attractor deals with a one-dimensional system given by two linear transformations. Despite the simplicity of the system the attractor is quite complicated.

Example 12.8.3. Let $X = R$ and

$$S_1(x) = \tfrac{1}{3}x \quad \text{and} \quad S_2(x) = \tfrac{1}{3}x + \tfrac{2}{3} \quad \text{for } x \in R.$$

Choose $A_0 = [0, 1]$ (see Figure 12.8.1). Then

$$A_1 = F(A_0) = S_1([0,1]) \cup S_2([0,1]) = \left[0, \tfrac{1}{3}\right] \cup \left[\tfrac{2}{3}, 1\right].$$

Thus, A_1 is obtained from A_0 by taking out the middle open interval $\left(\tfrac{1}{3}, \tfrac{2}{3}\right)$. Now

$$A_2 = F(A_1) = S_1\left(\left[0, \tfrac{1}{3}\right] \cup \left[\tfrac{2}{3}, 1\right]\right) \cup S_2\left(\left[0, \tfrac{1}{3}\right] \cup \left[\tfrac{2}{3}, 1\right]\right)$$
$$= \left[0, \tfrac{1}{9}\right] \cup \left[\tfrac{2}{9}, \tfrac{3}{9}\right] \cup \left[\tfrac{6}{9}, \tfrac{7}{9}\right] \cup \left[\tfrac{8}{9}, 1\right].$$

Again A_2 is obtained from A_1 by taking out two middle open intervals $\left(\tfrac{1}{9}, \tfrac{2}{9}\right)$ and $\left(\tfrac{7}{9}, \tfrac{8}{9}\right)$. Proceeding further we observe that this operation repeats and A_3 can be obtained from A_2 by taking out the four middle intervals. Thus, the set A_3 consists of eight intervals of length $\tfrac{1}{27}$. In general, A_n is the sum of 2^n intervals of length $1/3^n$. The Borel measure of A_n is $\left(\frac{2}{3}\right)^n$ and converges to zero as $n \to \infty$. The limiting set A_* has Borel measure zero since it is contained in all sets A_n. This is the famous **Cantor set**—the source of many examples in analysis and topology. □

Example 12.8.4. Let $X = R^2$ and

$$S_i(x) = \begin{pmatrix} \frac{1}{2} & 0 \\ 0 & \frac{1}{2} \end{pmatrix} x + \begin{pmatrix} a_i \\ b_i \end{pmatrix}, \qquad i = 1, 2, 3,$$

where

$$a_1 = b_1 = 0; \quad a_2 = \tfrac{1}{2}, \ b_2 = 0; \quad a_3 = \tfrac{1}{4}, \ b_3 = \tfrac{1}{2}.$$

Choose A_0 to be the isosceles triangle with vertices $(0,0)$, $(1,0)$, $(\frac{1}{2},1)$ (see Figure 12.8.2a). $S_1(A_0)$ is a triangle with vertices $(0,0)$, $(\frac{1}{2},0)$, $(\frac{1}{4},\frac{1}{2})$. The triangles $S_2(A_0)$ and $S_3(A_0)$ are congruent to $S_1(A_0)$ but shifted to the right, and to the right and up, respectively. As a result, the set

$$A_1 = F(A_0) = S_1(A_0) \cup S_2(A_0) \cup S_3(A_0)$$

is the union of three triangles as shown in Figure 12.8.2b. Observe that A_1 is obtained from A_0 by taking out the middle open triangle with vertices $(\frac{1}{2},0)$, $(\frac{1}{4},\frac{1}{2})$, $(\frac{3}{4},\frac{1}{2})$. Analogously each set $S_i(A_1)$, $i = 1,2,3$, consisting of three congruent triangles of height $\frac{1}{4}$ and $A_2 = F(A_1)$ in the union of nine triangles shown in Figure (12.8.2c). Again A_2 can be obtained from A_1 by taking out three middle triangles.

This process repeats and in general A_n consists of 3^n triangles with height $(\frac{1}{2})^n$, base $(\frac{1}{2})^n$, and total area

$$m(A_n) = \tfrac{1}{2} \left(\tfrac{3}{4}\right)^n,$$

which converges to zero as $n \to \infty$. The limiting set A_*, called the **Sierpinski triangle**, has Borel measure zero. It is shown in Figure 12.8.2d. Unlike the Cantor set, the Sierpinski triangle is a continuum (compact connected set) and from a geometric point of view it is a line whose every point is a ramification point. The Sierpinski triangle also appears in cellular automata theory [Wolfram, 1983]. □

In these two examples the construction of the sets A_n approximating A_* was ad hoc. We simply guessed the procedure leading from A_n to A_{n+1}, taking out the middle intervals or middle triangles. In general, for an arbitrary iterated function system the connection between A_n and A_{n+1} is not so simple. In the next theorem we develop another way of approximating A_* which is especially effective with the aid of a computer.

Theorem 12.8.2. *Let* (12.8.1) *be a hyperbolic system. Then for every* $x_0 \in X$ *and* $\varepsilon > 0$ *there exist two numbers* $n_0 = n_0(\varepsilon)$ *and* $k_0 = k_0(\varepsilon)$ *such that*

$$\mathrm{prob}(\mathrm{dist}(\{x_n,\ldots,x_{n+k}\}, A_*) < \varepsilon) > 1 - \varepsilon \qquad \textit{for } n \geq n_0, k \geq k_0,$$
$$\text{(12.8.18)}$$

where $\{x_n\}$ *denotes the trajectory starting from* x_0.

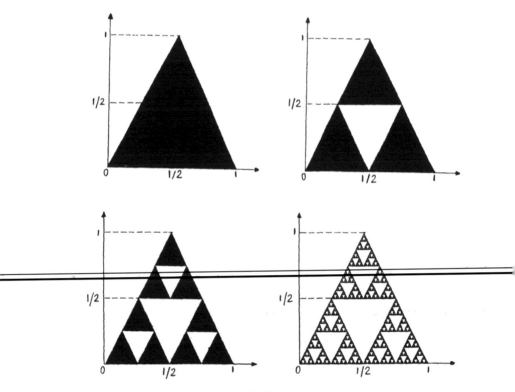

FIGURE 12.8.2.

In other words Theorem 12.8.2 says the following. If we cancel the first n_0 or more elements of the trajectory $\{x_n\}$, then the probability that a sufficiently long segment $x_n, \ldots, x_{n+k}$ approximates A_* with accuracy ε is greater than $1 - \varepsilon$.

Proof. Let $\varepsilon > 0$ be fixed. Choose a compact set $A_0 \subset X$ such that $x_0 \in A_0$ and $F(A_0) \subset A_0$. [From condition (12.8.4) and (12.8.6) it follows that such a set exists.] The sequence $A_n = F^n(A_0)$ is decreasing and $x_n \in A_n$. By Theorem 12.8.1 there is an integer $n_0(\varepsilon)$ such that

$$\text{dist}(A_n, A_*) < \varepsilon \qquad \text{for } n \geq n_0.$$

From this inequality, for every value of the random vector x_n there is $z_n \in A_n$ for which

$$|x_n - z_n| < \varepsilon \qquad \text{for } n \geq n_0. \tag{12.8.19}$$

This determines the number n_0 appearing in condition (12.8.18).

Now we are going to find k_0. Since A_* is a compact set there is a finite sequence of points $a_i \in A_*$, $i = 1, \ldots, q$ such that

$$A_* \subset \bigcup_{i=1}^{q} B_{\varepsilon/2}(a_i). \tag{12.8.20}$$

Pick a point $u \in A_{n_0}$. The set $\{u\}$, which contains the single point u, is compact and according to Theorem 12.8.1 there exists an integer r such that

$$\text{dist}(F^r(\{u\}), A_*) < \frac{\varepsilon}{4}.$$

The points of $F^r(\{u\})$ are given by $S_{\alpha_1} \circ \cdots \circ S_{\alpha_r}(u)$. Thus, for every $i = 1, \ldots, q$, there exists a sequence of integers $\alpha(i, 1), \ldots, \alpha(i, r)$ for which

$$|S_{\alpha(i,1)} \circ \cdots \circ S_{\alpha(i,r)}(u) - a_i| < \frac{\varepsilon}{4}.$$

This inequality holds for a fixed $u \in A_{n_0}$. When u moves in A_{n_0}, the corresponding value $S_{\alpha(i,1)} \circ \cdots \circ S_{\alpha(i,r)}(u)$ changes by at most $L^r c$ where $c = \max\{|u - v| : u, v \in A_{n_0}\}$. Choosing r large enough, we have $L^r c < \varepsilon/4$, and consequently

$$|S_{\alpha(i,1)} \circ \cdots \circ S_{\alpha(i,r)}(u) - a_i| < \frac{\varepsilon}{2} \quad \text{for } i = 1, \ldots, q, u \in A_{n_0}. \tag{12.8.21}$$

Now consider the segment $x_n, \ldots, x_{n+k}$ of the trajectory given by (12.8.1) with $n \geq n_0$. We have

$$x_{n+j} = S_{\xi_{n+j-1}} \circ \cdots \circ S_{\xi_n}(x_n) \quad \text{and} \quad x_{n+j} \in A_{n_0} \text{ for } 0 \leq j \leq k.$$

If the sequence $\xi_n, \ldots, \xi_{n+k}$ contains the segment $\alpha(i, 1), \ldots, \alpha(i, r)$, that is,

$$\xi_{n+j+r-1} = \alpha(i, 1), \ldots, \xi_{n+j} = \alpha(i, r) \tag{12.8.22}$$

for some j, $0 \leq j \leq k - r$, then (12.8.21) implies $x_{n+j+r} \in B_{\varepsilon/2}(a_i)$. The probability of the event (12.8.22), with fixed j, is equal to $p_{\alpha(i,1)} \cdots p_{\alpha(i,r)}$, and the probability of the opposite event is smaller than or equal to $1 - p^r$ where $p = \min p_i$. The probability that $\xi_n, \ldots, \xi_{n+k}$ with $k \geq rm$ does not contain the sequence $\alpha(i, 1), \ldots, \alpha(i, r)$ is at most $(1 - p^r)^m$. For sufficiently large m we have $(1 - p^r)^m \leq \varepsilon/q$. With this m and $k \geq k_0 = rm$ the probability of the event that $\xi_n, \ldots, \xi_{n+k}$ contains all the sequences $\alpha(i, 1), \ldots, \alpha(i, r)$, for $i = 1, \ldots, q$ is at least $1 - q(1 - p^r)^m \geq 1 - \varepsilon$. When the last event occurs, then for every point a_i there is a point x_{n+j+r} such that $|x_{n+j+r} - a_i| < \varepsilon/2$. In this case according to (12.8.20) every point $x \in A_*$ is approximated by a point of the segment $x_n, \ldots, x_{n+k}$ with accuracy ε. From this and (12.8.19) it follows

$$\text{dist}(\{x_n, \ldots, x_{n+k}\}, A_*) < \varepsilon.$$

The proof is completed. ∎

Theorem 12.8.3 gives a practical way of constructing a picture of A_* with the use of a computer. We simply must generate a segment of the trajectory $\{x_n\}$ according to equation (12.8.1). Neglecting the first segment $x_0, \ldots, x_{n-1}$, we assure that the remaining points $x_n, x_{n+1}, \ldots$ are in the ε-neighborhood of A_* where $\varepsilon = c_0 L^n$, and c_0 is a constant which depends only on the choice of the initial point x_0. Since the system is hyperbolic the sequence $\{L^n\}$ quickly converges to zero, and in practice, the value c_0 is irrelevant. We generate the points $x_n, x_{n+1}, x_{n+2}, \ldots$ until the picture no longer substantially changes. Theoretically, the approach to the set A_* does not depend on the choice of the probabilities $p_1, \ldots, p_N$ as long as $p_i > 0$ for $i = 1, \ldots, N$. However, the convergence to A_* may be slow if the value $p = \min p_i$ is small. Changing p_i does not change $A_* = \operatorname{supp} \mu_*$, but may change μ_*.

Example 12.8.5. Let $X = R^2$ and take S_i, $i = 1, 2, 3$ to be given by

$$S_1(x) = \begin{pmatrix} 0 & 0 \\ 0 & c \end{pmatrix} x + \begin{pmatrix} \frac{1}{2} \\ 0 \end{pmatrix}$$

$$S_2(x) = \begin{pmatrix} -r\cos\varphi & -r\sin\varphi \\ -r\sin\varphi & r\cos\varphi \end{pmatrix} x + \begin{pmatrix} \frac{1}{2} + \frac{r}{2}\cos\varphi \\ c + \frac{r}{2}\sin\varphi \end{pmatrix}$$

$$S_3(x) = \begin{pmatrix} q\cos\psi & -r\sin\psi \\ q\sin\psi & r\cos\psi \end{pmatrix} x + \begin{pmatrix} \frac{1}{2} - \frac{q}{2}\cos\psi \\ \frac{3}{5}c - \frac{q}{2}\sin\psi \end{pmatrix},$$

where

$$c = 0.255, \quad r = 0.75, \quad q = 0.625, \quad \varphi = -\frac{\pi}{8}, \quad \psi = \frac{\pi}{5}.$$

Choosing

$$n_0 = 100, \quad k_0 = 150000, \quad |x_0| \le 1, \quad p_1 = p_2 = p_3 = \frac{1}{3}$$

we obtain a representation of a tree as shown in Figure 12.8.3, as prepared for this volume by Dr. Z. Kielek. It should be noted that the "tree" has some natural asymmetry as usually appears in nature. □

The objects shown in Figures 12.8.1, 12.8.2, and 12.8.3 are called fractals, a name derived from the Latin word *fractus* meaning broken or partial.

One possible definition says that a fractal is a set which has a fractional dimension. To make this definition precise it is necessary to define a dimension applicable to a large class of sets. This is not an easy task and the several existing dimension definitions are, in general, not equivalent.

Here we give a simplification of the Hausdorff dimension proposed by N. A. Kolmogorov. It is called the **capacity** or **fractal dimension**. To understand the ideas that lead to this notion, define a d-cube of size ℓ by

$$K = \{(x^1, \ldots, x^d) : a_i \le x^i \le a_i + \ell \text{ for } i = 1, \ldots, d\}.$$

FIGURE 12.8.3.

This set K is evidently a d-dimensional object. The question arises how to derive the number d from the intrinsic properties of the cube, neglecting the trivial fact that the index i takes on d values in the definition of K. Assume first that $\ell = 1$ and that K is subdivided into cubes of size $\varepsilon_n = 1/n$. The number of these cubes is $N(\varepsilon) = n^d$. From this we obtain immediately

$$d = \frac{\log n^d}{\log n} = \frac{\log N(\varepsilon_n)}{\log(1/\varepsilon_n)}.$$

When the size ℓ of the cube K is arbitrary the calculation is a little more complicated. Namely, K may be divided into cubes of size $\varepsilon_n = \ell/n$ and the number of these cubes is $N(\varepsilon_n) = n^d$. Consequently,

$$d = \lim_{n\to\infty} \frac{d \log n}{\log n - \log \ell} = \lim_{n\to\infty} \frac{\log N(\varepsilon_n)}{\log(1/\varepsilon_n)}.$$

These calculations suggest the following.

Definition 12.8.3. Let $A \subset R^d$ be a compact set. For every $\varepsilon > 0$ denote by $N(\varepsilon)$ the minimum number of cubes of size ε needed to cover A. We define the **dimension** of K by the formula

$$\dim K = \lim_{\varepsilon \to 0} \frac{\log N(\varepsilon)}{\log(1/\varepsilon)} \tag{12.8.23}$$

if this limit exists.

Calculation of the fractal dimension by a direct application of Definition 12.8.3 is difficult. It may be simplified and the continuous variable ε replaced by an appropriate sequence $\{\varepsilon_n\}$. Namely, if for some $c > 0$ and $0 < q < 1$ we define $\varepsilon_n = cq^n$, and if

$$d_{cq} = \lim_{n\to\infty} \frac{\log N(\varepsilon_n)}{\log(1/\varepsilon_n)} \qquad (12.8.24)$$

exists, then the limit (12.8.23) also exists and $\dim K = d_{cq}$ [Barnsley, 1988; Chapter 5].

Using this property we may find the dimension of the attractors A_* described in Examples 12.8.3 and 12.8.4. First consider the case when A_* is the Cantor set. We have

$$A_* \subset A_n = F^n([0, 1])$$

and A_* can be covered by 2^n disjoint intervals of length 3^{-n} whose sum is equal to A_n. Since A_* contains the endpoints of these intervals the number of covering intervals cannot be made smaller. We have, therefore, $N(\varepsilon_n) = 2^n$ for $\varepsilon_n = 3^{-n}$ which gives

$$\dim A_{*\text{Cantor}} = \frac{\log 2}{\log 3}.$$

For the Sierpinski triangle the situation is similar. We have

$$A_* \subset A_n = F^n(A_0),$$

where A_0 is the initial triangle. The set A_n consists of 3^n isosceles triangles of height 2^{-n} and base length 2^{-n}. Every such triangle can be covered by four squares of size $2^{-(n+1)}$. On the other hand the vertices of these triangles belong to A_*. It is necessary to use 3^n different squares of size $2^{-(n+1)}$ just to cover the top vertices of these triangles. Thus, for $\varepsilon_n = 2^{-(n+1)}$ we have

$$3^n \leq N(\varepsilon_n) \leq 4 \cdot 3^n$$

and consequently

$$\frac{n \log 3}{(n+1)\log 2} \leq \frac{\log N(\varepsilon_n)}{\log(1/\varepsilon_n)} \leq \frac{\log 4 + n \log 3}{(n+1)\log 2}$$

which gives in the limit

$$\dim A_{*\text{Sierpinski}} = \frac{\log 3}{\log 2}.$$

Exercises

12.1. Let $X \subset R^d$ be a compact set and $C(X)$ be the space of continuous functions $f: X \to R$. Using the Weierstrass approximation theorem prove that in $C(X)$ there exists a dense countable subset of Lipschitzean functions.

12.2. Let $X \subset R^d$ be a closed unbounded set. Using the family of functions

$$f_{nw}(x) = w(x) \max(1 - n^{-1}|x|, 0),$$

where n is a positive integer and $w: R^d \to R$ is a polynomial, show that in the space $C_0(X)$ there exists a dense countable subset of Lipschitzean functions.

12.3. Let $X = [0, 2\pi]$ and let $\{\mu_n\}$ be the sequence of probabilistic measures with densities $(1/\pi)\sin^2 nx$. Find the weak limit μ_* of $\{\mu_n\}$. Is μ_* also the strong limit of $\{\mu_n\}$?

12.4. Let $S: R \to R$ be a continuous function such that $S(x) \neq x$ for $x \in R$. Show that for the operator $P\mu(A) = \mu(S^{-1}(A))$ there does not exist an invariant probabilistic measure.

12.5. Let $X = \{0, 1, \ldots\}$ be the set of nonnegative integers. Consider the iterated function system given by the two transformations

$$S_1(x) \equiv 0, \quad S_2(x) = 1 + x \qquad \text{for } x \in X,$$

and the probability vector $p_1 = p_2 = \frac{1}{2}$. Show that this system is strongly asymptotically stable.

12.6. Generalize the previous result and consider an arbitrary iterated function system (12.8.1). Show that if $S_1 \equiv 0$ and $p_1 > 0$, then this system is strongly asymptotically stable.

12.7. Let (12.8.1) be a hyperbolic dynamical system. Fix an arbitrary $x_0 \in X$. Prove that for every sequence $\{i_n\}$ with $i_n \in \{1, \ldots, N\}$ the limit

$$x = \lim_{n \to \infty} S_{i_1} \circ S_{i_2} \circ \cdots \circ S_{i_n}(x_0)$$

exists, and that the set of all such points x corresponding to all possible sequences $\{i_n\}$ is equal to A_* [Barnsley, 1988; Chapter 4].

12.8. Consider the hyperbolic dynamical system given, on $X = R^2$, by the eight transformations

$$S_i(x) = \begin{pmatrix} \frac{1}{3} & 0 \\ 0 & \frac{1}{3} \end{pmatrix} x + \begin{pmatrix} a_i \\ b_i \end{pmatrix} \qquad i = 1, \ldots, 8,$$

where (a_i, b_i) are all possible pairs made from the numbers $(0, \frac{1}{3}, \frac{2}{3})$ excluding $(\frac{1}{3}, \frac{1}{3})$. The attractor A_* of this system is called a **Sierpinski carpet**. Make a picture of A_* and calculate dim A_*.

References

Abraham, R. and Marsden, J.E. 1978. *Foundations of Mechanics*, Benjamin/Cummings, Reading, Massachusetts.

Adler, R.L. and Rivlin, T.J. 1964. Ergodic and mixing properties of Chebyshev polynomials, *Proc. Am. Math. Soc.* 15:794–796.

Anosov, D.V. 1963. Ergodic properties of geodesic flows on closed Riemannian manifolds of negative curvature, *Sov. Math. Dokl.* 4:1153–1156.

Anosov, D.V. 1967. Geodesic flows on compact Riemannian manifolds of negative curvature, *Proc. Steklov Inst. Math.*, 90:1–209.

Arnold, V.I. 1963. Small denominators and problems of stability of motion in classical and celestial mechanics, *Russian Math. Surveys*, 18:85–193.

Arnold, V.I. and Avez, A. 1968. *Ergodic Problems of Classical Mechanics*, Benjamin, New York.

Barnsley, M. 1988. *Fractals Everywhere*, Academic Press, New York.

Barnsley, M. and Cornille, H. 1981. General solution of a Boltzmann equation and the formation of Maxwellian tails, *Proc. R. Soc. London, Sect. A*, 374:371–400.

Baron, K. and Lasota A. 1993. Asymptotic properties of Markov operators defined by Volterra type integrals, *Ann. Polon. Math.*, 58:161–175.

Bessala, P. 1975. On the existence of a fundamental solution for a parabolic differential equation with unbounded coefficients, *Ann. Polon. Math.*, 29:403–409.

Bharucha-Reid, A.T. 1960. *Elements of the Theory of Markov Processes and Their Applications*, McGraw-Hill, New York.

Birkhoff, G.D. 1931a. Proof of a recurrence theorem for strongly transitive systems, *Proc. Natl. Acad. Sci. USA*, 17:650–655.

Birkhoff, G.D. 1931b. Proof of the ergodic theorem, *Proc. Natl. Acad. Sci. USA*, 17:656–660.

Bobylev, A.V. 1976. Exact solutions of the Boltzmann equations, *Sov. Phys. Dokl.*, 20:822–824.

Borel, E. 1909. Les probabilités dénombrables et leurs applications arithmétiques, *Rendiconti Circ. Mat. Palermo*, 27:247–271.

Boyarsky, A. 1984. On the significance of absolutely continuous invariant measures, *Pysica 11D*, 130–146.

Breiman, L. 1968. *Probability*, Addison-Wesley, Reading, Massachusetts.

Brown, J.R. 1976. *Ergodic Theory and Topological Dynamics*, Academic Press, New York.

Brunovsky, P. 1983. Notes on chaos in the cell population partial differential equation. *Nonlin. Anal.*, 7:167–176.

Brunovsky, P. and Komornik, J. 1984. Ergodicity and exactness of the shift on $C[0,\infty)$ and the semiflow of a first-order partial differential equation, *J. Math. Anal. Applic.*, 104:235–245.

Brzeźniak, Z. and Szafirski, B. 1991. Asymptotic behavior of L^1 norm of solutions to parabolic equations, *Bull. Polon. Acad. Sci. Math.*, 39:1–10.

Bugiel, P. 1982. Approximation for the measure of ergodic transformations on the real line, *Z. Wahrscheinlichkeitstheorie Verw. Gebeite*, 59:27–38.

Chabrowski, J. 1970. Sur la construction de la solution fondamentale de l'équation parabolique aux coefficients non bornés. *Colloq. Math.*, 21:141–148.

Chandrasekhar, S. and Münch, G. 1952. The theory of fluctuations in brightness of the Milky-Way, *Astrophys. J.*, 125:94–123.

Chapman, S. and Cowling, T.G. 1960. *The Mathematical Theory of Non-Uniform Gases*, Cambridge University Press, Cambridge, England.

Collet, P. and Eckmann, J.P. 1980. *Iterated Maps on the Interval as Dynamical Systems*, Birkhaüser, Boston.

Cornfeld, I.P., Fomin, S.V., and Sinai, Ya.G. 1982. *Ergodic Theory*, Springer-Verlag, New York.

Dłotko, T. and Lasota, A. 1983. On the Tjon–Wu representation of the Boltzmann equation, *Ann. Polon. Math.*, 42:73–82.

Dłotko, T. and Lasota, A. 1986. Statistical stability and the lower bound function technique, in *Proceedings of the Autumn Course on Semigroups: Theory and Applications* (H. Brezis, M. Crandall, and F. Kappel, eds.). International Center for Theoretical Physics, Trieste, *Pitman Res. Notes Math.*, 141:75–95.

Dunford, N. and Schwartz, J.T. 1957. *Linear Operators. Part I: General Theory*, Wiley, New York.

Dynkin, E.G. 1965. *Markov Processes*, Springer-Verlag, New York.

Eidel'man, S.D. 1969. *Parabolic Systems*, North-Holland, Amsterdam.

Elmroth, T. 1984. On the *H*-function and convergence toward equilibrium for a space-homogeneous molecular density, *SIAM J. Appl. Math.*, 44:150–159.

Feigenbaum, M.J. and Hasslacher, B. 1982. Irrational decimations and path integrals for external noise, *Phys. Rev. Lett.*, 49:605–609.

Foguel, S.R. 1966. Limit theorems for Markov processes, *Trans. Amer. Math. Soc.*, 121:200–209.

Foguel, S.R. 1969. *The Ergodic Theory of Markov Processes*, Van Nostrand Reinhold, New York.

Friedman, A. 1964. *Partial Differential Equations of Parabolic Type*, Prentice-Hall, Englewood Cliffs, New Jersey.

Friedman, A. 1975. *Stochastic Differential Equations and Applications*, vol. 1, Academic Press, New York.

Gantmacher, F.R. 1959. *Matrix Theory*, Chelsea, New York.

Gihman [Gikhman], I.I. and Skorohod [Skorokhod], A.V. 1975. *The Theory of Stochastic Processes*, vol. 2, Springer-Verlag, New York.

Gikhman, I.I. and Skorohod A.V. 1969. *Introduction to the Theory of Random Processes*, Saunders, Philadelphia. [Trans. from Russian].

Glass, L. and Mackey, M.C. 1979. A simple model for phase locking of biological oscillators, *J. Math. Biology*, 7:339–352.

Guevara, M.R. and Glass, L. 1982. Phase locking, period doubling bifurcations, and chaos in a mathematical model of a periodically driven oscillator: A theory for the entrainment of biological oscillators and the generation of cardiac dysrhythmias, *J. Math. Biology*, 14:1–23.

Hadamard, J. 1898. Les surfaces à courbures opposées et leur lignes géodésiques, *J. Math. Pures Appl.*, 4:27–73.

Hale, J. 1977. *Theory of Functional Differential Equations*, Springer-Verlag, New York.

Halmös P.R. 1974. *Measure Theory*, Springer-Verlag, New York.

Hardy, G.H. and Wright, E.M. 1959. *An Introduction to the Theory of Numbers*, 4th Edition, Oxford University Press, London.

Henon, M. 1976. A two-dimensional mapping with a strange attractor. *Commun. Math. Phys.*, 50:69–77.

Horbacz, K. 1989a. Dynamical systems with multiplicative perturbations, *Ann. Pol. Math.*, 50:11–26.

Horbacz, K. 1989b. Asymptotic stability of dynamical systems with multiplicative perturbations, *Ann. Polon. Math.*, 50:209–218.

Jabłoński, M. and Lasota, A. 1981. Absolutely continuous invariant measures for transformations on the real line, *Zesz. Nauk. Uniw. Jagiellon. Pr. Mat.*, 22:7–13.

Jakobson, M. 1978. Topological and metric properties of one-dimensional endomorphisms, *Dokl. Akad. Nauk. SSSR*, 243:866–869 [in Russian].

Jama, D. 1986. Asymptotic behavior of an integro-differential equation of parabolic type, *Ann. Polon. Math.*, 47:65–78.

Jama, D. 1989. Period three and the stability almost everywhere, *Rivista Mat. Pura Appl.*, 5:85–95.

Jaynes, E.T. 1957. Information theory and statistical mechanics, *Phys. Rev.*, 106:620–630.

Kamke, E. 1959. *Differentialgleichungen: Lösungsmethoden und Lösungen. Band 1. Gewönliche Differential-gleichungen*, Chelsea, New York.

Katz, A. 1967. *Principles of Statistical Mechanics*, Freeman, San Francisco.

Kauffman, S. 1974. Measuring a mitotic oscillator: The arc discontinuity, *Bull. Math. Biol.*, 36:161–182.

Keener, J.P. 1980. Chaotic behavior in piecewise continuous difference equations, *Trans. Amer. Math. Soc.*, 261:589–604.

Keller, G. 1982. Stochastic stability in some chaotic dynamical systems, *Mh. Math.*, 94:313–333.

Kemperman, J.H.B. 1975. The ergodic behavior of a class of real transformations, in *Stochastic Processes and Related Topics*, pp. 249–258 (vol. 1 of *Proceedings of the Summer Research Institute on Statistical Inference*, Ed. Madan Lal Puri). Academic Press, New York.

Kiełek, Z. 1988. An application of the convolution iterates to evolution equation in Banach space, *Universitatis Jagellonicae Acta Mathematica*, 27:247–257.

Kifer, Y.I. 1974. On small perturbations of some smooth dynamical systems, *Math. USSR Izv.*, 8:1083–1107.

Kitano, M., Yabuzaki, T., and Ogawa, T. 1983. Chaos and period doubling bifurcations in a simple acoustic system, *Phys. Rev. Lett.*, 50:713–716.

Knight, B.W. 1972a. Dynamics of encoding in a population of neurons, *J. Gen. Physiol.*, 59:734–766.

Knight, B.W. 1972b. The relationship between the firing rate of a single neuron and the level of activity in a population of neurons. Experimental evidence for resonant enhancement in the population response, *J. Gen. Physiol.*, 59:767–778.

Komornik, J. and Lasota, A. 1987. Asymptotic decomposition of Markov operators, *Bull. Polon. Acad. Sci. Math.*, 35:321–327.

Komorowski, T. and Tyrcha, J. 1989. Asymptotic properties of some Markov operators, *Bull. Acad. Polon. Sci. Math.*, 37:221–228.

Koopman, B.O. 1931. Hamiltonian systems and transformations in Hilbert space, *Proc. Nat. Acad. Sci. USA*, 17:315–318.

Kosjakin, A.A. and Sandler, E.A. 1972. Ergodic properties of a certain class of piecewise smooth transformations of a segment, *Izv. Vyssh. Uchebn. Zaved. Matematika*, 118:32–40.

Kowalski, Z.S. 1976. Invariant measures for piecewise monotonic transformations, *Lect. Notes Math.*, 472:77–94.

Krook, M. and Wu, T.T. 1977. Exact solutions of the Boltzmann equation, *Phys. Fluids*, 20:1589–1595.

Krzyżewski, K. 1977. Some results on expanding mappings, *Soc. Math. France Astérique*, 50:205–218.

Krzyżewski, K. and Szlenk, W. 1969. On invariant measures for expanding differential mappings, *Stud. Math.*, 33:83–92.

Lasota, A. 1981. Stable and chaotic solutions of a first-order partial differential equation, *Nonlin. Anal.*, 5:1181–1193.

Lasota, A., Li, T.Y., and Yorke, J.A. 1984. Asymptotic periodicity of the iterates of Markov operators, *Trans. Amer. Math. Soc.*, 286:751–764.

Lasota, A. and Mackey, M.C. 1980. The extinction of slowly evolving dynamical systems, *J. Math. Biology*, 10:333–345.

Lasota, A. and Mackey, M.C. 1984. Globally asymptotic properties of proliferating cell populations, *J. Math. Biology*, 19:43–62.

Lasota, A. and Mackey, M.C. 1989. Stochastic perturbation of dynamical systems: The weak convergence of measures, *J. Math. Anal. Applic.*, 138:232–248.

Lasota, A., Mackey, M.C., and Tyrcha, J. 1992. The statistical dynamics of recurrent biological events, *J. Math. Biology*, 30:775–800.

Lasota, A., Mackey, M.C., Ważewska-Czyzewska, M. 1981. Minimizing therapeutically induced anemia, *J. Math. Biology*, 13:149–158.

Lasota, A. and Rusek, P. 1974. An application of ergodic theory to the determination of the efficiency of cogged drilling bits, *Arch. Górnictwa*, 19:281–295. [In Polish with Russian and English summaries.]

Lasota, A. and Tyrcha, J. 1991. On the strong convergence to equilibrium for randomly perturbed dynamical systems, *Ann. Polon. Math.*, 53:79–89.

Lasota, A. and Yorke, J.A. 1982. Exact dynamical systems and the Frobenius–Perron operator, *Trans. Amer. Math. Soc.*, 273:375–384.

Li, T.Y. and Yorke, J.A. 1978a. Ergodic transformations from an interval into itself, *Trans. Am. Math. Soc.*, 235:183–192.

Li, T.Y. and Yorke, J.A. 1978b. Ergodic maps on [0, 1] and nonlinear pseudorandom number generators, *Nonlinear Anal.*, 2:473–481.

Lin, M. 1971. Mixing for Markov operators, *Z. Wahrscheinlichkeitstheorie Verw. Gebiete*, 19:231–242.

Lorenz, E.N. 1963. Deterministic nonperiodic flow, *J. Atmos. Sci.*, 20:130–141.

Loskot, K. and Rudnicki, R. 1991. Relative entropy and stability of stochastic semigroups, *Ann. Polon. Math.*, 53:139–145.

Mackey, M.C. and Dörmer, P. 1982. Continuous maturation of proliferating erythroid precursers, *Cell Tissue Kinet.*, 15:381–392.

Mackey, M.C., Longtin, A., and Lasota, A. 1990. Noise-induced global asymptotic stability, *J. Stat. Phys.*, 60:735–751.

Malczak, J. 1992. An application of Markov operators in differential and integral equations, *Rend. Sem. Univ. Padova*, 87:281–297.

Mandelbrot, B.B. 1977. *Fractals: Form, Chance, and Dimension*, Freeman, San Francisco.

Manneville, P. 1980. Intermittency, self-similarity and $1/f$ spectrum in dissipative dynamical systems, *J. Physique*, 41:1235–1243.

Manneville, P. and Pomeau, Y. 1979. Intermittency and the Lorenz model, *Phys. Lett.*, 75A:1–2.

May, R.M. 1974. Biological populations with nonoverlapping generations: stable points, stable cycles, and chaos, *Science*, 186:645–647.

May, R.M. 1980. Nonlinear phenomena in ecology and epidemology, *Ann. N.Y. Acad. Sci.*, 357:267–281.

Misiurewicz, M. 1981. Absolutely continuous measures for certain maps of an interval, *Publ. Math. IHES*, 53:17–51.

von Neumann, J. 1932. Proof of the quasi-ergodic hypothesis, *Proc. Nat. Acad. Sci. USA*, 18:31–38.

Parry, W. 1981. *Topics in Ergodic Theory*, Cambridge University Press, Cambridge, England.

Petrillo, G.A. and Glass, L. 1984. A theory for phase locking of respiration in cata to a mechanical ventilator, *Am. J. Physiol.*, 246:R311–320.

Pianigiani, G. 1979. Absolutely continuous invariant measures for the process $x_{n+1} = Ax_n(1 - x_n)$, *Boll. Un. Mat. Ital.*, 16A:374–378.

Pianigiani, G. 1983. Existence of invariant measures for piecewise continuous transformations, *Ann. Polon. Math.*, 40:39–45.

Procaccia, I. and Schuster, H. 1983. Functional renormalization group theory of $1/f$ noise in dynamical systems, *Phys. Rev. A*, 28:1210–1212.

Rényi, A. 1957. Representation for real numbers and their ergodic properties, *Acta Math. Acad. Sci. Hung.*, 8:477–493.

Riskin, H. 1984. *The Fokker–Planck Equation*, Springer-Verlag, New York.

Rochlin, V.A. 1964. Exact endomorphisms of Lebesgue spaces, *Am. Math. Soc. Transl.*, (2) 39:1–36.

Rogers, T.D. and Whitley, D.C. 1983. Chaos in the cubic mapping, *Math. Modelling*, 4:9–25.

Royden, H.L. 1968. *Real Analysis*, Macmillan, London.

Rudnicki, R. 1985. Invariant measures for the flow of a first-order partial differential equation, *Ergod. Th. & Dynam. Sys.*, 5:437–443.

Ruelle, D. 1977. Applications conservant une mesure absolument continue par rapport à dx sur $[0,1]$, *Commun. Math. Phys.*, 55:477–493.

Šarkovskiĭ, A.N. 1964. Coexistence of cycles of a continuous map of a line into itself, *Ukr. Mat. Zh.*, 16:61–71.

Schaefer, H.H. 1980. On positive contractions in L^p spaces, *Trans. Am. Math. Soc.*, 257:261–268.

Schiff, L.I. 1955. *Quantum Mechanics*, McGraw-Hill, New York.

Schwartz, L. 1965. *Méthodes mathémathiques de la physique*, Hermann, Paris.

Schwartz, L. 1966. *Théorie des distributions*, Hermann, Paris.

Schweiger, F. 1978. Tan x is ergodic, *Proc. Am. Math. Soc.*, 1:54–56.

Shannon, C.E. and Weaver, W. 1949. *The Mathematical Theory of Communication*, University of Illinois Press, Urbana.

Sinai, Ya. 1963. On the foundations of ergodic hypothesis for a dynamical system of statistical mechanics, *Sov. Math. Dokl.*, 4:1818–1822.

Smale, S. 1967. Differentiable dynamical systems, *Bull. Am. Math. Soc.*, 73:741–817.

Smale, S. and Williams, R.F. 1976. The qualitative analysis of a difference equation of population growth, *J. Math. Biology*, 3:1–5.

Szarski, J. 1967. *Differential Inequalities* (2nd Ed.), Polish Scientific Publishers, Warsaw.

Tjon, J.A. and Wu, T.T. 1979. Numerical aspects of the approach to a Maxwellian distribution, *Phys. Rev. A*, 19:883–888.

Tyrcha, J. 1988. Asymptotic stability in a generalized probabilistic/deterministic model of the cell cycle, *J. Math. Biology*, 26:465–475.

Tyson, J.J. and Hannsgen, K.B. 1986. Cell growth and division: a deterministic/probabilistic model of the cell cycle, *J. Math. Biology*, 23:231–246.

Tyson, J.J. and Sachsenmaier, W. 1978. Is nuclear division in *Physarum* controlled by a continuous limit cycle oscillator? *J. Theor. Biol.*, 73:723–738.

Ulam, S.M. and von Neumann, J. 1947. On combination of stochastic and deterministic processes, *Bull. Am. Math. Soc.*, 53:1120.

Voigt, J. 1981. Stochastic operators, information and entropy, *Commun. Math. Phys.*, 81:31–38.

Walter, W. 1970. *Differential and Integral Inequalities*, Springer-Verlag, New York.

Walters, P. 1975. *Ergodic Theory: Introductory Lectures*, Lecture Notes in Mathematics 458, Springer-Verlag, New York.

Walters, P. 1982. *An Introduction to Ergodic Theory*, Springer-Verlag, New York.

Wolfram, S. 1983. Statistical mechanics of cellular automata, *Reviews of Modern Physics*, 55:601–644.

Zdun, M.C. 1977. Continuous iteration semigroups, *Boll. Un. Mat. Ital.*, 14A:65–70.

Notation and Symbols

If A and B are sets, then $x \in B$ means that "x is an element of B," whereas $A \subset B$ means that "A is contained in B." For $x \notin B$ and $A \not\subset B$ substitute "is not" for "is" in these statements. Furthermore, $A \cup B = \{x : x \in A \text{ or } x \in B\}$, $A \cap B = \{x : x \in A \text{ and } x \in B\}$, $A \setminus B = \{x : x \in A \text{ and } x \notin B\}$, and $A \times B = \{(x, y) : x \in A \text{ and } y \in B\}$, respectively, define the **union**, **intersection**, **difference**, and **Cartesian product** of two sets A and B. Symbol $\emptyset$ denotes the **empty set**, and

$$1_A(x) = \begin{cases} 1 & \text{if } x \in A \\ 0 & \text{if } x \notin A \end{cases}$$

is the **characteristic** (or **indicator**) **function** for set A.

When $a < b$ the **closed interval** $[a, b] = \{x : a \leq x \leq b\}$, whereas the **open interval** $(a, b) = \{x : a < x < b\}$. The half-open intervals $[a, b)$ and $(a, b]$ are similarly defined. The **real line** is denoted by R, and the positive half-line by R^+. If A is a set and $\alpha \in R$, then $\alpha A = \{y : x \in A \text{ and } y = \alpha x\}$.

The notation $f : A \to B$ means that "f is a function whose **domain** is A and whose **range** is in B," or "f maps A into B." Given two functions $f : A \to B$ and $g : B \to C$, then $g \circ f$ denotes the **composition** of g with f and $g \circ f : A \to C$. If f maps R (or a subset of R) into R, and b is a positive number, then

$$g(x) = f(x) \qquad (\text{mod } b)$$

means that $g(x) = f(x) - nb$, where n is the largest integer less than or equal to $f(x)/b$. $\|f\|_{L^p}$ and $\langle f, g \rangle$, respectively, denote the L^p **norm** of the function f, and the **scalar product** of the functions f and g. $\bigvee_a^b f$ is used for the **variation** of the function f over the interval $[a, b]$.

The following is a list of the most commonly used symbols and their meaning:

a.e.	almost everywhere
$\mathcal{A}$	σ-algebra
$\mathcal{B}$	Borel σ-algebra
$d(g, \mathcal{F})$	L^1 distance between functions g and $\mathcal{F}$
$d_+ f / dx$	right lower derivative
$D, D(X, \mathcal{A}, \mu)$	set of densities
$D^2(\xi)$	variance of random variable ξ
$\mathcal{D}(A)$	domain of an infinitesimal operator A
$E(\xi)$	mathematical expectation of a random variable ξ
$E(V \mid f)$	expected value of V with respect to f
$E_i(x)$	exponential integral
$\{\eta_t\}$	continuous time stochastic process
f	an element of L^p, often a density
f_*	stationary density
$\mathcal{F}$	L^p set of functions, σ-algebra in probability space
$\{\mathcal{F}_t\}$	family of σ-algebras
$g_{\sigma b}(x)$	Gaussian density with variance $\sigma^2/2b$
g_{ij}^ϕ	Riemannian metric
$H_n(x)$	Hermite polynomial
$H(f)$	entropy of a density f
$H(f \mid g)$	conditional entropy of f with respect to g
I	identity operator
$K(x, y)$	stochastic kernel
$L^p, L^p(X, \mathcal{A}, \mu)$	L^p space
$L^{p'}$	space adjoint to L^p
$\mu(A)$	measure of a set A
$\mu_f(A)$	measure of a set A with respect to a density f
μ_w	Wiener measure
$\{N_t\}_{t \geq 0}$	counting process
ω	an element of Ω; angular frequency
Ω	space of elementary events
$(\Omega, \mathcal{F}, \text{prob})$	probability space
P	Markov or Frobenius–Perron operator
$P_\varepsilon, \overline{P}$	Markov operator
$\{\hat{P}_t\}_{t \geq 0}$	continuous semigroup generated by the linear Boltzmann equation
prob	probability measure
Prob	probability measure on a product space
R_λ	resolvent operator
S	transformation

$S^{-1}(A)$	counterimage of a set A under a transformation S
S_m	Chebyshev polynomial
S^1	unit circle
$\{S_t\}_{t\in R}, \{S_t\}_{t\geq 0}$	dynamical or semidynamical system
$\sigma(\xi)$	standard deviation of a random variable ξ
T	transformation
T^d	d-dimensional torus
$\{T_t\}_{t\geq 0}$	semigroup corresponding to an infinitesimal operator A
U	Koopman operator
V	Liapunov function, potential function
$\{w(t)\}_{t\geq 0}$	Wiener process
$(X, \mathcal{A}, \mu)$	measure space
ξ, ξ_i	random variables
$\{\xi_n\}, \{\xi_t\}$	discrete or continuous time stochastic process

Index

Applied Mathematical Sciences

(continued from page ii)

(continued on next page)

Applied Mathematical Sciences

(continued from previous page)